DECISION-MAKING AND AGRICULTURE

DECISION-MAKING AND AGRICULTURE

PAPERS AND REPORTS

SIXTEENTH INTERNATIONAL CONFERENCE OF AGRICULTURAL ECONOMISTS

Held at Nairobi, Kenya

26th JULY – 4th AUGUST 1976

Edited by
Theodor Dams, Institut für Entwicklungspolitik, Universität Freiburg, Federal Republic of Germany
and
Kenneth E Hunt, Agricultural Economics Institute, University of Oxford, England

University of Nebraska Press
Lincoln and London

Publishers on the Plains

UNP

Printed in Great Britain
First published in the United States of America
by the University of Nebraska Press, Lincoln, Nebraska 68588
Library of Congress Catalog Card Number 77-93962
ISBN 0-8032-7201-4

CONTENTS

AGRICULTURAL DEVELOPMENT AND ECONOMIC INTEGRATION

SYNOPTIC VIEW

A report on a number of sessions of the 16th Conference on the general subject of
Food and Population – Priorities in Decision-Making
will be published separately

PREFACE

The papers and discussion reports presented in this book record the proceedings of the XVIth International Conference of Agricultural Economists held at Nairobi, Kenya from July 26 to August 4, 1976 at the generous invitation of the Government of Kenya.

The Conference arrangements were the responsibility of the Kenyan Organising Committee under the chairmanship of the Honourable Maina Wanjigi, M.P.

Two special seminars were also held: one on Rural Development, organized by the Economic Development Institute (EDI), Washington, D.C., of the World Bank (IBRD), and the other on Post-Graduate Training in Agricultural Economics, organized by the German Foundation of International Development, Deutsche Stiftung für Internationale Entwicklung, DSE, Berlin. The Executive Committee of the IAAE has worked closely with both of these institutions, and the Association is very grateful for this coordination of these activities with the efforts of the IAAE to build up an efficient infrastructure of communications in Africa between the agricultural economists as individuals as well as to strengthen the international scientific cooperation in the field of agricultural economics.

The programme of the various sessions was built up by Professor Theodor Dams of the University of Freiburg, Federal Republic of Germany, and the book, therefore, owes its broad pattern to him and its more detailed content to the large number of individual participants who have contributed the papers and participated in the discussions.

The papers and the records of discussion were edited for publication by Mr. K. E. Hunt, Director of the Agricultural Economics Institute, University of Oxford, England.

The IAAE is deeply grateful to all of them for their assistance.

SAMAR R. SEN
President, International Association
of Agricultural Economists

EDITORIAL NOTES

The papers and summaries of discussions which make up this volume reflect the pattern of the central part of the conference concerned with the subject *"Decision-making and agriculture"*. This part of the conference programme was arranged as eight plenary sessions, in which the conference met as a whole, and special group sessions in which it divided into sections – usually four.

At the first plenary meeting Samar R. Sen Delivered the Presidential address on the conference theme and Theodore W. Schultz presented the First Leonard Knight Elmhirst Memorial Lecture entitled "On Economics, Agriculture, and the Political Economy". This lecture was established by the International Association of Agricultural Economists in memory of Leonard Elmhirst, of Dartington Hall, Devon, England, the Association's Founder President.

The six plenary sessions which followed were each devoted to a broad subject area within the field embraced by the theme. Some were introduced by a single paper, others by several.

The discussion at each of these plenary meetings was opened by invited speakers – usually two. The subject was then thrown open to general discussion and the session closed with the main speakers adding their concluding remarks.

The activities of these plenary sessions are recorded in this volume in the form of the Presidential Address, the Elmhirst Memorial Lecture, the other papers presented at plenary sessions, the contributions of those invited to open the discussions at these sessions, a summary of the general discussion and the concluding remarks of the readers of the main papers. The discussion summaries of the general discussion at plenary sessions were prepared by one of the editors (K. E. H.) from written statements provided by speakers and the taped record of the Proceedings. The names of all identified speakers are listed but, where appropriate, contributions from several speakers are combined and specific points are not attributed in the summaries to specific speakers. Rising printing costs have enforced the Association to abandon its practice of publishing a verbatim record of these discussions.

The Special Groups meetings which followed the plenary sessions to which they were linked were held simultaneously, the conference participants dividing among them according to inclination. The record given here follows a broadly similar pattern to that of the plenary sessions – main paper or papers, invited statements opening the discussion and a discussion summary (prepared by a rapporteur assigned by the Programme Organiser). Some slight variations in the pattern occurred reflecting convenience in arranging discussion. For example, one of the special groups was arranged to continue the discussion on M. Yudelman's paper in the plenary session devoted to Integrated Rural Development. The extended discussion was not summarised.

Among other activities of the Conference was a Symposium on agriculture in Kenya, for which the publication specially commissioned for the Conference (*Agricultural Development in Kenya: an economic assessment* by Judith Heyer, J. K. Maitha and W. M. Senga. (Nairobi, O.U.P., 1976) formed the introduction. A day was devoted to food and population problems on the basis of which the Association expects to publish a group of papers on the general subject of "Achieving a balance between population and food: priorities in decision-making". The 24 Contributed Papers read at the Conference have been published as the first of the new series of IAAE Occasional Papers.

SAMAR R. SEN

Decision-Making and Agriculture

The main theme of the Conference is "Decision-making and Agriculture" which really encompasses the decision-making processes "in" and "for" agriculture or, to put it differently, at the farm (or enterprise) levels and at higher (or policy) levels.

Economists have tried to develop concepts and tools which may help a rigorous analysis of the decision-making process in the context of supply and demand considerations and political scientists have tried to study the decision-making process in the context of the general relationship of the individual and the group with the state. But attempts made so far to study in depth the interaction of the decision making process of those who are actually engaged in agricultural operations and of others at different levels in the market, in the government and in the international spheres dealing with agricultural problems (in functional, planning or general policy formulation capacity) have been very inadequate.

1. MARKET ECONOMY AND SOCIALIST COUNTRIES

One can discern two distinct world trends in decision-making at the farm and higher levels in the agricultural sector.

In market economy countries, the basic decisions are taken by individual farmers and the market helps to coordinate them. But in practice, this co-ordination is often found to be unsatisfactory. In many instances the market fails to harmonise private and social interests, on the one hand, and short term and long term interests of the economy, on the other.

Many market economy countries have, therefore, found it necessary to circumscribe the individual decisions of farmers by advisory or regulatory actions at the local, market and national levels. The trend has, therefore, been towards a certain degree of centralisation.

In socialist countries, on the other hand, the basic decisions regarding agriculture are taken by the state acting on behalf of the society. Tasks are handed down to successive lower levels of government and individual farms do not often have much choice of their own in regard to inputs, outputs and techniques. The needed coordination is carried out not by the market but by the planning authorities. In practice, however, it has led in many cases to unuseful rigidities and inadequate scope for initiative and innovation at the farm level. This has resulted in serious inefficiency, especially as the central planners often do not, and cannot, have adequate knowledge about specific production conditions in different regions and farms.

Several socialist countries have, therefore, found it expedient to decentralise the decision-making process to varying extents and give greater autonomy with regard to specified decision-making aspects to the local authorities an individual farms.

For market economy as well as socialist countries, therefore, a scientific study of the decision-making process regarding agriculture at the farm and higher levels has become important.

There seems to be a converging tendency in the two. The market economy countries are paying increasing attention to problems of "public choice" and the socialist countries are giving greater consideration to implications of "private choice" than hitherto. There is also a number of countries which deliberately follow a policy of "mixed economy" and pay ernest attention to both, although the relative importance of the two varies widely in regard to issues like price determination, resource allocation and asset distribution.

Further, decision-making for agriculture is no longer confined to national authorities. Compulsions of international supply and demand situations as well as of environmental considerations are making it increasingly necessary to regulate national actions regarding agriculture through some sort of international consultations and even controls.

The UNCTAD meeting which was held in this very hall a couple of months back highlighted the growing international concern in this regard.

2. ECONOMIC AND POLITICAL REASONING

Through what process decisions are made "in" and "for" agriculture (in which agri-business and agro-industry may also be included) at the farm (or enterprise) and state levels and what is the role of intermediate institutions and international organizations is a subject the study of which must draw on the analytical tools of both economic and political sciences.

Decisions based on the operation of market forces, the role of the marginal unit, at what point it will opt out and what effect that will have on submarginal units and the equi-marginal approach to the allocation of scarce resources among competing purposes, represent some of the key elements of economic reasoning.

Decisions based on non-market considerations, inter-play and balancing of opposing views of constituent units, which seek to influence decision not through actual acts of opting out but through pressure of opinion, while striving to remain within the system, represent some of the key elements of political reasoning.

The discipline of agricultural economics, which is perhaps more aptly decribed as the Political Economy of Agriculture, has always been concerned, much more than other branches of economics, with practical problems and policy issues along with the pursuit of "pure knowledge". Therefore, it has necessarily to keep in view both economic and political reasoning in considering the decision-making process "in" and "for" agriculture.

Agricultural economics may in practice be doing a valuable service to both economic and political sciences inasmuch as it may prompt the development of new analytical tools which will be helpful in dealing with models which are much nearer the reality than the models developed so far.

3. THE INFORMATION SYSTEM

In any decision-making process the basic steps involved are (i) definition of the objective, (ii) identification of possible choices, (iii) collection of relevant information and (iv) drawing of appropriate inferences.

For the unsophisticated farmer in a subsistence type agricultural economy, who is a slave of tradition, is constrained by environment, action of neighbours and scarcity of resources and has very limited access to relevant key information, choices are very few, inferences are relatively simple and the decisions are usually un-innovative.

For sophisticated farmers in more developed economies and the public authorities in most countries, such constraints are much less, access to information is much better, choices are considerably greater, inferences are more sophisticated and the decisions are often of better quality and more innovative. But there is usually considerable scope for improvement even in the most developed economies.

A very important role that the agricultural economists can play in all these situations is to spell out carefully what kind of information needs to be looked into and developed at different levels and for different types of decisions, how it should be collected and presented, by whom and in what form, how it can be best utilised by the relevant decision-maker and what kind of research and educational facilities needs to be provided to achieve the optimum results in the kind of situation which is likely to prevail in the short run, and needs to be looked for in the long run.

Care, however, has to be taken that at each level of decision-making, the information to be considered is the most relevant and the minimum needed.

If the variety and complexity of the information are such as to be beyond the comprehension and the intellectual and material resources of the decision-maker, the result may be quite counterproductive.

There will be an advantage in quantifying and tabulating such information as can be so presented, even applying relative weights to as much of the unquantifiable information as possible and arranging the rest in some order of relevance and importance and with the utmost brevity. This will be helpful both for the comprehension of decision-makers and analysis by agricultural economists. These weights and arrangements will have largely to be based on judgement. How the quality of that judgement can be improved deserves special attention of research workers.

The information needed, the capacity for comprehension and the nature of analysis called for will naturally vary widely between the farm (or enterprise), market and public agencies at different levels and between subsistence, commercialised and socialised agriculture.

Anything more than simple farm accounts and extension advice may be too complicated for small farmers while agricultural policy makers may be able to make good use of a large array of information ranging from farm and market data and their projections, input–output matrices, R & D (Research and Development) findings and results of econometric analyses of various kinds. But to do the job properly, the policy makers may be well advised to have the assistance of economic/technological "bridgers" (as suggested by me at our last Conference), who can help "ensure that the several groups of specialists contributing to a decision place their information on the table not only in a reasonably fair manner but in terms which convey the significance of the issues to those who have to make decisions". With appropriate orientation, agricultural economists can well perform the function of such "bridgers".

4. DECISION-MAKING AT THE FARM LEVEL

Decision-making at the farm level is basically the function of the farmer or the enterprise manager.

His task is likely to be easier and efficiency greater, if he has a simple and clear objective like, say, maximising net financial return (or growth rate or whatever) and if he does not face any risk. It is likely to be somewhat more complicated if he has to modify such a straightforward maximisation criterion by the need for minimising the adverse effect of risks and uncertanties.

If in addition he has to modify such a criterion by broader economic or social considerations, his decision-making process is likely to be much more complicated, being subject to diffusion (or confusion) of objectives and therefore inefficient.

This difficulty can possibly be reduced if the farmer or the manager can follow a simple "maximisation" criterion and decisions about broader economic and social criteria are taken at other, say, market or public authority-levels and presented to him as given parameters, comprising a package of constraints and inducements.

As a citizen he may seek to get these parameters modified but as an operator he has to consider them as given.

Within these parameters, anything that may enable and encourage the farmer or the manager to be more knowledgeable, flexible, enterprising and innovative will make for better efficiency and economy in the short as well as the long run. Freedom to take decisions at the farm level and appropriate incentives for efficient management will help foster enterprise while provision of adequate research and information facilities will help promote innovation.

5. PRIVATE AND PUBLIC CHOICE

But it is precisely in this context that decision-making "for" agriculture will start having its impact on decision-making "in" farms (or enterprises).

Decision regarding the basic parameters, e.g., the package of constraints and inducements, comprises the main substance of decision-making for agriculture. And this basically involves public choice as distinct from private choice.

Generally speaking, if private choice is exercised at the operational level in the light of private costs and benefits and public choice at the regulatory level, keeping in view social costs and benefits, it should be operationally more efficient and economic. At least, it should be easier to apply the tests of efficiency and economy to the extent that the two can be viewed as two distinct processes.

The relative importance of private and public choices, the levels at which they will be decided and the way they will interact with one another will, of course, vary widely between the market economy and socialist systems. But there will necessarily be a combination of the two in differing proportions under both the systems.

For purpose of drawing inferences, however, there may be an advantage in considering them separately inasmuch as private choices will be based largely on economic reasoning while public choices will involve, in addition, a great deal of political reasoning. The nature of public choices will also differ significantly between market economy and socialist countries.

Since decision-making in and for agriculture will involve different combinations of private and public choices under different situations, systems and time horizons, the subject is naturally of great intereat to all agricultural economists. But while reasonable progress has been made in evolving analytical techniques for decision-making in agriculture, the concepts and tools available for decision-making for agriculture are yet most inadequate.

6. DECISION-MAKING "FOR" AGRICULTURE

I propose now to take a few more minutes in noting some of the issues which need to be kept in view in this context.

As I have mentioned earlier, unlike decision-making at the farm level, which involves relatively straightforward criteria, decision-making for agriculture involves a variety of criteria, some relatively simple and others complex, some of immediate import and others of long term significance. In addition, there are policy decisions which are made with a focus on quite other matters (e.g., general economic policy measures) but have their effects on agriculture.

Some of the less difficult, although quite important decisions for agriculture, would be those relating to improvement of the information and security systems, which would help decision-making at the farm level in becoming more efficient. These are considered less difficult, not because they are technically simple (in fact some of them may be technically quite complex), but because they are less likely to raise difficult political issues.

The information system would cover, as noted earlier, not only intelligence but also extension, education and research. How best the needed intelligence may be collected, processed and disseminated will raise important questions of methodology and costs. What extension and education systems would best suit the farm operators at different levels of sophistication would require careful study and involve quite difficult technical as well as administrative considerations. The type of research assistance needed by the farm operators will vary considerably between different farm situations and different stages of development.

Risks and uncertainties are closely associated with agriculture. These include not only natural hazards but also market fluctuations. Which of these can and should be covered, the relative efficacy and cost of security measures like insurance, price support, buffer stocks, etc., would raise rather complicated issues.

But the conflict of group interests and related political problems may not be serious in such cases.

Decisions by public authorities for agriculture aimed at ensuring appropriate supply, marketing, credit, irrigation and other infrastructure and, above all, land situations would be no less important for improving the quality of decision-making at the farm level. But they may raise relatively more difficult political issues, apart from quite complicated technical and administrative ones.

What is the appropriate economic return (as distinct from financial return which is of direct interest to the farm operator) to be aimed at, how it is to be measured, what package of constraints and inducements would help achieve it and how it can be best implemented comprise, however, really the key decisions for agriculture. Problems relating to the concept, estimation and application of accounting (or shadow) prices, wages, interests, etc. become relevant in this context. And political issues sometimes tend to become quite serious and complex in decisions of this type.

Then there is the whole series of issues related to agricultural "development" (as distinct from "growth"), role of backward and forward linkages and related structural changes. What are the criteria for considering whether the present trend and pattern of agricultural "growth" is optimum or not, in which directions changes are needed and how these changes can be most economically and efficiently brought about need consideration in this context. While agricultural economists will naturally have to focus their main attention on economic issues, they will have to take into account institutional and social factors also.

For example, in subsistence type of agriculture institutional and social factors are often more significant than economic factors. As agriculture becomes progressively commercialised, economic factors become increasingly more important.

With progressive commercialisation of agriculture, it is also increasingly influenced by international trade and finance. Supra-national considerations have to be kept in view by the national authorities in their decision-making for agriculture. They have also to consider whether the existing international arrangements are the most appropriate for objectives like "security" or "development" and, if not, what kind of change needs to be made and how.

It is obvious that as these different types of considerations are introduced in the decision-making process, it tends to become progressively more complex and difficult. Rational decision-making will be facilitated to the extent that the issues involved can be considered separately and at the level most relevant, and then coordinated step by step at other appropriate levels.

Similarly if the purely economic considerations are applied first at each step of analysis and social and political considerations are brought in later explicitly for ascertaining what kind and degree of modification may be desirable, the scope for diffusion (or confusion) of objectives will be minimised and the decision making is likely to be more rational and useful.

Those decisions which require detailed knowledge of and close contact with field conditions are best taken at operational levels. Others, which require a broader perspective and coordinating approach, may more appropriately be taken at successively higher levels, e.g., regional and national.

At operational levels there should be usually one prime objective but at regional and national levels there may be one or more main objectives and a number of sub-objectives. Traditional economic analysis will be more easily practicable at the operational levels. It will have to be supplemented by "trade off" types of analysis comparing the optimum levels of different sub-objectives and coordinating them with the main objective in such a manner as will give the optimum result from the overall standpoint.

The decision-making levels are arranged mainly in a hierarchical series in socialist regimes. The arrangement tends to be both hierarchical and lateral in market economy countries.

Sophisticated techniques of various kinds such as interactive iteration and successive approximation, econometric and programming models, general systems science simulation approach, etc. have recently been tried for developing some guidelines for decision-making but only with limited success so far, even for decision "aiding", not to speak of decision "making". It is too early to say if new approaches like "combinatorics" can play any useful role in this context.

Judgement reached after a systematic compilation and study of relevant key information and readiness to take some calculated risks and learn from experience seem likely to be the main basis of decision-making at most levels for quite some time to come.

In considering the decision-making process for agriculture, attention will also have to be paid to the decision-making structure and feedback arrangements.

The decision-making structure will vary considerably not only between different socio-economic systems but also between different stages of economic development within the same system. What kind of structure is likely to be the most efficient under specific socio-economic and development situations or at least what kind of considerations needs to be taken into account in deciding what would be the most appropriate structure, will require an objective and careful analysis of both social and economic factors.

Efficient feedback is important for any sound decision-making but the more hierarchical a decision-making structure is, the greater attention needs to be paid to the feedback arrangements. Adequate information flow and capability for prompt adjustment to changing situations are important in all decision-making structures. But appropriate arrangements for monitoring decision flow and interaction of decisions taken at different levels become increasingly important the more hierarchical the decision-making structure is.

7. DECISIONS REGARDING BASIC TRANSFORMATIONS

A tacit assumption of the decision-making processes contemplated so far is that the basic institutional or technological situation will not undergo a revolutionary change or basic transformation within a relatively short period. But if such a change or transformation is considered necessary or unavoidable, the criteria for decision-making will be much more ideological and political than is assumed in the foregoing observations. The process for bringing about such a basic transformation will also involve very different strategies. The difference then will be not of degree but of kind.

Basic transformation of the institutional type may be illustrated by a change over from subsistence to commercial farming in less developed countries, from labour intensive small farms to capital intensive large farms in developed countries, and from privately owned and operated farms to communes or collective farms in socialist countries. Such transformation is usually brought about by the joint operation of market and political forces in the first two and by the action of the state in the third.

Transformation through the pull and push of market forces sometimes takes the character of unbalanced but dynamic growth. Transformation through state action may also conceivably do the same but often it results in more balanced but less dynamic growth.

Basic transformation of the technological type may be illustrated by the many biological, chemical and mechanical innovations that have been introduced in the farms of both market economy and socialist countries in recent years. These innovations have been the result of R & D effort partly induced by market forces and partly promoted by the state in the former and mainly directed by the state in the latter.

The lag between R & D and extension varies widely from country to country. The lag is minimum and the progress in R & D is spectacular in the leading developed countries. In fact, R & D has become a massive self-feeding process in these countries.

The lag is the greatest and the progress in R & D is the mininum in the least developed countries. This lag is both a measure and a function of their underdevelopment.

If the innovative process, which obtains today in the developed countries, can be replicated in and the innovations, which are already known to them, can be adapted to the needs of the less developed countries, it can bring about a basic transformation in their economic situation.

The question is how can it be best done, what kind of decision making process will facilitate it, how soon can it be implemented and what kind of R & D activity needs special emphasis.

There are some who take a pessimistic view because they feel that a basic technological transformation of less developed countries is conditional on their basic institutional transformation, and serious social and political difficulties have to be overcome before such institutional transformation can take place.

There are others who take a more optimistic view on the ground that institutional transformation is often accelerated by technological transformation. They cite the fact that technological innovation is no longer dependent on the genius of an individual or accident of history but has become a well organised self-feeding process in the developed countries. The fact that the communication gap between developed and less developed countries is progressively decreasing, thanks to modern science and technology, is also cited by the optimists in support of their case.

The real issue is not so much the feasiblity of the technological and institutional transformation needed as the decision-making structure and process involved, related political and administrative problems and the time likely to be taken for bringing about the transformation.

The pessimists point to the "consumption explosion" taking place in the rich countries and the "population explosion" going on in the poor countries, the consequent rapid depletion of the exhaustible resources of the earth and increasing strain between the rich and poor countries. They hold that a crisis situation will be reached in 50 to 100 years time. The very wasteful use of exhaustible resources that is taking place under the present market conditions adds to their concern.

The optimists consider both the "consumption explosion" and "population explosion" to be a passing phase. They point to the accelerated growth of the conservationist and family planning movements in recent years. They believe that the "inflection" point, where the "geometric" growth (along an upward rising curve) of consumption and population will get transformed into a "logistic" growth (along a "S" shaped curve) will be reached in a couple of decades time. The projections based on the experience of the "pre-inflection" period of "pre-industrial" and "industrial" economies will turn out not to be true in the "post-inflection" period when what some of them call the "super-industrial" and "post-industrial" economy will prevail.

It is, of course, obvious that our world and its economy are today in a tremendous process of flux.

Technological change has become a massive self-feeding process propelled by an unprecedented knowledge revolution, capital accumulation and world wide interplay of dynamic socio-economic forces.

Agriculture itself is well on the way to changing its character dramatically, although the nature and pace of change differ widely from country to country.

The leading developed countries which already possess conventional technologies far in advance of others, have now set their sights on new revolutionary ideas like nutrient film techniques, controlled environment agriculture, genetic innovations, synthetic food factories and exploitation of non-conventional sources of energy.

The less developed countries are gradually moving from subsistence farming through commercial farming to agriculture based on science and technology that obtains today in the developed countries. There is a progressive spread fo the innovative spirit in these countries, which is likely to have very significant impact not only on food supply but also on population growth and consumption patterns over the next few decades, if it really catches on among the farmers and other rural people. If they become innovative about one thing, they should soon become innovative about other things as well.

Left to itself, however, this process of transformation will face serious ups and downs and be subject to considerable stresses and strains. The various hazards which the pessimists have highlighted are likely to pursue us throughout the transition period, especially during the rest of this century.

It is only through adopting the right kind of decisions in the realm of public policies that the adverse effects of these hazards and the transition period itself may be minimised and the more hopeful vistas that lie ahead may be reached.

Although there are quite a few such hazards, may I make a special mention of one? This is the hazard that is likely to result from the sharp difference in the growth of GNP, food production, population and per capita effective demand for food and other products in the developed and less developed countries during the transition period. There may be surplus of food in some of the former but there may also be acute shortage not only of food but also of purchasing power in many of the latter.

Concessional sale or donation of surplus food by the developed countries will not only have its limits but will also inhibit in the less developed countries the development of agriculture on which bulk of their growing population will continue to depend for employment and income for many years to come.

8. POPULATION AND FOOD

Here we come to the other important theme of this Conference namely, "achieving a balance between population and food supply".* The balance is urgently needed not merely from the global standpoint but also from the standpoint of specific regions in the less developed parts of the world.

Those who are forecasting the advent of a "post-inflection" era of plenty, first in the developed countries and then its spread by gradual stages to other countries, have to pay serious attention to the problems of distribution and related stresses and strains that will arise during the transition period when a large number of less developed countries will continue in the "pre-inflection" era of scarcity. Trade by itself will not be able to solve this problem as the imbalances will often be unduly large in a number of cases, giving rise to serious economic and social troubles. According to one estimate, in the next decade their import requirements may be two or three times as large as the present level. Proposed international food reserves may give some useful temporary relief in years of crop failure but will not provide a real solution.

It is mainly through a planned and massive effort at both national and

* A report on this part of the conference will appear as a separate publication by the International Association of Agricultural Economists (Edit).

international levels to bring about in the less developed countries such institutional and technological changes as would enable and encourage the masses, especially in rural areas, to (i) be informed, enterprising and innovative, (ii) modernise and improve progressively their agricultural production capability, (iii) make the most economic use of scarce factors, (iv) eliminate avoidable waste in production, storage and utilisation, (v) modify consumption patterns so as to make optimum use of not only the main products but also the bye-products some of which are wasted today, (vi) adopt effective family planning practices and (vii) improve their purchasing power, so that the serious economic and social hazards likely to be encountered, at least during the transition period, can be minimised and contained.

While the case studies and other papers prepared for this Conference throw some useful light on a few of these aspects, there is need for much more research to be done. For instance, a systematic study of the various elements mentioned above in a comparative manner for a sample of less developed areas, on the one hand, and for some sharply contrasting experiences in different patterns of production, consumption and development, on the other, could provide us with valuable insight.

One intriguing question is what would be the implications for global and regional food demand and supply situations during the next two or three decades, if the other countries followed, say, either the American or the Chinese pattern.

Another question is the likely effect of a substantial time difference in reaching the "inflection points" in regard to growth of population and per capita consumption of agricultural and non-agricultural products in developed and less developed countries and the implications for agricultural development and rural employment in the latter.

How best the possible adverse effects of the deceleration of population growth in both sets of countries, and of consumption growth in developed countries on the demand for agricultural products of the less developed countries, can be corrected by improving the per capita income of the poorer classes and the development of the secondary and tertiary sectors in the less developed countries themselves will also need careful consideration.

9. CONCLUSION

Be that as it may, there can be little doubt that if the very serious problems which the less developed countries are likely to face during the critical decades ahead are to be met, there will be urgent need for imaginative and effective public decision-making at international and national levels for the formulation and implementation of long term and short term policies and programmes for a more balanced development of the world's economy and its important sectoral and regional constituents. Intensive technological and economic research to meet the special requirements of different regions and spread of innovative education skills among the rural people will be as important elements of these policies and programmes as institutional reforms, trade promotion and transfer of resources. What kind of forward looking research and educational programmes should be taken up will need a special kind of perspective and strategic decision-making by the public authorities concerned.

These and other public decisions to deal with the emerging problems of a secular change may be significantly different from the kind of public choices for agriculture which I have mentioned earlier, and may be considered by some to be beyond the scope of conventional agricultural economics. But, if agricultural economists have to deal with the various problems which the impending transition in the world agriculture and world economy is likely to throw up, they will have to pay serious attention to this other kind of public decision-making that will be called for.

I hope that the deliberations in this Conference will throw useful light on some of these basic issues, indicate the directions in which further research effort needs to be concentrated and help evolve arrangements which would provide the intellectual, institutional and financial support that such effort will require.

SAMAR R. SEN

Inauguration of Leonard Elmhirst Lecture

We have gathered today to honour the memory of our Founder-President, Leonard Knight Elmhirst.

Leonard Elmhirst was a man of great vision and inititiative, a patron of the visual and performing arts as well as of the social sciences and a champion of the downtrodden and deprived all over the world. This last concern of his, in particular, prompted him to help the Indian poet-philosopher, Rabindra Nath Tagore, to initiate path-breaking experiments and studies in rural development in India, to undertake other valuable experiments in his own country and to establish the International Association of Agricultural Economists, of which he was the Founder-President for four decades.

I do not propose to take your time by going into the story of his unique life. A succinct account was published in the proceedings of our last Conference and many of you must be familiar with it.

I may briefly note, however, some crucial events in his remarkable career, which are of direct interest to our profession, namely, his first acquaintance with problems of under-developed agriculture during the First World War, his decision to study agricultural economics at Cornell University shortly thereafter, his close association with Tagore in setting up the Institute of Rural Reconstruction at Sriniketan, India, his many faceted experiments in the arts, crafts, education and social sciences at Dartington Hall in England, his initiative in organising and developing the International Association of Agricultural Economists and P.E.P. (Political and Economic Planning), his assistance for the International Institute of Agrarian Affairs (which was effectively our Association's base for 40 years) and the Oxford Institute of Agricultural Economics and his constant endeavour to facilitate communication between agricultural economists working under different social systems.

Leonard Elmhirst and his contributions for the promotion of the science of agricultural econmics will be remembered by the members of our Association with deep gratitude for ever.

He was a man of many parts with a special gift of friendship and inspiring people for great endeavour. He had a deep human concern and his constant exhortation to his fellow economists was that they should always keep "man" as the main focus of all their analysis. The message that he sent in August,

1973, to our last Conference held in Sao Paulo, Brazil, aptly expressed this basic approach of his. He wished "for lots of down to earth discussions at the grass roots with plenty of vision for the future, and with the boots of economists firmly rooted in the soil and their heads in the skies". "May the computors", he urged, "be kept properly in their place as servants of, and not masters over, humanity". He gave this Conference the broad humanist outlook that characterises it. What he sought to promote was the study not merely of the economics of agriculture but also of the political economy of agriculture.

It is not that all his widespread endeavours proved to be successful. He had his share of failures. But that never daunted him. He referred to his failures as "negative results", judging them as no less useful than "positive results" in any scientific experiment and cheerfully taking up new experiments. Even at the age of 80 when he died he was looking forward and not backward. He was one of the blessed few who are ever young at heart.

Your Executive Committee felt that the best way in which your Association could honour Leonard Elmhirst's memory would be to invite one of the world's outstanding agricultural economists to give in each of our Conferences a lecture to be known as the "Leonard Elmhirst Lecture", on some theme which would touch upon one of the key problems in the field of agriculture which was so close to Elmhirst's heart.

I have, therefore, the great privilege today of inviting the doyen of our profession, Professor Theodore W. Schultz, to deliver the first Leonard Elmhirst Lecture.

Professor Schultz was born in 1902 and obtained his Ph.D. degree from the University of Wisconsin in 1930. The same year he began his academic career as a teacher at the Iowa State College where he served as Professor of Agricultural Economics from 1934 to 1943. He joined the University of Chicago as Professor of Economics in 1943 and retired as the Charles L. Huchinson Distinguished Service Professor in 1972. He has served with great distinction in many U.S. Government and international assignments in the field of agricultural economics and policy and travelled widely in both developed and developing countries. He has been President of the American Economic Association and Fellow of the American Farm Economics Association and a number of other important learned societies. His contribution to the development of the science of agricultural economics has been outstanding and his defence of academic values courageous. He is a pioneer among agricultural economists in emphasising and analysing the role of human capital. Among his many publications, special mention may be made of his books: *Redirecting Farm Policy*, *Agriculture in an Unstable Economy*, *Production and Welfare in Agriculture*, *The Economic Organisation of Agriculture*, *Transforming Traditional Agriculture*, *Economic Crisis in World Agriculture*, *Economic Growth and Agriculture*, *The Economic Value of Education and Investment in Human Capital*.

The theme on which Professor Schultz will be speaking today is "Economics, the Farm People and the Political Economy of Agriculture". There could not be a more appropriate subject for this first lecture instituted to honour the memory of Leonard Elmhirst.

THE FIRST ELMHIRST MEMORIAL LECTURE

THEODORE W. SCHULTZ*

*On Economics, Agriculture, and the Political Economy**

The increase in the economic value of agricultural products since 1972 raises several important questions. Are the high world prices of these primary products here for the long pull or are they transitory prices? The answer, as I see it, depends in large measure on what governments do to agriculture. What then are the agricultural cost implications of the acts of governments? In this context, the issue of value and price can be put as follows: Are the underlying costs such that world agriculture could produce over the next decade and longer a supply of products that would be adequate to equate the demand at substantially lower relative world prices than the prices that prevailed during 1974 and 1975? I shall argue that the technical and investment opportunities to increase agricultural production are such that the answer to this key questions is in the affirmative. But this is not to say that these opportunities will be realized. It is not a prediction that governments will act appropriately.

I take it to be obvious that what governments do to agriculture differs greatly from country to country. It is also obvious that the economic performance of the agricultural sector differs markedly among countries. What is not obvious, however, is that much of the difference in the economic performance of the agricultural sector is a consequence of what governments do to agriculture. Most agricultural economists see this issue as too controversial and too unsettled for analysis. Some will say that it entails special values, welfare objectives and the maintenance of political authority and that these are matters that go beyond positive economics. It is my contention, however, that economics can deal with major parts of this issue, and that economists can evaluate the economic effects of what governments do to agriculture.

To suggest what I have in mind, consider the marked differences in the responses of agriculture throughout the world beginning in 1973 to the sharp rise in the economic value of agricultural products. The actual incentives that would warrant an expansion in production by farmers varied widely from country to country. A large part of the differences in these incentives is the result of actions taken by governments. Accordingly, the 1973–1975 period is akin to an "ideal" experiment to measure the influence that governments

* University of Chicago.

had on incentives and to determine the effects that these differences in incentives had on the production responses of farmers. I shall be returning to this "experiment" presently.

Using Viner's definition, agricultural economics is what agricultural economists do. Much of what they do is done well. The use of theory in empirical work has been justly praised by Wassily Leontief in his American Economic Association presidential address. He criticized the drift in economics toward theoretical assumptions with insufficient concern about observable facts. He noted, however, that "an exceptional example of a healthy balance between theoretical and empirical analyses and of the readiness of professional economists to cooperate with experts in neighboring disciplines is offered by Agricultural Economics".[1] Professor Leontief's assessment, however, is only part of the story. It is true that we who are agricultural economists know a good deal about soils, agronomy, crops and livestock, and we stay abreast of the contributions of agricultural scientists. We are not naive and simple-minded about the role of farmers as workers, capitalists, and entrepreneurs. We are skilled in using modern quantitative techniques, and occasionally there is one among us who contributes to their advances. Our studies of the management and production of farms are guided by the theory of the firm, and our macro models are designed to deal with agriculture as an integral part of the general economy. But what Leontief fails to see is the increasing opposition to economics in social and political thought, the debasement of economics by governments, and the unwillingness or inability of economists to challenge this adverse drift. That "healthy balance between theoretical and empirical analyses", attributed to agricultural economics, does not suffice to make this challenge.

This opposition to economics is not confined to either the low or the high income countries. Nor is it restricted to a particular type of government. It is not limited to agriculture, for clearly other parts of the economy are not spared. Most of the high priests of national and international politics, whether they speak for the first, second or third world, are at heart contemptuous of economics. But despite differences in political organization, in cartels, in marketing boards, in commodity agreements and in internal pricing of factors and products, *there is no free lunch*. The hard realities of the costs of producing goods and services are not abolished by either national or international politics. Herein lies not only the hope but the necessity of economics.

The easy road is to accommodate opposition. But the utility of economics is reduced by making it ever more permissive. Although corporations, labor unions, farmer organizations and consumer advocates perform useful functions, they are not innocent economic agents for they do conspire to exact benefits for themselves at the expense of others. To disassemble economics for the purpose of serving the special interest of these organizations is to sell economics short. While it is also true that governments are necessary, to make economics subservient to them regardless of what they do to the economy is to take the heart out of the utility of economics. What all this implies for us in agriculture is that it will not suffice for agricultural economists to take the particular economic goals of governments as given. For economists to proceed in this manner means that economics becomes hostage to government. When this occurs, and it can be readily observed, the result is that economists become "yes-men" in the halls of political economy.

I believe that the core of economics is sufficiently robust to evaluate the cost and welfare effects of various political-economic institutions. One need not become stranded on value judgments. No doubt I am now vulnerable; I obviously need a cloak of credibility. It is at hand from the pen of my colleague, Harry G. Johnson.

> One of the penalties or privileges of advancing age and professional maturation in an academic career in the social sciences is that one is forced to think in terms of progressively broadening frames of institutional and cultural reference. The young economist . . . comes out of the graduate student mill with a narrow range of specialized skills and the stamp of his teachers' ideas heavy upon him. He makes his way initially by applying his skills . . . in an institutional and social context taken as an immutable part of his environment. Only as he acquires self-confidence, on the one hand, and abrasive experience of the larger . . . world, on the other, can he afford again to accord himself the liberty to question society's institutions . . . .[2]

Needless to say, agricultural economists are not renowned for their critical evaluations of the economic effects of various political institutions on agriculture. At a more general level, the positive implications of value and price become weaker as economists become more permissive. To reach the Promised Land, never introduce value-judgments into your analysis, always screen your assumptions with great care to make absolutely sure that they are not contaminated!

There are two academic styles for dealing with economic assumptions and value judgments. From the twenties to the fifties it was the style to question the assumptions of theoretical analysis and dispute the value-judgments asserted to be implicit in the theory. It has been called, quite properly, "negative economics".[3] Since then the emphasis has shifted to whether or not the implications of the economic theory, given its assumptions, are validated by observable economic behavior. This approach has become known as "positive economics".

It is not difficult to question the simplistic assumptions of the early English economists. Nevertheless, particular implications of their theory are valid empirically, for example, that the removal of the tariff on imported wheat increases the real income of the working class. Marx accepted this implication of the tariff on wheat and called this tariff a "bread tax". Despite the criticism levelled against the assumptions on which Marx's theory rests, one of the implications of his theory is that technical advances in agriculture over time reduces the income share of landlords. The economic history of the high income countries strongly supports this implication. There are, however, other parts of classical theory and of Marxian theory that fail to meet the test of positive economics.

The analytical task that I deem to be possible and necessary need not be confined to current developments. Economic history is replete with political-economic experiments from which we can draw information. To gain an historical perspective, I shall begin with an account of the interactions between social thought and the institutional order to show the rise and decline of particular political-economic institutions and the apparent reasons for the instability of these institutions. I shall then consider some of the more important economic effects of various political-economic institutions on agriculture currently and in the recent past.

1. INTERACTION BETWEEN SOCIAL THOUGHT AND THE POLITICAL ECONOMY

I shall appeal to three assumptions, namely, that social thought consists of various social, political and economic ideas, that the dominant social thought shapes the institutionalized order of society (country), and that the malfunctioning of established institutions in turn alters social thought. The ideas that are embodied in social thought are of two historical types: those that rationalize and contribute to the codification of the prevailing order, and those that arise in protest to the established order which become embodied in social thought and then become strong enough to induce a real alteration in political-economic institutions.[4]

The mercantile economy that prevailed for at least a century, for example in England, prior to 1776 was buttressed by the established Church and the Law. It was also rationalized by the economic "literature" of that period which provided support for governmental restrictions on trade, on internal prices and wages and on migration – *restrictions which have once again acquired a modern ring.* The advocates of mercantilism had a theory which implied that keeping wages low increases the national product. It is this implication that provided support for the utility-of-poverty doctrine.

In retrospect, the utility-of-poverty doctrine of that period is startling not because it is no longer pursued by many countries but because it was so openly supported at that time. The doctrine held sway for a century in English history during which the poverty of the lower classes was declared to be desirable. Between 1660 and 1776, the mercantilism of England produced an intricate system of foreign and domestic policies, and it sought to rationalize the utility of poverty. Edgar S. Furniss[5] in his prize-winning Hart, Schaffner and Marx essay devotes a long chapter to "The Doctrine of the Utility of Poverty". The beliefs of illustrious individuals of that period should increase our ability to doubt the social thought of any period. Thomas Mun's view was that "penury and want do make a people wise and industrious". Arthur Young asserted that "every one but an idiot knows that the lower classes must be kept poor or they will never be industrious". John Law argued that "laborers were to blame for recurring high prices because of their insufferable habits of idleness contracted when food was cheap". William Petty joined in this chorus. There should be taxes on consumption and out-migration should be curtailed. Charity was thought to be the nursery of idleness. A larger population would keep the laborers poor, and immigration should be encouraged. George Berkeley, Bishop of Cloyne, proposed to reward parents of large families and to tax families with no children in support of the doctrine of the utility of poverty.[6]

One of the contributions of the classical economists was to show that, contrary to claims of mercantile economists, low wages imposed by governments reduce the national product. I view their analysis bearing on this issue as an excellent example of positive economics.

The ideas that are associated with liberalism emerged in large part as a protest to the adverse social and economic effects of mercantilism. The basic ideas of liberalism gradually developed into a strong internally consistent body of social thought, and over time liberal political and social institutions replaced mercantile institutions. 1776 has become a convenient birth year of liberal thought. What is often referred to as the bible of economics, namely the *Wealth of Nations*, carries this date. The argument of Adam Smith rests on the proposition that people, in responding to their own diverse self interests in their economic activities in an open competitive economy which is not fettered by private and public monopolies and which is supported by a political order in which the functions of government are greatly restricted, will maximize the social product. The economics of Smith became an integral part of the core of liberal social thought which, over the decades that followed, profoundly altered the institutionalized functions of many governments.

In Adam Smith's economy there is no room for the doctrine of the utility-of-poverty. The utility that matters is revealed by what people do in serving their own self interests. Smith's argument for an open competitive market implies free trade and a free labor market paying wages that are not fixed or restricted by government. It also implies that people are free to migrate from farms to towns and to leave the country if they choose to do so. Migration boomed in response to changes in institutions and to the economic opportunities associated with economic growth and the related growth in population which in turn also induced migration. Even in Russia of that era, we have August von Haxthausen's[7] account covering his 1843 tour into Russia of the liberal approach of that government in the opportunities that it accorded the Mennonites and Hutterites.

But the actual performance of the institutionalized liberal political economy in turn gave rise to protests which featured what was seen as a critical flaw of the liberal economic system, namely that it favored private property rights, and in doing so it gave rise to capitalism and to industrialization with insufficient protection of the rights of labor. The ideas advanced to correct this flaw are embodied in a body of social thought that calls for a political-economic order in which the functions of government are much enlarged, even more so than they were under mercantilism. As social thought it is viewed as socialism; as a political-economic system, it ranges from centrally planned economic development to a system of command economics. The ideas that emerged from these protests prior to those of Karl Marx called for various forms of socialism. The contributions of Marx, however, came to dominate the political and economic foundations that are required for socialism. The response to Marxian ideas has altered greatly the institutions of many nation states, and as noted, the economic functions of the governments of these states are much enlarged.

An essential part of the argument in support of socialism is that an open competitive system is blind and that the self interests of people must be directed and controlled in order to achieve efficiency and equity in economic development. Prices, wages and migration are instruments to be used by government in achieving its goals.

There now are signs of a nascent neo-liberalism. Because of the dependency of socialism on a vast increase in the functions of the government, which in some countries consists of strong authoritarian nation states, and because of the now widely observed adverse effects that the governments of these states have on personal freedom, protests akin to those of two centuries ago are once again the order of the day.

The historical record of the last three centuries is not inconsistent with the view that social thought and institutions are far from stable. The pattern of instability has some of the earmarks of a cycle. When institutions function badly they produce the seed for their decline. In my thinking, this cycle of instability is an indication of the incompleteness (inadequacy) of the social thought that shapes institutions.

2. AGRICULTURE AND THE POLITICAL ECONOMY

Much of the economics of agriculture that really matters is in the differences of the political-economic institutions. When markets are segmented by governments, the advantages of the extention of the market are lost. When imports and exports are controlled, agricultural prices do strange things. When marketing boards have a monopoly, farmers and consumers are well advised to beware. When governments authorize the procurement of agricultural products from farmers, the agricultural economy is placed in receivership. When the ministers of agriculture treat agricultural scientists as clerks, the agricultural research enterprise becomes stagnant.

The economics of agriculture in this context has both short and long run implications. Those which are short run are most readily put to test when a major change occurs. The events of 1973–75 came unexpectedly, and suddenly they resulted in large changes. I shall appeal to this period in commenting on particular short run implications. For the long run, I shall consider various governmental effects on agricultural production during the last three decades.

As I noted at the outset, the 1973–75 period on world agriculture is an instructive "experiment" in prices. We all know that the shortfall in food and feed grains enhanced the economic value of these products, but what is overlooked is the fact that this increase in value was greatly distorted by various governments. A major set of countries suppressed the economic signals that were called for by the shortfall. Consumers in these countries were spared from adjusting their consumption to accommodate the shortfall, and thus, the burden of the required consumer adjustment was increased in the rest of the world. Similarly, in agricultural production throughout the European Economic Community, agricultural prices were held in check and farmers did not have the required economic incentives to increase their production. In sharp contrast, in Australia, Canada and the United States, the food and feed grain markets were open and competitive and farmers responded sharply to the higher prices. So did the farmers in Brazil in the case of soybeans. In general, on this issue, many of the low income countries have a better economic record than the European Economic Community.

Does the control of agricultural prices in one part of the world increase instability in the other part? Looking at the price effects of the 1973–75 change in the supply–demand balance in food and feed grains, the answer is in the affirmative. Theory and evidence tell a consistent story on this point, namely, when a government achieves internal price stability by controlling agricultural exports and imports, it increases the price instability in the rest of the world.

The open market instability of agricultural prices that occurred in the early seventies was much greater than that of the sixties although the shortfall in grain production caused by poor crops in the sixties departed farther from the trend level in world production than did the shortfall in the early seventies. The reason commonly cited for this difference in the behavior of prices between the two periods has been the grain storage policies of the United States, Canada and Australia. But a careful analysis of the price effects of these stocks indicates that a large part of the greater price instability of the seventies remains unexplained.

D. Gale Johnson (1975)[8] has advanced the hypothesis that this unexplained instability of agricultural prices in the seventies is a consequence of the fact that "a much larger percentage of the world's grain production and consumption in the early 1970's than in the 1960's occurred in a framework of policies to achieve internal price stability through the controls of imports and/or exports". The evidence supports the Johnson hypothesis: fixed internal agricultural prices have become increasingly the rule in the Soviet Union, the European Economic Community and China; these countries account for about half the world's grain utilization in recent years. As a consequence, prices in about half of the world "do not serve the function of influencing either consumption or production when the world's demand–supply balance has changed". Accordingly, all of the adjustments must be made in the other half of the agricultural world. In the early seventies "these adjustments fell primarily upon two groups of countries – major grain exporters and the low income developing countries that imported grain". Theory implies that the increase in price resulting from a given shortfall of grains would be approximately doubled in that half of the world where prices are not fixed compared to the rise in prices that would have occurred if the other half had not fixed their internal prices of agricultural products. This important implication of the theory does not depend on value judgments; it is a testable implication, and in terms of positive economics, the evidence supports the implication.

In developing agriculture for the long pull, what types of public control machinery are on the recommended list of agricultural economists? A pointless question no doubt, since we have not made it our business to analyze and publicly certify the economic performance of alternative control devices. For my part in the spirit of the Mikado, I've got a little list of economic offenders who never would be missed – who never would be missed.

I know of no country in which the productive capacity of agriculture has been in fact increased by large P.L. 480 grain transfers from the U.S. to the receiving country. P.L. 480 is on my little list because when it comes to building agriculture, P.L. 480 is a liability.

The procurement of agricultural products from producers by the authority of the state is bad economics, and when it occurs it is a sure sign that the country which uses its police powers for this purpose is in deep agricultural trouble. Needless to say, command procurement is one of the major economic offenders on my list.

In marketing it is necessary to distinguish between cooperatives that are subject to competition and marketing boards that are vested with monopoly powers as agencies of the state. I find it odd that agricultural economists have so grossly neglected P. T. Bauer's excellent studies of West Africa's marketing boards. Marketing boards are always established for a noble purpose, e.g., to stabilize agricultural prices. But if the commodity is mainly exported, the board becomes a convenient device for raising revenue for the government; as the sole buyer it can pay a low price to farmers and sell abroad at a high price. It is a successful public device for pricing the best export crop to death! Then, too, some boards "employ" an inordinate number of people who are favored by the government – the well known practice of the "spoils system". A cogent and useful economic assessment of the various marketing boards in Kenya is at hand in *Agricultural Development in Kenya*.[9] I list state monopoly marketing boards as economic offenders.

While it is not difficult to detect the adverse effects of large P.L. 480 imports, of command procurement, and of state monopoly marketing boards on agricultural production, there are other and more important unsettled economic questions. Is it true that rents, interest and profits "never will be missed" once they have been eliminated? Why not substitute administered prices for competitive prices? When black markets prosper, what are the economic implications?

My own visits to the Soviet Union began in 1929 and I have extended them to observe agriculture in the other socialistic countries of Eastern Europe. Compared to most of the world, these are highly skilled, technically advanced and in general well-equipped countries. The puzzle is in the below par performance of agriculture. Despite the recent very large investments to increase agricultural production, it is evident that the cost of producing milk and meat is exceedingly high; the quantities that consumers are demanding are not forthcoming. Weather, climate and soils do not hold the key to this puzzle. Nor can the disappointing performance be attributed to a lack of competence in the management of agricultural production. In an analysis of the disappointing performance of Soviet agriculture since World War II, my colleague, D. Gale Johnson, finds that it is not primarily a consequence of the very large collective farms.[10]

The key to this puzzle is in the allocation of resources that are devoted to agricultural production. Economic inefficiency is the problem. Viewed as an allocative problem, it is not confined to the socialist countries. It is in fact pervasive throughout most of the low income countries, although the administrative devices that are used in allocating resources differ widely from country to country.

When the dominant social thought proscribes the use of rents, interest and profits and declares that market-oriented competition is blind, the accommodating political economy is severely handicapped in allocating resources for agricultural production. It may use all manner of devices – distributing fertilizer and other inputs by issuing quotas, controlling the migration of farm people from farm to farm and from farms to cities, segmenting the market for agricultural products within the country and allowing no movements between the different areas except as approved by the government, and even preventing farmers from buying and selling to each other.

The heart of the problem then becomes one of knowing what the real economic values are, values to be used in commanding production and consumption. In this context no matter who is making economic decisions, be they planners, managers of collective farms or small private farms, using the information that is thus provided, leads to economic inefficiency. This is not to argue that the supply of valid, usable economic information is costless. It becomes very expensive under the above circumstances for the available supply in many countries has two attributes: the quality of economic information is exceedingly low and, nevertheless, very costly. It should also be noted that improvements that are to be had from the work of the agricultural scientists in terms of better plants, animals, fertilizers and equipment – important as they are when they are used efficiently – in no way solve the problem of ascertaining the real economic value of agricultural products and factors. Nor is the solution to be had by simply using more advanced computer technology.

I return in closing to the question with which I began, what about future costs of agricultural products? In terms of technical possibilities and pure economic opportunities, the prospects for lower costs are good, but in terms of what is being done politically, the prospects are bad. Meanwhile, international food conferences produce a lot of weak reports, and social thought produces strong ideologies. But reports and ideologies do not produce food. Fortunately, plants and animals do not read reports nor do they discriminate against the ideology of any government. One thing is certain – what farm people will be able to do holds the key to our story. Would that agricultural economists would help them to write that story!

NOTES

[1] Leontief, Wassily (1971) 'Theoretical Assumptions and Nonobserved Facts', *American Economic Review* **69**, 1–7.

[2] Johnson, Harry G. (1975) *On Economics and Society,* University of Chicago Press, p. ix.

[3] This quote is from Johnson, Harry G. (1975) 'Learning and Libraries: Academic Economics as a Profession: Its Bearing on the Organisation and Retrieval of Economic Knowledge', *Minerva*, **13:4**, pp. 621–632.

[4] In part drawn from my paper, 'The Economic Value of Human Time Over Time', April, 1976, publication forthcoming.

[5] Furniss, Edgar S. (1920) *The Position of Laborers in a System of Nationalism*, Boston, Houghton Mifflin, see Chapter 6.

[6]See my 'Public Approaches to Minimize Poverty' in *Poverty amid Affluence*, edited by Leo Fishman, New Haven: Yale University Press, 1960.

[7]His *Studies on the Interior of Russia*, edited by S. Frederick Starr and translated by Eleanore L. M. Schmidt, University of Chicago Press, 1972.

[8]Johnson, D. Gale (1975) 'World Agriculture, Commodity Policy and Price Variability', *Proceedings of the American Journal of Agricultural Economics*, December.

[9]Heyer, Judith, Maitha, J.K., and Senga, W. M., (eds.) (1976) *Agricultural Development in Kenya*, Nairobi, Oxford University Press. See Chapter 10 especially.

[10]Johnson, D. Gale (1975) *'Theory and Practice of Soviet Collective Agriculture'*, Office of Agricultural Economics Research, University of Chicago, Paper 75:28, December. Unpublished.

GLENN L. JOHNSON*

Contributions of Economists to a Rational-decision-making Process in the Field of Agricultural Policy

In both socialist and non-socialist countries, governmental units use huge amounts of resources in amassing information, analyzing it, and reaching decisions on the problems before them. And, in private sectors, producers and consumers also use large amounts of resources to make decisions on, among other things, production, consumption, saving, and generation of human capital which are interrelated with public decision-making. Despite its obvious empirical and economic importance, the economics of public decision-making and management is relatively neglected; this neglect creates the opportunities to contribute discussed in this paper.[1]

An underlying unity emerges when we examine managerial and decision processes which transcends the developed/less developed, the micro/macro, the public/private, and, within the public sector, the centralized/decentralized spectra. This unity permits us to draw on a wide range of experience in studying public decision-making.[2]

In asking me to prepare this paper Professor Dams added the subtitle (The Position Between Adjustment Engineering and Critical Reflection on Values) and referred to a book about values and economists which I have co-authored.[3]

PART 1

Decision-Making and Management – A Summary Survey

This paper is written on the premise that it is important to be efficient – to be economic – in using the large amounts of resources and money devoted to public problem-solving concerning agriculture.

Management and decision-making are performed by more or less separate units or enterprises within an agency, firm or house-hold.[4] Enormous amounts of resources and inputs are used in these units.

* Professor of Agricultural Economics, Michigan State University. I am grateful to Drs. James C. Bonnen, Lester Manderscheid, Ulf Renborg and William Martin for reading and commenting on this and earlier versions of the manuscript. This paper is "Michigan Agricultural Experiment Station Journal Article Number 7640".

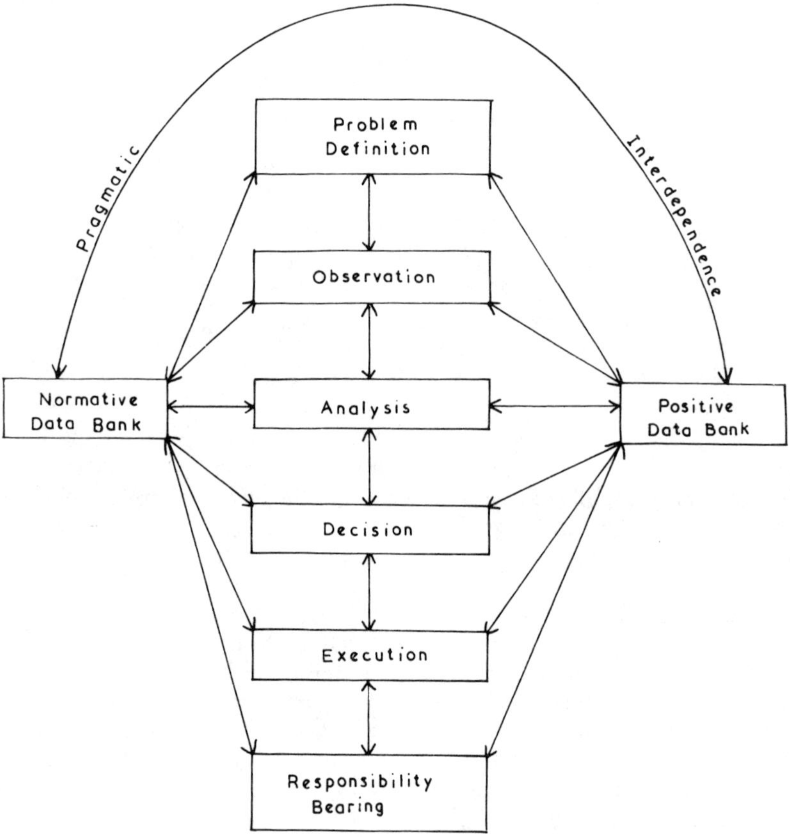

FIG. 1.[5]

The regulatory governments of the free enterprise, decentralized democracies devote high proportions of their budgets to decision-making and management. Socialized governmental units, which produce goods and services as well as regulate and control, devote smaller proportions of their budgets to decision-making and management but the absolute amounts are still large.

The processes whereby problems are defined and solved are behavioral. Hence, understanding managerial and decision-making units requires an understanding of human behavior. As in Figure 1, the behavior to be understood includes: (1) problem definition; (2) data and information acquisition, storage and retrieval; (3) analysis; (4) decision-making; (5) action, and (6) responsibility bearing. Each of these will be discussed in the following sections which will show how (1) economics contributes to performance of each function, and (2) economics can be used to increase the efficiency with which managerial units perform each function.

The crucial role of problem definition

Problem definition is crucial for decision-making units. Problems are defined in terms of both positive and normative information, the latter being both monetary and nonmonetary.[6] The existence of a meaningful problem implies some possibility, at least, that a decision rule could process the relevant positive and normative information into a prescription which would materially improve the existing problematic situation. In a study of the managerial processes of 1075 midwestern U.S. farmers (The Interstate Managerial Study, *IMS*), the structured questions used did not deal with problem definition or with the processes whereby farmers acquire and use normative information.[7] Fortunately, the IMS questionnaire contained sequences of non-structured, open-ended, probing questions. These revealed enough about problem definition for IMS researchers to see their error.[8]

Post IMS experiences with public decision-making units indicate that problem definition is of similar basic importance in their activities.[9] A related conclusion is that problem definition and solving is an iterative, adaptive, interactive process. When the investigator of a problem is not the decision-maker, the iterative, adaptive problem-solving process must also be interactive because decision-makers and affected persons are sources of information to investigators.[10]

Another important lesson growing out of the IMS experience is the distinction between kinds of information and problems. Initially, price,[11] production, new technology, institutional and human problems[12] were defined. Responses of the 1075 farmers to open-ended sequences of probing questions revealed, eventually, that these five "kinds of problems" were not problems at all; instead, they were kinds of information. Specific problems about which decisions are made, it turned out, typically require most of the five kinds of information for solution. Further, IMS research and subsequent experiences indicate there is probably no *stable* classification of problems for individual farmers or, for that matter, nations and regions, programs, and projects. The domain of each problem is described with a mix of information quite unique to each new problem.

Before leaving the subject of problem definition, we should discuss the word "paradigm" which is so fashionable these days. Kuhn, in his book on scientific revolutions,[13] noted that the problems before a scientific discipline sometimes change so drastically that a scientific revolution is required. As I see it, more or less minor paradigm revolutions are continuous for problem-solving researchers and decision-makers with each such revolution requiring a unique mix of contributions from various disciplines.

Whether a specific problem is macro or micro, public or private, etc., the efficiency of a decision-making unit depends on its ability to (1) define the problem and judge what information is required to solve it; (2) understand the statistical, technical, and human aspects of the process of acquiring, assembling, and storing the information; (3) convert normative and positive knowledge into problem-solving prescriptions; (4) execute decisions; and (5) monitor and bear responsibility for decisions made and executed.

Data acquisition (observation) storage and retrieval

Each type of information needed to solve a problem tends to be generated by a specific discipline such as, say, economics, sociology, political science, soils, crop husbandry, etc. IMS researchers found that midwestern farmers would use over 200 categories of information in solving problems of reorganizing and operating farms under the original five categories.

Further, it is now noted that there are three dimensions involved in the five general types of information considered. The difference between new technology and production with existing technology is in the *time* dimension, while the difference between the other four and price information is in the *positive/normative* dimension.[14] Basically, there were three broad categories of information (institutional, technological, and human), each existing in the past, present and future tenses and in the normative/positive dimension.

Kenneth Boulding, like IMS researchers, divides a decision-maker's positive image of his environment into the human, technical, and institutional and then sees superimposed upon that image an image of the values attached to the elements of the positive image. Further, he argued that the processes

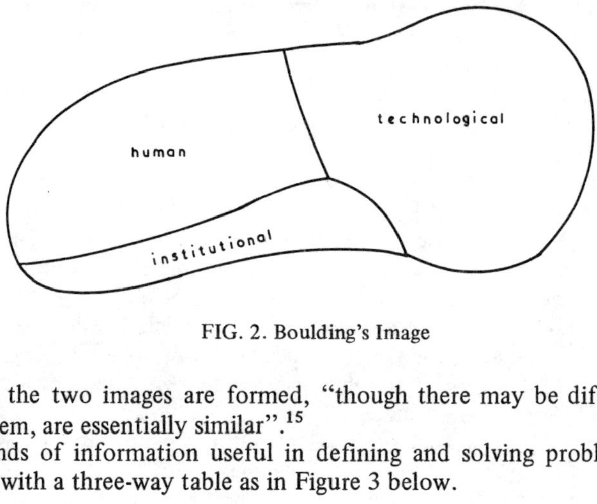

FIG. 2. Boulding's Image

whereby the two images are formed, "though there may be differences between them, are essentially similar".[15]

The kinds of information useful in defining and solving problems can be outlined with a three-way table as in Figure 3 below.

	Positive			Normative		
	Past	Present	Future	Past	Present	Future
Technical						
Institutional						
Human						

FIG. 3.

Prescriptive knowledge which is not diagrammed above[16] is functionally related to positive and normative information via decision rules (strategies). The normative includes both monetary and non-monetary values. Decision rules or strategies are considered in the Richardson, *et al.* paper on farm-level decision models presented at this conference. They, of course, deal mainly with the decision step in the total process of management or decision-making.

The IMS studied the processes whereby managers acquire, analyze and use information on prices or monetary values. It also measured utility cardinally.[17, 18] One emphases was on how information on past and present prices is processed into price expectations. A quickly learned lesson was that the price expectation models developed by statisticians and econometricians were far too specialized on statistics to cover what farmers actually do.[19] Midwestern U.S. farmers were so keenly aware of price-making forces that they took into account competing and complementing products, the influences on prices of governmental programs, inflation, changes in demand, war, labour union activity, monopoly, taxation and subsidy programs. The behavior of farmers forming price expectations is much too complex to be modeled with simple (or even more complicated Nerlovian type) distributed lag models.[20] M. Lerohl has developed price expectation models based on distributed lags *and* other kinds of information used by farmers; these have performance characteristics superior to or at least as good as distributed lag models.[21]

The attention given to how farmers form price expectations was not matched by comparable attention to how farmers form expectations about non-monetary values. This was a basic mistake which should not be forgotten by students of public decision-making and private managerial activity.[22]

The lesson immediately above leads to another. In order to understand fully the acquisition of knowledge by decision-makers and, hence, to be helpful to them in their quest for knowledge, one must be philosophically flexible. Decision-makers *do form both* normative and positive expectations. Philosophic positivsm (often incorrectly designated *the* philosophy of science) is based on the metaphysical proposition that there are no normative experiences to use in developing descriptive normative knowledge. Fortunately, there are other respectable philosophies which grant the possibility of objective knowledge of the normative and prescriptive. These include pragmatism and various forms of normativism. *Pragmatic* investigators believe that positive and normative information depend upon each other mutually in a problem-solving context. In U.S. agricultural economics, important productive workers who at least partially follow or followed this point of view include John R. Commons, L. Salter, Richard T. Ely, Kenneth Parsons,[23] John Timmons, Rainer Schickele, Erven Long and many more. The econometrician and theorist, now pragmatic, Georgescu-Roegan recently recognized[24] there is much to be said for the point of view that normative and positive information are interdependent. However, case studies indicate that normative and positive information are not always so interdependent that it is always necessary for positive truth to condition normative truth and vice versa.[25]

The IMS experience, the Nigerian and Korean studies,[26] and many consulting experiences indicate (1) the general inadequacy of assuming answers to normative questions in a conditionally normative way in order to proceed with problem-solving, but (2) that this approach works in many instances. Gunnar Myrdal followed this approach in Appendix II of *An American Dilema*[27] and, in so doing, made U.S. policy-making more rational. Kenneth Parsons fears the conditionally normative approach leads to "a reversion to the medieval view that the world of thought and action should be organized around assured values presented to mankind as dogma",[28] a view empirically supported by the existence of dictators (both right and left) and demonstrators (again both right and left) who try dogmatically to force others to embrace prescriptions based on values placed beyond the tests of logic and experience.

With respect to empirical normative knowledge, I find G. E. Moore's *Principia Ethica* (1903) most instructive, as did the late J. M. Keynes.[29] Moore demonstrates to my satisfaction the possibility of normative primitives to use (in the manner of the linguistic analysts) to convert analytic statements into synthetic (descriptively empirical) normative statements.[30]

We note that Figure 1 has (1) positive and normative information banks, (2) a pragmatic information feedback loop, and (3) feedback arrows to cover prices, political processes, press, and research and education as components of an information system. It is sufficiently eclectic to cover any philosophy, discipline, source of information or technique.

In IMS research, mechanistic expectation models of the type discussed earlier for prices were hypothized to be used by managers in forming expectations concerning technology, institutions and human behavior. This hypothesis did not survive pretest of the IMS field questionnaire. Expectation models for technological and institutional change used by midwestern farmers were related to past data and conceptual ideas and theories were used though not as systematically as in the case of price expectation models; apparently, sociologists and political scientists teach farmers less (or have less to teach) about institutional change than economists had to teach about price changes. Neither farmers nor academicians seemed to have highly articulated, conceptual systems for predicting technological change. Open-ended, probing questions failed to reveal much about how farmers form expectations concerning human behavior despite the fact that the questions were designed by a social-psychologist who devoted several unproductive months to tabulation and analysis of the answers.

To my knowledge there is no significant literature or research on how public decision-makers form expectations (normative or positive) with respect to technical, institutional and human change.[31] Presumably some of the above applies to public decision-makers as well as to farmers. My experiences with the Office of Technology Assessment, U.S. Congress, the U.S. National Academy of Sciences and the U.S. National Science Foundation and with non-U.S. agencies including FAO and IBRD indicate that study of their staff work and committe activity would cast some limited light on the public formation of technological, institutional and human expectations.

The price system is an important source of normative data and information for both public and private decision-makers. Other important sources of such data and information include political processes and the press where these are free enough to permit the origination and feedback of reasonably "true" information from affected persons to decision-makers. Governmental information systems serve both private and public decision-makers. Highly centralized economies, right or left, tend to prevent the market from operating a price information system, suppress political process, and control educational systems and the press to prevent them from feeding back non-monetary normative information. They also often fail to provide substitute ways of furnishing such information to their decision-makers and, hence, suffer from poor decisions. Many centrally organized economies use local organizations of farmers, laborers, youth military personnel, other groups, etc., for feedback and control purposes. Whether or not such organizations are feedback mechanisms for information or merely steering mechanisms by which the central establishment "runs" the system is questionable – perhaps, they generally play both roles with the need for the feedback roles by action agencies being reduced in economies with operating price mechanisms, free presses, and open political and educational processes.

In his Presidential speech at the American Agricultural Economics Association last year, Bonnen[32] drew heavily on the work of Edgar S. Dunn[33] on social information processing and statistical systems which is based, in turn, on the work of information theorists and cyberneticists. Information theorists and cyberneticists concentrate on positivistic information acquisition (monitoring), storage (banking) and what they call processing (which is similar to, but not the same as, analysis and decision-making) as conceived herein. Figure 4 presents an information theory point of view. The *monitor* in Figure 4 is the *observer* in Figure 1. The *information processor* is both the *analyser and decision-maker* while the *effector* is the *executive.* The feedback from the

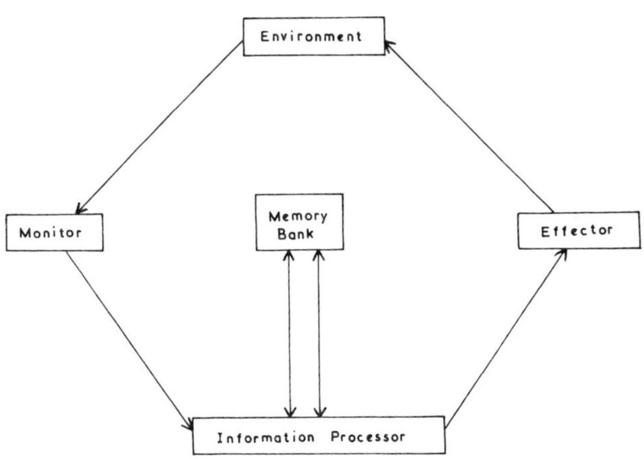

FIG. 4.

environment to the information processor via the *monitor* handles the *responsibility bearing function.* The memory bank does not distinguish between normative and positive knowledge. Dunn's discussion of information processing does not deal explicitly with how normative and positive information are processed into prescriptive information as he treats values and goals simultaneously without differentiating sharply between the normative (good and bad) and the prescriptive (right and wrong) in the C.I. Lewis sense.[34] While Dunn stresses the need to deal with the normative and prescriptive, his epistemological discussions are far from explicit.

There is an economics for deciding how much information of each kind is optimal in solving a particular problem. Knight's early but now obsolete distinctions among risk, uncertainty and certainty are relevant here.[35] A. G. Hart argued that Knight did not distinguish between risk and uncertainty on the basis of having *more than one* probability distribution to each of which is attached a probability in the case of uncertainty and *only one* in the case of risk. Hart noted that one could always compound two or more probability distributions into one by using the probabilities attached to each as weights but argued that it *might not pay* to compound at a point in time as an improved compounded probability distribution might be obtained later at an increase in cost less than the increase in value.[36] In so arguing, Hart anticipated Wald's work on sequential analysis which is now more or less subsumed under "pooled sample analysis".[37] In sequential choices between two alternatives, there are three alternative conclusions at any stage in acquiring information of a given type: (1) accept the first alternative, (2) accept the second alternative, or (3) continue acquiring information (learning). These three conclusions begin to focus attention on the economics of setting specifications for choices, i.e., on which alternatives are being considered and on what probabilities of errors of the first and second kind are acceptable. Setting specifications involves matching the cost of additional information against its value; *this is the economics of it.*

As some of both the marginal costs and marginal values are likely to be non-monetary, the economics of setting specifications in choosing between hypotheses is more akin to consumption and welfare than production economics. In any event, managerial and decision-making judgment with respect to acquisition of knowledge requires knowledge of the value (non-monetary as well as monetary in the context of the problem[38]) of the kind of knowledge being acquired.

Post-Wald risk is a situation in which specifications for a kind of knowledge have been met. Uncertainty, in turn, has to be divided into two parts, i.e., (1) a learning situation in which the specifications are not met but in which it pays to meet them and (2) situations in which it does not pay to try to meet the specification. The latter includes cases for which neither information acquisition nor decision takes place as well as situations in which outside circumstances force decisions. IMS researchers defined these situations as risk, learning, inaction and forced action.[39,40]

*Analysis (Use of Information)** – Another aspect of problem solving or function of management is the processing of primary data (both normative and positive) into secondary data and information. The domain of a problem tells us what is needed from what disciplines. And, the theories of the disciplines both tell us what data are needed and convert that data into information.

Economic theory also has the potential value of being helpful in defining optimal ways of solving a problem. Decision-makers economize when deciding, for example, that greater use of a particular biological theory has a greater pay off than, say, the use of a particular part of Keynesian theory.

Figure 1 shows two-way arrows among all the six functions of management, which indicates that the optimum amounts of observation and analysis are determined by an equilibrium among all six processes.

An orientation to economic theory both increases and decreases managerial efficiency. Efficiency is increased by bringing to bear the powerful analytical apparatus of economics on problems but may be decreased by the resultant possible neglect of other analytical apparatuses and concentration on the disciplinary *problems* of economics to the neglect of the practical problem.[41] Currently, economists are contributing their powerful analytical structures to the analysis of public problems involving human capital formation, family planning, and induced technological and institutional change;[42] hopefully, disciplinarians interested in these subjects will help economists make their contributions but prevent them (the economists) from uneconomically attempting to explain, originate and implement *all* human, technological and institutional changes when others may be able to do part of the job at lower cost.

Economists who use their theory both to define and solve problems must be sensitive to the problems which the market creates but cannot solve as well as to the problems which can be solved with market adjustments. Both of these are in addition to those originating outside the market. As a market system operates through time, those who can save get richer while those who must "eat up" capital to live, starve in a pareto-optimal way! In cases of inappropriate ownership patterns, the solution is a redistribution which hurts some in order to benefit others. In other instances, economies are simply out of economic adjustment and the resultant "problems" can be solved by mutually advantageous (pareto-optimal) trades among producers and consumers.

Dunn, in my opinion, unwisely designates progress through market solutions as economic and through non-market solutions as social. The distinction is unwise, because it is largely a distinction between private and public decision-making both of which have both economic *and* social significance.[43]

Among the problems which arise from market operations is the tendency of market-oriented agricultural economies to produce more output than can be sold at prices which equate marginal earnings with the marginal cost of acquiring resources.[44]

* Perhaps this function should be called synthesis rather than analysis as primitive terms (normative as well as positive) are used to convert analytic into discriptive information.

Decision-Making – The decision-making function of management is so important that management is often referred to as decision-making. A decision is a prescription of "what ought to be done" to solve a problem and, as such, is a function of both positive and normative knowledge, the functional relationships being the decision rule or strategy.[45] Some prescriptions are more risky than others. When decisions are viewed as above, it is difficult to use Knight's simple ideas of risk and uncertainty[46] even if modified to take account of Wald's sequential analysis[47] which tends to deal with one kind of positive rather than prescriptive knowledge. Risk is with respect to a decision, not one kind of knowledge going into that decision.

Interactive iteration is characteristic of problem definition and solution. Normative preconditions for maximization tend to be established in the process. These include: (1) establishment of a normative common denominator (CD) of all the goods and bads encountered in connection with the problem, this CD being necessary before maximization is mathematically possible, (2) if the decision will hurt some in order to benefit others as many housefold, family-farm and corporate private decisions and most public decisions do (at least in non-centralized economies).[48] an interpersonally valid CD, (3) establishment of the second order conditions mathematically necessary to assure the existence of a maximum in the CD, and (4) agreement on the decision rule to use. In the typical problem-solving exercise, a high proportion (over half) of the effort is devoted to establishing these normative preconditions and the positive constraints and consequences of alternate solutions. Initially the mode is not a maximizing one though it ultimately becomes so in most cases.

Attempts of economists to introduce maximization computations into decision-making are often premature and tend to foreshorten investigation of technical, institutional, psychological and social processes. The result is lost credibility for economists and economics – i.e., until the necessary normative and positive premaximization work is done, maximization cannot be reliable.

In order for a decision process and model to be credible, flexibility and generality are required with respect to: (1) types and sources of information – normative, positive and, eventually, prescriptive knowledge are required from any discipline (technical, institutional and humanistic) relevant to the domain of the problem, (2) philosophic orientations so as to permit work with normative, positive and prescriptive knowledge, (3) the use, non-use and delayed use of maximization, (4) use and non-use of such specialized techniques as linear programming, Cobb–Doublas analysis, input/output analysis (I/0), probabilistic estimation of the parameters of simultaneous equations from time-series and cross-sectional data, benefit/cost analysis, program planning and budgeting (PPB), program evaluational and review techniques, (PERT), deterministic adaptive models, etc., and (5) interaction with *people* (*decision-makers and affected persons*) involved as sources of positive and normative information which amounts to using indeterminate adaptive models.

General systems-science simulation approach (GSSSA) models, in my view, hold great promise for helping public and private decision-makers solve problems. This approach requires that generality and flexibility be maintained. I want to stress that GSSSA is an *approach*, not a technique. Recently substantial progress has been made with this approach in iteratively and interactively (adaptively) modeling the multidisciplinary domains of problems so as to utilize the computational efficiency of modern computers. Specialized techniques such as recursive linear programming or other recursive models and deterministic adaptive maximization models provide components which become parts of GSSSA models, if needed. Both formal (machine) and informal (committees or single person) components can be used as needed.[49] The same is true with respect to maximizing components. Utilization of computers is an incidental to the approach though the efficiency of the computer often frees both money and scarce professional resources for pursuit of flexibility and generality.[50]

In the following paragraph we deliberately discuss evaluations of models and decision efforts together as the truth and validity of a prescription depends on the truth and validity of the analytical and empirical content (or structure) of the model generating it.

Evaluation of a total problem-solving model or decision process in terms of its riskiness is far more difficult than relatively trivial evaluation problems in statistical texts. Complete models of problematic domains necessarily contain both human and less human (mechanical, mathematical and analytical and/or empirical) components. Decision-making efforts also involve such components.[51] Decision-making systems may be open in the sense that energy (more generally low level entropy) enters the system so that growth is possible. Among the instrumental resources which low-level entropy can produce are positive, normative and prescriptive knowledge. Growth and development result from the improved decisions based on such knowledge, some as a result of improved market decision – others as a result of improved non-market decisions. The output of a decision unit is a prescription based on (1) many kinds of positive and normative information both empirical and analytic, (2) adaptive (iterative and interactive with *people*) attempts to find a moderately usable CD to use in measuring disparate non-monetary costs and values conferred and imposed on different *people*, and (3) adaptive (iterative and interactive with *people*) attempts to arrange alternative acts in the order of their decreasing advantage as determined by an iteratively sought decision rule.

In general, the decision process and image (model) of a problematic domain involves an acceptable degree of risk to the decision-maker if:

(1) the decision-maker is ready, willing and able to act *and* if
(2) the costs of further improving the image or model (in the broad sense) exceed the value of contemplated improvements with respect to the eclectic tests of truth: coherence, correspondence, clarity, and workability.

Here is the key (1) to understanding a contribution of economics to rational decision-making and (2) for distinguishing between acceptable risk and uncertainty, the latter of which can be broken down into learning (more adaptive iteration and interaction), forced action[52] and inaction. There is an economics of matching added costs against added returns in setting specifications for decisions. How tight (conservative) or loose the specifications for an acceptable decision or problematic conceptualization should be is, in this view, influenced by how efficient the five functions other than decision-making are performed.[53]

The Executive or Action Function – Schools of business, military and public administration are likely to be more concerned about this function than economists. Traditional farm management men and farmers are likely to know more about it than production economists. And practicing public administrators, private entrepreneurs and housewives are probably the least likely to forget its importance.

Decisions which cannot be executed are of little value. More fundamentally, ability to execute and the required accuracy of decisions are interdependent. Still further, much execution is carried out by people who are essential sources of information in the managerial processes.

When the solutions of problems turn out to be technological, institutional and/or human change, execution requires special skills in these areas. Plant breeders and agronomists understand the processes of creating new varieties and cultural practices; political scientists, lawyers and sociologists – institutional changes; and educators and humanists – human changes. While changes which solve problems maximize human interest, in some sense, economics should not be expected to bear the burden of *explaining* the origin of such changes or of *seeing to it that they happen.*

Stress on administration to the exclusion of other managerial functions is often uneconomic. And, stress on one kind of administration to the exclusion of other kinds of administration and other managerial functions can be even more wasteful. We often find unproductive competition among administrative specialists rather than coordinated use of mixes of special administrative skills appropriate for executing the solution of specific problems based on all six managerial functions.

The roles of control and power in execution are important from the standpoint of executive efficiency. If complete objectivity with respect to normative, positive, and prescriptive knowledge could be reached, we would have consensus and there would be no need for power and control to *enforce* decisions which damage some in order to benefit others as everyone would agree.[54] However complete objectivity – i.e., agreement on positive and normative information, the preconditions for maximization including a decision rule, and on the prescription – is so expensive it is typically uneconomic. In the absence of complete objectivity, *use of force* – military, political, administrative, social, market power based on property ownership, religious – *is necessary to execute decisions.*

Imperfect knowledge and the need for executive power seem to be universal. Market decisions rest on political covenants which preserve the ownership of certain "rights" (and privileges) as private property. Non-market adjustments are based on other similar covenants or distributions of power among military and police officials, party members, civilian administrators and politicians.

Where the ownership of power and control is uncertain, there is danger of conflict from challenges which lead to tests of power on the field of battle, in the market, between personalities or in such political arenas as the national, local bureaucratic, academic, corporate, tribal, and familial (household and/or firm and extended). On the other hand, the ownership of power and control may be so definite that no challenge or conflict is possible and problems involving concentration of power go so long unattended that eventual change becomes catastrophic.

When the ownership of power is highly concentrated in the hands of decision-makers and executors while affected persons are powerless, meaningful feedback from affected persons is inhibited and the managerial process is deprived of an important interactive source of information. This occurs in authoritarian families, business and governments; it also occurs in free market economies with highly skewed, fixed distributions of property ownership. Prior to World War II, the Ford Motor Company suffered from this difficulty. Hitler's Germany and Stalin's USSR also suffered. In much of South America, property ownership is both skewed and highly stable. Authoritarian regimes (public, private, and mixed) are deprived of needed feedbacks if affected persons do not own the right to send messages. The ability to send messages depends on rights to participate in political activities, in the market, and to receive and distribute knowledge through open communication media and educational systems.

It is difficult to define and attain economic mixes (1) of centralized and decentralized ownership of power and control, and (2) the stability and flexibility in patterns of power ownership. Yet attaining such mixes is crucial to effective, rational decision-making.

Responsibility Bearing

Responsibility for decisions is borne by decision-makers and affected persons. Responsibility for the action actually executed is different than for the *intended action.* The executive may be held responsible for failure to do what the decision-making unit decided should be done.

The consequences of actions need to be monitored both positivistically and normatively for use in defining and solving the problems which emerge as the affairs of an entity are managed adaptively through time. Two important feedback loops for public decision-makers are political processes and the public press (if they are free to originate messages). Though the price system of a market-controlled economy is primarily a feedback mechanism among decentralized decision-makers, it is an extremely important source of normative information to public decision-makers; deprivation of public administrators of this source of information creates a need for information on unconsumed stocks in government warehouses, underemployment, rates of consumption, unfilled ration stamps, unmet quotas, etc. Where power is absolute, political monitoring may be neglected while the press is converted to governmental information systems to "feed forward" regulations and orders; to the neglect of the feedback function.[55]

PART II

Contributions Economists Can Make

With the above summary, we are now in position to consider the contributions economists can make – (1) as participants in decision-making, (2) as doers of subject-matter research, and (3) as doers of disciplinary research. All three are important and none are to be denigrated by me.

As Participants in Decision-Making – When an economist helps solve problems, he operates in a far different arena than when doing subject-matter or disciplinary research. The object is to help prescribe a solution. Interaction is required with decision-makers and affected persons possessing market, political, and/or military power.

A first and continuing task in problem-solving is to define the policy problem under consideration. Problem definitions determine the multidisciplinary mix required to understand the domain of the problem. Very few practical problems can be handled by a single discipline such as economics.

One of the important contributions which economists can make is their understanding of how markets operate and of the important role which price systems play in transmitting information among affected persons. Economists should also be able to (1) contribute an understanding of how market mechanisms can creat undesirable distributions of resource ownership and (2) help predict the consequences of changing ownership patterns.

In participating in problem-solving activities, economists represent but one of a number of disciplines. They are expected to contribute disciplinary excellence: theory, empirical knowledge and command over quantitative techniques for handling empirical information and for processing it into prescriptions. An economist will be looked to particularly for assistance in reaching maximizing (prescriptive) decisions. He will also be looked to for assistance in modeling the maximizing behavior of the producers, consumers, resource owners and public officials in the economy. In doing these things, he will be expected to join multidisciplinary teams and to recognize that the need for administration of problem-solving research supersedes the interest of particular disciplines, including economics.[56] If the problem-solving effort cannot be administered, the economist is well advised not to participate in it. And, if the economist is not willing to accept administration but wants, instead, to do his "own thing" as an economist, he should not join a problem-solving effort.

Great philosophic flexibility is required of economists participating in problem-solving exercises. It is reasonable to expect them to develop normative as well as positive descriptive information[57] and to be in command of economic logic for processing them into prescriptions. The normative information will be non-monetary as well as monetary.

Also, economists should be sensitive to the information obtainable interactively from decision-makers, executors of decisions, and affected people. Also they must be aware of the importance of property and power distributions as they affect decision and execution.

Great flexibility is required with respect to techniques. Decisions are reached on the basis of multidisciplinary conceptualizations of problematic domains. Such conceptualizations are difficult to verify and validate. The credibility of a prescription is determined essentially by the four tests of coherence, correspondence, clarity and workability. Economists should be prepared to use all four in problem-solving exercises.

Economists have had much experience in (1) developing models of the *structure* and the *conduct* of both private and public systems and (2) projecting the consequences of alternative actions in terms of various criterion or *performance* variables. Multidisciplinary projections have been relied on by public decision-makers in the past and will continue to be in the future; however, future projections can be far superior as the electronic computer has reduced costs. Common sense, flexible, multidisciplinary projections for problematic domains can now be made which would have been prohibitively expensive just a few years ago. One of the principle contributions which economists have to make to rational public decision-making is to bring computerized projections into much fuller use in assisting public decision-making. Such models include human components or are used interactively and adaptively with people – administrators and affected people.

Economists, in my view, have been slow in exploiting the general, systems-science simulation approach to making projections because (1) highly-trained economists are often unwilling to leave the narrow confines of economics to cooperate with other disciplines in using the approach, (2) of unduly specialized philosophic orientations, (3) of addiction to specialized quantitative techniques and (4) of cultural lag with respect to the general, systems-science simulation approach among older economists. Many older economists, having been burned in the past by the poor results from linear programming, econometric models, and other specialized quantitative techniques, now often react adversely to all quantitative models – even models which attempt to obtain enough generality in making projections to overcome the undue specialization which burned them in their youth and now blinds them in their maturity!

Michel Petit, Kim Dong Hi and Lars Folkeson will expand our thinking on modeling further tomorrow.

Subject-Matter Research – Such research is designed to bring together information on a specific subject which is relevant for solving a set of problems involving *that kind of information.* While not focused on a particular problem, subject-matter research is important for public decision-making. Examples include research on such subjects as energy, poverty, international trade, land-tenure, food, employment generation, environmental quality, etc. In doing such research, economists should recognize that they are working on a subject not a problem – on one kind of information needed to solve a specific large set of problems.[58]

Subject-matter research is relevant to sets of problems and, like research on a specific problem, is multidisciplinary. One of the first things to be done in mounting a subject-matter research effort is to define the *set of problems* for which a specific kind of information is being accumulated. Although subject-matter research has *some* of the characteristics of problem-solving research, the immediate objective is not prescriptive.

Economists are expected to (1) contribute disciplinary excellence to subject-matter research in the forms of economic theories, empirical information, and quantitative techniques and (2) cooperate with other disciplines in getting the necessary information together. However, the amount of administration required and which has to be accepted by economists is less than for problem-solving research.

Subject-matter researchers deal with data collection, storage and retrieval. They also obtain guidance from models and logical structures for, as Bonnen and others point out, it is theory which converts data into knowledge.[59]

For both the data and models, it is necessary for subject-matter researchers to recognize that:

(1) The subject involves a *set* of problems not a *single* problem, but
(2) That each problem in the set has a unique domain with which decision-makers using the subject-matter research will have to deal.
(3) The "practical out" is to:
  (A) develop models and data which will be useful, if not fully adequate, for each problem in the set of problems involved, and
  (B) which are componentized so that components can be added and subtracted to fit the specific domains of specific problems in the set of problems under consideration; and
  (C) which are also componetized so that they can be taken apart and recombined in configurations by decision-makers and problem-solving researchers to fit the specific domain of a specific problem being solved.

Disciplinary Research – Because many future problems are unknown, it is extremely difficult to classify disciplinary research as relevant and irrelevant. It can be classified as: (1) that known to be relevant and (2) that not known to be relevant. The concern here is with disciplinary research of known relevance. Often practical problems go unsolved because the necessary disciplinary research has not been none.

Doing disciplinary is much different than doing problem-solving and subject-matter research. Disciplinary research of economists, consists of attempts to improve the theories, quantitative techniques and data of economics. At times, problem-solving and subject-matter research require relevant disciplinary research as the theories, techniques and data of economics are often inadquate.

The contributions needed are both large and small in both statics and in dynamics, with respect to the theoretical, the descriptive and with respect to quantitative techniques.

The faddish word "paradigm" was discussed earlier in this paper. Some economists, particularly those concerned with non-market changes and dynamics, argue that what they call neoclassical paradigm needs to be abandoned and completely replaced by some other paradigm which addresses itself to a *new* problem. I doubt this. I believe, instead, that our paradigmatic changes should be mainly those which we encounter in going from one problem to the next. In doing this, we will find much from the neoclassical tradition, Keynesianism, modern welfare economics and various extant forms of dynamics to be useful. I believe we will evolve as a continually relevant discipline which helps solve each generation's problems. Our job as disciplinarians will be to extend and add to what we have so as to be able to model the economic aspects of the domain of the problems at hand.

In discussing needed disciplinary research in 1971, I wrote as follows:[60]

> The deficiencies . . . provide specific priorities for our disciplinary research. Some of the priority areas are: Investment–disinvestment theory; stock–flow conversion theory; user–cost concepts; benefit–cost concepts; the relationships between effective demand and production; the generation, saving, and investment of farm-produced capital; dynamic managerial theory; and premaximization concepts to use in establishing the necessary conditions for applying maximization theories. Other important deficiencies involve theories for dealing with the formation of and trade-offs among individual and group values. These and other subjects all need disciplinary research to make our problem-solving efforts and investigations more effective . . .
>
> While we still have difficulties in estimating the parameters of economic functions and sets of simultaneous equations, I believe that we have over-emphasized improving such estimates at the expense of failing to improve other more needed estimates and sources of information. In the last 20 years we have also made great efforts to improve maximization computations, particularly modifications and refinements of linear programming. I believe that this area of work has also been over-emphasized relative to the use of projections, simulations, and premaximization computations. Our quantitative techniques for dealing with normative questions . . . are inadequate and in need of much more disciplinary research.

What I wrote in 1971 still holds but I now add a more specific stress on dynamic economics. We saw earlier the need to understand the economics of performing the six problem-solving processes. A needed reorientation of risk and uncertainty theory towards prescriptive knowledge has scarcely started. The relationships among decision-making and execution, on one hand, and the stability and distribution of the ownership of power, on the other, are poorly understood although wars are fought about them. The essentially normative nature of production (that which is valued) and waste (that which is not valued or is valued negatively) needs to be researched. The dependency of the meanings of aggregate output and efficiency on the distribution of the ownership of rights and privileges needs to be better understood. Output and equity are not independent variables – the meaning of one depends on the other and the trade-offs between them are not simple indifference lines on simple, static, social utility functions. We require better techniques for working with nonmonetary values if we are to research (1) the non-pareto-better consequences of market activity in the presence of imperfect knowledge and transfer costs and (2) the economics of non-pareto-better, nonmarket adjustments in technology, institutions and people.

We also need disciplinary research on the economics of such information systems as market price mechanisms, political processes, the press, educational institutions including agricultural extension as well as public data and information systems.[61]

If disciplinary research is to be of known relevance then disciplinary researchers must be in touch with the problems for which it is relevant. Such contact is probably more important than administrative services insofar as disciplinary research is concerned. Cooperating with other disciplines and flexibility with respect to knowledge from other disciplines are much less important for disciplinary than for problem-solving and subject-matter research. Within the disciplines of economics, one can concentrate upon the positive, normative or pragmatic and hence need not be particularly flexible philosophically. However, if one wants to cover all contributions which economics can make to either subject-matter or problem-solving research, philosophical flexibility is required.

NOTES

[1]Dunn, E. S. (1970) *Economic and Social Development – a Process of Social Learning,* The Johns Hopkins Press, Baltimore; Mack, Ruth, *Planning on Uncertainty,* John Wiley, N.Y.; also, Dunn, E. S. (1974) *Social Information Processing and Statistical Systems – Change and Reform,* Wiley-Interscience, New York; and Arrow, Kenneth, *Social Choice and Individual Values,* John Wiley and Sons, Inc., and Arrow, Kenneth J. (1974) Nobel Laureate, *The Limits of Organization,* W. W. Norton & Co., New York.

[2]I will draw, in this paper, on the literature noted in these notes and on personal experiences in (1) the Interstate Managerial Study (IMS), a micro-level empirical study in seven U.S. states of the managerial activity of 1075 mid-west farmers, (2) administering problem-solving research at the Economic Development Institute, the University of Nigeria, and the Consortium for Study of Nigerian Rural Development, Michigan State University's Agency for International Development studies of agricultural sector, systems-science, simulation models, including two Korean projects which utilized such models, (3) consulting with foreign and domestic decision-making agencies in government and with individual farmers, and (4) researching problems of foreign and domestic decision-making agencies both public and private. These experiences are reported in the following: Johnson, Glenn L., Halter, A. N., Jensen, H. R., and Thomas, D. W., (eds.) (1961) *A Study of Managerial Processes of Midwestern Farmers,* Iowa State Univ. Press Ames. Also, Johnson, Glenn L., and Haver, C. B., (1953) *Decision Making Principles in Farm Management*, Kentucky Agric. Exp. Sta., Bul. 593. Johnson, Glenn L., Scoville, O. J., Dike, G. K., and Eicher, C. K. (1969) *Strategies and Recommendations for Nigerian Rural Development, 1969/1985 (CSNRD 33),* Consortium for the Study of Nigerian Rural Development with the support of USAID, Cont. AID/Afr. 264, MSU, E. Lansing, Mich. Also, Manetsch, T. J. *et al.* (1971) *A Generalized Simulation Approach to Agricultural Sector Analysis With Special Reference to Nigeria,* produced by the MSU Simulation Team under USAID, Cont. AID/csd-1557, MSU, E. Lansing, Mich. Rossmiller, G. E. *et al.* (1972) *Korean Agricultural Sector Analysis and Recommended Development Strategies, 1971–1985,* Korean Agric. Sector Study Team with support from USAID, Cont. AID/csd/2975, Ministry of Agric. and Forestry, Seoul, Korea and Dept. of Agric. Econ., MSU, E. Lansing, Mich. Johnson, Glenn L. (1952) *Burley Tobacco Control Programs,* Kentucky Agric. Experiment Station, Bul. 580, Feb. Also Johnson, Glenn L. and Zerby, Lewis K., (1973) *What Economists Do About Values – Case Studies of Their Answers to Questions They Don't Dare Ask,* Dept. of Ag. Econ., MSU, E. Lansing, Mich. Ch. II, 'The Problems of Improving U.S. Economic and Military Aid to Thailand',

Ch. III, 'Development Problems in Western Kentucky', Ch. V, 'Establishing a National Recreation Area in Mich.', Ch. IX, 'Research on Nigerian Rural Development', Ch. X, 'Selecting a Survey Sample Size'.

[3]Johnson, G. L. and Zerby, L. J., *Ibid.*, Chapters I and XI.

[4]This conception of managerial (decision-making) units makes it unnecessary to stretch our concepts of reality to the point of making management an input in, for instance, dairy production capable of generating marginal physical and value products instead of regarding it more realistically as deciding how much hay and grain a cow will get to eat to convert in marginal physical outputs of milk. In many analyses, management is inappropriately treated as a factor of production or "shifter" of production, supply and demand functions. Such "inputs", or shifters, are in effect *decisions*, not inputs, as to how factors of production and consumer goods (among other things) are used. See *The Management Input in Agriculture*, Agric. Policy Inst., Southern Farm Management Research Comm., Farm Foundation, April, 1963, Ch. I by Johnson, Glenn L., 'Methodology for the Managerial Input', and Eicher, C. and Witt, L. (1964) *Agriculture in Economic Development*, McGraw-Hill, New York, Ch. 6, 'A Note on Conventional Inputs and Nonconventional Production Functions'.

[5]Johnson, G. L. *et al.*, *Managerial Processes of Midwestern Farmers*, *op. cit.*, and *Research on Problems of Agrarian Change*, (tentative title) European Review of Agricultural Economics, The Hague. Authored by the TransAtlantic Committe on Agricultural Change with support from the Rockefeller Foundation, cooperating authors: Day, R. H., Heidhues, T., Petit, M., Renborg, U., and Johnson, G. L.

[6]See Friedman, Milton (1943) *Essays in Positive Economics*, University of Chicago Press, Chicago, for a presentation that fails to recognize the normative nature of prices.

[7]Definitions of positivism and conditional normativism are to be found in Johnson and Zerby, *op. cit.*, pp. 15f and 19f.

[8]Johnson, Glenn L., *et al.*, *op cit.*

[9]See Dunn, E. S., *Economic and Social Development*, *op. cit.*, Chapter V for an explanation of the neglect of the normative by economists and an exposition of why we must deal with the normative in decision-making.

[10]*Research on Problems of Agrarian Change* (tentative title), *op. cit.*, Ch. 5 by Johnson, G. L., 'Philosophic Foundations: Problems, Knowledge and Solutions', and Ch. 6 by Day, R. H., 'Towards an Adaptive Economic Theory', (forthcoming). Also see M. B. Mbithi's excellent paper on the human factor in farm management presented at this conference.

[11]Johnson, D. Gale (1945) *Forward Prices for Agriculture*, University of Chicago Press, Chicago.

[12]Katona, George (1951) *Psychological Analysis of Economic Behavior*, McGraw-Hill Book Co., Inc.

[13]Kuhn, Thomas S. (1970) *The Structure of Scientific Revolutions*, Vol. II, No. 2 of Foundations of the Unity of Science, University of Chicago Press, Chicago (enlarged edition).

[14]Johnson, *et al.* (1961) *Managerial Processes of Midwestern Farmers*, Iowa State Univ. Press, Ames, *op. cit.*, p. 26.

[15]Boulding, Kenneth (1956) *The Image*, The University of Michigan Press, Ann Abor, pp. 173. This quotation we note, as an aside, is at variance with Robbins, Lionel (1949) *The Nature and significance of Economic Science*, MacMillan and Co., Ltd, London, and Friedman, *op. cit.*

[16]See *Research on Problems of Agrarian Change*, *op. cit.*, Chapter V for a diagram including the prescriptive.

[17]Johnson, G. L. (1961) *Managerial Processes of Midwestern Farmers*, Iowa State Univ. Press, Ames, *op. cit.*, pp. 88ff and Ch. 7 by Halter, A., 'Utility of Gains and Losses'.

[18]Hicks, J. R. (1939) *Value and Capital*, Oxford University Press, London; Von Neumann, John and Oskar Morgenstern (1974) *Theory of Games and Economic Behavior*, Princeton University Press, Princeton, N.J. Halter, Albert N. and Gerald W. Dean (1971) *Decisions Under Uncertainty*, South-Western Publishing Co., Cincinnati,

Ohio; Friedman, M. and Savage, L J. (1948) 'The Utility Analysis of Choices Involving Risks', *Journal of Political Economy,* LVI, pp. 279–304.

[19]Heady, E. D. (1952) *Economics of Agricultural Production and Resource Use,* Prentice Hall, pp. 476ff.

[20]Nerlove, Marc (1958) *The Dynamics of Supply: Estimation of Farmers' Response to Price*, Johns Hopkins Press, Baltimore. Also see review by Johnson, Glenn L. (1960) in *Agricultural Economics Research,* USDA, January, Vol. XII, No. 1, pp. 25–28.

[21]Johnson, Glenn L. and Quance, C. Leroy (1972) *The Overproduction Trap in U.S. Agriculture,* Johns Hopkins University Press, Baltimore, Ch. 5 by Lerohl, M.L. 'Expected Product Prices', Also Petit, M. J. (1964) 'Econometric Analysis of the Feed-Grain Livestock Economy', unpublished Ph.D. thesis, MSU.

[22]See Dunn, *Economic and Social Development op. cit.,* Chapter V for a discussion of how widespread this mistake is among economists and in the scientific community, in general.

[23]Heady, E. O., *et al.,* (eds.) (1958) *Agricultural Adjustment Problems in A Growing Economy,* Iowa State College Press, Ames, Iowa, Ch. 18 by Kenneth H. Parsons, 'The Value Problem in Agricultural Policy'.

[24]Georgescu-Roegen, Nicholas (1971) *The Entropy Law and the Economic Process,* Harvard University Press, Cambridge, Massachusetts.

[25]Johnson and Zerby, *op. cit.*

[26]Manetch, *et al., op cit.;* Rosmiller, *et al., op. cit.*

[27]Myrdal, Gunnar (1944) *The American Dilemma* Harper Bros., New York, appendix II.

[28]Heady, *Agricultural Problems in A Growing Economy, op cit.,*

[29]Moore, G. E. (1956) *Principia Ethica,* Cambridge at the University Press, Cambridge, England.

[30]Carnap, Rudolph (1953) 'Formal and Factual Science' in Feigl, Herbert and Brodbeck, May, *Readings in the Philosophy of Science,* Appleton-Century-Crofts, Inc., New York, pp. 123–8.

[31]In response to a question by a discussion opener, Prof. Zohlnhoufer, a "strictly ordered preference function" is not assumed; instead, normative as well as positive learning is recognized as taking place to contribute to the normative data bank.

[32]Bonnen, James T. (1975) 'Improving Information on Agriculture and Rural Life', *AJAE,* Vol. 57, No. 5, December.

[33]Dunn, Jr., Edgar S., in *Social Information Processing and Statistical Systems, op. cit.*

[34]See Lewis, C. I. (1955) *The Ground and Nature of the Right,* Columbia University Press, New York, for a discussion of *right* and *wrong* goals, targets and actions as distinct from values. At the request of a discussant, Professor Zohlnhoufer, I add that *normative* concepts deal with the goodness and badnesses, *per se,* of conditions, situations and things. *Positive* concepts with the characteristics of conditions, situations and things other than their goodness or badness. The prescriptive deals with concepts about "what ought to be done" which depends functionally on both positive and normative concepts, the functional relationship between the two being a decision rule or strategy.

[35]Knight, Frank H. (1946) *Risk Uncertainty and Profit,* Houghton Mifflin Co., Boston. At the 1975 meetings of the American Agricultural Economics Association, a number of papers on risk and uncertainty did not up-date Knight's risk and uncertainty concepts. The paper by Richardson, *et at.,* at this conference ignores both the obsolete and modern distinctions and hence the economics of deciding how much information to acquire (learn).

[36]Hart, A. G. (1946) 'Risk, Uncertainty and the Unprofitability of Compounding Probabilities', *Readings in the Theory of Income Distribution,* The Blakiston Co., Philadelphia.

[37]Wald, Abram (1947) *Sequential Analysis,* John Wiley and Sons, New York.

[38]Acceptable positive truth is, we see, relative to the values sought and avoided in solving a particular problem.

[39]Johnson, G. L., *et al.* (1961) *Managerial Processes of Midwestern Farmers,* Iowa State Univ. Press, Ames, *op. cit.,* Ch. 3 by Johnson, G. L. and Lard, C. F., 'Knowledge Situations'.

[40]Hubbard, J. W. and Chastain, E. D., experienced difficulties with the classifications some of which may be related to the discussion in the next section. See Hubbard, J. S. (1960) 'An Appraisal of the Farmer Knowledge Situations Investigated by the Interstate Managerial Survey', *JFE*, Vol. XLII, May, p. 335f, and Chastain, E. D. (1956) 'An Empirical Study of the Decision-Making Process in Farm Management', unpublished Ph.D. thesis, Purdue University, Lafayette, Indiana.

[41]Johnson, Glenn L. (1957) 'Production Economics and the Field of Farm Management', *JFE*, Vol. 39, May, pp. 441f, and 'Stress on Production Economics', *Australian Journal of Agricultural Economics*, Vol. 7, No. 1, June 1963, pp. 12–26.

[42]I refer here to the highly productive work on these subjects at the University of Chicago and elsewhere in the last decade or so.

[43]Dunn, Jr., Edgar S., *Economic and Social Development, op. cit.*

[44]Johnson, G. L., *et al.* (1961) *Managerial Processes of Midwestern Farmers*, Iowa State Univ. Press, Ames, *op. cit.*, 'Implications of the IMS for Study of Responses to Price', Chapter 10; Edwards, Clark, 'Resource Fixity and Farm Organization', *JFE*, Nov., 1959; and Johnson, Glen L. and Quance, C. L., *The Overproduction Trap in U.S. Agriculture*, Johns Hopkins Univ. Press, *op. cit.* In addition, the reader's attention is also called to the unfortunate review of *The Overproduction Trap* by Dale Hoover, *AJAE*, Vol. 55, May 1973, pp. 354–5 which failed to detect the difference between the analysis in the appendix and earlier work by Cassells. The appendix indicates that supply responses would be expected to be symmetrically asymmetrical with respect to increases and decreases in product prices where Cassells concluded that they would be asymmetrical. The appendix indicates responses to price increases should not be expected to be the reverse of responses to price decreases with the same being true, symmetrically, for responses to price increases after price decreases. When farmers have been expanding investments in response to favorable prices they do not disinvest in response to small decreases in product prices. Similarly, when farmers have been contracting production as a result of decreases in produce prices they should not be expected to immediately expand investments when they experience small increases in prices. The mathematical appendix explains responses far different than postulated by Cassells. At a recent Agricultural Development Council Conference on Risk and Uncertainty held at the Centro Internationale Mejoriemente Mais y Trigo, March 9–13, 1976 papers and discussions dealing with market distortion due to risk and uncertainty were presented by Michael Lipton, P. B. S. Hazell and P. L. Scandizzo, David M. G. Newbery and Joesph Stiglitz. These papers were ground-breaking in general economics, if not in agricultural economics, for recognizing the distortion effects of risk and uncertainty; however, they failed to cover the market distortions for which *both* imperfect knowledge and acquisition cost/salvage price differentials are necessary but individually insufficient. These are covered in *The Overproduction Trap.*

[45]For a simple discussion of alternative decision rules see Clark and Schkade, *op. cit.*, pp. 415f. What I refer to as decision rules they call criteria for making decisions under uncertainty. Jock Anderson's paper presented at the ADC Conference on Risk and Uncertainty, March 9–13, 1976 in the CYMMIT, *op. cit.*, is a particularly good summary of alternative decision rules.

[46]Knight, Frank H., *Risk, Uncertainty and Profit*, Houghton Mifflin Co., Boston.

[47]Alternatively 'pooled samples'.

[48]Because mutually benefical decisions take place in the markets of decentralized economies but are centralized in centralized economies.

[49]We must stress the people content of the total model as anti-modelers often mistakenly assume that models exclude people.

[50]Manetsch, Thomas, *et al.*, *Generalized Simulation Approach to Agricultural Sector Analysis, op. cit.* and Rossmiller, G. E., *et al.*, *Korean Agricultural Sector Analysis and Recommended Development Strategies, 1971–1985, op. cit.*

[51]Piero Mini in a recent book on philosophy and economics distinguishes between a mechanical Descartian economics based on thought and logic and an empirical or synthetic economics based on experience with matter and sensation. The GSSSA under consideration here utilizes interactions between man and computer components and

both positive and normative primitives to map analytic into synthetic systems. As such it crosses the bridge between the two kinds of economics of concern to Mini. These GSSSA models and decision models are complex. Mini, Perio V. (1974) *Philosophy and Economics,* The Univ. Presses of Florida, Gainesville, Florida.

[52] A special case of added costs equal to added returns because outside forces either make added costs infinite or added returns zero.

[53] A paper by Jock Anderson at a recent Agricultural Development Council Conference at CIMMYT concentrated on decision-making without taking these crucial interrelationships into account.

[54] Completely ordered preference functions are not only not assumed, they are typically *uneconomic* to develop!

[55] Georgescu-Roegen, Nicholas, *op. cit.,* pp. 306–7.

[56] In this connection see my Fellows Address entitled 'The Quest for Relevance in Agricultural Economics', *AJAE,* Vol. 53, No. 5, Dec. 1971, pp. 728–39, Fellows Address, Summer Meetings, 1971.

[57] *Ibid.*

[58] In earlier work, I unwisely lumped problem-solving and subject-matter research together under the term 'issue-oriented, problem-solving work'. See Johnson, *The Quest for Relevance, op. cit.*

[59] Bonnen, James T., *op. cit.*

[60] Johnson, Glenn, L. (1972) 'Alternatives to the Neoclassical Theory of the Firm (NTF)', *AJAE,* Vol. 54, No. 2, May, pp. 295–303.

[61] In effect I extend Bonnen's plea to include information systems other than governmental statistical systems. See Bonnen, *op. cit.*

CHARLES W. CAPSTICK*

Agricultural Policy and the Contribution of Agricultural Economics Research and Analysis

I have been asked to set down in this paper some thoughts on the kind of research and analysis which agricultural economists might engage in if their work is to be of value to policy-makers. This is a subject which lends itself to the expression of personal opinions and the views below are offered as a contribution to discussion at this conference and should not be taken to represent in any way the position of the Department in which I am employed.

This topic is regularly examined by both agricultural economists and policy-makers and it is perhaps significant that the Association has chosen it as one of the subject themes for this conference. In spite of the efforts of policy-makers and agricultural economists, the pressures on our agricultural policies are no less great today than hitherto – some would say more so – whilst political considerations are forcing policy-makers to devise solutions which are complex both administratively and in respect of their economic consequences. To help overcome these problems, policy-makers are receptive to ideas capable of incorporation into their schemes and because of their experience and understanding of the agricultural and food economy it is to agricultural economists that attention inevitably turns. Perhaps, however, agricultural economists, particularly those outside Government Service, are a little too far removed from the policy-making scene to have sufficient knowledge at the time of all the issues to make a positive impact and may often appear to play the role more of critics after the event than contributors. For their part, agricultural economists may feel that they are consulted too infrequently or too late in the day, are expected to devote too much of their time to longer term issues and when commissioned work is completed, are not encouraged, since the results may be barely discernible in the policies finally adopted. In order to give structure to discussion of this diffuse subject, I will first expand on these impressions of the present contribution of agricultural economists in the policy field and then go on to consider whether agricultural economists in the Universities and research units might play a more effective role, what the subject areas are where more effort might be concentrated and, finally, how

* Head of Milk and Milk Products Policy Division (Formerly Senior Economic Adviser), Ministry of Agriculture, Fisheries and Food, London.

best these contributions might be fed into the policy making network.

1. IMPRESSIONS OF THE CONTRIBUTION OF AGRICULTURAL ECONOMISTS

There are, of course, many policy-makers and possibly almost as many agricultural economists and the generalisations above inevitably conceal the exceptions. Even so, I suspect that many agricultural economists outside those employed full-time, or as consultants, by governments or international agencies, would accept that there exists a gulf between their own research work and the support, market management, farm structure and other schemes which pour from governments almost daily. From the policy-makers' standpoint, the fact that the process of policy-making is both complex and protracted, spreading over several months and even years – with numerous possible schemes considered only to be rejected – makes it difficult for him to assess *ex ante* the contribution of individuals from only one of the professions to the package which finally emerges.

Policy-makers will, however, readily acknowledge the value of the comprehensive statistical series available on farm incomes, production levels, the changing structure of the agricultural industry, and so forth; information which exists largely as a result of the efforts of agricultural economists over the years. The same is true of the more obvious parameters, such as demand elasticities and the forecasts and projections which derive from the models which agricultural economists have created and are striving to improve. Data of this kind are essential to the policy-maker and complement the information flowing from his day-to-day involvement with representatives of trade, food manufacturing and producer organisations and extension and budgetary experts.

Although the precise arrangements may vary from country to country, it generally falls to the policy group to ensure that assessments, of an empirical kind, are carried out on the consequences of adjusting those policies for which it has responsibility. Usually in each department, working methods have evolved for analysing this material which strike a balance, taking account of timetable limits, between the use of involved methodology and straightforward desk calculations. In this process, it seems self-evident that the contribution of agricultural economists is invaluable, although because of its indirect nature, this seldom gains any overt recognition. Where economists, usually those employed by governments, are themselves personally involved with the assessment exercise, a more direct link between the profession and policy formation can be assumed but, even here, much of the basic material utilised will derive from the work of non-government economists. In carrying out their analyses, government economists usually follow the methods developed in the universities such as for the measurement of effective protection, clearly relevant in the GATT context, and of the balance of payments and welfare and resource implications of different price support systems. This is a further area where agricultural economists have played, and are playing, an important indirect role.

It is not my impression, however, that policy-makers would be as categorical in respect of the value of the larger statistical models. Most policy-makers, I believe, tend to regard the results from such constructions, particularly if they have not been tested over a lengthy period, with a good deal of suspicion and probably prefer to rely on their own experience and a less scientific approach. The difficulty is that whilst mathematical techniques have developed along with the computer and plenty of data about agriculture is there for processing, the models still seem reluctant to yield a range of plausible outputs. Confidence is not increased over-much when different result patterns emerge after adjustments have been made. Large scale models can, of course, help us to understand the relationships existing within the agricultural sector and indicate the broad impact of policy changes such as price and subsidy adjustments. But for them to provide more than an occasional back-up service, the policy-maker would, I think, demand to see how the model has performed over a period of time.

Turning to the question of published analyses relating to major policy issues, I would argue that, apart from occasional examples, policy decisions have not been greatly influenced by studies which analyse existing policies and go on to volunteer assessments of the economic consequences of feasible alternative policy options. Some economists here with longer experience than I may not accept this and I would concede that there are many examples of such work which are impressive. I can quote a domestic example of an analysis by Professor Britton of the United Kingdom Cereals Market and there are the reports of workshop-type exercises relating to commodity, structure and marketing issues. At the international level I will, for illustrative purposes, mention only the recent study by D. Gale Johnson on World Problems and Prospects, which included discussion of a proposed world grain reserves policy. Yet in spite of the substantial flow of published material of this kind, it is my impression, which admittedly derives largely from contact with the European scene, that there have been relatively few occasions when they can be linked to a change in the direction of policy – or have influenced retention of the status quo. Certainly this seems so in relation both to the number and scale of the policy issues and to the effort and volume of published material that are devoted, for example, to further refinement of statistical measuring techniques.

This is in contrast to the achievements of agricultural economists at the micro level. Where a problem has been clearly defined, whether it be in the field of the creation of new industries based on agriculture in the developing world or at the farm planning level, the profession has made a significant contribution. It is, therefore, somewhat disconcerting to arrive at the above conclusion in respect of the contribution at the macro level, particularly as remarkably close co-operation exists between agricultural economists, extension specialists, and others concerned with the industry – including policy-makers. Some years ago Wassily Leontief remarked that agricultural economics offered "...an exceptional example of a healthy balance between theoretical and empirical analysis and of the readiness of professional economists to co-operate with experts in neighbouring disciplines...". This must surely remain valid, yet as I have suggested, one can turn to only a limited number of influential studies into the aims and effects of policies pursued by governments and international organisations and offer alternatives. It is difficult to pinpoint why this seems to be so. Agricultural economists are kept fully stretched, the Journals have difficulty keeping pace with their output and their advice is widely sought both on an *ad hoc* basis and as consultants. Perhaps it is that when their attention focuses on the policy field, it tends to be devoted to the wider issues, such as the fundamentals of the EEC Common Agricultural Policy or developments in trade in agricultural and food products. The agricultural economist, positioned some distance from the canvas, is well placed to turn his attention to "over-view" studies of this kind and provide us with an impartial view. This is an important role and such work is often commissioned by governments and international organisations to indicate the direction policy might follow. Usually, however, these studies are too generalised to serve other purposes and by the time they arrive on the desk of the policy expert, grappling with his more specific problems, may be of only marginal value.

I have suggested in this survey of impressions that there are areas in the policy field where agricultural economists are making only a limited impact of a direct and clearly identifiable kind. This may be partly the fault of policy departments in failing to indicate clearly their requirements but it might be that economists themselves may choose to avoid too close an involvement with the policy area. It is the case, however, that policy-makers welcome and are able to make use of information from non-governmental sources on the benefits or costs flowing from existing policies, on the methods that might be used to measure these and, above all, ideas and suggestions about what is to follow. Agricultural economists working as individuals and as members of team projects, can obviously meet more of these demands but to do so, decisions would have to be made to address their work and present the results for this particular purpose. A massive switch of research effort is hardly called for, it is simply that a small number of economists might profitably turn their attention to this area to add to those who already do so.

2. POSSIBLE SUBJECT AREAS

Agricultural policy, like any other policy be it governmental or commercial, has a pyramidal structure. At the top we have the broad policy objectives and strategy, whilst in the next tier are the policies for the commodity groups, land development, capital, regions and so on, and further down the individual schemes which are the interface between the policy and farmers, traders and processors.

As regards the broad objectives of agricultural policies, there are striking similarities as between countries. Generally, as for example does the Treaty of Rome – which lays down the aims of the Common Agricultural Policy – they refer to the living standards of farmers, regard for consumers, efficiency, security of supplies and market stability. Where differences arise, it is in the emphasis given to thse diverse aims. Whilst it is for governments to decide on such priorities, economists can, of course, claim the right to examine the overall policy with the aim of measuring, in terms of resource use within the economy, the implications of the particular balance that has been struck at a point in time. This, however, is taking us into the area of overall economic management and, whilst more precise knowledge is needed on the effect of, say, a change in the overall output of agriculture on activity in the food and other sectors and on economic growth, our attention might be better concentrated on less grand issues.

Within the agricultural sector most governments influence, and in several instances determine directly, the balance between the outputs of different commodities and of different regions. This is generally achieved via price or subsidy levels or by the fixing of production quotas. The difficult decisions reached on these matters are based on both economic and political criteria and it would seem unfruitful for agricultural economists in the universities and research units to attempt, via published work, to become involved in these annual exercises. However, study of the input/output pattern that has emerged as a result of these decisions and assessment of the economic implications would provide guidance on the priorities that might be followed in future and hence is an appropriate subject area. It is to assist with this work that further development of large scale models can be justified and indeed, why so much attention is devoted to them. But again, this is an example of wide ranging work, usually handled by small teams of economists, which tends already to be supported by governments and international bodies.

In the middle level of policy formation, i.e., for commodity groups and specific schemes which spread across the commodity boundaries, the greatest scope would seem to exist for agricultural economists to increase their contribution. To do so, however, the economist must specialise and acquire more than a superficial knowledge of the commodity group or subject area. In my view, the economist loses credibility with the policy-maker if in his published work he has ignored or overlooked factors to which the policy-maker has had to attach great weight. Conclusions and prescriptions which pass easily over political and legislative constraints, external trade rules and the practicalities of marketing and manufacturing, would not be regarded as helpful.

The economist does not, of course, have to become an authority on these matters, but concentration on, say, a commodity group enables him to communicate more fluently with the policy-maker and others concerned with the commodities, and his analyses will gain in plausibility. For example in my own particular field of interest, milk products, the policy-makers would not regard it as helpful to be confronted now with an analysis of the situation in the European Economic Community which simply prescribed a reduction in the target price of milk at the next price fixing round. The existing arrangements and the complexities of the problem are such that the Community is seeking ways of arriving at a satisfactory price level over a period of time. Consequently, the question is not whether prices should be reduced but rather, what array of measures might be implemented which would enable the price goal to be ultimately achieved.

For the agricultural economist to do his job effectively in cases such as this, a situation parallelled in most countries, he must, of course, concentrate on the task of measuring the effects of his prescriptions to set against those of others. Knowledge of the commodity area would, however, enable him to place the results in perspective, as it would assist him to select the measures in the first place. And if it is no easy matter to devise solutions to be assessed even with a fair background knowledge of the commodity in question, it must be infinitely harder without it. I am, therefore, suggesting that rather more economists might concentrate on these commodity and other specific policy areas. Usually economists have little difficulty in criticising existing policies – the situation in Europe is not exceptional – and wider discussion of alternatives, which would presumably be more economically rational, must improve their chances of gaining acceptance. An example of this kind is that many countries now operate schemes to assist small-scale farmers to leave the industry and certainly the publications and seminars by economists some years ago, helped them gain acceptance. In those countries with large numbers of agricultural economists, these second tier policy areas are probably well covered but in the case of the Common Agricultural Policy, which is a relatively recent creation, there is a case for much more concentration on the parts as against the whole.

The detailed schemes which apply the commodity and other policies are subject to day-to-day adjustment. Changes of this kind derive from market pressures or the demands of farmers and food processors. Decisions are taken at short notice and it is unlikely that non-government economists can here make much of a contribution. The most that the economist can do is note the way policies are being operated and use this information as part of his data base.

At the implementation level there is, however, one area of considerable involvement by economists, namely farm planning, which is worth mentioning. Considerable effort has been devoted to the development of elaborate programming techniques for handling farm data and assumptions about future price levels. There are, however, relatively few reports of success stories on the use of the more complex systems, largely because they have not been found acceptable or usable by extension workers and farmers. By far the major part of planning advice continues to be based, in spite of the economists' efforts to produce computerised systems which embrace the whole farm situation, on the use of budgets, cash flows, loan repayments and other partial analyses. One therefore questions the need for able economists to devote their energies to further elaboration of techniques of doubtful acceptability by the end users. Without acceptability the exercises can hardly have been cost-effective.

The question, therefore, is whether agricultural economists should disengage from this area and leave it to the extension worker, as was the case some decades ago, after budget techniques had been pioneered and developed by agricultural economists.

3. DISSEMINATION OF RESULTS

The impact of studies by agricultural economists in the policy field is enhanced when reports are drafted with the non-specialist reader in mind, a dictum of which agricultural economists are fully aware as reference to workshop reports will confirm. There is a place for description of the technical side of the analysis but as such material is largely for other economists to examine, it is best separated from the main messages which the economist wishes to transmit. Although obvious, it is worth stressing that the more succinct the report, the greater the chance it will be read at the higher levels in the policy-making hierarchy. Senior policy-makers are inundated with briefs and reports directly calling for their comment and little time is left to consider material not requiring their immediate attention. Apart from the definitive reports of studies, there is also a place for summaries published in non-academic journals and the presentation of the results in seminars. These are important additional outlets, generally well used already, which reach an audience different from the readership of the Journals of our agricultural economics societies.

An example of the difficulties faced in disseminating the work of agricultural economists which is of interest to policy-makers, is found in the European Economic Community. Economists in the member states are studying various features of this policy but as the reports may be in any one of five languages, the transmission of ideas is considerably restricted. As I have indicated, agricultural economists have the responsibility of dissecting this policy, the problems of which have both domestic and international repercussions, and producing constructive solutions to overcome them. And if their work is to have much impact across the Community, there is a clear need for a close identity of aims between economists and policy-makers and also between economists themselves in the various member states. In the latter context, it is of course up to agricultural economists in the EEC to organise their affairs accordingly and also to arrange for their findings to be made more readily accessible to policy-makers in the member states and in the central policy-making body. Although this example relates to a European problem, it nonetheless serves to illustrate that policy studies which are not made accessible to policy-makers themselves, or those who brief them, are largely irrelevant. If I may adapt the quotation, if agricultural economists do not contribute to the solutions, are they not part of the problem?

DISCUSSION OPENING – George Bublot, *Belgium*

In opening the discussion on Dr. Johnson's paper I feel some discomfort, not only because I have not been able to grasp it in the depth it deserves, but also because such an approach to the subject is not very familiar to agricultural economists. In basic terms, Dr. Johnson's paper is the fruit of a very great intellectual effort, an essay on the fundamental logic of decision-making. One can hardly help but agree with the six stages of decision-making behaviour suggested in the introduction.

But the distinction between the normative, positive and prescriptive dimensions of the problem does not appear as clearly as one would like. For example, the term "positive" seems to have been inspired by the title "positivism" given to the philosophical ideas of Auguste Comte. But in another human science – law – the term "positive" is the opposite of "natural". In algebra, "positive" is "greater than zero", and the opposite of "negative". There are other interpretations of the word "positive" which it is not possible to dwell on here.

This little exercise in linguistics is, as far as the terms used are concerned, far from being over. It reveals different senses of a word between disciplines, and even in English and French usage. The sense in which these terms are used therefore needs to be defined.

To conclude this preliminary comment, I would only underline the need to study in depth, in addition to Dr. Johnson's text and the ideas he introduces, the numerous articles, papers and pieces of research cited as references, which formed the basis of his paper.

My second reflection is more pragmatic in scope, since it is inspired by the conditions in which the decisions of everyday life are taken. Let us consider, for example, the determination of an optimum production plan for a farm. It is a decision at the micro-level, well-known to agricultural economists, which can be reached by a variety of different techniques. But the carrying out of this decision implies a sequence of secondary decisions on how to use the available means to put that decision into effect. For example, the allotment of a particular area to a given crop has to be matched by second order decisions on production techniques to be used (choice of variety, fertilizers, timing of harvest . . .). This underlines both the pyramidal structure of decision-making, and the all-embracing, global nature of the whole make up of the decision itself together with the organization of the means by which it is carried out.

This examination of the rationality of a decision imposes on us the choice of an objective, which gives an aim to the decision. But this view considers rationality in its narrowest sense, as the best compromise between the available means and the aim pursued. On a larger scale rationality can be seen as the definition of the aims which conform to the scale of values of those who take the decisions, or those whom the decisions concern. Rationality in this context appears very subjective, tied to cultural and historical heritages, and totally outside the competence of the agricultural economist.

Agricultural decision-making can also be technical, economic or political. Technical decisions seem easy to take, since they are based on easily made observations or experiments, which are easy to carry out. Economic decisions have to bring in technical aspects, but because they involve the future, and as we can neither master nor know the factors in the future which affect them, they seem very difficult, because of their unavoidable association with uncertainty. Notwithstanding the importance of this latter factor and all the work which has been done on it, there is an ever-widening gap between the attempts made to define uncertainty and the extent to which it is actually taken into account.

Decision-making in the policy field involves other human factors besides the economic ones. It is elementary to state that land law, cultural heritage, pressure from certain groups, basic concepts of income distribution, or the role of the state in social life . . . influence political decisions and define the whole framework of social values into which economic decisions fit, the latter being completely subordinated.

In conclusion to this second point it seems to me that, to be operational in the field of decision-making, one must be aware from the very beginning of the actual conditions in which decisions are taken. This approach perfectly complements that adopted by Dr. Johnson. It may help to put decision-making in its place in the general context of human behaviour.

My third and last thought relates to the contribution which the economist can offer to the making of rational policy decisions. Here one cannot but agree with Dr. Johnson's views. The economist (1) will intervene as a participant in decision-taking, (2) will suggest research subjects, and (3) will be involved in interdisciplinary research.

It is fundamental in this context to remember that he will know the very wide range of methods which could contribute to a rational decision. In addition, these methods will be used to solve problems which correspond to the very object of the discipline of economics, i.e. the allocation of scarce resources which could have a variety of uses. This implies that the economist should be aware of the limits of his competence.

The task of the economist will always be incomplete, because the policy decision cannot be limited solely to its economic dimensions. But it can be made easier through liaison, or even better, co-operation, between the thinker, the theoretician or planner of the decision, and he who actually bears the responsibility for it.

At the academic level, the problems of decision-making could well form the subject of a university course. Is not decision-making the synthesis of all the constituents of human knowledge?

The many contributions to be presented on this theme during this conference should be, for agricultural economists such as ourselves, both an invaluable basis for consideration and research, and a stimulus to exploration in this very wide field.

DISCUSSION OPENING – Werner Zoehlnhoefer, *Fed. Rep. of Germany*

In his paper Prof. Johnson gives us a comprehensive review on how economists may contribute to rational decision-making in the field of agricultural policy. He does this by dividing up the decision-making process into different stages (which are, of course, separable only analytically) and by transcending the differences which may be due to the fact that there are various *kinds* of decision-makers.

By viewing decision-making as a process, Prof. Johnson certainly chooses an appropriate way of dealing with the subject. Treating decision-making as such – without making a difference between, e.g. governments on the one hand and private farmers on the other hand – seems to me less fortunate. It is true, in both cases choices are to be made, formally speaking. But the structure of interests involved, the kinds of information needed, the distribution of power and the decision rules – that means practically all factors conditioning structure and outcome of the decision-making processes – are quite different. The building up of the necessary consensus, for instance, is much more important and, at the same time, much more difficult in political than in economic decision-making. As a consequence, the contributions which economists may be able to make to public decision-making should be quite distinct in *kind* and *perspective*, not only in amount.

Another difference I consider to be significant in dealing with the problem in question is that between routine decisions and innovative decisions. Again, I think, the informational needs connected with these basically different decisions are so distinct as to demand specific treatment. Quantitative decision models with maximizing behavior on the basis of a common denominator, for instance, may be a realistic and helpful approach to certain types of routine decision-making by private firms. By contrast, innovative decision-making by governments poses quite different problems and, therefore, calls for different contributions from economists.

Closely connected with these arguments is a more basic point: I refer to Prof. Johnson's concept of rationality. From what I understand, his decision-makers are all maximizing an objective function. They are, in other words, all supposed to possess a strictly ordered preference scheme and to collect as much information as is economically justified to make optimal decisions.

Meanwhile, however, representatives of a behavioral theory of decision-making, like Braybrooke and Lindblom[1], Herbert A. Simon[2] or Cyert and March[3] have argued quite convincingly – to me, at least – that maximizing is practical only under very restrictive conditions. Even in the area of private economic decision-making only routine decisions about well structured problems are susceptable to maximizing in an *economically meaningful* way and – accordingly – made by use of algorithms.

In general, however, comprehensive or *absolute rationality*, as Simon calls it, in a highly complex, uncertain and permanently changing environment exceeds men's computational capacities (including those of man-made computers). And, what is even more important, the search for optimal solutions presupposes a strict preference ordering, that means: a quite comprehensive and detailed consensus on values which is generally missing. In modern societies this is particularly true with respect to political decision-making where so many diverging interests and value orderings are involved.

If, then, public agencies do not and cannot make decisions according to the *synoptical ideal* of comprehensive rationality, as Braybrooke and Lindblom call it, the question arises, how rational decision-making by governments is possible at all.

Braybrooke and Lindblom argue that, as a rule, public decision-making – at least in a modern democracy – is characterized by a multitude of actors making decisions about matters of their respective competence which involve only incremental changes of policy. There is no comprehensive ex-ante co-ordination; the minimum of "negative" co-ordination (R. Jochimsen) necessary to secure gradual goal attainment is achieved by continuous processes of mutual adjustments among the different decision agencies.[4]

This so-called strategy of "*disjointed incrementalism*" has been criticized and further developed by Amitai Etzioni.[5] He shares the objections against the synoptic ideal. At the same time, however, he also finds the disjointed incrementalism to be unable to secure even the (limited) goal attainment possible in a pluralistic society. Therefore, he suggests what he calls the strategy of *mixed scanning* for problem-solving. This strategy involves a multi-step approach to decision-making. At *first*, a global survey over possibly relevant problem dimensions and policy alternatives is designed to discover those areas which seem to be of primary importance. They, therefore, deserve closer scrutiny. In a *second* step, then, these problems and policy alternatives selected are to be studied in detail. The transition from global scanning to detailed analysis may, of course, go on more gradually, i.e. in many more steps.

At any rate, this sort of mixed scanning is supposed to allow rational decision-making without exceeding either men's limited computational capacities or society's limited consensus on values.

It is neither possible nor necessary to discuss the relationships between these strategies of decision-making and the contributions of economists to public choice in detail *here*. From a *global* point of view, it seems to me highly probable that, if decision-makers in general and public agencies in particular, for good reasons, are striving for *satisfactory* rather than *optimal* solutions to agricultural problems, the contributions of economists to rational decision-making will be somewhat *different* from what Prof. Johnson's models imply. *Some* kind of information he considers important may be quite *irrelevant*, while, at the same time, there will be *questions* of high significance to *policy-makers* which economists do not deal with – simply because of their inadequate concept of political decision-making. To mention only one but a very significant example of the latter sort of problems: Economists have so far hardly taken any notice of the fact that political decision-making is basically determined by the desire of the decision-makers to stay in office – up to and beyond the next general elections.[6]

Generally speaking, it seems probable that the contributions of economists to rational decision-making significantly depend on the strategy of decision which political decision-makers apply. As a consequence, economists are able to make their full contribution to rational decision-making only to the extent, to which they take into account the conditions, under which political decision-making takes place.

Therefore, I agree with Prof. Johnson that we do not need to abandon the neo-classical paradigm completely and replace it by some other. I definitely would argue, however, that marked paradigmatic changes in the direction indicated above will be essential, if economists are to fully use their potential of improving economic policy in general and agricultural policy in particular.

NOTES

[1] See D. Braybrooke and Ch. E. Lindblom, *A Strategy of Decision*, New York 1963.

[2] See Herbert A. Simon, *Models of Man*, New York and London 1957; *Administrative Behavior, A Study of Decision-Making Processes in Administrative Organizations*, 2nd ed., New York 1961.

[3] See R. M. Cyert and J. M. March, *A Behavioral Theory of the Firm*, Englewood Cliffs, N.J., 1963.

[4] See Ch. E. Lindblom, *The Intelligence of Democracy, Decision-Making through mutual Adjustment*, New York and London 1965.

[5] See A. Etzioni, *The Active Society, A Theory of Societal and Political Processes*, London and New York 1968.

[6] See A. Downs, *An Economic Theory of Democracy*, New York 1957; and G. Tullock, *The Politics of Bureaucracy*, Washington 1968.

Report of the General Discussion

Though there were numerous references to concepts and philosophies the greater part of the discussion centred on decision-making situations in the centrally planned countries, Europe and the Third World.

The view was expressed that Professor Johnson had not correctly presented the decision-making processes in socialist countries. It was emphasised that social and economic policy goals are basic and the elaboration of decision depends firmly on broadly based economic research, carried out in a network of research institutions; efficiency and consensus have an important place in the socialist practice. The definitions of these concepts attracted attention from other speakers. In some circumstances it appeared that efficiency was concerned with maximising food supply while minimizing costs but other factors were involved, too, and the criteria varied with the level in the structure of the decision-making.

In the EEC context there was some debate on whether Capstick's appraisal conveyed a balanced impression. The decisions in relation to the European milk policy were political decisions and these necessarily were of the nature of compromises between economic and political considerations and the individual policy tendencies of all the individual member countries. It was also pointed out that there were many agricultural economists advising the Commision, either as members of its advisory committees or through study programmes to which university staff contributed. In some cases, at least, the influence of agricultural economists was limited by their unduly narrow perspectives, either geographically, or over time. Furthermore, it was argued that the decision process, by taking a commodity approach along the Capstick's line, could be misleading, because commodity and structural issues interlock and decisions on the one necessarily imply decisions on the other whose implication should not be left out of account.

The developing country situation did not figure largely in the discussion. At present there seemed to be rather slight communication between agricultural economists and policy-decision-makers in these countries. It would increase if the decision-makers saw a role for economic analysis and if institutions were developed which concentrated on policy orentated research. Moreover, advisers must take account of the national philosophy of the country concerned. If self sufficiency or production by the masses – not mass production – were central to national policy the adviser must accept this situation and advise accordingly – if he wanted to be listened to.

What seemed to be the contention of politicians and civil servants, namely, that economists were ineffectual because they usually did not work on the problems which occupied policy-makers and, even if they did, they were poor communicators, started several lines of comment.

One simply argued that economists should not mould what they do to match what politicians see as the important current problems. It was for the economist, as economist, to identify key problems and not to compromise his professional purity by picking politically interesting issues. Perhaps one might associate with this view the argument that there was need for a section in large government services whose job it was to take the output from economic analysts and translate it into language to which decision-makers are accustomed. However, it was also argued that it was misconceived to think in terms of politicians having problems which economists could use their discipline to solve. If the politician was clear as to what the problem was, then he would not find it difficult to reach a solution consistent with practical politics. The economist's contribution was to consult with the politician and help him to identify the problems – then, unless specific requests for help were made, to leave him to solve them.

In still more basic terms it was argued that an economist is trained to treat decision-making as value neutral – but it is not. The desire to exercise the power of decision-making can both influence who makes decisions and what is decided. The issue in many countries is whether governments or private agencies should make decisions.

Arguably we should examine objectives other than economic and try to handle the resulting complex; but we may then be less able to be useful in other directions. Some speakers thought that economists may simply have to content themselves with pointing out the cost, in reduced efficiency, of giving weight to other than economic objectives – accepting that if they do, they may not greatly influence decisions.

Participants in the discussion included: J. A. Akinwumi, *Nigeria*; P. C. Baillet, *France*; P. C. Bansil, *Zambia*; H. F. Breimyer, *U.S.A.*; I. D. Carruthers, *U.K.*; R. Khan, *Pakistan*; V. I. Nazarenko, *U.S.S.R.*; P. C. van den Noort, *Netherlands*; D. Paarlberg, *U.S.A.*; J. F. van Riemsdijk, *Netherlands*; G. Schmitt, *Fed. Rep. of Germany*; F. G. Sturrock, *U.K.*

Glenn L. Johnson (in reply)

First I am very grateful to both the openers, Drs. Zohlnhoefer and Bublot, and the commentators from the floor. Their comments and suggestions are most helpful to me as I attempt to clarify my presentation. The need to clarify is evident from my agreement with comments purported to be in disagreement with me, but with which I do agree. Obviously, I have not communicated well.

The formulation presented in my paper is not based on an assumption of strictly ordered preference functions; hence, I cannot disagree with Zohlnhoefer who feels such an assumption is unwarranted. The normative data bank in Figure 1 along with the discussions of normative observation and analysis and the objectivity of normative knowledge indicate that preference functions emerge out of an interactive, iterative problem-solving process.

As to rationalism, much depends on how we define it. My assignment was to discuss "rational decision-making". To some, reliance on custom, habit, tradition and "power covenants" when the cost of the errors made in so doing is less than the cost of making decisions is rational – to others, any reliance on tradition, custom, habit and authoritarian power distributions is irrational.

The comments on definitions of terms are well taken. I should have included the glossary I have published elsewhere.*

Of course there are hierarchies of problems and decisions. There are decisions about the amounts of information to acquire and analyse in solving a problem. There is an infinite regression of decisions about the decision rules to use in solving a hierarchy of problems. Also in execution there are sub-decisions, sub-subdecisions, etc., as to how to attain the prescriptions decided on in the decision-making step.

Decision-making to solve problems is not mechanical. Models of decision processes need not be mechanical and models need not be mechanically used. This is what interactive, iterative, adaptive decision-making is all about – interaction with people, singly, in groups, and combined into organizations. In problem-solving, interactions among investigators, administrators and affected people are important. In fact, they are a source of both positive and normative information and, especially, of creative innovative ideas as to possible problem-solving courses of action.

The observation that making a decision or participating in the making of a decision has value is crucial and important. Indeed, that value is a basic force behind women's liberation, worker alienation, student unrest, etc. The discussion on this point provides an example of how normative experiences provide normative primitives to use in transforming logical systems into objective descriptive normative knowledge.

Comments about feedbacks in planning Soviet farms are interesting and informative. Soviet accounts do provide much information to the "upstairs" from the "downstairs" and descriptions about how this is done is valuable. Yet there are questions about how certain kinds of normative and positive information flow in Soviet state farms in the absence of two important information systems: (1) the market price system and (2) an open political system. The Stalinist period is characterized by many Soviets as well as external observers as one in which it was difficult for ordinary "downstairs" Soviet citizens to send messages "upstairs". And, in Soviet agriculture, the problem was no less severe than in Soviet urban centres. I am very interested in Soviet mechanisms for transmitting to the "upstairs" the kind of information which goes upstairs through the market price mechanisms and the open political systems of, say, the Western European countries.

In closing, I am most grateful for the comments and criticisms I have received. I will use them both, to add clearly identified, clarifying comments and footnotes to my paper and in my future writing on this and related subjects.

* See G. L. Johnson and L. K. Zerby, pp. 11–12, cited in the paper, and G. E. Rossmiller, *et al.*, *Korean Agricultural Sector Analysis and Recommended Development Strategies*, Department of Agricultural Economics, Michigan State University, East Lansing, Michigan, 1972, pp. 34–5.

C. W. Capstick (in reply)

Various speakers have understood me to be implying that agricultural economists have not made a contribution in the policy field. On the contrary, I acknowledge that they have made many significant contributions. Rather, I was trying to underline the nature of the policy decision-makers' situation in relation to such contributions. The number of policy decisions which have to be made in a given time is very great. There are many commodities and many subsidiary policies. Administrators – like university economists – are very busy. They are each faced with a series of responsibilities. Communication between academic agricultural economists and policy decision-makers must take that into account.

The view has been expressed that governments do not sufficiently bring in agricultural economists where they might be useful. Perhaps that is so, but, in my view, the realities of the government process is such that if agricultural economists wait for highly positive government interest they will have to wait a long time. With the pressures as they are, administrators are just not in a position to sit back from current business and sketch out research projects whose results might be helpful in making decisions next year or the one after that. If they are to be heard where decisions are taken, agricultural economists must show that they have something pertinent to say.

It has been pointed out to me that there have been innumerable papers on milk problems in the EEC – with which I agree; that decisions have been taken quite at variance with these analyses; and the view expressed that I blame agricultural economists for the situation. I do not! Rather, I was trying to look farther ahead and hoping that agricultural economists would give a little attention to these problems in future. If their work is publicised the pointers will work their way through gradually and result in changed attitudes to the handling of problems. One of the problems is that the CAP is relatively new and perhaps agricultural economists have not yet adapted themselves to the breadth of the issues.

Finally, the difference in government philosophies was referred to. I did not refer to this. It would certainly take a long time for agricultural economists to change the direction of thought of a government which is intent on following its chosen line. They would certainly have to take the realities into account.

MICHAEL PETIT*

The Role of Models in the Decision Process in Agriculture

Any decision process is a link between thinking and doing, between reflection and action. In order to assess the role of models in the decision process in agriculture it therefore is essential to have a clear idea of the relationship between reflection and action. Put in such general terms the question is not restricted to agriculture. Indeed, most of the concepts and hypotheses presented here probably would apply to any decision process. Yet the expression "in agriculture" has been kept in our title essentially for two reasons. First, as agricultural economists we are mainly interested in problems arising in agriculture. The main theme of our conference is precisely the decision process in agriculture. Second, the experience from which the concepts are derived as well as the examples given below all stem from agricultural economics research.

This paper contends that models, often very informal models, are an intimate part of, and play a crucial role in, all decision processes. Formal models can be useful if they refine this process, i.e., make it more effective. If this premise is accepted, the role of models and of model-builders is clear. They must help the decision process. To support this argument the paper is organized in four sections. The first presents a general model of adaptive behavior which emphasizes the close interrelationship between reflection and action seen as two parts of the same adaptive process. In the second section, two classifications of models built and used by agricultural economists are discussed in the light of the general adaptive behavior model. The study of these classifications leads us to argue that the usefulness of models depends more on the role given to them by the analyst rather than on the type of model used. Accordingly a conception of the relationships between decision-makers, models and model-builders is derived in the third section. All students of decision processes must face two major analytical difficulties (model specification and model validation). These are discussed in the last section, on the basis of the general thesis of this paper.

Before proceeding to the first section we should specify exactly what we mean by the word model. Let us state that a model is a representation – necessarily incomplete – of reality in our mind.[1] Such a broad definition is

* ENSSAA, Dijon, France.

needed because it captures the essential epistemological nature of models and permits assessment in proper perspective of the long-term efforts of economists at building more and more formalized decision models.[2]

1. THE GENERAL MODEL OF ADAPTIVE BEHAVIOR

The purpose of this model is to establish close interrelationship between reflection and action. In other words, the analysis of the decision process requires a model of the decision process itself.

1.1 Agent, environment, goal, perceptions:

A synthetic presentation[3] will be made here starting from fundamental concepts and relating them together. Decisions are made by individuals attempting, through their actions, to modify their situations vis-a-vis their environment. Two interrelated concepts emerge here: the *agent* and *his environment*. These are two subsets of the universe. Because only a fraction of the universe is relevant to the decision of any agent, one would not need to include the whole universe in the environment of the agent at any one time. But it is preferable to do so because one important aspect of the behavior of any agent is the search for information beyond the border of his immediate environment. To fix *a priori* the outer limit of the agent's environment would not permit the model to capture this essential dynamic feature of the decision and action process. As to the border line between the agent and the environment, it is sufficient to postulate that it exists and that it may vary. For instance, it will be useful to consider that a farmer's environment includes, among other elements, sometimes his farm and his family, at other times only his farm; in the latter case the agent then is made up of the whole family without considering the interrelationship within it between, say, husband and wife, or parents and children.

Agents act in order to modify their situation with respect to their environment. Their action is thus postulated to have a purpose. They have *goals*. A major question pertains to the ways in which goals are formed and modified. Before investigating this point further, it must be pointed out that decisions to act are made by agents on the basis of how they perceive their environment and themselves, including their goals and their possibilities to act. Two new concepts have appeared here: their possiblilities to act, i.e., the *feasible domain of actions* and *the agent's perceptions*. Models play a crucial role in these perceptions. Essentially the broad definition of models given in the introduction leads us to argue that these perceptions are models, some vague and nonformalized, others more formalized.

1.2 Perceptions, learning and acting

The agent's perceptions depend upon his past experience, his memory, and the flow of new information reaching him all the time, originating in particular from his actions. Thus learning, which is the modification of one's perceptions, and acting are two intimately related processes. Conscious efforts to learn, such as reading a book or listening to someone, can even be viewed as actions;

this proposition strengthens the argument for a close interrelationship between learning and acting. Indeed, they can be viewed as parts of a single process; we can even learn about learning.

The postulate that all actions, including learning, have a purpose may appear much too rationalistic. After all, we often make mistakes and often learn without trying ("involuntary learning"), for instance, for the simple reason that we did not even know that there was anything to learn. But this criticism does not invalidate the postulate. The purposive nature of action implies only that when taking a decision, the agent does so to reach some goal. It is true that he can be mistaken both concerning his goal (once he has reached it, he discovers that he should have pursued another) and the relevance of his action (it did not permit him to reach his goal). But actually this general uncertainty is taken into account and adaptation to it is an essential feature of the model. Any perception is recognized as tentative, and a wise man often agrees to revise his position. Any action ought to be *controlled*, i.e., its impact should be monitored and due corrections should be made in case of errors or of changes perceived in the environment. A simple example: I have decided to drive to my office – my goal. I may revise this decision if I find out that the road is too dangerous today. I am alert to possible, unpredictable obstacles and prepared to make a detour. In both cases, the action to drive can only be understood if one recognizes that it is taken with the purpose of reaching the office.

1.3 Goal formation and modification:

The nature of goals can best be understood by taking into account the hierarchical relationships between categories of actions. The example of the automobile trip can best illustrate this hierarchy. When someone decides to drive from A to B, his goal is to arrive at B. If he gets there, he clearly will have modified his situation vis-a-vis his environment. His action is to drive there. But this action entails a host of elementary sub-actions: slowing down, accelerating, turning, etc. Each sub-action itself can be split, conceptually *ad infinitum*, into more elementary sub-actions: lifting the foot, changing gear, moving one's arms . . . At the other extremity of the spectrum the goal "to reach B" can be interpreted in terms of higher order goals. The agent wants to go to B because he has work to do there. He has work to do there because he is engaged in a given project, because he pursues this career, because etc. Any decision to act fixes a goal for the sub-actions necessary to implement this decision. Thus a goal really is an intended action with an expected outcome.

The choice of a goal is a decision. It depends upon higher order goals and the perception which the agent has of his environment and of his possibilities to modify it. Goals can be revised when these perceptions are modified. But where do goals of a higher order come from? Clearly from yet higher order goals. How far up can we go in this hierarchy? Conceptually *ad infinitum* again, but is this a satisfactory answer? Given that goals are intended actions, higher and higher order goals relate to more and more general actions. Is it wise to assume that agents have such very general action projects? Conceptually perhaps, but these are at best very vague. Is it not more useful to consider that given the impossibility of formulating an infinite hierarchy of articulated goals, agents resort to a temporary, revisable set, and to some general rules of behavior? These reflect the set of values held by the agent. (For the distinction between goals and values see Lewis (1955).) In this perspective, values differ from goals because they are general whereas goals, even when of a higher order, refer to a very specific individual in a given situation at a given point in time. Values transcend situations. They also transcend individuals as one speaks of social values. Note here, however, that the model does not assume that values are fixed once and for all, nor that the agent knows them precisely; nor does the model assume that values are restricted to influencing general, grandiose plans of action. On the contrary, they also influence the most trivial actions. For instance, one uses an extra blanket during the winter simply because it is comfortable to be warm in bed. It is not necessary to look for a higher order goal. The ways in which values influence the formation of goals deserve to be specified further, but this article is not the appropriate place. Let us remember only that our model remains vague on this point.

1.4 Adaptive behavior and the role of models:

The model of decision-making elaborated so far can be summarized as follows:

At any point in time the agent has a perception of his situation (vis-a-vis his environment), of his goals, of his possible actions and some idea of the values he holds, including, possibly, contradictions among these values.

All these perceptions are changing because the objective situation of the agent changes and because he continuously learns more and more about these various elements. His behavior is a continuous adaptation process. He acts and thus adapts to changes in his environment and in his perceptions. In the same process, he learns more and adapts his goals and his perceptions. At all times, he knows that his perceptions, his knowledge, are provisional and subject to revision. In a way, he is continuously in the process of revising his knowledge, he destroys old models and builds new ones. Such a view of knowledge and actions as a continuous process appears in complete agreement with the findings of Bachelard, a philosopher of science and Piaget (1968), a student of the elaboration of cognitive processes among children.[4]

2. CLASSIFICATIONS OF MODELS BUILT BY AGRICULTURAL ECONOMISTS

Several classifications have been used to sort out the various types of models used by economists.[5] In agricultural economics two classifications seem to have played an important role in the discussion of the relative merits of various types of models. These discussions on the merits of models are useful to reveal the role assigned to them by those who build and use them. Some authors have distinguished between positive and normative models. Another useful classification is based on criteria relative to the mathematical structure of the model or to the source of data used to estimate their parameters.

2.1. Normative versus positive models

The vogue of programming, particularly linear programming, models of the farm as a production unit has led to considerable discussion concerning the use of such models. In the U.S., large research projects undertaken to study the impact of new agricultural technologies and farmers' supply response (Dairy Lake State Study, NC54) were based on linear programming models of representative farms. Similar efforts were undertaken in numerous other countries. All these large models produced highly unrealistic results which were considered as such by the researchers involved. As a result, the question of the relevance, and hence of the real nature of these models, which had consumed large amounts of research resources, had to be faced. It then was often argued that such models were "normative", i.e., that they indicated what farmers should do. A somewhat more sophisticated variant of this interpretation calls such models "conditionally normative" (if farmers really want to maximize their income, this is what they should do). This interpretation is closely linked with the use of these models in farm management advisory work. The existence of an objective function to be maximized (or minimized) leads one to speak of optimization and hence to present the results as norms[6] to be fulfilled by the farmers. Such models are contrasted to other models which do not postulate an optimizing behavior, i.e., which are viewed as representing how farmers actually behave, rather than how they should behave. Examples of the second category are econometric supply models which postulate the existence of a supply function without any specific optimizing assumption concerning the producers' underlying behavior. Such models are then called positive or descriptive models.

Glenn Johnson (1976) has argued strongly that these labels are not appropriate because they reflect only a popular version of nineteenth century philosophy. An outright positivistic philosophy of social sciences is not tenable today. Social facts and social values are known to be intimately related, hence values cannot be excluded from the realm of social science investigation, and hence purely positive model of economic behavior do not exist. They are always based on some assumptions – perhaps implicit – relative to values. Conversely, the assumption that farmers maximize income does not make the model purely normative, as is well illustrated by the label "conditionally normative". Johnson argues further that values must be investigated and that an objective knowledge about them – including agreement on what is good and bad – can be obtained. Without pushing this argument further, we will use the expression optimization model instead of normative model and put the word "positive" between quotation marks. Let us examine how this distinction between these two types of models stands in the light of the general adaptive behavior model adopted in this paper.

First, it must be emphasized that the use of optimization models does not necessarily imply the production of norms (or goals). In farm supply analysis, for instance, the rationale for building a model of a representative farm, as illustrated directly by the word representative, is to capture the essential features characterizing the economic workings of the farm as a production unit. Consequently, a linear programming model of a farm built to understand farmers' supply behavior must be viewed as a behaviorial model. As any model, this one is based on an assumption to be tested, in this case that farmers seek to maximize one objective[7] subject to a set of constraints. Such a model does not imply that farmers are only worried about income; constraints can render an account of other objectives, such as the search for leisure or for security. Furthermore, the results of such models must not necessarily be viewed as goals. If the represented farmer does not behave as suggested by the results of the model, he is not necessarily wrong. Rather the model is not an accurate representation of what it purports to represent.

Whether or not a farmer should modify his decision and would find it advantageous to behave as indicated by the model results is an important question in farm management work. This can be decided only by the farmer, possibly after a discussion with the model-builder, once he has reassessed his goals, the situation of his farm and his environment, and his feasible domain of action. Such a prescription would be supported wholeheartedly by any experienced extension worker and is in full accord with the general adaptive behavior model. Thus, if one speaks in terms of norms, the main purpose of an optimizing model is not to produce goals ("norms") but rather to question the "norms" implicitly, or explicitly, accepted by the decision-maker.

What about the so-called "positive" models? Do they play a different role in the decision process? To be of any use they must be a fairly good representation of some segment of the economy. Hence, they should be useful to one or several decision-makers in understanding his, or their, environment or, more precisely, situation vis-a-vis environment. It is true that the goals of the decision-makers do not appear as such in a "positive" model. But, from the general adaptive behavior model, we know that goals are projected actions with expected outcomes. A "positive" model can be very useful in assessing the likelihood of such expected outcomes. As such, to continue to speak in terms of norms, the model will be useful in questioning the "norms" of the decision-maker. This fundamental similarity of functions of these two types of models, distinguished earlier precisely on the basis of their purpose, arises from the fact, pointed out by the general adaptive behavior model, that goals are intimately related to the situation of the agent.

2.2 Classification of models according to their mathematical structure or to the source of their data

In this section, three broad generations of models which have appeared one after another in our profession will be discussed briefly. The emergence of a new generation is proof that previous ones were not wholly satisfactory. The fact that no new generation eliminated the previous ones shows that there is not one privileged category of models. To distinguish only three generations is of couse to paint a picture with very broad strokes of the brush. This will, however, support the main thesis of this paper: whatever the model used, what matters most is the role given to it in the analysis which it supports. Ultimately, this role must be viewed in the general context of the decision process(es) served by the analysis.

The three broad generations to be discussed here are econometric models based on statistical inference, programming models, and general simulation models. These labels admittedly can be questioned. They are meant to help clarify the discussion.

Econometric models based on statistical inference appeared first in the 1920's. They can accommodate fairly robust hypotheses of economic behavior and have the advantage of permitting the reader relatively easy access to the data. For instance, an interested reader could, if he wished, check the computations or trace the consequences of a change in the original set of hypotheses. Hence these models have the appeal of some objectivity. This advantage should not be overstated. It is this author's personal experience that considerable judgment is involved in building such models.

Programming models, first developed in the 1940's, began appearing in the agricultural economics profession in the 1950's, and were in full vogue in the 1960's. Two major features distinguish them from the first category: the source of their data and their mathematical structure. Statistical inference is not used to estimate their parameters which are built up from a variety of sources. Since, furthermore, they entail a much greater number of parameters, peers'* control of data and of computation procedures is made almost impossible. The existence of an objective function forces the model-builder to specify further his hypothesis relative to the behavior of the economic agent studied (say the farmer) or the set of relationships between economic agents within an economy (as in an inter-regional competition model). This loss of robustness can of course be viewed as a disadvantage. On the other hand, programming models are very efficient instruments to organize coherently a huge mass of information; the very specificity and explicit character of the assumptions on which they are based lead to their thorough discussion – a definite advantage.

General simulation models appeared in the late 1960's. The word simulation may lead to confusion. The other two categories of models can also be used to simulate the functioning of what they are supposed to represent, but in the latter category this is the unique objective. Their advantage is their flexibility both in terms of data used and of mathematical structure. They are really an outgrowth of system analysis and they have been used mainly as a way to simulate the behavior of a system through time on the basis of some assumed relationships between variables.[8] Their purpose is not to test such relationships directly. Yet they may provide an indirect test, as when model results are compared with data on the actual behavior of the system and differences are interpreted.

System analysts, and scientists who have worked in collaboration with them, are convinced of the superiority of these general simulation models in assisting decision-makers. The outstanding publicity given to the book *Limits to Growth*, which was based on such a model, tends to support this claim. This book has been severely criticized by other scientists for lack of rigor in the conduct of the research. I feel that such a bold attempt at modelling the world was bound to produce only very rough results yet its great merit lies

* See previous para - (edit).

precisely in the controversy it engendered. It called public attention to important issues and demonstrated that we really know little about them.

A general conclusion about these various types of models stands out. No type is inherently superior to another in assisting decision-makers because decision makers use all types of available information. We should not restrict ourselves to specific kinds of data. At the same time we should be conscious of the uncertainty surrounding the data we use and of the relevance of this uncertainty. In this respect it may be useful to denounce the slogan often heard in our profession. "No model can be better than the data which go into it". Such a statement seems obvious yet it is meaningless. Criteria for assessing the reliability of data conceivably can be phrased in terms of accuracy statements[9] but the results of the model should be judged in relation to the situation of the decision-maker: What is it that he needs to know about the phenomenon under study? What are the other sources of information? Is the value of the added information provided by the model worth its costs? Given the numerous externalities involved in model construction and utilization, estimating model benefits and costs is very difficult indeed but one should at least recognize this as the ultimate criterion.

It appears from his discussion that the role of models in the decision-making process does not depend upon their mathematical structure or the type of data they use. What really matters is the underlying "philosophy" of the analyst. What are his goals in building a model? What issues does he want to deal with? How well thought out are they? What light does the model throw on these issues? What questions remain open? Why? The choice of the most appropriate model depends upon the answers given to these questions.

3. DECISION-MAKER, MODEL, MODEL-BUILDER

The role played by the analyst in the conduct of the analysis in the previous section has been emphasized. Though no novelty, it deserves to be repeated. The purpose of this section is to better specify the role of the analyst, his relationship with decision-makers, and the possible utility of models in this relationship.

3.1 Place of models in the decision-making process

In the adaptive behavior model, an economizing principle is implicit. To assert that actions are taken in the pursuit of goals implies that agents are interested in reaching their goals. Since this always costs something, i.e., sacrifice in terms of attainment of other goals (there are trade-offs between goals), it is safe to assume that agents seek to minimize such costs. But as already emphasized, this behavior takes place in a climate of general uncertainty where information has a cost. Models thus are used so as to act more efficiently. Provided that they can be relied upon, elaborate models will be better than rough ones because they are more efficient as ways to organize a multitude of information in a relevant manner. The proviso, of course, is essential; keeping this in mind, progress can be seen as leading toward greater model formalization[10] as a way to reach better coherence between goals and actions. Such is the fundamental role played by models. But the argument applies here to the models which are in the mind of the agent, which form his "perceptions" of his environment and of himself. What happens when the formalization process has gone so far that a division of labor is necessary? How will model-builders relate to decision-makers?

3.2 Relationship with decision-makers

In order to build realistic and comprehensive models, analysts will find it advantageous to work in close collaboration with decision-makers. Experience shows, however, that this is easier said than done, perhaps for want of a clear understanding of each other's respective decision and action domains. Furthermore, one can justly argue that social scientists, such as agricultural economists, seeking to sort out the fundamental economic problems of agricultural development, play a useful role even though they do not relate their work either directly or indirectly to any specific category of decision makers. One should, however, take care not to draw fallacious implications from such an argument. The usefulness of models obviously depends upon the quality of the new knowledge which they help to produce. The social relevance of this new knowledge must ultimately be judged from the point of view of the decision-making processes to which it contributes.

3.3 Non-neutrality of social scientists

The claim that social scientists can be fully objective, meaning here that they are neutral with respect to the phenomena which they analyze, can no longer be supported, If they have any value, models will be useful to some decision-makers. They will help some to understand their own situation and action possibilities better than others. Hence they will influence their actions. Furthermore, with any model of society can be associated an underlying implicit ideology which it strengthens. This has an impact on social values and hence on the "social game". We thus reach the conclusion, which is becoming increasingly accepted, that pure, scientific objectivity is an illusion, at least in the social sciences. But Myrdal's recommendation (1969) that social scientists should spell out their own values is not adequate to the problem at hand here. No one fully knows his own values. Any discursive presentation always sounds too noble, probably because it does not take sufficient account of internal contradictions within the set of values held by any individual. Actually our values are revealed by our actions. Thus social scientists, whatever the degree of sophisitication of the model which they use, should be conscious that they are ideologically involved, and that a high degree of sophistication may obscure the real values held by the analyst. Hence the least which can be expected of model-builders is that they clarify in their own minds their position in the "social game" as best they can and make it as explicit as possible. This is where discussion among scientists having different ideological biases can be very useful. In general, the choice of the boundaries and of the elements of the system which the analyst chooses to represent is often very revealing of his ideological bias. For instance, the set of interrelationships between social groups in the development process is often neglected by model-builders. For a conscious effort to analyze it in the case of the Kosi region in Bihar (India), see Biggs (1973).[11]

Marxists have a different conception of scientific objectivity. They argue that the evolution of society obeys the objective laws of class struggle. In this case the model-builder's objectivity consists in accepting this postulate and his task is viewed as that of prescribing the consequences for actions of these objective laws. For instance, a planning model must yield the optimum set of actions to be undertaken. This presentation may not do full justice to our Marxist colleagues' position. Whether the adaptive behavior model, which in my view applies also to the case of a centrally-planned economy, is compatible with Marxism or not remains to be seen. The views on scientific objectivity, however, appear impossible to reconcile.

But we share one conviction with Marxists: The analyst is not neutral in the social game. If he works directly with decision-makers, he reinforces their power, at least in the short run. If he does not work directly with them, he stengthens the underlying ideology of his own position, particularly if he does not attempt to expose the internal contradictions of his own position. Besides the ideological question, model-builders face analytical difficulties which are discussed in the following section.

4. MAJOR ANALYTICAL DIFFICULTIES

Any model-builder faces difficulties in at least two phases of his work: the specification and the validation of his model. Since these questions arise for all models, they will be discussed here.

4.1 Model specification

The first question to be answered in constructing a model is: What level of approximation is accepted? The answer, of course, depends upon the purpose of the model. Two conflicting pressures must be reconciled here: the model should be as close a representation of the studied phenomenon as possible; but too intellectually ambitious a project may be less fruitful than a more modest one.

The following unfortunate experience of this author illustrates this point. Research on the obstacles to the adoption of intensive forage production techniques by dairy farmers in Eastern France led to the hypothesis that technical uncertainties in the production process itself were such that many dairy farmers could not take the risks associated with these techniques. A simulation model was built with the purpose of testing this hypothesis. A programming model was deemed to be inadequate, since what mattered in that case was the cumulative effect of uncertainties in the sequence of elementary production processes (grass, planting, growth, grazing, growth again, grass and other foodstuff consumption by dairy cows, milk production . . .).

A production model synthesizing available biological information was built to represent the production process of a specific cowherd on an ordinary farm for which data on total daily milk production were available and were used for testing model results. After much effort, a reasonably good model simulating the first hundred days of the grazing season in the spring and early summer was constructed. But this was a far cry from what was needed to test the general hypothesis on which the whole project was based. The model really pointed out large gaps in biological knowledge. For instance, the relationship between rainfall and grass growth is both obvious and impossible to specify accurately. Hence, it was impossible to derive the probability distribution of milk production from data on rainfall variations. The cumulation of such obstacles led us to abandon the project.

This failure clearly was due to an over ambitious project. However, even though no formal test of the hypothesis could be conducted, the familiarity which we acquired with the phenomenon studied in the process of building and refining the model convinced us of the validity of our hypothesis. But how to communicate such a conviction? How to submit it to the criticism of our peers? How to specify the hypothesis so that its domain of validity could be assessed more precisely?

These questions call for building new models at various levels of specification. Which ones exactly, we do not know. This doubt should at least be sufficient to illustrate that "overkill" is a danger in model building. The level of approximation to be accepted, and hence the specification of the model, depends upon the quality of the knowledge pertaining to the phenomenon studied and its relevance to the issues to be clarified.

4.2 Model validation

To validate a model is to accept the hypothesis on which it is based. If one accepts that new knowledge is gained when old mistakes are corrected, it is the validation process rather than the resulting validated model which is really fruitful. For the purpose of action, however, it is a valid model which is needed. Hence the validation process serves two related purposes regarding learning and doing.

With these purposes in mind, what criteria of validation should be used? Of course, one expects that models will be logically consistent. Precisely, one advantage of mathematics and formal logic is to insure internal logical consistency. But to what extent do theoretical concepts correspond to real characteristics of the phenomenon studied? Are those available good measurements of the theoretical concepts? The internal logical consistency of the model enhances but does not guarantee the logical consistency of the analysis conducted with the help of the model. Yet it is the latter which counts.

The above questions lead to a second criterion of validation: How can one compare the model with the reality it purports to represent? Without entering into a philosophical discussion on the nature of reality, let it be accepted that a model is never compared with reality but with another image of reality; which image and which comparison? The first criterion which comes to mind is: Consistency with observation; model results should be compared with direct observations of the studied phenomenon. The model will be valid if its results are reasonably close to the observation. Two questions arise in this respect: What is reasonably close? What is the epistemological nature of these observations? Actually the closeness of the fit can only be judged subjectively. Even statistical models giving a range for a parameter estimate at some level of significance cannot escape this. As L. Mandersheid has shown very well, there is nothing absolute about a 5 percent or a 1 percent level of significance. Hence, the question to ask is whether or not the model renders an account of observed variations in the phenomenon studied. Whether this account is judged satisfactory or not depends upon the use made of the model, i.e., ultimately on the questions which it answers for one or several decision-makers. As for the epistemological nature of observations, it has been argued that no observation is strictly objective for it depends upon the observer. What he observes depends upon his ideas of what should be observed, hence of his *a priori* knowledge. This point is certainly well taken and one should probably speak of *consistency with experience*. To be valid a model must produce results (logical consequences on one's hypotheses) consistent with one's previous knowledge which had led one to make specific observations. In case of inconsistencies, hypotheses must be questioned and revised to render account of all observations. Experiments – seldom possible in the social sciences, it is true – are for some disciplines powerful ways to generate observations relevant to putting previous models to a serious test. But in this view they are only one category of experience.

Pragmatist philosophers have also suggested the test of workability. A model is valid if it works, i.e., if it leads to the right action. Such a criterion will readily be accepted if one accepts the general adaptive behavior model for which learning and doing are intimately related. But a blind application of this test can be dangerous. A model may "work" for a given situation and yet be invalid in all others while, to be useful, knowledge must have some degree of generality. A moment's reflection will also show that workability is a special case of experience. Hence, the test of workability can be accepted as a necessary, but not sufficient, condition of model validation.

Ultimately these three criteria of model validation (logical consistency, consistency with experience and workability) establish the degree of reliability which can be given to the knowledge thus generated and made available to decision-makers. As reiterated in this paper, knowledge must be viewed much more as a process rather than as a state; a model is not valid once and for all. In the pursuit of knowledge, the more a model raises pertinent questions, the greater its utility. Thus, somewhat paradoxically, the most useful models are those which have the greatest chances of being invalidated. Invalidation, the rejection of hypotheses, leads to correcting one's previous errors. It is the essential step in the learning process.

5. CONCLUSION

If one accepts the proposition that learning and acting are two interrelated phases of the same fundamental process of adaptation, the role of models in decision-making is very clear. Models are the simplified representations of reality formed in the agent's mind, his perceptions of himself and of his environment. Decisions to act are based on these perceptions. Conversely, as a result of his actions, the agent is led to revise his perceptions: he learns. Hence the agent is constantly adapting his perceptions and his actions which are themselves closely interrelated. He may also learn involuntarily as new information reaches him. In this perspective, models play a crucial role, even though most of them are very informal.

Model formalization can be a source of progress, i.e., of a better coherence between ends and means, provided that it produces reliable and sufficiently flexible models at a reasonable cost.

The role of model-builders is to help decision-makers in building such flexible and reliable models. This can be done in a variety of ways, as illustrated, for instance, by the agricultural economists' numerous modes of operation. A plea can be made here, however, for an "analytical" use of models.[12] By this expression we mean that the analyst should always consider the model as a tool for the analysis of the reality which is represented by the model. For instance, a model is a tool designed to analyze the situation of a decision-maker vis-a-vis his environment. Model results can never be taken as *prima facie* goals. They are only the logical consequences of the hypotheses, either explicit or implicit, and must be critically evaluated as such.

It should perhaps be emphasized that the adaptation process described above does not apply only to individual decision-makers. Collective action is not ruled out; on the contrary the concepts developed in this paper can be very useful to analyze why groups are formed, why they act, why they change membership, why they survive or why they disappear. The limitations of the approach in terms of decision processes must, however, be recognized. To place emphasis on economic decision-making processes implies that one views the economy, and therefore society, as a complex set of decision-makers influencing each other and linked together by a large number of interrelationships. In this perspective, conflicts are normal and recognized as such. The adaptive behavior model does not tell us how conflicts are solved. It does not even tell us which conflicts are important and which ones are secondary in the evolution of the economy and of society. As such it falls short of being a global theory of social change.

NOTES

[1]Some authors would argue that such a definition of the word models is much too broad. They would prefer to distinguish among hypotheses, theories and models. Hypotheses are essentially tentative propositions about reality and as such simplified representations of parts of it. A theory is a structured set of interrelated hypotheses. Models are still more specific. In a way they are specified versions of theories, often taking a mathematical form with specific values of the parameters. Such distinctions are often useful particularly in the process of building econometric models. They can help the analyst to become more aware of what he is doing. For instance, an uncertain estimation of a parameter does not necessarily imply that the mathematical structure of the model should be revised, the failure of a model to capture essential aspects of reality does not necessarily invalidate the theory. Perhaps, one should then wonder what kind of procedure is needed to put theories to crucial test. But this question, albeit important, would take us far away from the main objective of this paper.

[2]One could argue that all economic models are not decision models. Some models are used to represent a set of interrelationships between economic agents without specifically attempting to help any decision-maker. But even those are, in some sense, decision models. In order to be of any value, such models presumably should clarify some issues

for someone interested in those issues, i.e., a decision-maker who will use the information generated by the model in his decision process. Admittedly, this loop is not always closed.

[3]No attempt will be made to acknowledge the numerous intellectual debts incurred in building up my own version of this model. Most of these can be found in Day (1976), Petit (1975 and 1976), and Renborg (1976).

[4]Kuhn's thesis on scientific revolutions (1970), although it emphasizes the discontinuous nature of such progress, is fully consistent with the adaptive behavior model inasmuch as it argues that scientific discoveries occur through the destruction of old paradigms and the construction of new ones.

[5]We will not attempt here a review of the literature on models, not even decision models. For two excellent reviews see Dillon (1971) and Anderson and Hardaker (1972).

[6]In this context the word norms is synonymous with goals and thus differs from values.

[7]This objective may be profit or a farm income or a function of several objectives, as in the case of a utility function.

[8]For a good illustration, see the famous *Limits to Growth* (1972).

[9]For instance, the probability that x falls within $\bar{x} \pm t\sigma$ is α.

[10]As indicated above, model costs also must be taken into account here.

[11]In a letter to this author Biggs wrote: "Models developed for decision making processes in agriculture have not addressed: (a) policy issues of income distribution, (b) technology (in its broader sense to include more than HYV's), (c) accumulation of savings – by whom and invested in what specific items of capital, (d) the broader range of socio-political issues which came into real world policy decision-making (academics can avoid these things but not the civil servants and other policy-makers) . . . ". This judgment may be a little harsh but it clearly makes a point: The choice of phenomena judged to be important enough to be represented in a model depends upon the model-builder's ideological position.

[12]The responsibility for this expression is shared with my colleague, Jacques Brossier, who has written several good papers on this subject. See Brossier (1973).

REFERENCES

Anderson, J. R. and Hardaker, J. B. (1972) 'An Appreciation of Decision Analysis in Management', *Review of Marketing and Agricultural Economics*, **40**(4): 170–184.

Bachelard, G., *La formation de l'esprit scientifique*, Paris.

Biggs, S. D. (1973) *A Programming Model for the Agricultural Socio-Economic System to Investigate Output and Income Distribution Implications of Rural Policies: Kosi Region, Bihar, India*, IDS, Brighton, 232 pp.

Brossier, J. (1973) 'Un essai de liaison entre recherche formation et action', *Recherches d'economic rurale*, Paris, INRA.

Day, R. H. (1976) 'Toward an Adaptive Economic Theory' in Day, *et al.*, Agricultural Change and Economic Method, *European Review of Agricultural Economics*, special issue, to be published.

Dillon, J. L. (1971) 'An expository Review of Bernoullian Decision Theory in Agriculture: Is Utility Futility?', *Review of Marketing and Agricultural Economics*, Vol. 39, pp. 3–80.

Johnson, G. L., 'Philosophical Foundations: Problems, Knowledge, and Solutions', in Day *et al. op. cit.*

Kuhn, T., 'The Structure of Scientific Revolutions' in *Criticism and the Growth of Knowledge*, Lahatos and Musgrave (ed.).

Lewis, C. I. (1955) *The Ground and Nature of the Right*, Columbia University Press, New York.

Meadows *et al.* (1972) *The Limits to Growth*, Universe Books Publishers, New York.

Myrdal, G. (1969) *Objectivity in Social Research*, Pantheen Books, New York.

Petit, M., 'Farmers' Adoption of Technical Innovations', in Day *et al. op. cit.*

Petit, M. (1975) 'L'adoption des innovations techniques par les agriculteurs Plaidoyer

pour in renouvellement de la theorie economique de la decision', *Pour.* 40.
Piaget, J. (1968) *Epistomologie genetique*.
Renborg, U., (1976) 'Steering Economic Decisions' in Day *et al. op cit.*

DONG HI KIM*

Coordination in Models Representing Different Decision Making Levels in Agriculture

1. INTRODUCTION

Quantitative models for the agricultural sector or for regional planning have become increasingly important in the overall framework of economic and social development planning in a number of nations, both developed and developing. For example, large scale models for agricultural sector planning have been developed in some countries in Asia including the Republic of Korea, Thailand and the Philippines.

However, the progress made in model-building and in the utilization of formal models for planning and decision-making in agriculture varies a great deal from country to country. At present it appears that the developed countries have reached a much higher level of sophistication in developing and utilizing formal models than most of the less developed countries.

A number of reasons can be listed to explain these differences. They can generally be grouped into the following four categories: (1) differences in the availability of qualified human resources and facilities, (2) differences in the availability of reliable statistical data, (3) differences in the degree of perception and credibility accorded to models by decision-makers and (4) differences in the institutional and organizational barriers.

Of these factors, the first and second – relating to human resources and data – seem to be most relevant to the fact that most developing countries lag behind in the use of formal models for economic planning.

In order to make formal models more useful in agricultural planning, coordination is essential. Generally two types of coordination are called for: vertical coordination or linking the micro and macro models, and horizontal coordination among the related submodels. The primary focus of this paper is on coordination in model formulation and multilevel planning in agriculture. This will be followed by a discussion of the Korean experiences in this field.

2. COORDINATION IN MODEL FORMULATION IN AGRICULTURE

It has been frequently observed in many countries that the macro plans

* Director of the National Agricultural Economics Research Institute, Ministry of Agriculture and Fisheries, Republic of Korea.

formulated by planning experts have not been effectively implemented and have failed to produce the outcomes expected by the central planners. The gap which is evident between what macroeconomic planning models predict and what is actually occurring in the economy has been the focus of concern of both policy-makers and model-builders. This gap may be due to a variety of factors such as aggregation bias, insufficient statistical data, unrealistic assumptions which oversimplify the resource structure and behavior of economic decision-makers, inadequacy of the model employed, poor systems of implementation and random variables such as weather. These factors can be grouped into three categories as such problems of modeling or model formulation, problems of implementation and random variables.

In the following, analytical problems related to model-building, the nature of coordination and general adaptive systems approach as a most promising approach to well coordinated decision model-building will be taken up in order.

2.1 *Analytical Problems*

With respect to modeling techniques, remarkable progress has been made with mathematical tools which can be applied to improve decision models in agriculture, though there still exists inadequate statistical data base in terms of quality and quantity especially in developing countries.

The task of coordination is generally concerned with both the process of model specification and model validation which are usually confronted by a model-builder.

The modelling effort begins with a theoretical framework for which relevant knowledge is extracted from all related disciplines and experiences, and then the further step to mathematical modelling is taken by which the relationships between enlisted functions or variables and the constraints are determined to answer the specific questions raised.

Most of the analytical tools used in economic analysis are flexible and general in nature so that their application is unlikely to be restricted by the types of problems to be tackled. On the other hand, models are relatively inflexible in as much as they are problem-specific. As such, they must be modified if they are to be used to examine different problems.

In reality, even though no two models are exactly the same, their basic structure can be almost identical if certain assumptions are satisfied. Differences in the goals and objective functions of different decision-making units often make the models or even the structure of models inconsistent. If the goals of objective functions are the same among the different levels, the structure of the models can be made similar. If a consistent set of goals and objectives can be defined, then a system of models can be developed which will be consistent.

A micro planning model is often used for decision-making on farm firms in order to optimize the allocation of farm resources. This type of decision model is constructed under a set of prices, factors and products expected during the production period. Such a model, however, may not be very meaningful unless both forward and backward linkages are developed between the micro and the more macro regional and national planning models. These linkages are essential since they provide the means through which individual farm units are supplied with information on the aggregate consequences of their production and consumption behaviors.

The nature and scope of linking the macro and micro decision models in agriculture appear to be closely related to the type of economic system under which the agricultural sector is organized and operates. For example, in a centrally planned economy the task of coordinating the macro and micro models and their linkages is relatively easier than in a market economy where the decision-making power and risk taking are delegated to the individual firms.

In a market economy, the goals or objective functions of decision-making units at different levels are usually so diverse that the structure of models and variables to be used should differentiate among them. For example, the macro goals may be to increase national wealth, increase aggregate consumption, increase the rate of economic growth, or to attain an equitable income distribution. On the other hand, the individual farmers' goals may be to maximize profit or family labor income, or to maximize family utility by balancing income and leisure.

A question arises on how to transfer readily the macro goals of an agricultural plan to the micro sphere, or on how to find the common denominators between goals of various decision-makers at different levels. A form of ranking function by which various goals are transformed into a set of objective functions being weighted by priority ranks would be a solution to this problem. However, this is a philosophic issue as well as a methodological question in economic analyses. For example, the use of multi-criteria variables without attempting to reduce them to a single objective function is also acceptable in formulating a policy model (Rossmiller *et al.*, 1972).

From the methodological point of view on linking multilevel models or vertical model coordination, it may be useful to examine the following three cases separately.

Case I: The mathematical forms for equation specifications of models are differentiated among various decision-making levels.

Case II: The mathematical forms employed are the same, but the variables included are not identical among different level models.

Case III: The mathematical forms and the variable names included are the same, but only their sizes are differentiated among the levels.

Case I is often due to behavioral divergences of variables at different decision-making levels, while Case II is caused by varying assumptions underlying the different models. However, the form of combining these two cases is more frequently confronted by the analysts in the process of model-building.

It is generally noted that what are taken as "given" in the micro models, e.g., prices, are the values which are to be determined in the macro models. Also, the types of constraints applied to the contrasting levels of decision-making are different. For example, at micro and regional levels, the decision-makers are faced with uncertainties both on yield and prices of products, whereas the central model-builder is primarily faced with yield variance because regression models for the central planning authority can provide reasonable predictions on prices.

Case III appears to entail the least complex problems in linking multilevel models. The main problem in this case, however, is the consistency of aggregation or disaggregation of variables, as well as the consistent estimation of parameters at the different levels. This consistency is guaranteed only under certain constraints (Kim, 1971). Decomposition of an agricultural sector model into subcomponents based on commodities and geographical areas may be one way to tackle this problem.

Furthermore, since the central or regional planner should acquire knowledge on how the low level decision-making units respond to those elements of a plan which alter the planning coefficients of each of the target systems, he needs a predictive model which will indicate the aggregate consequences of individual behaviors.

Some other important constraints imposed on modelling agricultural decision-making come from the spatial and temporal characteristics of agriculture. Agricultural production is a biological process located in various areas under diverse physical, technological and economic conditions. Thus the model structure and variables included in more general cases must be differentiated from one level to another. And the variations in the structure and in the parameters of a model over time should also be taken into consideration, especially when a medium range or longer time horizon is called for in planning.

2.2 *Nature of coordination*

Coordination in modelling for decision-making includes two components: a formal interactive linkage between submodels and cooperation between the model-builders and decision-makers concerned. In general, there are two contrasting types of macro models: those models which are built from the "bottom up" and those which are built from the "top down". The former describes the micro behavior of farms and then aggregates to the macro level, while the latter starts at the macroeconomic level. However, these two types are unlikely to be independent of each other. For an effective vertical coordination in multilevel agricultural planning, some method of horizontal coordination is essentially required between the model-builders and relevant units of decision-making especially at the central and regional levels.

In the case of planning for actions, a possible gap between a plan and its realization exists due to an incongruity between the boundary of planning areas and that of effective decision-making.

In some instances, it has been observed that the parties concerned with planning have no authority to make effective decisions required for implementation. This implies that the acts of modelling for prediction (positive modelling) and of modelling for planning actions need to be distinguished and to be adequately linked with each other as the need arises.

This further indicates that the roles of formalized planning models need to be clearly defined in view of the political, economic and social settings under which the modelling efforts are made. By this process, the roles of coordination in modelling, which is instrumental and interactive in nature, can specifically be defined.

The planning parties, including experts and decision-makers, are required to maintain close rapport for coordination in the process of model-building and planning in such tasks as setting up the objective functions, conceptual framework, structure and variables of the models to be employed and in generating the data to be used.

It may be argued that the validation in economic analyses is strictly a matter of simple application of general criteria of validation by professional experts, rejecting an involvement of non-professionals in the process even if they are directly concerned. However, this is unlikely to be true especially in a practical application of social science as in formal decision models for economic activities. A test of logical consistency of an analysis would be an easy part which can be done without participation of other than experts. But testing consistency of a practical decision model with experiences and workability requires coordination between the model-builders and the users. Thus the analysts and the users of models at various decision-making levels must keep open a workable communication channel and to reach a common understanding about norms and facts relevant to model validation. If they are placed in separate agencies, an insititutional linkage is necessary for effective coordination between them.

2.3. *General adaptive systems approach*

The methodological problems in building quantitative decision models have been discussed in a number of expert seminars and professional papers (Heady, 1971; Cocks and Carter; Thorbecke). Until recently various forms of optimization models, such as linear programming and recursive programming, and consistency models, such as input–output models, have been extensively employed for agricultural sector planning or regional modelling.

Regression equations have also frequently been incorporated in building decision models. A model with only regression equations, however, often fails to consider the spatial and structural features of agriculture. A good example of welding a regression analysis and a programming model for resource allocation would be the case of quadratic programming (Takayama and Judge).

Nevertheless, the comprehensiveness, consistency and generality, which are desirable characters of a general adaptive planning model, appear to be reasonably insured in a systems simulation approach. This is because in this approach various analytical tools can be mobilized and linked together to deal with the complex problems and components such as dynamic behaviors, policy variables, uncertainties and interactions between subcomponents in agriculture.

Simulation modelling uses a building block approach by which the relevant components are separately modelled and then linked together one by one to form a comprehensive system of models. This approach has many advantages, i.e., it is time-saving in operation, flexible in the use of analytical tools for building the individual component submodels, and easy to coordinate in spite of the fact that a test of validity of the whole model is often extremely difficult.

This approach can also offer a further advantage in a case when lack of statistical data restricts reasonable estimation of certain parameters. Because of these advantages, recently this approach has been increasingly introduced in a number of fields of inquiry including agricultural planning.

3. THE KOREAN EXPERIENCES

3.1. *Use of programming models*

As Korea has experienced a rapid economic growth primarily through industrialization, the agricultural sector has also steadily been modernized, demanding more formalized approaches to agricultural planning at various decision-making levels. In practice, the linear programming technique has been applied to farm planning at sectoral and farm levels. For example, the National Agricultural Economics Research Institute (NAERI) has developed a regional linear programming model to compare profitabilities of different types of farming. In addition, an interregional linear programming model has been developed which generates optimum solutions to farm resource allocation questions encompassing 50 commodities and 9 regions. The model using alternative assumptions on biological technology and grain price policy has recently been employed to provide information to the planners engaged in developing the nation's Fourth 5-Year Plan (Choi and Kim, 1975).

In this process of modelling, cooperation has been maintained to a reasonable degree between economists, agronomists, extension agents, farmers and policy-makers in generating information to be put into the model.

In spite of the fact that ordinary linear programming models have various constraints which limit their extensive application, Korea's experiences have in general demonstrated the usefulness of this sort of analytical tool in providing the decision-makers with economically feasible solutions to the resource allocation problems. As this approach is expected to be included in the Fourth 5-Year Development plan (1977–1981), a further geographical decomposition will be followed.

3.2. *The Korean agricultural sector study*

In an attempt to produce more information useful to policy-makers, the Korean Agricultural Sector Study (KASS) has been undertaken as a joint effort between Michigan State University in the U.S. and the National Agricultural Economics Research Institute in Korea since September 1971 (Abkin *et al.*, 1976; Byerlee and Halter; Lee, 1975; de Raen and Lee, 1972; Rossmiller *et al.*, 1972).

Systems simulation is the major tool for modelling and has been employed together with other conventional quantitative analytical techniques.

The problem definition underlying the Korean Agricultural Sector Model (KASM) is a broad one encompassing the major concerns of agricultural policy-makers in Korea. These include increasing farm income; achieving self-sufficiency in the major foodgrains; improving agriculture's trade balance; and reducing the contribution of food prices to inflation. Major emphasis is placed on production and consumption of agricultural crop and livestock commodities, while little attention is paid to the marketing sector, although this is of increasing concern to policy-makers. KASM is designed for use in medium and long-range (5–25 years) analyses.

The initial KASM model, which was developed in 1972, had three agricultural regions; upland, single cropping paddy and double cropping paddy regions based on cropping patterns, in order to reduce aggregation errors in production models. But the results of this approach were later found not to be significant as expected, calling for better classification of regions and improved quality of statistical data. Therefore, spatial decomposition has been tentatively abandoned in the current KASM model.

The components of KASM, their major interactions, the policy inputs considered and some of the main outputs (performance indicators) are shown in Figure 1. The interactions, shown on the figure by lines and arrows, are labelled with abbreviations which are explained in the key in the upper right-hand corner of the figure. Policy inputs are indicated by Ⓟ. Although some of the components shown in Figure 1 are still in various stages of development, they are shown here to give a complete picture of the guiding view of the system.

The principal components and their inputs and outputs are:

(1) *Population*, which projects farm and non-farm population levels using a cohort survival model and accounts for farm–non-farm migration. Its main *input* from the sector model is non-agricultural labor demand (NLD), which is used to determine migration. The population component *outputs* are the agricultural labor supply (ALS) and population levels (POP).

(2) *Demand–price–foreign trade*, which projects farm and non-farm consumptions patterns, consumer and producer prices and agricultural imports and exports using linearized consumption functions and finding a simultaneous solution for prices and consumption while satisfying an income constraint. *Inputs* to demand component include domestic supplies (QS), income (I), the non-food price index (CP), world food prices (WP) and price and trade policies. Principal *outputs* include producer prices (PP) and demand for non-food commodities (QD).

(3) *Yield/input demand*, which uses a production function model to project crop yields and input application rates resulting from government programs in land and water development and in crop improvement research and extension. *Inputs* include agricultural input price indices (IP), producer prices (PP) and government investment policies. *Outputs* are input application rates (IAR) and yield levels (Y).

(4) *Farm resouce allocation*, which uses a recursive linear programming model to allocate land, labor and capital to crop and livestock production. Inputs include yields (Y) and input application rates (IAR), input prices (IP), producer (PP) and the agricultural labor supply (ALS). *Outputs* projected are land allocation among the crops, desired livestock herd sizes and feedgrain import required.

(5) *Crop* and *livestock production*, which compute as outputs total production and quantities supplied for human consumption (QS).

(6) *Fisheries production* is not currently modelled but is an input projected from outside the model.

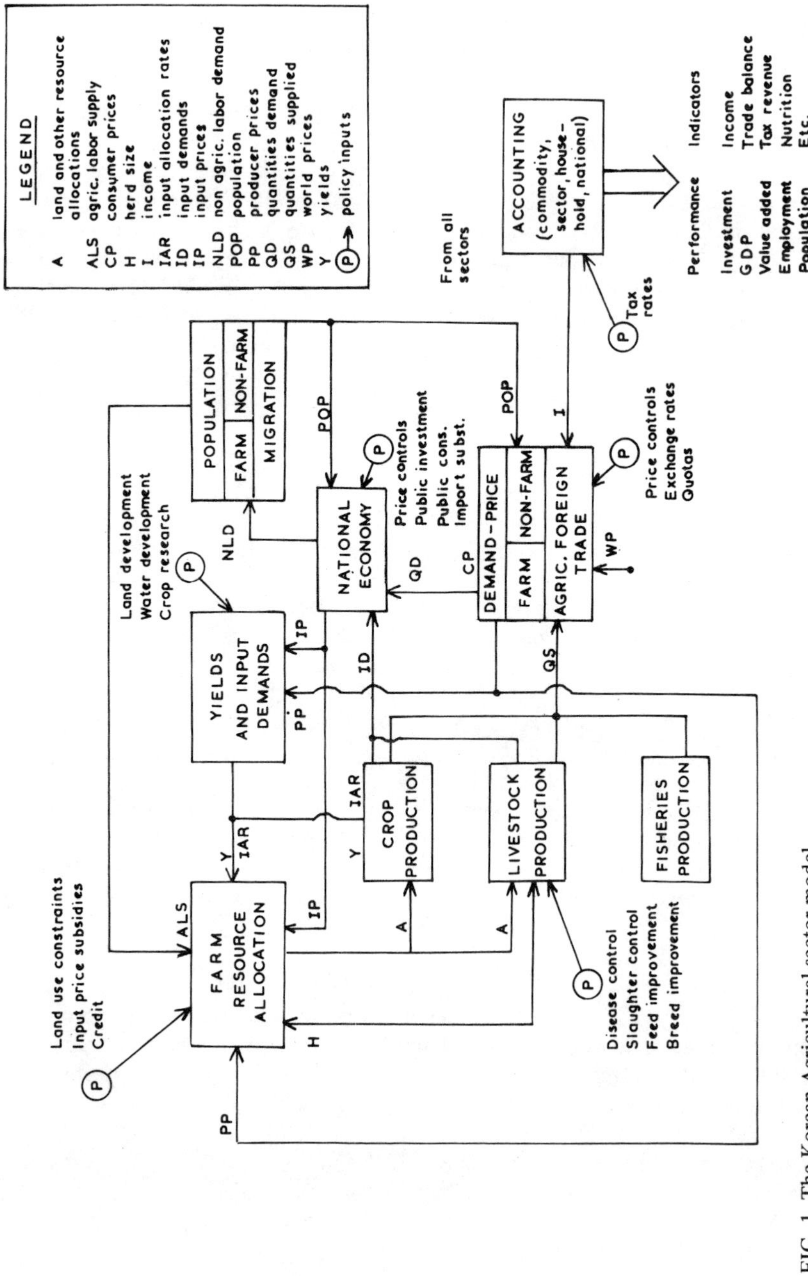

FIG. 1. The Korean Agricultural sector model.

(7) *National economy*, which links the agricultural and nonagricultural sectors through a 16-sector input–output model. *Inputs* from the agricultural sector model are expenditures on nonfood commodities (QD), population (POP) and demands for agricultural inputs (ID). *Outputs* to agriculture include the price index of consumer goods (CP), agricultural input price indices (IP), and nonagricultural demand for labor (NLD).

(8) *Accounting*, which takes variables determined in the rest of the model and computes such indicators of system performance as value added, employment, population, nutrition, income, trade balances, etc.

Although the KASM is near completion, many problems have still to be solved for institutionalization of this approach in the practice of planning and decision-making in agriculture. The major problems facing the Korean analysts lie in such areas as quality data to represent economic and behavioral variables related to agriculture, increasing interests of the decision-makers in what they want measured and analyzed, keeping will-trained analysts and qualified economists in the research or planning positions for reasonably long periods of time, and closer cooperation between researchers and decision-makers at various planning levels. The problems listed above seem to be similar to those facing most of the contemporary developing countries.

As a matter of fact, the Korean government has recently taken a series of actions favourable to solving these problems, such as strengthening agricultural statistical service, expanding training programs for civil servants and private industrial managers at various decision-making or planning levels, balancing the salary scales for civil servants with that for professionals in the private sector, and organizing working parties and committees at various levels for better communication and coordination, horizontal and vertical. As a result, the working environment for professional analysts in Korea has been significantly improved.

Nevertheless, it is recognized that continuous efforts should be made on all fronts to bring about better performance in the steering of good planning and decision-making for agricultural growth and farmers' welfare.

4. CONCLUDING DISCUSSION

A building block approach, based on systems simulation, appears to be a methodologically promising one for the efficient coordination of multilevel planning in agriculture. This is because the systems simulation approach is comprehensive in scope and general in the use of analytical tools and data, and thus provides both the analysts and users with flexible means of coordination by which behavioral components can be well accommodated.

Yet there are several problems associated with the approach which must be confronted at the outset of development. The systems simulation approach is characterized by high initial costs, a rather long development period and requires the services of both qualified analysts and a well developed statistical data base on the technical and economic features of the agricultural sector and the behavior of individual farmers.

In many instances it has been observed that this approach is too sophisticated to be understood by decision-makers at various levels, especially in a developing country where budgeting methods or simple analytical tools have been dominantly employed. As a matter of fact, the adoption of such a sophisiticated planning tool as system analysis and computerized simulation method would be an important innovation in policy formulation, calling for step by step efforts on the part of the initiators toward the full scale acceptance by the potential users.

At the national level it would be desirable to develop a systems simulation model which predicts the aggregate consequences of various constraints, such as weather and factor supply, and varying policy alternatives. However, this effort should be supported by a number of micro-oriented behavioral studies. The KASM has suffered from the shortage of behavioral information relating to lower level decision-making units.

It is also desirable to develop an inter-regional competition model based on regionalized data on production resources and product markets if they are spatially differentiated. The regionalized data, of course, are to be aggregated from micro data by a clearly defined procedure. The underlying assumptions of perfect competition, free mobility of farm resources within a region and balance in resource structure between farms should be carefully examined.

Then a question arises on how a farmer builds his own decision model for optimum allocation of farm resources. In a market economy, it would be impossible to build formal farm planning models for all farm units. Instead, representative farm models for various sizes and types of farming can be built on the basis of physical and technological data available and price expectations which are drawn from the equilibrium prices generated by the sector model under alternative policies and resource conditions.

During the two-way process of modelling: upward for data generation and downward for decomposition of the large model, it is necessary that the limitations of the information generated by a model thus developed should be clearly understood by all parties concerned.

The model-builders or analysts need to be warned not to oversell their products to the users. Even though a more comprehensive sector or regional decision model is desirable, it is often observed that a simplified model is much more workable than a sophisticated one in aiding the policy-makers in the practice of decision-making. The greater need in the whole process of modelling is to facilitate close interactions and communications among the model-builders and the decision-makers or users.

A decision model will by no means replace the decision-maker. The model is simply an aid to the practice of decision-making. It would not be appropriate to see the act of coordination in modelling as a mechanical process of linking submodels or multilevel models. Instead, it should be viewed as a multisided, dynamic and evolutionary process, which is conditioned by types of decision-making institutions and by refining analytical tools and data, and by improving the understanding of the need for, and ways of, coordination among all parties concerned. Therefore, coordination should be institutionalized throughout the whole process of model formulation including generation of information, selection of theoretical and empirical model framework and verification of the analytical work.

REFERENCES

Abkin, M. H., *et al.* (1976) *The KASS Approach and Agricultural Policy Analysis*, Issue Paper.

Byerlee D. and Halter, A. N., 'A Macro-Economic Model for Agricultural Sector Analysis', *Am. J. Agr. Econ.*, Vol. 56, pp. 520–523.

Choi, H. H. and Kim, D. H., 'Resource Allocation on Sericulture-Grain Farms in Kimjae Plain Area', *Korean J. Agr. Iron.* Vol. 15, pp.

Cocks, K. D. and Carter, H. O., 'Micro Goal Functions and Economic Planning', *Am. J. Agr. Econ.*, Vol. 50, pp. 400–410.

Heady, E. O. (ed.) (1971) *Economic Models and Quantitative Methods for Decisions and Planning in Agriculture*, Proceedings of an East-West Seminar, The Iowa State Univ. Press.

Kim, H. T. (1971) *Effects of Aggregation on Errors of Prediction of Macro-Variables*, Unpublished Ph. D. Thesis, Pennsylvania State Univ. Chapter II.

Kim, S. J., *et al.* (1975) *A Study on Agricultural Production Allocation Between 9 Regions in 1981: A Linear Programming Model*, NAERI, Seoul, Korea.

Lee, J. H. (1975) *Projections of Agricultural Land, Yield Level and Intensity by Coops: A system Simulation Approach*, NAERI.

De Raen, H. and Lee, J. H., (1972) *Dynamic Model of Farm Resouce Allocation for Agricultural Planning in Korea: Application of Recursive Programming within a General Systems Simulation Approach*, Working Paper, MSU, East Lansing, U.S.A.

Rossmiller, G. E., *et al.* (1972) *Korean Agricultural Sector Analysis and Recommended Development Strategies, 1971–1985* KASS Team.

Takayama, T. and Judge G. G., Spatial Equilibrium and Quadratic Programming, *J. Farm Economics*, Vol. 46, pp. 67–93.

Thorbecke, E., 'Sector Analysis and Models of Agriculture in Developing Countries', *Food Research Institute Studies*, Vol. 12, pp 73–89.

L. FOLKESSON*

The Role of Models in the Formulation of Agricultural Policy

1. INTRODUCTION

A well-known definition states that a model is a simplified description of reality. It is then safe to assume that agricultural policy, as well as economic policy in general, is always based on models. It is obvious, however, that the theme of my paper contains many more aspects that this particular one. To display at least some of these aspects, I have found it necessary to set certain limits to my presentation.

First, I will discuss quantitative models, based on so-called modern mathematical and econometric methods. The building of such models has now for rather a long time been a subject at the research frontier of our field of science.

Second, within the field mentioned I will especially discuss macroeconomic models. This does not mean that in my view, microeconomic models have no role in policy formulation. The need for knowledge of a microeconomic nature, when constructing macroeconomic models, is also fully recognized.

My third limitation is that I will keep the economic and policy-making systems of the developed market-oriented economies particularly in mind. I am aware that there is a very interesting development of quantitative macroeconomic models in several countries with centrally planned economies and that this may also be true for some developing countries but my experiences do not permit me to speak on these issues. I can therefore only hope that those who have experiences of model applications from centrally planned or developing economies will share these with the rest of us in the discussion which follows.

The particular question identified is thus the role of macroeconomic models in the formulation of agricultural policy within developed market economies. This question is controversial. It is controversial not only at the policy-making level, but also among us agricultural economists ourselves. To support this statement with a number of quotations would be easy. The only generally accepted viewpoint seems to be that, at least so far, in the vast majority of cases agricultural policy is formed without reference at all to the models we

* Professor of Agricultural Marketing and Policy of the Swedish College of Agriculture.

are interested in. It is therefore a matter of great importance to develop a better common understanding of the models, among policy-makers as well as among ourselves. In my opinion, this task is even more important than the development of more sophisticated and complex models.

If so, how can we identify the specific role that the models could and should play? How can we judge whether or not the models we develop fulfil this role? And, above all, how much of our limited time should we allocate to the building of quantitative macroeconomic models?

These and related questions are, as far as I know, not discussed very much in our literature or scientific journals. Neither does this paper claim to present the answers. My hope is, however, that it will point to at least some of the issues involved.

2. FORMULATION OF AGRICULTURAL POLICY

From our literature and scientific journals it is obvious that the making of policy recommendations is one of the major objects of our science. At the same time, we do not seem to spend very much time on analysis of the policy formulation process itself.[1] What did initiate a particular political decision? How much time was available before a decision had to be made? What information material was collected? How was the material analysed? What influence did the political and administrative framework have on the decision? How was it followed up?

I want to stress the lack of such analyses. Some of you may feel that questions of this type should be left to the political economists or to the political scientists. But is it realistic to expect those professions to take an active interest in agricultural policy? Others of you may feel that you already know the answers to the questions raised, due to active personal participation in the policy formulation. My question to you is then: Why don't you make more efforts to document your experiences, so that they also can be made available to those of us who want to promote the use of models?

Behind my argument is a firm conviction that knowledge about the policy formulation process itself is of vital importance, if the aim is to develop models which are intended to play a role in this connection. For how can we, in the absence of such knowledge, define the role of models or judge whether or not the models developed fulfil this role? This can hardly be done, if we have no clear picture of the alternative or complement to model use, or if our only assumptions are that the existing policy formulation is "highly informal". that it is "unsystematic" and that it is based on "hunch and intuition".

Let me, therefore, make a few remarks about the policy formulation process. My starting point is then the description of it in the theory of economic planning. This may or may not be a good picture of how it actually works under real-world conditions.

I am aware that the very term economic planning is controversial, in particular when, as here, it refers to macroeconomic or central planning. So it is necessary for me to be rather precise on this point. There are several definitions of economic planning to be found in the literature. The one I have chosen has been provided by Professor Leif Johansen of the Oslo University (Johansen, 1974). I quote:

> Macroeconomic planning is an institutionalized activity by or on behalf of a central authority for:
> (a) the preparation of decisions and actions by the central authority and
> (b) the coordination of decisions and actions by lower units of the economy
> for the purpose of governing the development of the whole economy and its constituent parts so as to achieve certain goals for the economy and to harmonize the development with broader non-economic goals.

This definition claims to be valid for centrally planned as well as market economies. The important thing is, as is pointed out by Professor Johansen, the weight that is put on "decisions and actions of the central authority" relative to "coordination of decisions and actions taken by the lower units of the economy".

By lower units of the economy is meant, e.g., households, farms and industrial enterprises. It is then obvious that in market economies, the policy formulation aims much more to the "coordination of decisions and actions taken by the lower units of the economy" than to "decisions and actions of the central authority itself". All the important elements of the agricultural policy underline this statement.

Even so, if we accept Professor Johansen's definition, it is quite relevant to speak about the existence of central economic planning in the market economies.[2] It is then also relevant to speak about a corresponding planning model. This model, i.e., the planning model, must not be mixed up with the quantitative macroeconomic models of interest here. I will come back to this very important point later.

Another aspect of the policy formulation process is, again according to the theory of economic planning, its sequential or stepwise character. This specific feature is described by different authors in slightly different ways (see, e.g., Kirschen *et al.*, 1964, and Tinbergen, 1956). The following steps are, however, usually identified:

(1) recognition of a policy problem, initiation of the policy formulation process,
(2) analyses and investigations,
(3) consultations with, e.g., political parties and different interest groups,
(4) governmental proposal to parliament, discussion and decision, and finally
(5) execution of the policy measures agreed on.

This description is obviously especially valid in case of a major policy issue. It is then normal, at least in my own country, that the inititation takes the form of a governmental directive to a specially appointed *ad hoc* committee. Members of this committee may, again at least in my own country, be members of parliament and/or governmental administrators as well as experts from different non-governmental organizations and "independent" experts. The committee is, moreover, charged with the responsibility to report its findings to the government.

Needless to say, the policy-making process may take several different forms under real-world conditions. This may not only depend upon its being a major policy issue or not. Other factors such as if the policy issue at hand is of a repetitive nature or not, the time available for the decision and the degree of political agreement of different levels, may also be of importance in the connection. These features are fully recognized in the literature on economic planning.

If we look at the content of the agricultural policies of the developed market economies, we find a number of common features. As we all know, there is also a number of specific features for each country as to the formulation of goals and as to the choice of means.[3] The specific features are, among other things, related to the foreign and trade policy, to the relative size of the agricultural sector and to the prevailing structural conditions within agriculture as well as within other sectors of the economy.

There is one common feature that I want to stress in particular. It is that the goals of policy are in most cases expressed in quite general terms. I mention this because I now and then get the impression that many model-builders argue for more precise goal formulations. They are of course completely free to do so. They should then, however, be aware of the fact that there may be very rational arguments behind the choice of quite general goal formulations. One such argument, trivial but nevertheless important, may be that it is not possible to obtain a needed political agreement about more precise goal formulations.

There may also be even more important arguments behind the goal formulation strategy referred to. These arguments are directly related to the basic concept of central economic planning within a market-oriented economy.

The major role of this planning is, as I have already pointed out, to coordinate the decisions and actions by the lower units of the economy. It is then quite obvious that the more precisely the overall policy goals are expressed, the stronger will be their influence on the decisions and actions by the lower units. This effect may sometimes be in agreement with the prevailing political and planning principles. In other cases, it may very well be in conflict with these principles. This is the case when it is felt that more precise goal formulations would limit too much the role of the decision-making at the lower level.

Another argument for general goal formulations, also related to the basic concept of central economic planning within a market-oriented economy, can be derived from the fact that there is always a considerable uncertainty about the future development. I have only to refer to the international market development, or to the supply response, in order to support this statement.

Let us, with the uncertainties in mind, assume that the policy-making authorities issued very specific goal formulations. Let us, moreover, assume that they really were prepared to take actions to secure the achievement of the goals. Depending upon what the future developments actually turned out to be, the achievement of the goals might require the use of, e.g. far-reaching quantitative regulations of the foreign trade and/or of the supply. The use of such means may again be felt to limit too much the role of decision-making at the lower level. If so, it is again quite rational to stick to the general goal formulations.

3. QUANTITATIVE MACROECONOMIC MODELS

I will now add a few words about the macroeconomic models. It is outside the scope of this paper to present a full catalogue of all such models, even within the field of agricultural economics (see, e.g., Heady, 1971). More important is to identify the common features of the models and to make some general observations in this connection.

The models of interest describe the whole or at least the major part of the agricultural sector. This description is, in the typical case, quite disaggregated in terms of the number of outputs, production opportunities and inputs. A further level of disaggregation, with respect to regions and/or types of farms, is also found in many cases.

Another feature in common is that the models are composed of a number, often quite large, of mathematical relations or equations. Each equation is then assumed to represent a particular aspect of the underlying economic system. Examples of such aspects are different balances of a physical or of a monetary nature, different technological relationships, different assumptions about the behaviour of the lower level economic units and different institutional relationships such as, e.g., the existence of market regulations.

A third feature in common is that the construction of the models and the estimations of the data needed are in most cases major undertakings in terms of time required and costs involved. At the same time, the calculations are in many cases easily carried out by the use of a computer. Under such circumstances, the models often lend themselves very well to the generation of a number of alternative solutions from alternative specifications of different model parameters. This feature is of considerable importance for our discussion.

There is quite a large number of different models that fit the description given so far, with respect, e.g., to how the time factor is dealt with, to the types of data used, as well as to the mathematical techniques employed to generate the model solutions. I will not, as I have already said, go very deeply into these questions. There are, however, two observations that I want to make.

My first observation is related to the purpose of the model calculations. As is well known, we may distinguish between models for forecasting the development of the economic system being studied, on one hand, for determining a preferable state or a development path of the system, on the other.

In the first case, when the purpose is to make forecasts, the development of the system studied is viewed as a function of factors which cannot be controlled by the policy-makers as well as of the application of different policy instruments. Moreover, it is implicit in this approach that the evaluation of the alternative forecasts generated and, consequently, also the choice of values of the policy instruments, has to be made outside the models.

In the second case, when the purpose is to determine a preferable or optimal state of the system studied, we have to introduce an explicit representation of the relevant policy goals. This is done via the introduction of an objective function, which is to be maximized or minimized, and sometimes also via the introduction of specific restrictions into the equation system of the model.[4]

In this case, as well as when the purpose is to make forecasts, we also have to make assumptions about the factors which cannot be controlled by the policy-makers. A difference is, however, that as a rule we cannot determine the values of the policy instruments, if our purpose is to determine the state of the system which is optimal in relation to the policy goals specified.[5] As a consequence, the determination of the policy instruments that lead up to the calculated optimal state usually has to be made outside the models.

Although there are important differences of principle between the two approaches mentioned, there are also a number of similarities. The choice of approach is therefore a practical question that has to be determined in view of the policy problem at hand.

Against this background it is a little surprising how we model-builders often debate quite intensely the pros and cons of, e.g., econometric models in relation to programming models. This discussion is perhaps even more surprising since, as a rule, it does not give many references to the underlying policy problems.

Behind the debate I think one can trace the fact that the whole field of quantitative macroeconomic models is a large one. We, who are interested in the development and application of such models, therefore tend to specialize within one particular sector of the whole field. As a result, we develop vested interests in the promotion of the models that belong to our own particular sector.[6] Our discussions of the different models are then not seldom a reflection of the power struggle within our field of science and it is this reason that I argue for a better common understanding among ourselves about our models. It is hard to see how the models can play a proper role unless such an understanding can be developed. Perhaps this conference could play a role to this end.

My second observation is related to a question of terminology. We often present our programming models, when they include explicit assumptions about the policy goals, as so-called planning models. But in my experience this frequently creates a lot of misunderstandings when the models and their results are discussed with administrators and policy-makers.[7]

An example may illustrate the possible reasons for the misunderstandings. Let us assume that the policy-making authorities are considering the introduction of active measures to promote a certain regional development within agriculture. Let us assume, further, that they ask an agricultural economist to study the problem and that he does so by means of an inter-regional programming model of the standard type. His results would then consist of one or several location patterns, together with the corresponding cost estimates.

If the economist then claimed that he had prepared a "plan" for the location of agriculture, there would be good reasons for misunderstandings.

First of all the policy-makers could feel that it would be completely outside the prevailing policy-making framework to prepare such a plan. Secondly, the policy-makers would probably remind the economist that if such a plan is to be prepared at all, several other types of information in addition to those represented in an inter-regional programming model would have to be made use of. Thirdly, the plan that the policy-makers themselves had in mind would probably be quite specific about the actions to be taken, rather than about the expected "end" results. They would therefore not understand the economist who claimed that he had prepared a "plan", if he did not show at the same time how it could be achieved.

So, in order to avoid misunderstandings, I think it would be very wise not to use the word "planning" in connection with our macroeconomic models. Though I am not certain which terminology we should use instead and am open to different suggestions. I am, however, quite sure that we should reserve the term "the planning model" for the policy-formulation process itself.

4. ROLE OF THE MODELS

Let me now return to the questions I raised in my introductory remarks. The first one was: How can we identify the specific role that the models could and should play in policy formulation?

In an interesting paper (Johansen, 1971) on planning theory Professor Johansen, whom I have already quoted, has defined what he calls "the maximum function" of a model. A model is said to satisfy this function if it is used "to calculate which is the best possible decision and if the decision then is implemented".

Is this "maximum function" what we have in mind, when we develop models about different aspects of agricultural policy? Or, to carry the argument one step further, do we like Simon in his book *The Shape of Automation: For Men and Management* foresee a situation when we will be able to gather all relevant data in a gigantic information system? And be able to develop the models needed to analyse all the data? A situation when the information and model systems are supervised by a group of "managers", sitting in front of computer display units, and whose closest collaborators are model and programming specialists?

This vision of "the maximum function" is certainly not that of Professor Johansen. Nor, for that matter, is it my own vision – not even if we limit our discussion to policy issues of a "pure" quantitative nature. I base my view on several arguments, some of which I have already indicated. One of these is that what can be expressed by numbers and equations is only part of the information needed for policy decisions. Moreover, the character of the decisions that have to be made is changing all the time, due to the dynamics of the technological, economic, social and political developments. On the other hand, the construction and introduction of new information and model systems take a considerable time. When a policy-decision has to be made, would it not then by very irrational to rely only on the particular results coming from a formal information and model system?

So it is safe to assume that our models could hardly satisfy "the maximum function" mentioned. As a personal observation I would like to add that even if they could, they should not – not be allowed to, if you wish – play such a role. For this would mean that we would have to go over to a policy-making system that certainly would be much more authoritarian than the present, unless every citizen obtained a PhD in econometrics and computer programming, which seems to be highly unlikely.

Which are then the arguments that we model-builders usually use, when we want to promote the role of our macroeconomic models in policy formulation? These arguments are, as far as I have been able to find out, of the following types:

(1) that the models permit a large number of relationships to be analysed simultaneously,

(2) that the models permit even large volumes of data to be analysed systematically and

(3) that the models permit the generation of alternative model solutions within a relatively short time and at low costs,

and moreover, as a consequence of the properties mentioned:

(4) that the models can increase knowledge of the planning situation and also that the models can improve decision making on the basis of existing knowledge.

To be able to evaluate these claims, I have looked for comments from people holding responsible positions in policy formulation at the national and international levels. I have found several remarks about the role of our science in policy formulation. Some of those remarks, e.g., that agricultural economists should take more interest in real-world problems and that we should use a more understandable language, may have a connection with our model-building activities. But I have certainly not found any comments that take up directly the claims that we use to justify our work. How should this silence be interpreted? It seems reasonable to assume that the responsible policy-makers are not unaware of the fact that model-building is an important activity of our science.

Perhaps the policy-makers, as well as some of ourselves, are still uncertain if our claims are justified? This assumption does not seem unreasonable in view of the fact that model-building is after all a relatively new activity.

Or is it that the preparation of quantitative forecasts, and quantifying the consequences of alternative decisions, are tasks that are in practice carried out at lower administrative levels? And that we model-builders are wrong in claiming, as we sometimes do, that the results from our calculations are of specific interest at the highest policy-making level?

One feature common to our claims seems to be that they particularly stress quantitative aspects such as the many relationships included, the large volumes of data handled and several alternative solutions possible. Nobody can deny that many of our model studies are quite large in these dimensions. But how strong is the correlation between, for example, the number of relationships and data on one hand, and the quality of the results on the other? Not very strong, I think, if it exists at all.

And moreover, do not the very properties mentioned sometimes tempt us to build large models, very detailed within areas where data are easy to obtain, but with a very limited scope? Are we not tempted to present very detailed results, which may give a wrong impression as to their exactness? Do we not sometimes calculate so many alternatives, that it becomes impossible even for ourselves to identify the main findings?

Moreover, could not claims also be made for the real-world planning model? I assume that it is designed so that the insight and experiences accumulated within different administrative agencies are made use of in a proper manner. And that persons with knowledge about different spheres of economic and social life are actively engaged in the different steps of the policy formulation.

By now I may have given some of you the impression that my own view is that the quantitative macroeconomic models have no role to play at all in policy formulation. As I have indicated earlier, this is not the case. I have only tried to argue in the spirit of the Swedish author Strindberg, who told us that we must demolish the old structures to let in air and light.

My own view is that quantitative macroeconomic models certainly could have a role of their own to play in policy formulation, as one source of information among several others. They could be a tool to identify which of the assumptions made – about policy goals, instruments and non-controllable factors – are especially important for the outcome of the analysis. Such sensitivity analyses could in my view very well lead to a more efficient use of the total resources available for data collection, analyses and investigations.

The models could moreover form a basis for bringing together information from several different sources and thereby maybe lead to new formulations of real-world problems. However, this role is certainly more modest than the one some of us seem to have in mind when the claim "to improve decision-making" is made. After all, what is a wise decision?

If you accept my own view on the role of the models, channels are essential for communications between various agencies, especially those whose task it is to prepare material for policy decisions and the model-builders. This has been pointed out by several authors. Some authors have suggested that there should be certain persons entrusted with the specific task of acting as communication links. The motive for this suggestion is that a good ability to develop models, and a good ability to communicate with other people, are not always found together.

I must confess that I find the idea of such intermediaries rather peculiar. The reason may be that I come from a small country where we cannot afford too many types of specialists. At the same time I feel that having such intermediaries would only add to the present confusion. How can a model-builder who is not prepared or able to make personal efforts to obtain insights into policy issues develop models of an applied nature? And how can he do this, if he is not prepared or able to explain in an understandable way how the models are constructed and what the findings are?

My second introductory question was: How can we judge if our models fulfil their role or not? In the literature one can find quite a lot of suggestions concerning the evaluation of forecasting models but very little about the evaluation of programming models, designed to illustrate the choice between policy alternatives. Professor Schultz made a rather grim remark on this issue when he said: "I am regretful we have not even tried to resolve this problem". Personally I am not sure that the situation is much better when it comes to the forecasting models. I have in mind the tests that could show if our models do play a role in the policy formulation or not. I, moreover, assume that we are satisfied if our models serve as one source of information, among several others, for policy formulation.

One idea could then be to test the models in relation to "the maximum function" earlier mentioned. But such a test would most likely be negative in almost all cases. If the test on the other hand turned out to be positive, this would, as I see it, rather be the result of pure coincidence than of anything else. So this type of test would not be helpful. The tests that I would like to propose would rather be to ask questions of the following types: Was it possible to explain the construction of the model to the administrators and the policy-makers? Because if it weren't, they would be unlikely to pay attention to the findings. Did they ask for modifications of the model and for more alternative calculations? Did they spend a considerable time discussing the findings? Did they explicitly refer to the model study in their own report? If they did, my own conclusion would be that the model has played its proper role.

Maybe the ultimate test of this type would be the following: Did the administrative agencies want to take over the model and use it themselves for further studies? Or did they claim that after all, the model was their "own baby"? They would certainly do so if they thought that the model was a very useful one because they would feel that access to the model could strengthen their own role in the policy formulation.

The type of tests now proposed should, however, by no means be a substitute for our usual professional tests. I have in mind, for instance, if the assumptions and the models were clearly described, if the study could be repeated by someone else, if its validity were described and if the results were presented independent of prejudice. I rather think that these tests should be applied more vigorously than they sometimes are.

The two types of tests argued for could occasionally give different results – for instance, where a heavily promoted model study, which was quite poor from the professional point of view, reached the attention of agencies with little earlier experiences of such studies. But this would probably only happen once. So I am convinced that the two types of tests would need to yield the same result, if our models are to play a role in a longer perspective.

My last introductory question was: How much of our limited time should we agricultural economists allocate to model-building activities? The answer is related to our personal aims, so the question is a very general one. Let me assume, however, that our aim is to put ourselves in a position that enables us to make policy recommendations. What are then the alternatives open to us, as agricultural economists in particular?

There are then several alternatives open to us. Among them are the identification of new sources of statistical data, the description of the current development in a consistent way, the forming of new theories about the factors behind the development, the discovery of new policy instruments and, of course, also the building of quantitative models that could summarize existing knowledge. Which alternative should we then choose? That is a decision that I think each one of us has to take on a personal judgement of his or her comparative advantages, after making a careful analysis of pros and cons of each alternative in relation to the current state of knowledge.

5. CONCLUDING REMARKS

I will not try to summarize my discussion, so my concluding remarks will be very brief.

I am aware of the fact that I have not been able to give very precise answers to the questions I raised in the beginning of my paper. Nevertheless, I feel that my approach has been justified. For before one can give the answers, one has to formulate the questions.

When thinking about my concluding remarks I have looked to our literature for a good way to characterize the particular activity of developing quantitative macroeconomic models. The best I have found has been formulated by Jan de Veer. I can only agree with him when he has said (de Veer, 1971) that "the building and operation of these models is itself an act of policy-making".

NOTES

[1]One of the few works that I know of, where not only the models but also their policy-making framework are dealt with, is Miller 1973. Another pioneering work, which gives a description of the Swedish approach to the formulation of agricultural policy, is Anderson 1972.

[2]The term "indicative planning" is sometimes used in this connection.

[3]See, e.g., OECD, 1975.

[4]I have in mind the case when the goal structure includes at least one goal dimension which has the character of a flexible target in Tinbergen's terminology. This is, as far as I have been able to find out, the typical case for the agricultural policies of most of the market-oriented economies.

[5]There may, for example, be a number of alternative values of the policy instruments that can be expected to lead to the same state of the system studied.

[6]In this respect, I look at myself as a rather typical "programming man".

[7]The same observation has been made by a Norwegian colleague of mine. See, Reisseg, 1971.

REFERENCES

Anderson, A. (1972) Översiktlig planerig i lantbruket. (Perspective Planning in Agriculture), *Lantbrukshögskolans meddelande* **B17**. Uppsala.

Heady E. O. (ed.) (1971) *Economic Models and Quantitative Methods for Decision and Planning in Agriculture. Proceedings from and East–West Seminar*. Iowa State University, Ames.

Johansen, L. (1971) *Planlegging og Spill*, (*Planning and Games*), Oslo University of Oslo, Institute of Economics, Reprint Series No. 76.

Johansen, L. (1974) *Lecture Notes on Methods of Macroeconomic Planning*, Oslo, University of Oslo, Institute of Economics. Mimeo.

Kirschen, E. S. *et al.* (1964) *Economic Policy in our Time. Vol. 1: General Theory*, North Holland Publishing Company, Amsterdam.

Miller, T. A. (1973) Economic Adjustment Research for Policy Guidance: An Example from Agriculture, in, Judge, G. G. and Takayama, T. (ed.). *Studies in Economic Planning over Space and Time*, Amsterdam, North Holland Publishing Company.
OECD. (1975) Review of Agricultural Policies: *General Survey*, Paris.
Reisegg, F. (1971) *Extent of Gaps between Plan and Realization*, In: Heady, E. O. (ed.), (*op. cit.*).
Simon, H. A. (1965) *The Shape of Automation: For Men and Management*. Harper & Row Publishers, New York.
Tinbergen, J. (1956) *Economic Policy: Principles and Design*. North Holland Publishing Company. Amsterdam.
de Veer, J. (1971) *Extension and Administrative Needs in Application*, in: Heady, E. O. (ed.), (*op. cit.*).

DISCUSSION OPENING – Q. B. O. Anthonio, *Nigeria**

Dr. Petit started by explaining that the general model of adaptive behaviour was designed "to establish close interrelationship between reflection and action". After defining agents, environments, goals, feasible domains and the agent's perception, the author lucidly analysed the close relationship between perception, learning and acting. This was followed by a full discussion of adaptive behaviour and the role of models which opened with the – albeit controversial statement – that "at any point in time the agent has a perception of his situation" (not necessarily) . . . "of his goals" (not always) . . . of his possible actions, and some of the values he holds . . .".

In the discussion of classification of models the author seems to reject the notion of the normative versus positive models on the basis of Glenn Johnson's (1976) dictum and evolved the "better" phraseology of "optimization model" instead of normative model and preferred the word "positive" to be in quotation marks. I feel that this is rather hair splitting. The crucial issue really is that "optimization" is, after all, a normative concept no matter how one wants to redefine it and inclusive of the fact that other objectives (e.g. income or consumption needs) impose constraints on any one objective.

At one stage of the discussion on model classification, the paper becomes rather muddled. To be specific, the author asserts that "Thus, if one speaks in terms of norms, the main purpose of an optimizing model is not to produce goals ("norms") but rather to question the "norms" implicitly, or explicitly, accepted by the decision-maker". (p.) I am totally at a loss in reconciling "norms" and optimizing. I would have assumed from the author's own definition of goal, ("Agents act in order to modify their situation with respect to their environment. Their action is thus postulated to have a purpose. They have *goals*".) that optimizing is, of course, a goal and thus consistent with a type of "norm". Whether or not linear programming correctly and adequately represents this "norm", is the key question. And if it does, whether farmers behave in such a way as to optimize a particular constraint or objective is, to my understanding, again, a different question.

The argument that "the usefulness of models depends more on the role given to them by the analyst rather than on the type of model used", is not

* At Joint ECA/FAO Agriculture Division, Addis Ababa. Ethiopia

tenable, since the capacity to operate the model itself is highly dependent on the type of model, its versatility, and relevance.

In most developing countries, and particularly in areas where computer facilities can be a constraint, over-complicated models make a mockery of the concept of models as an aid to the decision-making process. They may become too cumbersome, too costly and time consuming.

I am frustrated by the assertion that "failure of a model to capture essential aspects of reality does not necessarily invalidate the theory". (p. 74n). The question then is, what help is the model giving to decision-makers in the light of this type of uncertainty, especially at macro level analysis which often builds within it a whole series of aggregative misfunctioning (or errors)?

Professing that "all economic models are not decision models" is really begging the question. To my understanding of this exercise, the choices to be made have to be obvious and in fact to precede the hypothesis to be tested, the thesis to be confirmed; the models give the decision-maker the opportunity of choice by illustrating expected consequences of alternative decisions.

The author took us through a kind of structural classification of models. Discussing the advantages and disadvantages of these models (econometric) (1920), programming (1940's), and simulation models (1960's), the author believes that the best of the lot is the simulation model since it is a form of system analysis and permits of a very wide range of uses. I agree with this, but mention should be made of the fact that because of the lack of scientific rigour in this model, it permits a whole lot of errors which in most cases cannot be crosschecked and the unwary believer can easily be led astray. For policy-decisions based on such a model, these errors can give rise to gross misallocation of resources and an outright unsavoury position. This situation should not be overlooked.

Our profession is constantly being criticised by decision-makers on the basis that we allow slipshod work to pass, using data of doubtful authenticity to build up magnificent designs that are not only expensive but of limited practical use. As scientists, even social scientists, we have to search more and more for useful models with built-in verification. The underlying "philosophy" is important, but so are the data and model that correctly explain this "philosophy". Personally, I consider these three aspects, philosophy (or hypothesis), model used, and the data incorporated, constitute an eternal triangle of interrelated "variables" essential in any meaningful analytical economic model.

Perhaps the frightening portion of the paper is in the conclusion that "scientific objectivity is an illusion". If that is correct, and I doubt if it is, what is the whole exercise about models for? Is it not to be more rational and to use more scientific concepts to improve decision-making? What are optimization, maximization, cost minimization, all about – scientific or mathematical concepts and functions are surely intended to illustrate rational optimum positions.

There is no doubt whatsoever that models are very useful in decision-making. Unfortunately, the types of models which our professionals have produced, to the best of my knowledge, have in no way engendered confidence in the minds of decision-makers, or even solicited it. Far too often, unfortunately, it is the mechanics of such models that are given the greater attention. In this respect, I agree with the author and state finally that "model formalization can be a source of progress . . . provided that it produces reliable and sufficiently flexible models at a reasonable cost".

DISCUSSION OPENING – P. Reider, *Switzerland*

Dr. Folkesson raised all the important questions relating to interregional models but they are too many to answer fully. Regarding the relationship between model-builders and decision-makers, the former is too narrow a term. The builder must be deeply knowledgeable about agricultural policy and politics. With such experience good, problem-orientated, country or region specific, models can be built.

Decision-makers are not interested in methods so it is not a question of "selling" models but of using suitable methods for solving problems. A model-builder who is an agricultural economist with a thorough understanding of agricultural and economic problems can give useful and effective advice to the decision-makers out of his experience of applying models. He knows the problems of government and what government decision-makers must know to be in a position to make the best decisions.

A search for features common to a number of models may not be very useful. How to build a model is often a matter of what is practicable. The question of objective functions cannot be reduced to simple terms either. Since each policy may have several goals, different objective functions must be considered. But it is not possible to treat the whole complex of agricultural goals with mathematical models. Because of this complexity, I think the best approach is usually to apply the model to calculating the outcome of different alternatives – for instance, the effect on agricultural income distribution of different support measure.

Finally, I believe that only the builder understands his model well enough to evaluate it. Consequently it is only to be expected that evaluations in the literature will be scarce.

Report of the general discussion

The importance of models in promoting rigour in policy analysis, the specifically LDC problems in using models, and the pros and cons of simple models were aspects common to several contributions.

On the rigour aspect the importance of developing a framework of thought for the policy-maker, the value of simple models, the contribution of models to greater discipline in analysis, seemed to meet general agreement in principle. Where a simple model, could be developed involving a small number of variables, and showing clearly the economic mechanism and the behaviour of economic institutions, the alternatives were easier for the policy-maker to grasp.

Though simple models might have such attractions it was felt that, in fact, decision-making in agriculture was by its nature complex and, since models contributed only in so far as they contributed to a knowledge of reality, it was unlikely that simple models would be realistic. Departures from realism were seen to take several forms. Assumptions about economic relationships and the decision-making framework – often that of North America or Western Europe – with the resultant models, were liable to be applied in a much too imitative way in developing countries. Application of sophisticated models is pushed ahead before theory has been modifed to reflect local conditions. Into what model, for instance, can the decision process be fitted which placed a Nigerian cocoa farmer in a position now of great socio-economic influence because, as a young man, he spent his first income on buying a chieftaincy instead of buying land?

Distortion and irrelevancy were traced to deficiencies in data – sometimes directly from error, sometimes because absence of the necessary data forced modification and simplification of models beyond a reasonable relation to reality.

The less specifically economic features needed more attention in model-building than, so far, they had received – for example, political and administrative considerations. The significance of the increasing impact of agricultural producers and workers on decision-making at all levels was also stressed; so too was its importance.

Apart from this general core of discussion, various specific questions were posed. For example, since agricultural economists could be presumed to be scientists and concerned with facts, was not the positive their natural approach; the normative seemed more suited to priests or politicians. And if so, any contribution in a normative vein invited frustration. Or again, what meaning was to be attached to "simple" in relation to models? Who was in mind? – Presumably non-economists. But in that case it was "operational" that should be stressed – an operational model could, through its results, be understood by administrators and politicians.

Participants in the discussion included: J. A. Akinwumi, *Nigeria*; A. T. Birowo, *Indonesia*; P. C. van den Noort, *Netherlands*; L. A. Odero-Ogwel, *Tanzania*; K. Prasad, *India*; D. Tomic, *Yugoslavia*.

M. Petit (in reply)

It has been questioned whether it was reasonable to assume that at any point in time the agent has a perception of his situation, of his goals and of his possible actions. The question is put whether this does not assume too rationalistic a pattern of behaviour? A similar point is raised by the question whether a model could take account of the decision by a cocoa farmer to buy a chieftaincy rather than land. My claim is precisely that the general adaptive behaviour model provides us with an approach to building more specified models capable of taking account of such behaviour. Saying that our agent has a perception of his situation, his goals and his possible actions does not imply that these perceptions are clear or even free of inconsistencies. Understood in this manner, I do not feel that these assumptions are unrealistic.

About the distinction between normative and positive models, it is precisely because I feel that the distinction is misleading that I have used the expression "optimization models". For reasons already indicated, I do not agree with the claim that optimization is normative.

The failure of models to capture essential aspects of reality does not invalidate the theory on which these models are based. I am surprised that this should surprise anyone. For instance, there are many good modern agricultural economists in Western Europe using Marx's theory in spite of Marx being totally mistaken about the evolution of agriculture in Western Europe. Similarly the numerous failures of market mechanisms do not prevent us from using neo-classical market theories.

I would concede that I may have overstated my case in emphasizing that the usefulness of a model depends more on the role given to it by the analyst than on the type of model used. This might imply that the type of model has no importance. Of course such is not my position. I have, in fact, written that some models will be adequate in some circumstances and other models in other circumstances.

The point that it may serve African colleagues to have a paper incorporating models of North American, or West European origin published in international journals – even though they are of little practical use to Africa – is very well taken. I submit that the general adaptive behaviour model is general enough to raise good questions about the behaviour of farmers everywhere, permitting economists to conduct "concrete analyses of concrete situations" and thus follow Lenin's excellent prescription.

Finally I must record two further points of disagreement. I do not believe that we should confuse philosophy and hypotheses. The former deals here with the scientist's relationship with society; the latter with the way society, or more precisely in our case the economy, is working.

I maintain that pure scientific objectivity is our illusion. Our main task as social scientists is, in my view, to increase everybody's understanding – including our own – of social problems – in other words to make knowledge more objective, knowing full well that it will never be fully objective.

Dong Hi Kim (in reply)

Agricultural economists should pay more attention to what we do not know about how to sell the products of our research to potential customers in their diverse situations. It seems to be a necessity for successful co-ordination in modelling for agricultural development to maintain a minimum balance in basic knowledge and attitudes toward the problems and analytical approaches between the analysts and the decision-makers.

On adaptability, simplicity and operational feasibility of a decision model, the concept of adaptability of a planning or a decision model can be expanded so that a practical decision model in agriculture includes the behavioural variables of the micro decision-making units. Secondly, the theoretical frameworks and analytical models which have been developed mainly in the western industrialized countries should be made adaptable to meeting the current problem-solving needs of the underdeveloped countries.

The operationality, and the simplicity, of a decision model is important so far as it fulfils the specific needs of an analysis or a planning situation. However, a simplified model, in terms of structure and variables covered, can hardly be meaningful for the explanation of a complex situation or for the planning of a bio-sociological system such as agriculture. By contrast, a rather sophisticated model can better serve for comprehensive analysis or for workable planning in agriculture.

L. Folkesson (in reply)

I would agree with Dr. Rieder that my paper raises more questions than it answers. I mentioned this in the paper. But my paper has served its purpose if the questions taken up are the relevant ones.

My definition of a "model-builder" has been held to be too vague. When I began to write the paper, I intended to be more specific but it turned out to be difficult to draw a clear picture of a "model-builder" without, at the same time, drawing a picture of an agricultural economist – and an economic advisor in general, for that matter.

As regards the role of a model as a tool to obtain a better understanding of practical policy issues, I fully agree with this statement if it is read as better understanding among policy-makers. I, myself, also think that the sociological aspect is a very important one and had it in mind when I stressed the role of our model as a link between different experts as well as between experts and decision-makers.

We have confined our discussion to the role of models in relation to policy-making, but a valuable alternative would be to look at them as educational tools. This question is very well worth a seminar of its own.

R. A. RICHARDSON, J. B. HARDAKER AND J. R. ANDERSON*

Farm-level Decision Models for Developed Agriculture

1. INTRODUCTION

The focus of this paper is decision-making by farmers in the so-called "developed" countries. We have directed our attention to the modelling of farm-level decisions in commercial agriculture. It is through the construction and use of models that agricultural economists can better understand the options open to farm decision-makers, can appreciate the effects of farmers' attitudes on the choices they make, and can help farmers to make better decisions.

Farmers, like other people, are frequently confronted with situations requiring a choice. Many choices are made more or less habitually or randomly without much consideration of alternatives. However, some decisions are seen as more important than this, requiring careful evaluations. It is in regard to these more weighty decisions that the analyses of agricultural economists are potentially useful.

Farmers are faced with deliberative choices in regard to production. They must decide what techniques of production to use and how to allocate their resources of land, labour and capital among the alternatives open to them. Investment decisions, characterized by consequences extending far into the future, can be regarded as a special kind of production decision. Farmers must also make decisions about marketing. They must decide not only what to produce, but how to sell it. Such decisions involve the method of pricing, choice of selling outlet and the integration of production and marketing activities. It seems that agricultural economists have tended to neglect this important class of decisions in their analyses in the past. Finally, there is a large class of personal decisions which farmers face relating to choices about lifestyle, family matters, and so on. While we recognize the importance of such personal decisions, we do not deal with them explicitly, since our concern is with the business decision of commercial farmers.

Farm-level decision models can be either normative or positive. That is, they can be directed towards assisting the farmer to make a rational choice,

* University of New England, Armidale, Australia.

or they can be designed to aid understanding and prediction of the choices farmers make. Petit (1976) examines this dichotomy between normative and positive in the framework of a model of general adaptive behaviour. He seeks to classify models on the basis of their mathematical structure or source of data. However, we have chosen to classify models in terms of the embodied assumptions about the decision makers' beliefs and preferences. Hence in the following sections we turn to an examination of how a farmer's degree of knowledge about the possible consequences of his decision and his preferences for those consequences, bear upon his choice. From this preliminary discussion we develop a taxonomy of farm-level decision models. We use this taxonomy to highlight areas of emphasis and neglect in the recent literature. The paper concludes with a discussion of possible developments in farm-level decision modelling in developed countries.

2. THE DECISION ENVIRONMENT

Many models are founded on assumed certainty, in the sense that they are built using single valued parameters. If the inevitable uncertainty of the real decision environment is recognized at all, it can usually be accommodated in such models only very imperfectly, perhaps by some form of sensitivity analysis. The dangers of using deterministic models in a stochastic world have been emphasized in the context of planning by Anderson (1967a) and Upton and Casey (1974) and in the context of appraising efficiency by Pasour and Bullock (1975).

With improved computational resources, it has become more practical to incorporate dimensions of risk into farm-level decision models. In this context risk is taken to mean uncertain events to which probabilities are attached and it is argued that all probabilities are subjective, reflecting the degree of belief the decision-maker holds in the occurrence of relevant uncertain events (de Finetti, 1972). Objective data bearing on the uncertainty, such as relative frequencies of past events, are seen merely as aids in formulating probability judgements. In this view, the classical distinction between risk, where probabilities are "known" and uncertainty, when they are not, is invalid and all cases of uncertainty for which possible events can be enumerated are categorized as (subjective) risk.

While the subjectivists' view appears to be gaining ground in decision modelling, some authors have argued that subjective probabilities by themselves may not capture all a decision-maker's feelings about the uncertain environment. On some occasions he may be more confident about his subjective probability judgements than on others. The concept of ambiguity has been introduced as a measure of degree of confidence in probability estimates (Ellesberg, 1961; Bernard, 1974), and some attempts have been made to include ambiguity in agricultural decision models (e.g., Maruyama and Kawaguchi, 1971). However, there is by no means universal agreement on the need for this further dimension of uncertainty and our sympathies are with those who have argued strongly against it (e.g., Raiffa, 1961).

Because it is impossible to account for the inconceivable, all decision-making embodies a dimension of ignorance. The existence of ignorance is recognized in models of an adaptive nature that incorporate feedback procedures for modifying solutions in the light of new information (Petit, 1976). However, by its very nature, ignorance cannot be formally incorporated into a decision model. Hence in our taxonomy of models we will ignore ignorance, along with ambiguity and uncertainty without probabilities, and focus on the simple dichotomy between certainty and subjective risk for describing planning environments.

3. PREFERENCES AND DECISION RULES

Unless made at random, which would be a rather happy-go-lucky approach in positive terms and quite inconsistent from a normative view, decisions imply preferences. In turn, the implementation of preferences implies decision rules within our models. Normatively, the most attractive decision rules are those based on subjective expected utility (SEU) theory. The cornerstone of this theory is the SEU theorem which is based on axioms of choice. The theorem tells us that for a decision-maker whose preferences do not violate the axioms, there exist: (i) a subjective probability distribution for the set of outcomes associated with any choice alternative under uncertainty; and (ii) a utility function U whose subjectively expected value gives a measure of attractiveness for each of the available alternatives under uncertainty. The alternative with the highest SEU is the optimal choice.

Categorization of preferences, and hence of the function, U, can be based on whether or not preference depends on only one or more than one attribute of the decision outcomes. Preference functions for single attributes are easily estimated and imply $U = U(X_1)$. Multiple attribute preference implies $U = U(X_1, X_2, \ldots, X_n)$. In so far as either of these cases results in some attributes not being considered, the SEU approach is lexicographic since these other attributes are not taken into account.

The single attribute case needs little elaboration. It is well covered in the literature (e.g., Anderson *et al.*, 1976, or Halter and Dean, 1971). The multiple attribute case is more complicated and also more realistic, since decisions are rarely based on a single attribute. Complications arise in the multiple attribute case because of the necessity to assess preferences in more than a single X_i dimension simultaneously and to allow for trade-offs between the several dimensions. A variety of methods has been developed in recent years for eliciting multiple attribute utility functions, as reported by Keeney (1973) and Kirkwood (1976). They all involve intensive interrogation procedures that are worthwhile only for sizeable decision problems. In most cases, a reasonable pragmatic approach for multiple attribute problems is to assume a utility function of the additive form $U = \Sigma_i k_k U(X_i)$. The weights k_i $(i = 1, 2, \ldots, n)$ are fairly easily determined and the additive function is generally adequate to discriminate between available alternatives.

In terms of the specification of preferences, the most complicated decision problems are those involving multiple attributes over time, i.e., where X_i $(i = 1, 2, \ldots, n)$ has a time sequence of uncertain outcomes $X_{i1}, X_{i2}, \ldots, X_{it}$

over the time periods $1, 2, \ldots, t$, Conceptually, such problems can be logically viewed as multiple attribute situations where there are $(n)(t)$ attribute dimensions. As a practical matter, however, elicitation of such functions, except for two or three attributes over a period of two or three years, is an impossible task. The analyst is forced back to the patently illogical but comprehensible approach of maximizing the utility of present (financial) value. In other words, decision costs dictate a second-best approach.

Though not so obvious, the question of decision costs is also relevant to any application of the SEU approach. For normative analysis, the SEU model should only be used if its use gives benefits beyond its cost of application, on *a priori* appraisal which has its own decision costs. Hence an infinite regress of prior appraisal is implied. To those not prepared to act on judgement, this infinite regress is seen as a telling criticism of the SEU approach. It is also suggested that real-world decision-makers may not follow the SEU model because they prefer other decision rules such as safety-first criteria. As yet, however, there is no substantive evidence on just what decision rules may best describe the behaviour of decision-makers as they express their preferences in actual choices.

Day (1971) has attempted to represent in models of decision-making behavioural factors that allow for the interaction of values and experience through what he calls environmental feedback. A conclusion from such work is that decision rules should perhaps allow for "flexible" utility and, accordingly, the results can only be useful for quite short time horizons. Goals may be represented by a lexicographic ordering, especially if the decision-maker is viewed as a satisficer in search of acceptable alternatives. Over successive time periods, one feasible approach to modelling this type of behaviour is recursive programming which, in its most recent incarnation, Day (1976) call "cautious sub-optimizing".

4. A TAXONOMY OF MODELS

There have been so many attempts (see, e.g., Throsby, 1974) at inventing an instructive taxonomy of models used in agricultural economics that a further essay in the saga must risk being of negative marginal value to the profession. However, we brave this danger in the hope of more succintly depicting recent progress in the invention, innovation and diffusion (to follow De Benedictis' 1974 suggestion) of farm-level decision models.

At the outset, it should be made clear that we can only usefully categorize modelling approaches that are relatively formal. In thus banishing informal or intuitive methods to the unclassified wilderness, we are conscious that the great bulk of *actual* farm-level decision-making is based largely on intuition. We can try to mime such intuition through positive models and to improve on it through normative models, but classification of intuitive methods seems impossible.

From the decision theoretic point of view that we espouse, the most useful primary taxonomic device is the assumption made about the preference of the decision-maker. In this regard, the fundamental distinction is drawn between orderly Bernoullian utility functions (to be categorized as regular preferences) and "other" categories, designated here as irregular preferences. Under the latter heading we herd together lexicographic orderings (e.g., Dean 1974–5), *ad hoc* rules of thumb (such as criteria for games against nature), arbitrary satisficing procedures, safety-first rules, etc.

The primary classification is presented in the left-most tree in Figure 1. A three-way bifurcation of preferences (and corresponding models) is indicated. As well as their regularity or otherwise, preferences are distinguished as involving one or more than one attribute, and as reflecting consequences over single or multiple periods.

Several of the branches in the left-most tree of Figure 1 lead to model subsets that are so close to being empty of important theoretical and/or operational models that they can be virtually ignored in this short review. In fact, most important developments can be captured by examining only branches A (regular, single period, single attribute preferences) and H (irregular, multiple period, multiple attribute preferences). This will do some but not too much violence to the reality of modelling. For instance, when we speak of linear programming methods on branch A, we recognize that while a single attribute is being optimized in the objective function, linear programmers in practice accommodate additional or subsidiary attributes of farmers' preference functions in the constraint of the planning matrix. In terms of our classification, such efforts thus fall more correctly on branches C or G, which are not explicitly reviewed here.

To examine models in branches A and H further, we resort to a more detailed level of classification as shown in the right-most tree of Figure 1. The first basis of finer classification is the recognition, or otherwise, of uncertainty in the planning environment. Models are then distinguished by whether they permit the identification of optimal decisions, or are either non-optimizing or exploratory. These first two next taxa are of greater consequence than the third, which merely styles the mathematics used in solving the models.

The taxonomy introduced in the trees of Figure 1 is now exploited in reviewing recent work in farm-level modelling.

5. DESIGN AND USE OF FARM-LEVEL DECISION MODELS

It is difficult to know what, on a world-wide scale, is the current state of application of farm-level decision models. Probably a minority of farmers in developed countries use even the simplest formal models to aid their decision-making. The use of methods such as budgeting may be more widespread in farm advisory agencies but more complex models, such as the methods of decision analysis or computer-based models, appear to be used only rarely.

Analysts of pragmatic or unsophisticated bent tend to confine most of their empirical work to heuristic, exploratory computations in a presumed certain environment. On branch A (Figure 1) this work would most often be single period "hand" budgeting. On branch H it would be budgeting over

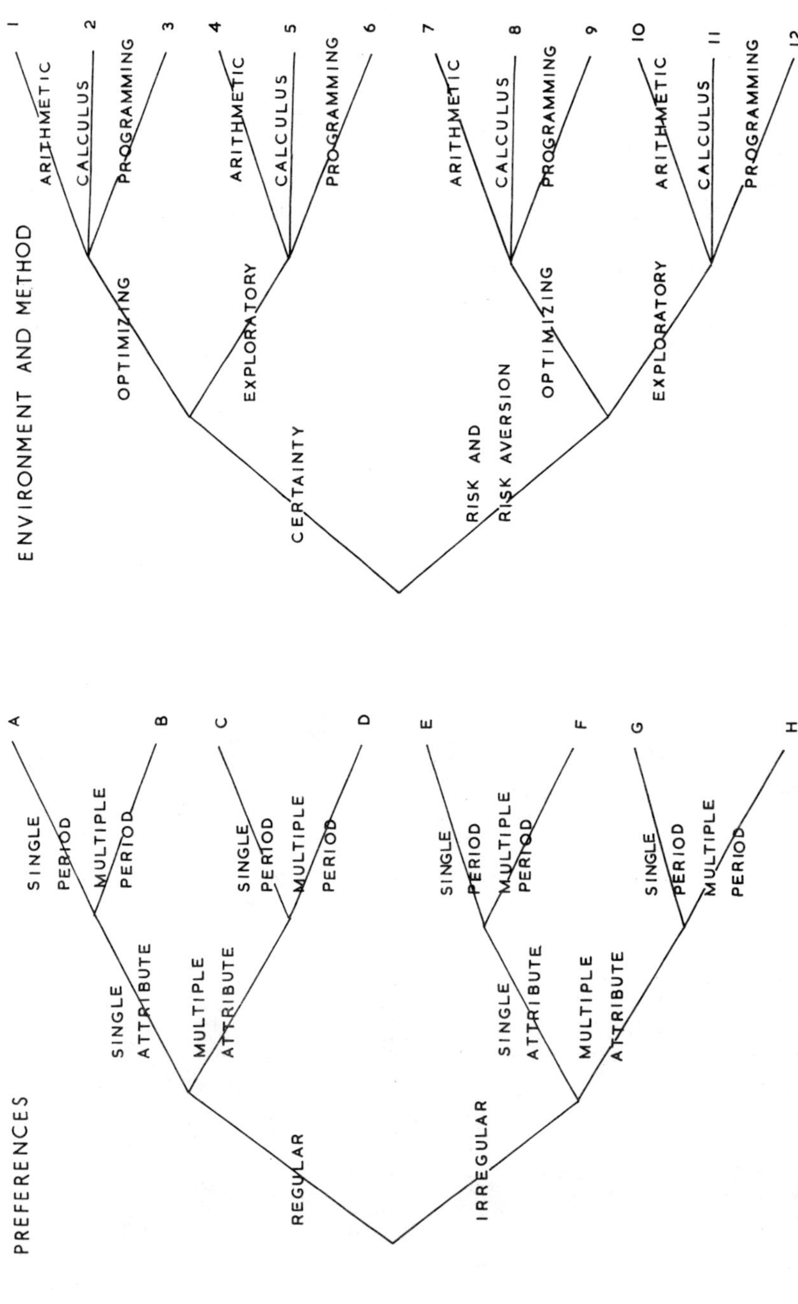

FIG. 1. A taxonomy of models.

many periods with consideration of several criteria of success, such as net present value, ability to meet financial obligations, post-tax earnings, etc. These relatively simple (some would say artistic) methods embody the great majority of influential planning work for decision-making at the farm level in developed (and less developed?) countries.

Budgeting methods are demanding of expertise and experience, but not of the even scarcer computing resources and skills of applied mathematics and statistics often required to pursue alternative and seemingly more appropriate methods. Choice of budgeting is often rationalized on grounds of low costs of analysis. This may be defensible in the case of one-period budgets, but the high labour cost inherent in long-run poly-period budgeting is receiving greater recognition as evidenced by the growing attention that institutions are paying to computational assistance, including minicomputers and programmable desk calculators.

Having now acknowledged that most of the "practical" farm level decision models are of categories A4 and H4 (to combine the trees of Figure 1), it behoves us to review briefly some of the more research-oriented recent activities of agricultural production economists.

Deterministic, static linear programming models, located in category A3 of our classification, have been quite widely used in agriculture, both for whole-farm planning and for such more specific purposes as the formulation of least-cost diets for farm livestock. Three main developments in the use of programming models for the analysis of farm-level decisions can be identified. These are: (i) procedures to reduce the costs of using such models for individual farm planning; (ii) improvements in the availability, reliability and capacity of computer algorithms for mathematical programming, including non-linear programming; and (iii) methods for incorporating risk into programming models. Each of these developments is now briefly considered.

Although estimates vary widely, it seems that it usually takes about two weeks of an experienced analyst's time to collect the data to construct a linear programming model of an individual farm, to have the model solved by computer and to interpret the results to the client. This represents a cost which, at commercial rates, few farmers can afford, and an input of time which few extension services can justify providing free of charge. Consequently, attempts have been made to make the process more economical, usually by seeking to spread the costs of model development over several farmers. At least three approaches are followed. First, "standard" models may be developed which are intended to be broadly descriptive of all the farms of a given type in a given area. When planning for a particular farm, only certain key coefficients in the standard model are adjusted to reflect the individual features of that farm. Second, a special matrix may be developed for each individual case using a matrix-generating program. That is, the task of model construction is formalized to the degree that the necessary processing and arrangement of information from the case farm can be undertaken by computer. Third, some sort of group planning may be undertaken, usually based on a "representative" farm and often making extensive use of parametric programming procedures. Advice to a particular farmer is formulated by modification and interpretation of the results for the representative farm.

Developments in the computer world have had a number of implications for the use of programming models in agriculture. With increasing availability and accessibility of computers, the potential for use of programming models has increased. Some of the computing problems, such as limits on matrix size, which bothered analysts a decade or so ago, are now largely eliminated. Thus, the solution of large problems, for example for long-run whole farm planning, is now possible if not routinely practicable. Workable algorithms to handle non-linear problems, notably integer programming, quadratic and separable programming, are now much more widely available, permitting more realistic formulations of farm planning models. The literature reflects the increased use of such algorithms, at least in research applications.

Finally, there is evidence of a widespread recognition of the need to account for uncertainty and risk aversion in programming models used for farm planning (Kennedy and Francisco, 1974; Lin, Dean and Moore, 1974). Approaches that have been suggested range from the use of linear and quadratic risk programming to forms of stochastic programming (Anderson *et al.*, 1976). The main thrust of these developments occurred a few years ago and it seems that more recently the emphasis has been on refining and extending the application of the methods that have been proposed. For example, Barry and Willman (1976) have used quadratic risk programming to determine optimal forward contracting arrangements, one of the few recent studies relating to farmers' risk-efficient marketing decisions (see also Heifner 1972).

There has also been increasing explicit recognition of risk and risk aversion in modelling other than mathematical programming. Developments include innovations in (category A8) response analysis under risk (Anderson, 1973), production under risk (Blandford and Currie, 1975) and marketing under risk (Ward and Fletcher, 1971). Analogous developments have been witnessed for non-optimizing exploratory models. We are likely to see a proliferation of (category A10) techniques such as stochastic budgeting (Mill and Longworth, 1975) and related Monte Carlo planning techniques which permit consideration of many constraints (Anderson, 1975; Hardaker and Tanago, 1973), especially as access to computers and software improves (Candler, Boehlje and Saathoff, 1970).

In terms of preferences, much innovative work has been directed at moving the emphasis from branch A towards branch D but, as noted above, the difficulties inherent in specifying and manipulating multiperiod multiple-attribute preference functions have been such as to relegate much of this endeavour to the more pragmatic approaches of branch H (Anderson, 1976b). Thus there have been several studies of growth of the farm firm, often using lexicographic objective functions (e.g., Dean, 1974–5) and almost inevitably using the methodology of systems simulation (Eisgruber and Lee, 1971; Harle, 1974; Johnsson, 1973). For a more spacious critique of such methodology in agricultural economics, see Anderson (1974).

6. FUTURE DEVELOPMENTS

It seems clear that the organizational structure of agriculture in many developed countries is inhibiting to the more extensive use of decision models. A situation of large numbers of relatively small farm businesses, geographically dispersed and independently managed, is not conducive to the widespread use of advanced management aids. Few farm businesses are large enough to justify employing operations research staff, even on a part-time or consultancy basis. Some government extension services undertake a certain amount of decision modelling in an advisory context, but the ratio of advisers to farmers is seldom sufficiently favourable to make such an approach workable for the majority of farmers. In these circumstances it is not difficult to see why decision modelling is not more widely used. Similarly, it is hard to be optimistic about the chances of any dramatic change in future.

Two circumstances give hope that the existing range of decision models may be put to somewhat better practical use in future. First, there is a trend in most developed countries to larger farms and for farming to be regarded more as a business and less as a way of life. Larger farm businesses may be expected to make more comprehensive use of such management aids as are available. Second, the progressive improvement in computer and associated telecommunications facilities will reduce the problems of spatial isolation of farms and will bring the latest computer aids literally into farm kitchens or offices, to the finger-tips of farmers or their advisers. In speculating about the future, therefore, we can imagine a situation of slow but steady growth in the use of farm-level decision models. Such models will be computer-based (Blackie and Dent, 1974), perhaps including interactive computing facilities so that, through a dialogue with the computer the farmer, perhaps aided by an adviser, can develop a model that suits his decision problem and is consistent with his beliefs and preferences.

REFERENCES

Anderson, J. R. (1973) 'Sparse Data, Climatic Variability and Yield Uncertainty in Response Analysis', *American Journal of Agricultural Economics,* Vol. 55, No. 1 (February 1973), pp. 77–82.

Anderson, J. R. (1974) 'Simulation: Methodology and Application in Agricultural Economics', *Review of Marketing and Agricultural Economics',* Vol. 42, No. 1 (March 1974), pp. ?–55.

Anderson, J. R. (1975) 'Programming for Efficient Planning Against Non-Normal Risk', *Australian Journal of Agricultual Economics,* Vol. 19, No. 2 (August 1975), pp. 94–107.

Anderson, J. R. (1976) 'Essential Probabilistics in Modelling', *Agricultural Systems,* Vol. 1, No. 2 (1976a), (in press).

Anderson, J. R. (1976) 'Modelling Decision-Making Under Risk', mimeo, University of New England, Armidale, (1976b).

Anderson, J. R., Dillon, J. L. and Hardaker, J. B. (1976) *Agricultural Decision Analysis,* Iowa State University Press, Ames (in press).

Barry, P. J. and Willmann, D. R. (1976) 'A Risk Programming Analysis of Forward Contracting with Credit Constraints', *American Journal of Agricultural Economics,* Vol. 58, No. 1 (February 1976), pp. 62–70.

Bernard, S. (1976) 'On Utility Functions', *Theory and Decision,* Vol. 19, No. 2 (1974), pp. 205–42.

Blackie, M. J. and Dent, J. B. (1974) 'The Concept and Application of Skeleton Models in Farm Business Analysis and Planning', *Journal of Agricultural Economics,* Vol. 25, No. 2 (May 1974), pp. 165–75.

Blandford, D. and Currie, M. (1975) 'Price Uncertainty – The Case for Government Intervention', *Journal of Agricultural Economics,* Vol. 26, No. 1 (January 1975), pp. 37–51.

Candler, W., Boehlje, M. and Saathoff, R. (1970) 'Computer Software for Farm Management Extension', *Americal Journal of Agricultural Economics,* Vol. 52, No. 1 (February 1970), pp. 71–80.

Day, R. H. (1971) 'Rational Choice and Economic Behaviour', *Theory and Decision,* Vol. 1, pp. 229–51.

Day, R. H. (1976) 'Cautious Suboptimizing', mimeo, University of Wisconsin, Madison, (1976).

Dean, G. W. (1975) 'Firm Theory Incorporating Growth and Risk: Integration into Farm Management Research', supplement to *International Journal of Agrarian Affairs,* double number 1974–5, pp. 110–20.

De Benedictis, M. (1974) 'Comment', in *The Future of Agriculture: Technology Policies and Adjustment,* Int. Assoc. of Ag. Economists, Oxford, pp. 183–8.

De Finetti, B. (1972) *Probability and Induction: The Art of Guessing,* Wiley, New York.

Eisgruber, L. M. and Lee, G. E. (1971) 'A Systems Approach to Studying the Growth of the Farm Firm', in Dent, J. B. and Anderson, J. R. (eds.), *Systems Analysis in Agricultural Management,* Wiley, Sydney.

Ellsberg, D. (1961) 'Risk, Ambiguity and the Savage Axioms', *Quarterly Journal of Economics,* Vol. 75, No. 4 (November 1961), pp. 643–69.

Halter, A. N. and Dean, G. W. (1971) *Decisions under Uncertainty,* South-Western, Cincinnati.

Hardaker, J. B. and Tanago, A. G. (1973) 'Assessment of the Output of a Stochastic Decision Model', *Australian Journal of Agricultural Economics,* Vol. 17, No. 3 (December 1973), 170–8.

Harle, J. T. (1974) 'Further Towards a More Dynamic Approach to Farm Planning – A Technically Based Model of the Farm Firm', *Journal of Agricultural Economics,* Vol. 25, No. 2 (May 1974), pp. 153–64.

Heifner, R. L. (1972) 'Optimal Hedging Levels and Hedging Effectiveness in Cattle Feeding', *Agricultural Economics Research,* Vol. 24, No. 2 (April 1972), pp. 25–36.

Johnsson, B. (1973) 'Uncertainty in the Environment of the Farm Firm: A Simulation Approach', *European Review of Agricultural Economics,* Vol. 1, No. 4 (1973), pp. 391–413.

Keeney, R. L. (1973) 'Concepts of Independence in Multiattribute Utility Theory', in Cochrane, J. L. and Zelany, M. (eds.), *Multiple Criteria Decision Making,* University of South Carolina Press, Columbia, pp. 62–71.

Kennedy, J. O. S. and Francisco, E. M. (1974) 'On the Formulation of Risk Constraints for Linear Programming', *Journal of Agricultural Economics,* Vol. 25, No. 2 (May 1974), pp. 129–45.

Kirkwood, C. W. (1976), 'Parametrically Dependent Preferences for Multiatributed Consequences', *Operations Research,* Vol. 24, No. 1 (January–February 1976), pp. 92–103.

Lin, W., Dean, G. W. and Moore, C. V. (1974) 'An Empirical Test of Utility versus Profit Maximization in Agricultural Production', *American Journal of Agricultural Economics,* Vol. 56, No. 3 (August 1974), pp. 497–508.

Maruyama, Y. and Kawaguchi T. (1971) 'An Approach to Farm Planning under "Ambiguity"', in *Policies Planning and Management for Agricultural Development,* Int. Assoc. of Agric. Economists, Oxford, pp. 347–55.

Mills, S. J. and Longworth, J. W. (1975) 'Stochastic-Computerized-Activity-Budgeting for Sheep Enterprises', *Australian Journal of Agricultural Economics,* Vol. 19, No. 2 (August 1975), pp. 108–18.

Pasour, E. C. Jr., and Bullock, J. B. (1975) 'Implications of Uncertainty for the Measurement of Efficiency', *American Journal of Agricultural Economics,* Vol. 57, No. 2 (May 1975), pp. 335–9.

Petit, M. (1976) 'The Role of Models in the Decision Process in Agriculture', Paper delivered at this *Conference,* July–August 1976.

Raiffa, H. (1961) 'Risk, Ambiguity and the Savage Axioms: Comment', *Quarterly Journal of Economics,* Vol. 75, No. 4 (November 1961), pp. 690–4.

Throsby, C. D. (1974) 'New Methodologies in Agricultural Production Economics: A Review', in *The Future of Agriculture: Technology, Policies and Adjustment,* Int. Assoc. of Ag. Economists, Oxford, pp. 150–69.

Upton, M. and Casey, H. (1974) 'Risk and Some Pitfalls in the Use of Averages in Farm Planning', *Journal of Agricultural Economics,* Vol. 25, No. 2 (May 1974), pp. 147–52.

Ward, R. W. and Fletcher, L. B. (1971) 'From Hedging to Pure Speculation: A Micro Model of Optimal Futures and Cash Market Positions', *American Journal of Agricultural Economics,* Vol. 53, No. 1 (February 1971), pp. 71–8.

DISCUSSION OPENING – Jacoubus T. Van Riemsdijk, *Netherlands*

This paper gives a good general appreciation of the current situation but the authors do not go far with regard to possible future developments which might eventually lead to a substantial improvement of the state of the art. This seems to be because the paper only covers models in the area of farm planning in the strict sense and omits the important problem area of farm operation.

We all know that farmers proceeding with the execution of similar "plans" under similar conditions may well end up with farm incomes which differ by over 50 per cent of the total production cost per farm and per year. Problems concerning farm operations are therefore very important and should no longer be neglected by those concerned with farm-level decisions. This calls for more than merely analysing farm accounts. We should follow the farmer's activities carefully throughout the year. Farm operating is, in my opinion, nothing else than a long sequence of decisions made in a very uncertain environment. We should feel ashamed to have devoted so much effort to farm management problems and yet still not be able to identify the factors which make one farmer highly successful whereas another fails badly.

Agricultural economists in co-operation with colleagues in other fields of agricultural sciences can, and should, help farmers to improve their score.

When projects fail we should not be content to say that they were too ambitious (see the paper by Michel Petit) when the failure is due to large gaps in knowledge relating to farm operation. We should try to close the gap by finding out the real content of the learning processes which a successful farmer can accomplish properly. The result of a project of my former staff members indicates this to be possible. Such projects, and others along the line of the work of Nils Westermark, would supply important feedbacks to model-builders and simulation specialists, to the rank and file of farmers and – most important indeed – to classroom-educators of the generation to come. This comment is meant not as a criticism of the authors but as a stimulant to a discussion on means of improving the state of the art.

DISCUSSION OPENING – A. Hjörtshoj Nielsen, *Denmark*

My comments will be concerned with the commercial farm – often a family farm – of a developed country with a market economy, not with models for use on large scale farms in centrally planned economies.

Any critical remarks bear more on the state of the art than on the paper which I think accepts things as they now are.

In relation to the opening paragraph of their paper, I feel that we should concentrate on how we can help farmers to make better decisions but I am not convinced that the results of these papers will be readily used in action. I am not against basic research and research carried out to gain new knowledge but I feel that agricultural economics is an applied science. We can hardly justify our work if we do not contribute to the solution of today's problems in our societies.

We seem to have a dichotomy in our efforts to establish ourselves in a unique and sophisticated science and, at the same time, to contribute to solving practical problems. This has been said before but it merits being said again.

There is a remarkable number of papers – including some of my own – in our profession's journals which concluded with words something like these. "We have demonstrated the usefulness of this powerful new technique which very easily can be programmed on a large computer. The example shown, however, is very simplified and has no direct bearing on real practical problems".

I agree that in research we should use techniques well ahead of those used in practical farming and should apply sophisticated models but why do we not develop these models in connection with actually solving real world problems which somebody has asked us to solve? If we do that we are sure to have a demand for our results.

Turning to some more specific points – at the beginning of the paper it is said that agricultural economists have tended to neglect decisions on marketing problems for individual farms. I have some doubts whether aspects of pricing and selling outlets represent relevant problems for most farmers in western economies. If problems exist in these areas they can hardly be solved by individual farmers but will require some institutional action either by farmers' organizations or by governments.

Next, I feel that the difference between success and failure in farming is often a question of good or bad day-to-day management. In other words, it seems to be very important to be efficient in the use of resources, to have an efficient conversion rate in feed consumption, to have high yields in the field, etc. To help the farmer to have control in his business we need to develop information systems – accounting procedures – which signal to the farmer whether he has done well or not. These problems seem to be somewhat neglected in the paper. You may say this class of problem is outside the scope of decision models. I think they are integral parts of decision models.

Another thing which seems to be neglected in much work is the problem of the effect of tax and depreciation rules on the economic results and liquidity of farms. In most western countries it is absolutely necessary that we take taxes into consideration.

While I agree with the author that computers will give us better possibilities for modelling the real world, we must be aware that we do not live in an orthogonal world – parameters cannot be varied independently. We cannot raise, say, milk prices by 20 per cent and assume this would have no effect on the technical coefficients in our model. This recognition of the non-orthogonal world is often neglected.

Let me end by joining the author in his conclusion that we may expect farmers to make more use of management aids in the future but let us help to solve their problems not create ones of our own!

RAPPORTEUR'S REPORT – J. Brossier, *France*

This report considers first technical, and then more philosophical questions which were covered in the discussion with final sections on the role of models and means of making them more useful and efficient.

The technical section was concerned with the classification of models and their construction. First – what use is a classification? If it is generally agreed that all classifications provide some useful guidance, it is difficult to go beyond the one presented here. Some may consider it useful; others did not regard it as essential to approach these problems at the farm level. However, we have to recognise that Petit and Brossier's model certainly would not fit within it – it is not a mathematical model and is set on a different plane. The classical distinction between "risk" and "uncertainty" was challenged. Some progress in research in this field has been made by introducing the concept of "subjective probabilities".

The philosophical questions surrounding "normative" or "positive" received much attention, including Johnson's and Petit's written papers. Various philosophers (Weber, Popper) were cited to illustrate the debate. Some speakers felt that "normative" and "positive" are, and should be, completely separate concepts. The economist should confine himself to the positive; in other words – he should stay at the factual level) and thus remain neutral and objective, leaving decision-making, for example, to the politician. Others saw some aspects as positive, others normative, and thought that social scientists could not leave the latter out of account because it provided a basis for deciding between right and wrong and thus formed the basis for giving advice. To those who took this view the normative had to be included because it provided an essential part in understanding the workings of the economy. The decision-maker could not be helped to make better decisions if economists left that aspect out of the analysis. This group criticised Richardson's classification because it considers objectives as exogenous to the models even though they are an integral part of the decision-making process.

The whole group felt that all economic analyses (positive or normative) should be subjected to validation criteria – consistency, feasibility, communicability, (as Petit set out) together with criticism by scientists and the opinions of decision-makers. There was general agreement on the importance of efforts to this end.

On role of models, and problems of relationships between farmers and decision-makers, while there was unanimous agreement that the economist's role is to help farmers and others to make good decisions, the question of methods, and in particular the role of models, remained open. Following on from the example of the car driver in Petit's paper, two points were made. The objective is not that the farmers should ponder every decision, but that they should be able quickly and automatically to assess its consequences. On the other hand if we, as economists, could not be sure that, when they are better informed, farmers and other decision-makers would make better decisions, there would be no point in us or our discipline.

The debate on models developed into one between simple and sophisticated models. While there was no consensus, a majority feared that too sophisticated models may complicate problems rather than solve them. Some regretted that economists set up their own problems to analyse and do not take the real problems which face farmers. It was felt that what is actually needed is the means to make day-to-day decisions, using simple records.

All agreed, however, that models are useful teaching tools. This is already recognised at the university level; but models are also proving a vehicle for discussion with decision-makers, thereby aiding the building and improving of the models themselves. But what about the qualities of the analyst? That question recurred in the discussion. It became clear however, that there was no universal model that would suit all needs.

As regards the gulf between theory and practice, the question was posed whether the way ahead lay with, on the one hand, a greater use of computers, or, on the other, with a radical change in our methods – for example, by going to see the farmer's wife who is involved in decision-making. Ideas put forward to bridge the gap included:

(i) taking more explicit account of the time factor in decision-making,
(ii) studying the detailed role of prices and markets: does the farmer have a real choice, for example?
(iii) exploring the role of training in decision-making, with more research into the learning process,
(iv) taking more account of different decisions (for example, there may be conflict between the farm business and the farm family which affects farm business decision) and the various non-economic factors. (For example, the farmer may act sometimes as an economic man; sometimes he is swayed by emotion.)

Many disagreed with the optimistic view of Anderson *et al.* on the prospective use of models by farmers. There was also dispute about the inevitability of the trend towards the larger farm and concentration on the farm as a business and the exclusion of the idea of farming being, to a significant extent, a way of life.

Q. B. O. ANTHONIO*

The Dynamics of Decision-making and Policy Formulation for Agricultural Improvement in Low-income Countries

1. INTRODUCTION

Research and studies in decision-making for agricultural development have been neglected in the past primarily because of the expense and the large number of well-trained personnel from various disciplines required to carry it out. That this neglect persisted for so long, however, in spite of the pivotal role that decision-making plays in economic development generally, and in agricultural improvement in particular, cannot be easily explained. Clearly a thorough knowledge of the decision-making processes helps to remove serious bottlenecks, to reduce persistent production constraints, and to expedite sustained over-all agricultural improvement.

In most countries, contemporary developmental agricultural policies tend to incorporate extension education as a key area, if not the only major instrument for the transfer of "improved" technology to farmers in low-income countries. This apparent emphasis on extension education is based on the assumption that education leads to improved decision-making at the farm level and that this will automatically improve agriculture. This is not without the realization that there are other important constraints, institutional, social, and political, that tend to impede decision-making for agricultural development.

The complicated nature of agricultural improvement in low-income countries stems from the fact that most governments, in an effort to meet the aspirations of their citizens, are compelled to accelerate the pace of agricultural improvements. If left to natural evolutionary processes they would take far too long. But, unfortunately, in trying to help or accelerate the phenomenon of improvement, most planners underrate the major limitations imposed by the human factor: labour, skill, organization, and decision-making. The human factor is by far the most important element in agriculture, but it does not receive the emphasis it deserves; rather it is erroneously assumed as given. In explaining farmers' responses to an innovation, it is

* Joint ECA/FAD Agriculture Division P.O. Box 3001 Addis Ababa, Ethiopia. The views expressed in this paper are the personal views of the author.

quite obvious that the most significant factor is the type and nature of decisions being made. In other words, the motivation of agricultural improvement really has to start with improving the nature of decision-making not only on the farm but in all areas affecting the progress of agriculture.

One obvious area of neglect in research and the literature is the theory of decision-making in agriculture. For the purpose of this paper, decision-making is broadly defined to incorporate all conscious thoughts and actions from thought conception to the eventual step of taking action based on the decision made. The essence of the final decision being made is clearly interpreted (or understood) from exhibited behaviour or reports emanating therefrom. In other words, according to Thornton, (1962), decision-making "consists of that chain of mental activity which follows either the reception of stimuli from the outside world or the growth of a desire for change within the individuals".

The improvement of decision-making, however, presupposes clear and thorough understanding of how the different categories of decision-makers arrive at their decisions. Such an understanding helps not only to demarcate the areas of the different levels of decisions but even more to identify which decisions are best taken and by whom. It is useful, for instance, to appreciate the over-riding role of objectives and the dual problems arising from apparent conflicts in them. Obviously these issues have to be carefully analysed and understood before any effective agricultural improvement can be achieved.

In this exposition, agricultural improvements (a more easily defined term) is used in the sense of increased productivity* that yields the producers increased or improved real income or welfare. This increased productivity can be a higher output per man, or per hectare of land, or per unit of capital or management invested in production, measured in money terms.† Defined this way, the value terms of increased productivity have to be appraised in the light of the economy and changes in the demand and supply variables in each case. In a restrictive (or marginal) sense, agricultural improvement will be easily assessed on the basis of the return per unit of the most limiting resource. In most parts of Africa, agriculture is restricted in the broadest sense by capacity or skill in farm organization, which is also influenced by decisions being made at all stages of the improvement processes.

The next section of this paper examines the classified orders of decision-making. The simple classification need stems from the realization that social institutions *and their antecendent or concomitant constraints set limits to the range of effectiveness of the decisions taken by the individual farmer and other decision maker.* This classification also assumes that the society has the

* That is, showing significant average and marginal increase simultaneously when the point of agricultural improvement has been attained. In other words, the decisions being analysed in this paper are, for the sake of simplicity, limited to those that result in progress to a better situation than the one that prevailed before, all things considered.

† Measurement of net return on capital can employ the basic budget analysis or the gross return approach, while return to managerial skill can be assessed on the basis of profit level relative to the technology being employed and the type of enterprise.

responsibility on behalf of all its members to regulate, oversee, and reward the use of resources for achieving desirable economic ends.

In the following sections the constituent members involved in decisions and policy-making are discussed. Emphasis is placed on the need for timely and relevant information for better decision-making in low-income countries. A summary of conclusions appears in the final section.

2. DYNAMIC ORDERS OF DECISION-MAKING

The exposition of the dynamic orders of decision-making in a sense starts with a categorization of the basis of who and what group of people are making what decisions. Categorization, for the sake of simplicity, is made into first, second, third and the higher orders of decision-making. It is clear that to understand the processes involved in the positive and dynamic essence in decision-making towards increased agricultural improvement, these orders of decision-making have to be fully classified, especially with respect to the limitations encountered by the decision-makers in their efforts to achieve the goals of agricultural improvement. These constraints have to be fully appreciated *ex ante* to suggestions towards the modifications, necessary changes, and improvements that are to be introduced and hopefully be subsequently made operational and sustained as such. In other words, according to Okran (1948), "It is only a matter of re-arrangement". The alternative to a judicious analysis of the situation can only lead to the typical folly of "the man who for reason of slow speed and progress re-arranges his horse and cart; he then places the cart before the horse. All hopes of progress are therefore checked. Pandemonium displaces peace. The cart misunderstands the horse, and the horse misunderstands the cart. The man distrusts them both: but mark you, the horse, the cart and driver are well known and recognised means of progress" (Okran, 1948).

First order decisions

First order decisions are those over which one individual has absolute authority. The effective scope of the decision may, however, have forward and backward linkages beyond the individual. In traditional agriculture, there are very few decisions of this order that affect farm production or the owning, acquisition, and use of farm resources. The disposition of the labour available to the individual farmer, however, falls within this category. This is, in fact, the most important simple (though not easy) decision the farmer has to make: how best to use his labour.

Agricultural extension, for decades, concentrated on convincing the farmer or group of farmers about the merits of a new technology or innovation without due regard to the facets of decision-making processes involved and especially the labour constraints and reward uncertainties (and risks) the farmers have to face in arriving at the decision whether or not to accept and use or adopt the new practice.

Hence, the extension agents' strategy and preoccupation with advising farmers on new technology can be more effective if advisory agents appraise first and foremost the additional labour requirements of any new practice being recommended. Beyond this, efforts for improvement should concentrate on the farm-family as a unit rather than on the individual farmer, as at present. This is principally because the decisions involved in adopting innovations are often being made at the farm family level. Resources in agriculture that are within the use and absolute control of the farmers are severly limited, and farmers are very reluctant to accept innovations that may call for resources not within their disposal: that is making for decisions beyond the first order category.

The second order decisions

These sets of decisions are invariably made by a defined small, but homogenous, group (or unit) of people bound together temporarily or otherwise by a common and easily identified objective or relationship. The family or a firm is an example of such a unit. The resulting action from such a decision primarily affects those making the decision even though the long-run impact may have other implications even beyond the unit or group. Examples of such decision are those that are made by the family, a business group, society or a board of directors.

In farming enterprises, the farm family make second order decisions with respect to some aspects of farm operations. Second order decisions can be differentiated from the first order decisions in two main respects: in the second order decisions, there is usually more than one decision-maker and the nature of decisions to be made are such that the members may individually have different or even conflicting views and the impact of such decisions may be different for the different members of the decision-making group. Furthermore, when an individual is empowered by the group and entrusted with the execution of the decisions, there is usually the need for subsequent ratification or reports on progress being made. Conflicting effects on members of the groups of such decisions reflect some inconsistency, and may affect the unformity and effectiveness or otherwise of the execution of such decisions. Conflicts have to be firmly resolved to improve this order of decision.

In traditional agriculture, this order of decision-making is perhaps one of the most important. Given a set of natural and/or induced situations, the group directly concerned or affected by the availability, ownership, and use of farm resources and products, have to take decisions on the appropriate cause of action. As much as possible, conflicts have to be resolved to the satisfaction of most if not all members of such groups failing which either the decision becomes inconsistent with the group objectives and is subsequently rejected or the group becomes ineffective or moribund. Tardiness and delay in arriving at decisions are typical of this category of decision-making. It is in this particular frame of reference that most government policies aimed at improving agriculture in the past have either not been able to meet with expectation, or, as with some, failed woefully. This has been particularly so when policies fail to take cognizance of how the different types of these second-order decisions are arrived at, and in particular how they can be improved, and made more internally consistent and effective.

The third order decisions

Decisions clearly outside the scope of any one individual or any related homogenous groups can be classed as the third order set of decisions. Decisions on the use of common resources, on land ownership, price of resources and products, to name a few, fall within this category. In this category of decision, the farmer as an executive is still actively engaged in decision-making – participating in influencing decisions. The main difference between the third order decisions and the next category, the higher order decisions, is that, whereas with the third order decisions the participation of the farmer both in formulation and execution of decisions is paramount with the higher order decisions the farmer is only remotely referred to and not actively involved in decision making. In essence, the higher decisions are rather "imposed" on him. Unfortunately, these decisions, unless well thought out may impose constraints and hardship on him and disrupt his sets of first order decision rather than help him to improve his agriculture, and his living standards. Under such circumstances, his being able to influence such decisions in a direction conducive to his welfare should be considered marginal.

Irrespective of the nature of the governmental set up, existing political structures have implicit effect on agricultural improvement. Political and social structures and institutions within the society can sometimes impose serious constraints on the decision-making processes especially those referred to as the second and third order series. Improvement in the economic, social and political structures of the society gives a fair chance for everyone to have a full knowledge of and participate in decision-making. In cases where the farmer's participation is expected before or after the policy is formulated and executed, his decision to participate can only be of benefit to him when such policies are fully understood and actively supported by him; and, of course, in harmony with the sets of decisions of the first and second order he is likely to face.

The higher order decisions

This is the category where policy decisions become appropriate and relevant. The higher order decisions by implication affect everyone within and beyond the agricultural sector. Directly or indirectly, they also influence the other sets, level or orders of decisions. Clearly, here too, the requirements of objectivity and consistency are equally important. Two groups of conflicts are easily observed. First is disharmony resulting from conflicts between the decisions made within the different orders of decisions and policies. Inconsistency or even dissonance between second and first order decisions will certainly give rise to a conflict in the mind of the farmer, or the farm-family, and the effectiveness of either set of decisions will not be assured.

Secondly, policy dissonance may also result from a conflict between one set of policy decisions and another even within the same order. Both sets of conflicts have grave consequences for agricutural improvement. It is therefore necessary for effective improvement of farm operations and productivity that these conflicts are minimised, if not eliminated. For most farm problems, several alternative solutions arc invariably put forward, but according to Heady (1961, p. vi), "progress towards a solution rests on the resolution of conflicts in goals and values". Goals, defined as ends or objectives towards which behavior may be directed, are likely to be achieved only when decisions are made on the "*best*" way to utilise available means.

Policy

Many people are often not quite clear about what is the nature of policy or who makes policies. Invariably, they assume that only the executive council or Ministers, legislators and administrators formulate policies. This misconception stems from ignorance in assuming that because government decisions being actively made by these group of men tend to have the force of law, they are the only policy-makers. On the contrary, policy in a dynamic sense of the word is, or should be, constantly formulated as the point of concensus of individual decisions and in consistent harmony with other relevant sets of decisions. In other words, the day to day decision of every individual influence and eventually add up to a general policy. Good policies therefore grow out of different sets of decisions and are aggregates of them.

With respect to agriculture, Schickle (1954) asserted that "agricultural policy must be judged in the context of the cultural and social environment of which it is a part" (p. 1). In essence, good agricultural policy is part and parcel of the social milieu. Consequently, any effort to introduce policy, or aggregates of higher and other sets of decisions, aimed at improving any aspect of human endeavours in any society should commence with improving the institutions through which decisions are aggregated and crystallised into policy. This is of particular significance for agriculture where, to the farmer engaged in subsistence farming, emphasis is basically on survival. The implication is obvious since for any effective improvement in the agricultural sector policies have to be evolved with these objectives in mind. In the final analysis the decisions that are crucial are made by the farmer – and his family – that is, they are first and second order decisions. And until these are also improved, the third and higher order decisions and policies will probably only have marginal effects.

Clearly, a poorly formulated farm policy has very little chances of achieving its objective even with the best of executing agency. There is a large body of experiences in most low income countries to emphasise the fact that poorly formulated policies, resulting largely from confused objectives, values, and decisions about means and ends with respect to agricultural improvement, have plagued these countries for decades, if not centuries (John C. de Wilde, 1967). With mounting pressures for agricultural improvement under the more complex economic and political situations we now witness, rapid and lasting improvement in policy making is vital. Necessary improvements can be assured only by creating improved procedures for securing improved policy.

3. IMPROVED POLICY FORMULATION

Improved policy formulation is a necessary step towards achieving a consistent and meaningful policy for development. In any society where there are limitations to resources, human and non-human, all efforts and policy guidelines aimed at improving agriculture have to state clearly and in rather convincing and categorical terms what objectives or results are desired, what resources are to be expended and the means (as well as techniques) to be employed in such an exercise. It is only when this is done that one can be in a position to judge whether a good policy is being produced and how far it can be achieved. A good policy when assessed has to be regarded as satisfactory if, firstly, the stated objectives are desirable and consistent with other objectives and other policies of interest to the agricultural sector, the society and the nation as a whole. Secondly, the policy when critically appraised, has to meet the condition of its being harmoniously operational, and the ends to justify the means. Far too often, most agricultural policies in low-income countries have in the past failed to meet these two tests.

Whether a policy will pass the test of consistency and feasibility is largely influenced, if not predetermined, by at least four important facets of policy formation: namely,

(a) consistency in the objectives of the policy;
(b) the quality of participants in policy-making;
(c) amount and quality of information available to decision-makers; and,
(d) processes by which a consensus of opinion is arrived at.

If policies are to improve in their essence and feasibility these four areas have to be considerably improved. It is in fact unavoidable that without improvement – and improvement simultaneously in all these four facets of policy-making – the quality of decision and policy will be poor and will fail to achieve the desired agricultural improvement in low income countries. If they remain ineffective and confused the aspirations for agricultural improvement will remain as dreams: dreams that may take for ever to come true.

Consistent objectives

Objectives can be defined simply as what achievement is being aimed at, whether stated or assumed, by carrying out a particular line of action as explicitly stipulated in a decision or policy. Two kinds of objective have to be clearly defined. The *normative* objectives constitute a class of objectives based on value judgement as good and which are broad in scope and of general or universal desirability. On the other hand, the *positive* objectives are the more operational though more limited in scope compared to the former. Furthermore, positive objectives are of spectifc relevance to policy. Operational objectives form the core of achievements to be aimed at or accomplished, given available resources and time dimension within the policy that is being produced or formulated. Whereas the normative objectives such as increased social product, improved welfare, increased income, greater national wealth, etc., are areas of little or no controversy it is very difficult to gain full support of political will for the positive objectives such as farm income, vis-à-vis urban income, etc. This difficulty stems from differences in goals and values between individuals, and between different societies and economic sectors of the same society.

Goals and values need not be completely uniform, nevertheless. Agreements and decisions on the desirable ends can be suitable and equitably worked out provided three major areas influencing policy are adequately taken care of. These areas are that:

(i) public interest overrides or takes precedence over individual interest;

(ii) educational (not schooling) quality improves so that values are consistently made more efficient; and that;

(iii) communication between and among the different partners or sections (or factions!) of the society can be assumed to be relatively honest, reliable and in harmony by constantly exchanging views with all concerned.

Finally, the point has to be stressed that without a clear-cut objective in the first place, policy decisions may prove ineffective. In essence, if people cannot agree on a common ground with respect to their desires and anticipated achievements and how these can be attained, then, they do not deserve and should not expect that the mere uniting of any policy will improve their lot.

Quality of participants

The quality of participants in decision- or policy-making can be improved: (a) in the choice of participants themselves and (b) the methods by which they are selected. With respect to all the different categories of decision-making and policy discussed earlier, quality of participants can be improved along the following four important lines: (i) relevance, (ii) adequacy, (iii) representativeness, and (iv) quality of participants.

(i) *Relevance.* The member participant selected has to be directly or indirectly relevent to the issues being discussed. In low-income countries, this is a difficult issue to resolve, especially as people tend to have a tendency merely to acquire power as a means of ensuring that they are involved in all processes of decision-making, sometimes at the risk of efficiency and their own health, instead of helping to achieve better decisions. Far too often, senior administrators and chief executives have been justifiably accused of failing to delegate authority, or trying to cope with too much. Consequently, decision-making is slow, long winded, and invariably of poor quality.

Relevance is essential in agriculture. In particular, the thrust of improved innovation in agriculture through improved extension services well be more effective if the individual being so educated is the right person to make the decision. In other words, the sets of decisions to be made for effective implementation of the adoption processes being encouraged will be more fruitful if the strategy of extension is made more relevant to the specific decision-maker.

(ii) *Adequacy.* The number of participants is a reflection of size, nature of decision and the optimum number of persons required to make the relevant decision. "Two heads are better than one", but too large a group is not always easy to work with. Being adequately represented gives credence to the decisions being made as well as ensuring that enough shades of relevant opinions are represented, and discussed, before any final decision is taken.

(iii) *Representativeness* of participants cannot be denied as vital for enhancing the quality of decisions. It is essential to have relevant and adequate representation to ensure that all shades of relevant opinion are represented and that everyone feels a sense of belonging as part of the policy-making machinery, thus creating a sense of commitment to all concerned, and therefore more willing to ensure the success of the decision.

To achieve representativeness for agricultural policy participants, *the* list needs to consist of the *farmers' representative* (selected by or with the approval of the farmers themselves), *administrators* to co-ordinate and carry out the essentials of contact and reporting (Anthonio, 1974); (3) *technicians* of one type or the other essentially to advise on the techno-economic and social aspects and supply other basic intelligence of a technical nature.

(iv) *Quality of participants* Can be defined on the basis of the attributes of personal competence, knowledge of the issue, experience, and the ability to appreciate public need. These qualities have little or no relevance to the more socially dominant factors such as education, religion, wealth, age, and tribal grouping, which unfortunately are some of the attributes that today tend to dominate the selection of participants in agricultural policy making in several low-income countries.

It has to be admitted that defining the quality of participants implies some subjective judgements. Nevertheless, it is clear that as the quality of the participants improves, simultaneously with the other features, the quality of the decision made is bound to be better. Therefore, the selection process for appointing, electing or nominating participants should be such as to ensure that the "best" appropriate criteria for selection processes are evolved in the first instance. This is a fundamental issue. A poor selection of participants is not likely to secure a good decision and bad policies are not only meaningless, but sooner or later lead to more serious problems and greater frustration.

Quality and quantity of information available

Firstly, subject to the limit of costs, there is no alternative to making sure that all members involved in a decision or in policy-making have the opportunity of access to full available information before the decision or policy is made. Secondly, an infinite array of inaccurate and irrelevant information is meaningless and certainly not as useful as a small but well documented amount of correct, reliable, relevant and timely information. Relevant information is that which helps to improve available knowledge about a particular issue, and likely to improve the decision being made.

The four main factors conducive to the improvement of available information for decision-making are as follows:

(i) quantity – to be just sufficient

(ii) quality – reliable and useful;
(iii) presentation – simple, intelligible and easily transmitted;
(iv) timeliness – to meet the desired goal or be available to the right person at the right time relative to the decision timescale.

(i) *Quantity* should "be enough for the use" to be made of it. One of the most limiting factors to procuring full information is cost. Gunnar Myrdal (1967) believes that people tend to operate within the frame of rational ignorance because of the cost of securing information. There are different types of information as well as different methods of presentation – in figures, words, graphics, photographs and a host of others. The means of distributing them are just as many and the area of communication is wide, important, and easily abused.

The contention being made here is that, provided particular information is relevant, reliable and correct, then the more of it supplied the better. The supply method should also be appropriate, ensuring the most feasible assimilation by those concerned.

(ii) *Quality* refers to the positive contribution such information will make in improving the quality of the decision or policy being made. Quality of information will vary depending on source, method of collection, care taken in its processing, and method of presentation. The more reliable the information, the more acceptable and the more useful it is to decision-making.

It is true that most low-income countries plan the development programme without adequate facts and sometimes with the wrong ones. This is most unfortunate. Hence if agricultures is to develop, policy making has to be improved. To ensure this, in future there will need to be more data with improved reliability, relevance and acceptance.

(iii) *Presentation* of information, for quick comprehension and without deliberate bias, is absolutely essential. There is no alternative whatsoever to making sure that the presentation of the available and reliable information is such that it meets the purpose of supplying the information. There are quite a number of ways of doing this, but the criterion of selecting the best method is the test of understanding. That is, the presentation of information is such that the different categories of users are given the best opportunity, via the presentation, to quickly grasp, assimilate and be in a position to use the information.

(iv) *Timeliness* of information is important, especially as decisions have to be made within a time limit. In any case it is meaningless to present useful information only after a decision has been made. All relevant and accurate information should be adequately presented well ahead of time for effective use by all concerned as a necessary adjunct to making a good decision. Any other strategy can only generate confusion, distrust, and disharmony in decision-making and these should be avoided at all costs.

Finally, it has to be mentioned that the critical aspect of assessing the quality of a decision lies in both the acceptance of operationality and effectiveness of such conclusions as well as the available alternatives with respect to organizations or institutions for making necessary adjustments, corrections, or even repudiating (with alternatives) the previous decisions. More especially, when the manifestations of decisions have long gestation period, as in the case of development plans or agricultural investment, it is probably more relevant and important to correct past mistakes.

Consensus of opinion
With the foregoing three facets fully discussed, the area of consensus of opinion remains largely a socio-political question. It is an area where altruism seems to hold that "a country gets the government it deserves". The way and manner in which opinions are exchanged, analysed and a consensus arrived at are influenced by the people themselves, their goals, values, and the means at their disposal. Their expression is translated into their governments and institutions which govern the major part of their lives. In this paper, it will suffice to add that a more representative consensus, as represented by reduced conflicts, is generally a better position to be in.

4. SUMMARY

Perhaps the most neglected and yet most crucial area of policy for agricultural development in low-income countries is how decisions are made at all stages and for all categories of decisions. This apparent neglect has arisen purely from a genuine lack of understanding of the inter-relatedness of the political, economic and social interactions of farmers and policy-makers in low-income countries. It is true to say that research studies and analysis of the processes by which decisions are made are far from being strictly economic. Rather, other variables such as power, social status and prestige are as important as income and wealth in the decisions being made. Hence, the processes of decision-making seem rather confusing but very vital and consistently important in understanding participants' objectives. Ideally, therefore, the improvement of decisions will, to a very great extent, be influenced by a number of factors. Some of these factors, discussed previously include the following:

(1) Consonance and consistency in perspective and objective with regard to the sets of decisions to be made and those responsible for making the decisions. Far too often, decisions are made by those not directly involved or affected by the objectives to be achieved.

(2) Nature, selection and relevance of participants in the decision-making structures.

(3) Information available – to be of good quality, timely, relevant, accurate enough and freely available to all.

(4) Improved communication within and between categories of decision-makers to achieve overall harmony.

(5) Decisions to be made must clearly be within the effective competence and scope of the person(s) expected to make such decisions.

(6) An effective organisation or institution for identifying development priorities of the society and the resources requirement to meet the priorities must be identified with a view to working out realisitc and operational decisions.

The variables listed above are by no means exhaustive or exclusive and should not be regarded as such. Nevertheless, within the context of the low-income countries, these forces, more than any other, influence both the quality and the speed with which these decisions are taken.

Effective improvement in decision-making is a necessary precursor to any useful operational national agricultural policies which can lead to improved agricultural practices at the farm gate. Furthermore, the fact also has to be accepted that as the level of decision to be made becomes more complicated and more involved, so is the need for a better set of decisions, precisely because of the larger area, scope and greater implications of such decisions. The higher order decisions, for instance, are the more intractable. Furthermore, there is a general weakness in the processes by which decisions are made and until these serious weaknesses are resolved, agricultural improvement in low income countries will for ever remain in files, textbooks and development blue-prints. Whether or not the society is prepared to act is a crucial question: whether the future of agriculture in low income countries will bring the "green" or the "red" revolution will largely depend on how soon these improvements in decision-making are expedited.

REFERENCES

Anthonio, Q. B. O. (1972) 'Rural-Urban Inbalance In Nigeria and the Adebo Commission Report', *Qtly, Journal of Administration,* Jan 1972, pp. 147–155.

Heady, E. O. (1965) *Agricultural Policy under Economic Development,* Iowa State University Press Ames, Iowa, U.S.A.

Myrdal, Gunnar (1967) *Asian Drama: An Inquiry into the Poverty of Nations,* 3 Volumes, Pantheon Press, New York.

Okran, J. A. (1948) *The Need of the Farmers of Nigeria,* (Private publication on the Nigerian Agricultural Ordinance) Lagos.

Schickele, R. (1954) *Agricultural Policy,* McGraw-Hill, New York, U.S.A.

Thornton, D. S. (1962) 'The Study of decision-making and the relevance to the study of farm management', *The Farm Economist,* Vol. X, No. 1, 1962 and Vol. X. No. 2, pp. 40–56 and 71–84 respectively.

de Wilde John (1967) *Agricultural Development in Tropical Africa,* Volume II, ('The case studies', The International Bank for Reconstruction and Development). The Johns Hopins Press, Baltimore, Maryland, U.S.A.

PHILIP M. MBITHI*

Farm Decision-making With Respect to Social Psychological Elements and the Human Factor in Agricultural Management

1. INTRODUCTION

Decision-making for the farm family is the settlement of questions which arise in the day to day, season to season, operation of a farm. It implies mental confrontation with the structure of ideas, problems, or even formless information, and the settlement of these issues into concrete action guidelines or actionable opinions. It involves taking into account all factors within the farm's production and social environment; making choices, discriminating on the basis of feasibility, and hence identifying consequences for alternative actions. However, the farm family makes decisions often on limited information and limited understanding of its environment.

In the past there has never been any great disagreement that farmers, even peasant farmers, make decisions. The great debate has been; "How do they make decisions? What logic do they use? Are they rational decision-makers? Who/what influences their decisions?"

This paper is not an attempt to resolve any of these questions. It is based on the assumption that farmers make decisions. It attempts to evaluate the social and management environment of farm decision-making and shows how this has led to serious social psychological problems.

The farm is defined as the average peasant farm in East Africa, with a modal acreage of about 3·5 hectares in the case of Kenya. This farm is directly linked with various information systems such as extension services, agro-service firms, and also to the "rural community" in as far as this forms the socio-cultural and political environment of the farm.

Our studies (Mbithi, 1974) on the farm decision-making process in the area of transfer of technology indicate that farmers make many types of inter-related decisions:

(a) Specifically decisions on farm activities (what, when, how, for whom) on cash crops, food crops, livestock and off-farm activities.

(b) Effects of such activities on family goals – educating children, paying

* University of Nairobi, Kenya

self-help dues, treatment of illness, social debts (such as reciprocal payments of bride-wealth).

(c) The relations of such decisions to past events and the serious influence of past actions of change agents, marketing agencies, etc. In one case, heavy pressures by government agents for farmers to grow cotton were seen by farmers as dangerous precedents especially since the agents were meddling in allocation of farm family labour "and pretty soon will take over our homes".

(d) The influence of such decisions, and the resultant acitivities, on one's "standing" with one's peers, with one's social group. Buying a new plough may bring social prestige or lose one's closest friends. Investing in new noisy equipment may attract the attention of witch-doctors and those with an evil eye. It may therefore mean that before such investments are made, one may have to decide whether one's Christian principles will be compromised, if one brought in a medicine man to insulate the farm family and property.

2. THE FARM HOUSEHOLDER DECISION-MAKING PERSPECTIVE

Decision-making model builders in the field of farm management are aware that farm householders in rural East Africa operate an essentially indivisible complex reality (Webster, 1970; Heyer, 1969, Hunt, 1973). This reality constitutes what social scientists divide neatly into areas of farm decisions – farm management, general production economics, political parameters, social inter-action patterns and "institutions", public relations, religious beliefs and behaviour, health beliefs and treatment behaviour. Thus, to social scientists, peasant farm development occurs as a result of discrete activites of specialized institutions such as agriculture, community development, health, the Church, agroservice agencies, etc. Each institution, operating any one of a combination of activities, assumes that its activities are necessary and sufficient to effect total change to the farm household. As this assumption is critical in ensuring institutional survival, image projection, it often leads to its logical absurdity when agencies compete in their attempts to reach the farmer and sometimes sabotage each other's efforts.

As Mbithi and Mutiso (1974) show, the independent decision-making framework of the farmer is that of a responsive actor and adaptive actor in a teaming, dynamic and fluid environment. His social decision-making environment is described by his culture, of which he is a product, and whose complex values, cognitions, beliefs, and experiences are important components of his decision-making frame of reference. Technical and economic factors are extremely important, but nor necessarily the only factors and in some decisions not really critical. This becomes more important when one realizes that farmers are not trained engineers or economists but products of their social background, inheritors of well-proven indigeneous practices.

It is against this background that the results of a recent study to test the hypothesis that change agents are not really critical in farmer decision-making did not surprise scholars at the University of Nairobi. Table 1 shows the findings.

The communities which have high percentages on consultation with the

TABLE 1. *"When you wish to move into a new farm activity, for example adopt hybrid maize or the new drought escaping maize, whom do you consult?"*

Community & Province	People Consulted					
	Chief/ Agricultural Officer	Local Farmer	Kinsman/ friend	Himself	Total	Sample Size as % of Population
Tetu, Central	54	30	10	6	100	15
Machakos, Eastern	30	25	30	15	100	10
Karaba, Eastern	56	20	22	2	100	10
Ishiara Eastern	50	10	35	5	100	10
Shiakago Eastern	20	25	36	19	100	100
Ngoliba Central	20	28	32	20	100	30
Average	37	23	28	12	100	–

chief and agricultural officer are those with a strong farm cash crop economy. Tetu is in the high potential coffee and dairy farming agricultural areas of Central Kenya; Karaba a pioneer settlement with a strong cash crop production of green grams, corriander and maize, and borders the famous Mwea Rice Irrigation Scheme; Ishiara an intensive peasant-holder irrigation scheme.

The findings in the above-mentioned study, that in any one day a farmer may be visited by a health visitor who may advise on the need to reduce cash crop acerage to increase food crop acreage and the nutrition of the family; that some time later on the same day he may be visited by the crop specialist who reprimands him for not expanding his acreage of macadamia nuts, and he may yet, if he is near the road, be visited by the District Officer who threatens to "take him in" if he does not pay his taxes. These findings throw light on the confusion and frustration these encounters may generate. In certain areas, they appear to have confused farmers to a standstill.

3. THE CHANGE AGENTS PERSPECTIVE

Leonard (1974) and Mbindyo (1974) found that grassroots workers are very susceptible to commands from up the extension agencies hierachy. They define their duty as essentially responding to, and following the directives of, the professional. The professional has an extremely technical approach to the farm. His approach tends to follow the sequence outlined on the next page.

This approach assumes the following:

(a) That the rural farm can be seen as a responsive, expectant production firm which can be organized on true scientific principles according to prescriptions from experts, whoever they may be.

(b) That the farm is operating in a knowledge vacuum and there is not need for the expert to integrate existing knowledge, or pay attention to

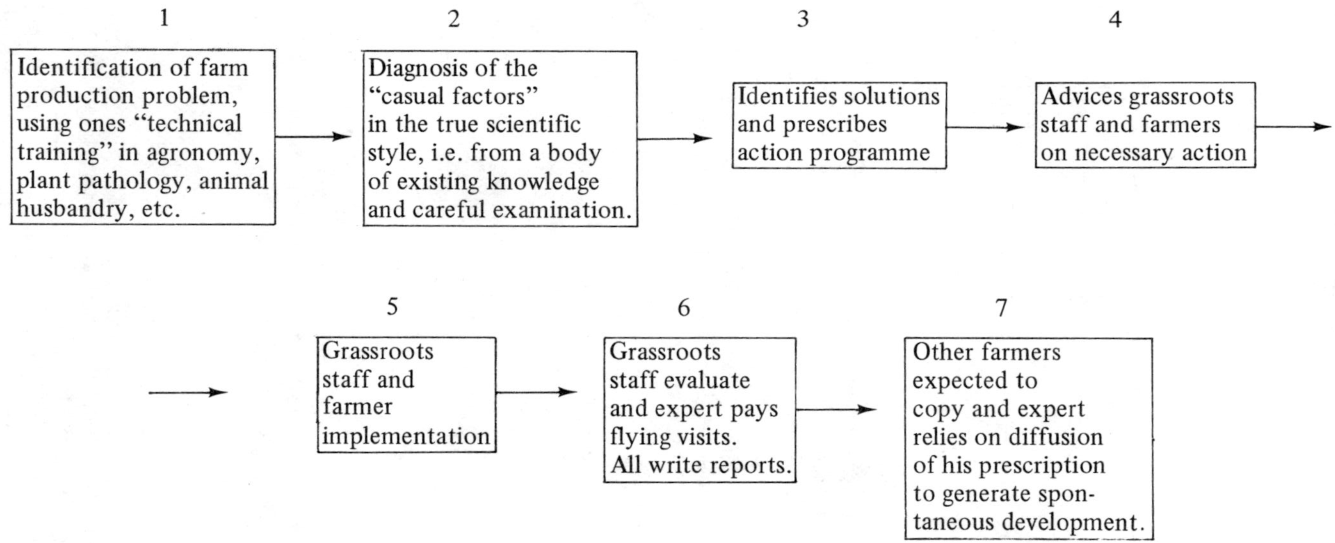

SOURCE: Mbithi, P. M., *"Integrated Social Science Approach to Development Planning", UNESCO,* September 1974.

other factors such as historical or social environmental, or to pay attention to the personality of the farmer. This assumption is very useful for medical doctors, although even then the need to "treat the mind" and the use of "bed-side manners" reduces its negative and alienating influence.

(c) That the change agent is operating with all necessary information: that he is well trained to assess local micro-level situations and resource patterns and has a battery of alternative prescriptions to suit each case or that he has originality, adaptability and foresight to bridge any gaps in knowledge or prescription. Given that most training in agriculture tends to rely on teaching materials from Europe and America, from temperate agricultural regions and given that little local systematic research on farm inputs, pests, diseases, resource management has been done, it is rather difficult to accept this image of the professional change agent.

(d) That the farmer understands and appreciates the technical nature of the expert prescription and has the resources to implement them.

Careful studies have shown that there are many things wrong with these assumptions:

(a) The change agent in East Africa is a member of the educated elite, who became alienated from his illiterate parents when he went to school at the impressionable age of between 6 and 8, spent at least fifteen years "learning" acquired book-keeping, western value systems, and life styles. He shares little with his rural folk and has no empathy with their thought processes. He may be arrogant or paternalistic but he does not recognize them as independent managers of their holdings.

(b) Alternatively, the rural farmer has, for the past seventy years been subjected to systematic brainwashing, through ethnocentric western religion, colonial administration, and all other contacts. He has been told repeatedly that he is backward, heathen, ignorant, illiterate and heading straight to "hell". As Mbula (1974) argues, this orientation, exalting the supremacy of western technology, western value systems, western religion, saps self ascertiveness, self identity from the humble rural farmer, and even makes him apologize for daring to see things differently. This approach generates a schoolboy mentality among rural adults who are very willing to suspend their world view, abandon it and embrace that of the expert, who may have just recently scraped through an agricultural exam based on a narrowly designed curriculum.

4. SOCIAL PHYSOLOGICAL OUTCOMES

4.1. *Dependence*

The extension approach which aspired to subjugate the farmer to the control of the change agent was based on the assumption that farm management decision-making could be understood and initiated, reinforced and directed by the expert.

The change agent ignored the fact that in East Africa change agents reach only about 10–15% of the farmers and, even with increasing extension saturation, the agent–farmer ratio was 1:1500. By undermining the farmers' self-assertiveness the agent undermined the farmers' creativeness and originality. In certain cases, those defined as worthless "laggards" retreated into reactive exhibitionist traditionalism; others became hesitant, vague and extremely dependent on "advice". It is not unusual to be told that a farmer has not adopted such a recommendation because the agent has not come round to clarify this or that, or because the agent has not supplied the commodity free or at the farm gate.

A strong agent influence, reinforced by a from top-down planning and resource flow pattern, has created dependency among farmers by enhancing the perception that government will, and can, do everything, i.e., will organize the farm inputs, subsidize them, transport them, send someone to advise, offer loans, take risks on failure, write off loans, organize marketing and give you the money in a bag not through an account. It should be clear that many of the members of this Conference would be strongly tempted to fake helplessess to receive such "service". It should also be clear that governments in developing countries cannot spread such "service" very far, as the financial resources and technical manpower is limited and farming populations constitute over 80% of their total populations.

The need for farmers to be self-reliant is therefore obvious. Farmers must become inquisitive, creative and must experiment with new methods on their own. They are more conversant with the peculiarities of their farm resources and their needs, and must become adaptive and independent.

4.2. *Alientation*

By ignoring local world view, local practices and social control, to which farmers are continuously exposed, extension agents become alienated and isolated from the general thrust of community life. By using national resources to win a few "progressive" farmers, the majority "non-progressive" are pushed to the periphery of development investment. They are thus alienated from the innovation process and resource flow process, and increasingly lag behind. The progressive farmers in turn increase their production and incomes. In certain cases they begin to expand by buying land from the non-progressives, expanding into non-farm activities, and increasingly control community labour, land and capital.

Apart from the implication of exploitation, the ego problems of the non-progressive are critical. He is openly ignored as a laggard, illiterate, poor and receives very little reinforcement. Roling (1971) argues that such neglect in the face of increasing materialistic aspirations for a better life, education of one's children, etc., leads to acute frustration. He argues that the major sources of frustration in rural areas is the ever increasing aspiration generated by a bombardmant of new life styles, by mass media and the demonstration effect of urban elite achievements. However, these rising aspirations are not matched by concomitant improvements in material possessions. The gap between aspirations and achievements constitute a measure of frustration. This frustration leads to demonstrations of energy where, though some may innovate, the majority may retreat in apathy or escape into pseudo-religion, drunkenness, and pseudo problem solving such as magic.

4.3. *Resource immobility*

Many students of rural development are convinced that many farm and community resources are locked into immobility due to the dependency problem and the poor personality non-rejuvenating, ego suppressing, morale sapping, rural development approaches of the past decades.

When farmers rely on government agents, they do not exploit their long-term grasp of farm production issues. They are relatively less imaginative and are willing to go on using the same old tools, following the same practices, year in, year out.

However, more significantly, they normally do not release all physical resources. In East Africa, in non-cash cropping areas, men do not perform most farm tasks. If they were committed, enthusiastic and excited creative actors, their participation would be assured, and meagre farm capital would not be diverted to non-production activities.

An example of release of resources by local villagers is discussed by Mbithi and Rasmus (1975) where they show that through Harambee self-help activities, grassroots investment in rural development activities accounts for about 30% of the total national investment in rural development. They show, through careful analysis of labour and capital contributions, that families contribute such items as eggs, beads, designing working songs, farm produce, and that all cash contributions which are not in kind are in very small denominations. They show that, however, in spite of the size or quality of donation, the contributor is praised, publicly mentioned and made to feel he/she belongs, is important and should contribute whatever he/she has. In this manner, rural people contribute over 90% of the total investment in Harambee self-help as compared to governments 6% and foreign donors 0·3%. The other 3–4% comes from big businessmen and rural elite.

4.4. *Suppression of indigeneous adaptation techniques*

Current international studies on drought and famine have become critical of traditional stop-gap relief measures and also of the extreme multi-million dollar large scale development programmes. The argument is that relief measures are too little, come too late and are too short run whereas large scale projects threaten the delicate ecosystems. The need to study the rural man and his responses to his environment and study his systems of adaptation has been articulated by social-ecologists. They claim that much wisdom is enshrined in traditional institutions and practices in social organizations in folklore and oral tradition etc. Recent studies by Mbithi (1974) in patterns of technological adaptation to marginal farming environments confirm the seriousness and worth of this orientation.

5. POSSIBLE RE-ORIENTATION

The emergence of a strong ethno-scientific approach to rural development has entrenched the necessity to utilize local systems of knowledge. Efforts to obtain a coherent local world view which is the basis for farm and community decision-making are gaining momentum. According ot W. F. Thomas, "what men define as real, is real in its consequence".

Thus, local decision-making frameworks can be seen as specific socio-cultural, socio-technical and ecologically bound knowledge models. They incorporate:

(i) the priorities as defined by local communities and farmers, (ii) an approach to a problem in its total context, (iii) within East Africa, a strong component of reinforcing the individual and collective personality in its capacity to tap local opinion, contribution and involvement.

In this context there is a strong need to:

(a) Re-open discussion on farm management and production models whose formulation was earlier dominated by agricultural economists.

(b) Re-discuss the concept of technology to shift it from a sterile scientific outlook into the psychic world which is dominant in indigeneous agricultural decision-making criteria. In our studies in Kenya we found that agricultural practices such as time of planting, harvesting and storage, were governed by strict observances, rituals and taboos.

This is not unique to rural folk. In launching a ship the Queen of England breaks a bottle on its side. At a Christian Wedding the couple is showered with rice. The role of witchcraft in football in East Africa is becoming a big worry to sports administrators.

The professed sterility and objectivity of fertilizer applications, for example, is often treated with scepticism. At one session, a young agricultural officer was asked whether he knew what he was talking about when he said that fertilizers were not European medicine. Had the Europeans told him it was not and had he naively believed them?

(c) Reject totally the "trickle-down effect theory" and its justification for the progressive farmer strategy in agricultural extension which is now known to lead to rigid barriers in access to national farm resources for the majority of farmers and perpetuates unacceptable income disparities.

(d) Accept that for rural East Africa, and possibly other developing countries, the displacement of the less able farmers from agriculture, in the true European tradition is no longer desirable or realistic given the acute unemployment and slum developments in urban areas. Thus farm management planners will need to rediscuss their assumption of a bottom ceiling on land size and their love for the progressive farmer who systematically displaces all "incompetent" farmers, saves extension costs and diffuses technology.

(e) Support current search for an integrated rural development strategy where the aim is to: (i) integrate goals of agricultural and rural development, (ii) integrate those rural poor who have been left out of the development, process, (iii) integrate the programmes and agencies of rural development.

REFERENCES

Ascroft, Joseph, *et al.* (1972) 'Does Extension Create Poverty in Kenya?' *East African Journal IX* 3 (March), 28–33 Nairobi.

Heyer, Judith (1968) *Peasant Farming in Lowland Machakos,* Occasional Paper, Institute for Development Studies, University of Nairobi, 1968.

Leonard, David (ed.) (1973) *Rural Aministration in Kenya,* East African Literature Bureau, 1973.

Mbindyo, Charles (1974) *The Effects of Extension Workers Role Orientation on Rural Development,* unpublished M.A. thesis, University of Nairobi, 1974.

Mbithi, P. M. (1971) 'Issues in the Transfer of Useful Knowledge in Agricultural Development in Kenya', *Journal of Agricultural Administration,* Reading, 1971.

Mbithi, P. M. and Rasmusson, R. (1975) *The Structure of Grassroots Harambee in Kenya,* University of Nairobi, 1975.

Mbithi, P. M. and Mutiso, R. (1974) *Integrated Rural Development, Popular Participation,* July 6–9, Nairobi.

Mbula, Judith, (1974) *Penetration of Christianity into Akamba Traditional Family,* unpublished M.A. thesis, Nairobi.

Roling, Niels (1971) *Some Socio-psychological Aspects of Development,* unpublished paper, Institute for Development Studies, University of Nairobi.

Webster, Paul (1970) *Computer Simulations of Farm Decision-Making: The Kabayolo Model Farm,* Department of Rural Economy, Makerere.

Wills, Jane (1969) *The Peasant Farm and the Theory of the Firm,* RDR No. 68, Makerere Institute of Social Research.

RAPPORTEUR'S REPORT* – John Cleave, *I.B.R.D.*

Although each takes a different perspective of agricultural decision-making, largely in an African context, the two papers presented to the Group do complement each other; Dr. Anthonio's paper suggests a heirarchy of levels of decision-making against which the process of reaching decisions, and proposals for improvements at all levels, can be analyzed. Dr. Mbithi's paper focuses on the environment in which the "lower-order" or farm-level decisions are made, suggests some social-psychological problems which results from it, and proposes new approaches to avoid these problems.

It was generally agreed that improved decision-making at all levels is an important element in agricultural development, and that an understanding of the process of arriving at decisions and the constraints which influence decision-makers is an essential first step in aiding such development. In discussion, it was emphasized that Dr. Anthonio's heirarchy (from "first order" decisions over which the individual has complete control; through "second order" decisions made by a homogeneous group with common objectives; and "third order" decisions made by a heterogeneous group not necessarily bearing the consequences of its actions; to "higher-order" or policy decisions, also outside the control of farmers) is a continuum of interdependent and interacting decisions. Not only do higher-order decisions (touching, for example, on crop prices, market organization, or regional resource allocations) have enormous impact upon the setting in which lower order decisions have to be made, but equally, if they are to be effective policy decisions have to take account of the responses, goals and values which go into farm-level decisions. Moreover, agriculture cannot be treated in isolation from other sectors nor, in an open economy, in isolation from the decisions of other countries. There

* The texts of opening statements were not available

was wide agreement also on the desirability of improving the means of reaching a consensus of opinion in policy-making and avoiding the predominance of sectional interests.

Speakers emphasized the need to improve the involvement of farmers in higher-order decisions, both through insititutional improvements to enhance communications between groups at all levels and by improving the quantity, quality and timeliness of information on which decisions are to be based.

There was general agreement with Dr. Mbithi's characterization of extension agents as alienated technicians who tend to underrate the experience and knowledge of farmers, and to over-rate their own understanding of the farmer's production environment and operational constraints. Following from this, there was considerable discussion on the need to find ways to improve extension services. The importance of improving the training of extension workers, and especially of enhancing their understanding of the farmer's resource constraints and conflicts so that advice could be modified to fit the needs of the individual farmer, was generally agreed upon, although the question of how to do it when extension staff are few and poor in quality was not answered. The desirability of organizational changes to associate extension workers more closely with farmers by drawing them from villages and de-emphasizing the "Progressive Farmer" approach was also suggested.

The group did, however, take issue with the characterization of the farmer as a dependent, subjugated, and largely helpless pawn. Speakers emphasized the rationality of farmers, the breadth of their knowledge derived from experience – both good and bad – and their sharp understanding of their resource constraints and the risks under which they operate. Indeed, the very independence of farmers has frequently been bemoaned by those who see irrationality in their refusal to adopt innovations which are, in fact, often ill-adapted to the farmer's circumstances.

Speakers suggested that new technology frequently increased the risks with which farmers are faced, including the risks attendent upon greater dependence on unreliable government input, marketing and advisory services, and on government policy-level decisions which do not adequately take account of their effects on farm operations.

Related to this point, a speaker stressed the need to include high-level decision-making as an element in farm-level models, and a useful characterization of the farm-level decision-making process was made by Dr. I. J. Singh. He said that the time had come to move away from the static "neo-classical model" of the decision process, which had proved useful in the past, but which viewed farmers as optimizing single-valued goals, with full information and without risk or learning, and adjusting fully and instantly to changes in their external environment; to a dynamic "adaptive model" wherein farmers can be viewed as "satisfying" multiple and heirarchial goals and operating without full information in a risky environment and adapting only cautiously over time to external changes, and then only in the light of their own experience and learning ability.

It is clear that decision-making on the African farm involves a complex interaction between goals, constraining factors, resource limitations, enterprise choice and techniques, on which higher-level decisions impinge critically. The papers presented by Doctors Anthonio and Mbithi, it was agreed, provided a useful focus on a range of issues and the factors involved in the decision-making process.

A third, contributed, paper presented at the Conference (John H. Cleave *Decision making on the African farm*) further analyzes the nature of farm decisions and provides illustrations of the way African farmers modify their production strategy within the limits of their resources, objectives, and constraints, and could be read in conjunction with those discussed at this session.

JOSEPH SEBESTYÉN*

On Regional and Interregional Models of Planning for Agriculture

The spatial aspect of economic activity has been a topic of discussion and studies for a long time. Agriculture has always been an attractive field for such studies. Starting with Thünen a long list of authors dealing with problems of space in agriculture or other sectors of the economy could be enumerated. For brevity, I draw attention in the reference list to a paper by Weinschenck, Henrichsmeyer and Aldinger [1].

1. NATIONAL AND REGIONAL PLANNING

In many non-socialist countries there is a regular activity of national planning. Local and regional plans are also prepared but not always in connection with the national plans. There are efforts in various countries to elaborate national plans in a regional setting. As an example refer to the recent French REGINA model [2].

In the Soviet Union Lenin stressed already in 1918, i.e., at the start of the economic planning activity, the need to develop national plans in a regional setting, as we read in [3]. He emphasized the importance of a rational distribution over the country of the productive forces, the importance of a rational spatial division of social labor.

The Soviet experience, gained in the course of the steps toward the realization of these goals, was helpful for the other socialist countries where the need for consideration in national planning emerged soon after taking the first steps in building up a planned economy. This was true with Hungarian planning too.†

Important principles in the traditional methodology have been comprehensiveness and consistency. The very detailed system of balances required correspondence between figures for different sectors in the national economy as a whole and in various regions. Of course, due to weaknesses in planning –

* Research Institute for Agricultural Economics, Budapest, Hungary.

† The present author has good memories of the book entitled *Territorialnoe planirovanie*, by L. Volodarsky, from which he learned about the methodology of regional planning in 1949–50, and he still feels that his mathematical model building activity profited greatly from that traditional regional planning methodology.

or rather of the planners – these basic requirements have often been only partially met. However, mathematical model builders have still much to do in developing a system superior in every respect to traditional methodology.

1.1. *Types of problems in planning over space*
The spatial aspect of decision-making in economic activity, or of decision-making in the planning process, touches a number of problems. A few of them, characteristic ones, will be listed here.

(a) Finding a correct location for a unit of economic activity (e.g., a plant or a complex).

(b) Finding correct distribution over a space of a type of economic activity, along with resources and flows (i.e., sectoral location).

(c) Finding correct structure of economic activity within a region under, or without influence of, national plans or results from solutions of problems of type (b).

(d) Finding a correct distribution over space of economic and other activities and resources, in the form of activity structures (comprehensive plans) held appropriate for units on different levels of the hierarchy, and the flows between them.

Policies of a financial character (pricing, taxation, subventions etc.) often appear in problems (a)–(c). Problem (d) involves policies not only of financial character but also policies of social–cultural and environmental conditions, population movements, etc.

Various mathematical models and procedures have been developed for handling these problems. The mathematical apparatus used most frequently involves different types of functions and linear algebra (in the form of Leontief analysis and mathematical programming).

I note that the important task of establishing industrial (or agro-industrial) complexes touches points (a)–(d). As to treatment in relation to problems (a)–(c), I refer to several Soviet and American studies [3–6]. An excellent treatment of pricing and allocation over space can be found in Takayama and Judge [7].

1.2. *The choice of regions*
Depending on the geographical picture, its size and on the social system of a country, the considerations for marking out the regions may be different. Let us refer to only three items from a rich literature: Fox and Kumar [8], Kostetsky [9] and Alexandrov [10]. Difficulties with the number and size – to which there may be related also a problem of the degree of non-homogeneity – can be reduced by the application of a multi-level planning procedure which may reconcile the advantages and disadvantages of a microregional and of a macroregional approach in model building.

I would like to stress that the boundaries of regions should encircle complete administrative units. One reason for it is that the data basis, or at least an important part of it, is generally available for administrative units. Another reason is that the leaders of these units have to represent the interests of the population of entire communities and are often responsible for important decisions concerning the respective units as entities.

1.3. *Planning downwards or upwards?*

No doubt serious regional planning cannot be imagined irrespective of central (government) decisions and information. On the other hand, the government also must be informed about needs and possibilities in their distribution over the country. Thus, a continuous exchange of informations between government and units on lower levels of the hierarchy is necessary for any type of planning.

We can speak of a centralistic type of planning if every important decision in the course of planning, particularly those concerning targets and resource allocations, comes downwards to a unit from the next higher level of hierarchy. Given official omniscience, it would be very efficient. In its very rigid form it already belongs to the past. As practised at present we see it in a looser form – as certain segments of the planning procedure in socialist countries and, for example, as the procedure corresponding to Tinbergen's concept [11, 12] of "planning in stages".

"Planning upwards" means that the units belonging to each level of the hierarchy make most decisions for themselves in the process of planning and the next higher level of hierarchy has to respect these decisions in setting up its own plan. Of course, the decisions which are necessarily made at appropriate higher levels are respected. This type also can be found in the practice of socialist countries, in a looser form. In the present author's view [13], regional and national plans can be built up in this way as will be discussed later.

2. TYPES OF MODELS AND THEIR USE FOR PLANNING

Section 1 has a rather general character although our efforts are directed to improving agricultural planning. But we can help agriculture more efficiently if we deal with its problems set in a general socio-economic planning process. Thus purely agricultural models should be considered as specific elements of a broader framework, or as precursors of such elements.

Now we shall review four types of models as means of analysis for planning purposes or as tools of planning.

2.1. *Single functions and systems of functions*

(a) *Production functions.* Production functions by regions or sub-regions appear very often in the literature, particularly if we also count here the crop response functions for different soil types. Crop and livestock response functions can help in studying possibilities of shaping yields and technologies for regions with different conditions. The result of such calculations may be useful in constructing the system of coefficients for Leontief analyses, for optimizing and simulation models.

Functions describing the dependence of total farm output on various types of inputs also offer productivity, substitution, etc., comparisons between regions for considerations about policies to be included in planning models. By integrating regional production functions into a system, effects of different policies can be studied in each region and on the national level.

Interregional and international comparisons of resource efficiency in agriculture derived from production functions can be found, e.g., in Heady and Dillon [14]. From the Soviet Union I mention Zakumbayev [15] who estimated Cobb–Douglas functions for sectors, including agriculture, in a regional setting, for republics, and within Kazakhstan for macro and micro-regions, to analyze development.

The present author has also made calculations based on production functions in the preparatory stages of both the 4th and the 5th Five-Year Plan, with functions computed on the sector and on the regional level.

(b) *Demand functions.* The models known to me include demand either in fixed quantities – eventually in variants or in values changing according to a parameter – or in the form of demand functions. Some interregional models consider national demand only, others calculate regional demand figures from total national demand proportionally to population. Pant and Takayama [16] derived regional demand functions by substituting regional per capita income in the national per capita demand function and multiplying this formula by regional population. Kottke [17] estimated regional demand functions from data by states.

(c) *Supply functions.* As elements of a model, e.g., of a simulation model, they are generally estimated by variables like prices, area and inputs as well as proxies for technical change and eventually for natural conditions. Often we see them derived from results of optimization as normative supply functions.

2.2 *Leontief analysis*

The interindustry analysis has become not only a tool of national planning but has also found a wide application in spatial planning. We find input-output tables for cities, and for smaller and larger regions throughout the world, in countries having quite different social systems. These regional Leontief analyses have generally been of a static nature. As a new development in this field should be mentioned the elaboration of dynamic regional interindustry models for the Baltic Republics of the USSR [18].

Another important type of interindustry model is that reflecting spatial interdependence. Riefler [19] gives a good survey, with a long reference list, so citing examples here seems unnecessary. A specific type of these interindustry–interregional models is that emphasizing agriculture, with a disaggregation by commodity groups of the agricultural sector in a regional setting. The spatial-sectoral interdependence has generally been the focus of the studies, including projections. A further step is represented by primary factor commitment projections like the study by Schluter and Heady [20] presenting agricultural employment projections on the basis of an interindustry–interregional model emphasizing agriculture.

If we focus our attention on the interdependence within agriculture, the connection between local, or regional, and national tables has to be organised. Of the present author's multipurpose studies the one to be mentioned is that which, based upon the tables calculated for each large-scale farm in a 2·5% sample, furnished distributions by regions of direct and cumulative, inverse, coefficients, also [21].

2.3. *Mathematical programming*

Spatial location of agricultural production had been mentioned among the fields of application by the very first paper [22] on the procedure which later became widely known under the name linear programming. Optimizing models are abundant and show a great diversity in relation to time and to coverage. The majority have aimed at finding an optimal final state for the spatial distribution of production. The simplest ones optimize only the location in space of an activity or groups of activities. Others include meeting demand, not only on the national level but in a regional setting too. The production–transportation models for agriculture often serve a multiplicity of purposes. As examples, I mention recent studies which consider, among others, policies for pricing, water use and pollution restriction [23, 24].

Another type of model deals explicitly with time. One of them, described in [5], is fully applicable to agriculture, although it was originally constructed for purposes of industrial development and location in Siberia. Here the variables and constraints have a yearly timing within the planning period. The interregional recursive programming models elaborated at the USDA as aids to agricultural policy-making have a cobweb character. (See, e.g., Miller [25] and Schaller [26]). For an application to a single region I refer to a paper by Singh [27].

The programming models also differ heavily in coverage. The variables may cover the whole agriculture or only part of it, may specify types of farming from technical as well as a social point of view, and may include links to non-agricultural activities. The constraints are also detailed in varying degrees. Some models include mainly commodity balances and technical constraints like a Polish model with high activity coverage [28], others include financial and behavioral equations along with detailed technical constraints over a deep spatial disaggregation like CHAC, having a much smaller coverage of Mexican agriculture [29].

The models of partial coverage, either in terms of activities or of constraints or both, may throw light on special problems. They may help in building more correct and more comprehensive models and may also furnish considerations for traditional, and more comprehensive, planning.

2.4 *Simulation*

This procedure may use different mathematical tools. For a single region, Singh [27] applied recursive programming while Miernyk *et al.* [30] the dynamic input–output analysis. The USDA models [25, 26] are based on the interaction of demand functions and optimizations through linear programming by regions. Systems of equations or functions characterize the simulation models constructed for Nigeria [31] and South Korea [32]. Both of them include macroregions and connections with other sectors of the national economy. With a broad coverage and an appropriate choice of coefficients, such simulation models may contribute much to a more realistic planning. Running policy alternatives by using such models may help the decision-makers to construct strategies for development and policies serving the realization of the strategic goals.

3. THE PLANNING PROCESS

The success of plans depends to a great extent on how seriously planning has been taken by a government of a country first of all, by the leaders of the planning units and, last but not least, by the planners themselves. Development in many countries would require changes in the economic, social and political structure. To carry out these changes in a relatively short time, or even abruptly if necessary, needs a firm resolution of the government and its agencies, not only at the start but in the process of realization, also, when difficulties have to be faced day by day.

3.1 *The planners*

In any stage of the planning process, the planners must take stands and their preferences and dis-preferences may strengthen or weaken the intentions and efforts of the government. This is true for mathematical planning too: in model building and interpretation of results, attitudes play a role which should not be under-estimated. For example, in respect of the relation of regional development to national economic growth, Courbis [2] stressed that the conception of the REGINA model shows an explicit acknowledgement of the non-neutrality of spatial factors and regional policy to national development. And this occurred not independently of the attitudes of the planners and model-builders. Let us bear in mind, even a purely technocratic attitude in modelling can work against a development conceived on the basis of social justice and the result finally touches not only certain regions or groups of population but, also, the sound growth of the whole national economy.

In many cases, regional–national planning has been assigned to a central group – at least, many publications do not mention planners on lower levels of the hierarchy. Thus, the role, if any, left to the lower levels, might by that of consultants. At a very early stage, e.g., in a stage of developing models by researchers, it may be accepted. However, if we assume that planning means a set of decisions with a multiplicity of consequences, those who are interested should participate, at least through their representatives. Thus, planning agencies should include regional and local authorities, representatives of social groups and experts in various fields available in a certain region. The planning agencies at various levels of hierachy may have keen discussions with each other but they should respect the right to make certain decisions which have necessarily to be made at a higher or lower level.

3.2 *Goal setting*

Government intentions concerning the socio-economic development of the country necessarily involve a spatial aspect even at the stage where the plan is taking shape, the natural resources to be used, problem spots to be eliminated, etc. In order to improve this first crude goal setting, a series of partial studies should be carried out.

These partial studies, focusing problems of agriculture, should cover at least the following areas at the national as well as at regional level.

(a) Population and labour force projections, by social and professional groups. Needs and possibilities for mobility (interregional and interprofessional) should also be studied under the present and possible future situations characterized by social system, educational and economic opportunities.

(b) Development of sectors other than agriculture, including various types or services. Projections, estimates and decisions have to be investigated from the point of view of demand and supply, employment opportunities and effect upon living conditions, with special regard to relations with agriculture.

(c) The formation of agrarian structure – a realistic picture assessing needs and possible directions of change, the forces in favor and against, the time span of realization, the factors and measures needed to make the change a success. Estimates about the effects on shorter and longer term.

(d) Agricultural production and technology should be surveyed. The balance of demand for products and the production potentials should be investigated, considering efficiency with changes in the agrarian structure, supply of inputs, general and professional education, diffusion of new farming techniques, incentives for increasing and modernizing production, financial sources of development.

(e) Income distribution should be studied by groups of population, according to the levels needed, the sources of income and the feedback to production.

(f) Government intervention with direct and indirect effects of policy variants should be surveyed in the field of legislation, finances, a state sector of production as a moving force, social-cultural development, population and employment policies.

An evaluation of such studies would show, how ambitious goals might and should be thought of as feasible, in the qualitative and quantitative sense, serving for starting points for the second stage of planning.

(a) *Mathematical models in the preliminary studies.* Mathematical models can be widely used in the course of these preliminary studies. Their application is helpful not only because of the information furnished by the solutions but at least as much through their effect upon thinking. The keen discussions in this stage about model-building, evaluation and assessment of results, can bring benefits for the later stages.

Demographic models may be used for studying and forecasting movements of population and labor force.

Production functions and Leontief tables offer good possibilities. A special form of the latter is an intra-sectoral table in a heavy disaggregation, the input coming from other sectors being linked as a premultiplying matrix. It must be noted that intra-sectoral intermediate demand must justify the construction of such tables well suited to different types of organization, levels of technology, and connectable to more aggregative national Leontief models.

The behavior in connection with technical progress in practice should be studied, in its dependence on time and other factors, to evaluate coefficients for Leontief tables, MP matrices and equations in simulation models.

The simulation models used for the evaluation of the effects of different policies characterized quantitatively for future periods can cover either more narrowly a sector, as in an investigation made by Reynolds, Heady and Mitchell [33] or include to a certain extent the behavior of the economic environment of the sector in question as reflected by the above-mentioned studies of the Michigan team [31, 32].

The optimizing models should be used on both regional and interregional levels, for single sectors and for the ensemble of several sectors of the national economy, since computer time and cost would hardly allow running models of complete comprehensiveness in so detailed a form as is necessary in this stage.

As a result of the solution of a considerable number of such models, the spatial structure of economic activity offers information about employment, income distribution, living conditions, demand and supply, resource use, investment for different purposes etc., depending on the policies and socio-economic conditions reflected by the models.

(b) *The evaluation of the results.* The different models offer, of course, because of their character, different figures for the same categories. Comparisons of the results furnished may lead to acceptance or rejection of certain assumptions of the model-builders.

In the course of an assessment of results from various models, one can set up systems of balances on the regional and national level, for separate sectors and for the whole economy. We can see which policy, with how much effort, would lead to attainment of certain goals, which goals would be overfulfilled and which ones not attained. A projected population structure would need a certain food supply, an income structure: how do the food production and the income distribution meet these requirements according to solutions received from the models? How do the results of models of manufacturing industries correspond to the demand in input goals derived from models for other sectors? Does capital formation keep pace with the investment activity as it is distributed over time?

A long series of such questions must be asked and the staff of planners and their consultants, going through the results, item by item, must identify disproportions, find promising policies and change (replace) assumptions, coefficients, relationships in models in the course of this analysis. After this the important national and regional goals may be fixed for the second stage of planning.

3.3 *Planning over space and time*

Starting from the preliminary studies, aided by mathematical models as mentioned above, the national goals should be tested by Leontief analysis, accompanied by projects involving policies touching the socio-economic sphere. It is preferable to set up several variants of the goal system, representing strategies of development, to allow the legislators to choose the path to be followed.

In the case of setting up the regional–national plan by using mathematical methods, one should first decide upon the direction of the planning procedure: going downwards or upwards.

(a) *Planning downwards.* In the first case, one might use an interregional Leontief analysis but this choice might give rise to questions touching serious problems. Even in the case of a static model, the spatial nomenclature of sectors should be fixed over the whole period and one should decide about interregional flows by fixing the coefficients and their eventual changes over time. More problematic would be an interregional dynamic Leontief model.

Choosing a normative approach for downward planning, a simultaneous solution for all regions should be considered. Here we have to deal with the case mentioned under 1.1(d) and we must consider, because of indivisibilities a mixed integer programming approach (with general integer and continuous variables). The objective function may represent different economic categories. The system of variables and constraints should be chosen in such a way that they would not eliminate possibilities important for the next step, the optimization for subregions. I stress that lower bounds of income for different social strata should be fixed for each region, along with requirements for social–cultural and infrastructural development. Foreign trade, balance of payments, financial balances, migration, and eventually transportation, should also be explicitly treated.

As far as optimization for the subregional level is concerned, the solutions for the respective regions would serve as components of the constraint systems.

Planning downwards is an efficient procedure but it does not allow much freedom of decision for lower levels as either responsible leaders or owners.

(b) *Planning upwards: building up a national plan from below.* With the exception of managing the economy in a rigid centralistic manner, decisions in planning for future action should be made by the representatives (leaders) of the respective units (firms, municipalities, regions, etc.). These units on various levels of hierarchy are assured of an economic independence by the legal framework prevailing in the country which not only allows, but also requires, their self-contained decision-making.

Under such circumstances, building up the national plan from plans of units belonging to different levels of hierarchy seems to be preferable. This would also correspond to the principle of an active and wide participation in the planning process. A basic requirement of such planning is that the totality of plans prepared by the units on a level of hierarchy must be compatible with the plan of the unit they belong to on the next higher level of hierarchy.

Since the aspirations and the response to the socio-economic environment, including incentives and other types of government intervention, of the individual planning units may be different, compatibility with the plan of the unit on the next higher level can be more easily reached if there are variants of plans from which the higher unit can choose, one for each lower unit, in such a way that the higher unit can also develop its plan in a number of variants.

As was mentioned in 3.1 and 3.2, the preliminary studies, including investigations by partial mathematical models, and the discussions in the course of the evaluation of them, furnish rich material for the planners at all levels. Based on them and on the information received from higher levels, variants of plans can be developed by each unit, by using either traditional or mathematical methods. A system of goals with more modest or more ambitious aspiration levels elaborated according to different rates of change over time, considering the occurrence of more or less favourable situations with response to possible government actions, may lead to a useful number of variants which can be submitted to the unit (authority) on the next higher level.

The choice from the variants received from the units at the next lower level occurs according to the specific tasks and goals of the respective unit on the higher level (e.g., of a regional authority), in relation to a criterion. The same procedure can be followed in setting up the national plan – also in variants.

This procedure can easily be formalized as a mixed integer programming approach for several levels of hierarchy. The structure of the models had been described in [13], along with results from testing their workability. For the test, a problem of location of agricultural production had been chosen, with price response, the minimum necessary levels of subventions by regions as influenced by increasing requirements in earning foreign currency. (The results from running the extended more complex problem with new computer facilities confirmed the earlier judgment about the workability of such type of planning).

A main characteristic of this multi-level planning over space is the assumption that the plans of any unit on any level of hierarchy are organic entities representing strategies relative to a structure, an interdependent system, thus the variants cannot be mixed. Linear combination being excluded by the acceptance of political indivisibility, the higher level must choose in a zero – or one – manner. With technical indivisibility also appearing in the plans, the optimization problem has 0–1 integer, general integer and continuous variables. The variables which can assume a value equal to either one or zero represent the plans of the units on the next lower level as indivisible entities.

Comprehensiveness and consistency are enforced by the nature of this approach, in a degree obtainable at the respective level, from both social and economic points of view. Consistency of ends and means on a level of hierarchy can be assured by a proper formulation of the models for lower levels. Only that information can be included into the model of a unit on a higher level which can be derived from the solutions for the units on lower levels. I would like to stress that not only production but other categories too (e.g., income distribution, employment, migration, housing, transport facilities, capital formation, credit, investment, etc.) should be also investigated in their timeliness, in their intra-unit (intraregional) and interunit (interregional) relations, too, not only on the national level.

3.4 *Planning at the international level*

In the present world, the usual international trade alone can less and less satisfy the requirements for international economic relations. The great differences in resource endowments, in conditions of rational use of national recources stemming from the degree of socio-economic development, the population pressure and the prospects of food production etc. press for a change in attitude. The idea of international planning has also emerged – let us refer here only to one book [11]. Mathematical economists have made efforts to set up models covering to a certain extent several countries [34] and an international effort for linking national models has emerged [35].

The procedure formalized in [13] and verbally treated in Section 3.3(b) of the present paper may also be one of the means helping to make a step forward. If we consider that countries can develop medium – and long-term – plans in several variants representing different strategies for socio-economic development, one might try to select those variants which can be connected with a common benefit from coordination. The procedure proposed by the present author would assure, as with the case of interregional planning within a country, that even the weaker nations should not accept, as a result of international coordination, plans which were less advantageous than the most modest of the variants elaborated (possibly optimized) by themselves. Any type of pressure for the sake of non-mutual benefits is regarded as a topic beyond the scope of this paper.

Since even models constructed in highly developed countries display considerable differences, much effort would be needed for the introduction of requirements in model building to get a data system for a model of international choice.

4. SUMMARY

This paper has tried to show how a wide variety of mathematical models has been used for regional–interregional analysis and planning of separate sectors or of the national economy. Of course, only a small fraction of the literature could be cited, mainly recent publications characteristic of types of problem treatment. This abundance of models also made it possible to avoid formal presentations. The paper has been aimed at insisting on a treatment of agricultural planning, not separately, but as an organic element of national planning. By making use of the already existing partial models, the possibilities of planning in a multi-level system were shown, alluding to the advantages of the upward procedure with emphasis upon the rights to independent decision-making on the lower levels of hierarchy and upon the reconcilability of interests and goals of the different levels of a hierarchy.

REFERENCES

[1] Weinschenck, G., Henrichsmeyer, W. and Aldinger, F., 'The theory of spatial equilibrium and optimal location in agriculture: a survey', *Review of Marketing and Agricultural Economics,* Vol. 37, No. 1.

[2] Courbis, R., 'The REGINA model. A regional–national model of the French economy', *Economics of Planning,* Vol. 12, No. 3.

[3] Nekarasov, N. N. (1975) *Regional Economics,* (in Russian), Izd. Ekonomika, Moscow.

[4] Albegov, M. M. (1972) 'Regional models in an automated system of planning', Rome Congress, 1971, *Papers of the Regional Science Association,* Vol. 29.

[5] Aganbegian, A. G. *et al.* (1974) *Economic Problems of Development of Siberia,* (in Russian), Izd. Nauka, Novosibirsk.

[6] Vietorisz, Th. (1967) 'Industrial development planning models with economies of scale and indivisibilities', Lund Congress, 1963, *Papers of the Regional Science Association.* Vol. 12.

[7] Takayama, T. and Judge, G. G. (1971) *Spatial and Temporal Price and Allocation Models,* North-Holland, Amsterdam.

[8] Fox, K. A. and Kumar, T. K. (1966) 'Delineating functional economic areas' in *Research and Education for Regional and Area Development,* Iowa State University Press, Ames.

[9] Kostetsky, A. U. (1974) 'A method of solution for the rational regionalization of the country', in *Mathematical Aids for the Problem of Location of Production,* (in Russian), Izd. Nauka, Moscow.

[10] Aleksandrov, Yu. K. (1974) 'Application of random search in problems of regionalization', in *Mathematical Aids for the Problem of Location of Production,* (in Russian), Izd. Nauka, Moscow.

[11] Tinbergen, J. (1967) *Development Planning,* World University Library, London.

[12] Mennes, L.B.M., Tinbergen, J. and Waardenburg, G. (1969) *The Element of Space in Development Planning,* North-Holland, Amsterdam.

[13] Sebestyén, J. (1973) 'On models of planning regional development in Hungary, in Judge and Takayama (eds.), *Studies in Economic Planning Over Space and Time,* North-Holland, Amsterdam.

[14] Heady, E. O. and Dillon, J. L. (1961) *Agricultural Production Functions,* Iowa State University Press, Ames.

[15] Zakumbayev, A. K. (1975) *Methods of Assessment of the Degree of Economic Development in Republics and Regions of the USSR,* (in Russian), Izd. Nauka. Alma Ata.

[16] Pant, S. P. and Takayama, (1973) 'An investigation of agricultural planning models; a case study of India's food economy', in, Judge and Takayama (eds.), *Studies in Economic Planning Over Space and Time,* North-Holland, Amsterdam.

[17] Kottke, M. (1973) 'Allocation of milk through space and time in a competitively-mixed dairy industry', in, Judge and Takayama (eds.), *Studies in Economic Planning Over Space and Time,* North-Holland, Amsterdam.

[18] Kossov, V. V. (1973) *Interindustry Models,* (in Russian), Izd. Ekonomika, Moscow.

[19] Riefler, R. F. (1973) 'Interregional input–output: a state of the arts survey', in, Judge and Takayama (eds.), *Studies in Economic Planning Over Space and Time,* North-Holland, Amsterdam.

[20] Schluter, G. E. and Heady, E. O. (1975) *Projected 1980 Overall Agricultural Employment by Regions, Commodities and Related Sectors,* CARD report 59, Iowa State University, Ames.

[21] Sebestyén, J. (1969) 'Problems of rational location in Hungarian agriculture', Budapest Congress, 1968, *Papers of the Regional Science Association,* Vol. 22.

[22] Kantorovich, L. V. (1939) *Mathematical Methods of Organization and Planning of Production,* (in Russian), Izd. Leningrad State University, Leningrad.

[23] Meister, A. D. and Nicol, K. J. (1975) *A Documentation of the National Water Assessment Model of Regional Agricultural Production, Land and Water Use and Environmental Interaction,* CARD, Iowa State University, Ames.

[24] Dvoskin, D. and Heady, E. O. (1975) *U.S. Export Capabilities Under Various Price Alternatives, Regional Production Variations and Fertilizer Use Restrictions,* CARD report 63, Iowa State University, Ames.

[25] Miller, T. A. (1973) 'Economic adjustment research for policy guidance: an example from agriculture', in Judge and Takayame (eds.), *Studies in Economic Planning Over Space and Time,* North-Holland, Amsterdam.

[26] Schaller, W.N. (1968), 'A national model for production response', *Agricultural Economic Research,* Vol. 20, April.

[27] Singh, I. (1973) 'Recursive programming models for agricultural development', in, Judge and Takayama (eds.), *Studies in Economic Planning over Space and Time,* North-Holland, Amsterdam.

[28] Gomulka, Yu. (1976) 'A preliminary model for the location of agricultural production', in *Bulletin No. 22. of the Center of Coordination of CMEA Countries for Scientific Research,* (in Russian).

[29] Duloy, J. H. and Norton, R. D. (1973) 'CHAC, a programming model of Mexican agriculture', in, Goreux, L. M. and Manne, A. S. (eds.), *Multi-level Planning: Case Studies in Mexico,* North-Holland, Amsterdam.

[30] Miernyk, W. H. *et al.* (1970) *Simulating Regional Economic Development,* Heath, Lexington.

[31] Manetsch, Th. J. *et al.* (1971) *A Generalized Simulation Approach to Agricultural Sector Analysis,* Michigan State University, East Lansing.

[32] Rossmiller, G. E. *et al.* (1972) *Korean Agricultural Sector Analysis and Recommended Development Strategies, 1971–1985,* Agricultural Economics Research Institute, Seoul–Michigan State University, East Lansing.

[33] Reynolds, Th. M., Heady, E. O. and Mitchell, D. O. (1975) *Alternative futures for American Agricultural Structure Policies, Income, Employment and Exports: a Recursive Simulation,* CARD report 56, Iowa State University, Ames.

[34] Balassa, B. (ed.) (1975) *European Economic Integration,* North-Holland, Amsterdam.

[35] Ball, R. J. (ed.) (1973) *The International Linkage of National Economic Models,* North-Holland, Amsterdam.

DISCUSSION OPENING – E. A. Attwood, *Ireland*

We must congratulate Dr Sebestyen on the breadth of his knowledge of the literature and for the way in which this knowledge has been summarised into an integrated assessment of the subject.

There are two major issues which arise from the paper. The first is that most of the work has been of a basically theoretical nature and some further examination of the problems in the day-to-day application of this work would be useful. Let me illustrate this by reference to the discussion on "Planning downward" in which the use of an inter-regional dynamic Leontief model is considered. This discussion is in terms of what should be done, rather than in terms of what actually happens in practice. There is a need for a further and more detailed examination of the subsequent outcomes of the many regional models specifically referred to in this paper. Many of these studies are of a predicitive character and these predictions should be considered in relation to the developments which took place after the models had been completed. There are, of course, many valid reasons why the real world does not evolve in the way predicted in economic models but, even so, it is a useful discipline for model-builders to have their models checked against the subsequent outcomes. Perhaps some additional consideration of this aspect would have been useful, though clearly time was limited.

The second point is that in the paper there is an implied, rather than explicit, discussion of the entirely different situation in planned and in market economies. In practice there could be a valid distinction in a horizontal and a vertical plane – between market and planned economies (with various degrees of combination of these two different economic systems within a spectrum covering the whole range of actual economic systems in different countries). At the same time it would be desirable to distinguish according to the sheer size – mainly in terms of area – of different countries. Regional and inter-regional models in very large countries, such as the U.S.A. and USSR, are of a very different character from those which would apply in small countries – such as, for example, in Ireland. If a common methodology and uniform data were available it would be possible to consider regional and inter-regional planning models for a number of countries – and perhaps for a whole continent – but in practice attention is focused primarily on models applying within national frontiers, and these cover countries of very widely differing size.

Finally, a more general comment; the problems of the inter-relationship between regional and national models of the development of agriculture and of the economy as a whole are of critical importance, and these are generally not considered in the same detail as the models for the individual sectors. This is certainly true for my own country and my impression from this paper is that this is true in many other countries as well. Yet if we are to see the maximum impact of the work of agricultural economists in the decision-making process this aspect needs to be given very full attention.

DISCUSSION OPENING – Claude Baillet, *Offices EEC Commission*

In opening the discussion on Dr. Sebestyen's paper I shall confine myself to a few general comments on certain aspects connected with the novelty, diversity, limitations and usefulness of regional models for agriculture.

(a) Judging from the literature to which the speaker referred, drawing-up regional and inter-regional models for agriculture would seem to be a relatively new departure by agricultural economists. The approaches and methods are many and various; even the objectives are different and in course of evolution from one country to another and even from one region to another; that is a feature of developing disciplines.

(b) These regional models all have one thing in common, however; they constitute the inevitable link and an essential correlating factor between the micro-economic models conceived at farm level and the macro-economic models constructed at the national level. In this respect, they are an indispensable complement to the planner's set of tools, whatever the type of planning envisaged.

(c) Regional and inter-regional agricultural models are becoming more important both in planned economies in which they constitute a privileged instrument and in market economies in which they represent an essential basis for forecasting. They are also a valuable tool to the regional authorities, especially where the latter are seeking more far-reaching responsibility. In this connection they could be a significant factor in greater regional independence in a system of inter-dependent regions which have a greater awareness and are better organised. They could in this way play a decisive role in achieving more democratic management of the economy and thus help to reinforce the feeling of solidarity between the various economic sectors of a nation.

(d) The development of regional agricultural models encounters one basic problem, however – the frequent absence of adequate statistical data. Despite the considerable efforts made in many countires, there are large gaps in most cases; these gaps discourage, or even thwart, attempts to construct such models.

The complexity and troublesomeness of the calculations which was formerly a major obstacle to the construction of models in general, and agricultural models in particular, has been solved by the prodigious, universal development of data processing. It is now the lack of statistical base which is the main limiting factor.

(e) Regional agricultural models also presuppose an explicit and quantified formulation of the general and specific agricultural policies and an accurate knowledge of agricultural prices at the regional level. This is probably one of the reasons why they are more highly developed in countries with planned economies than in those with free market economies.

(f) The conditions required for the correct operation of regional agricultural models still apply, of course, when the notion of region is taken beyond national frontiers. At the international level, however, these conditions are even more difficult to create than within one country. None of the existing international economic unions, even the advanced ones, seems to have produced any real agricultural development models so far or even to have attempted or managed to integrate models produced for the individual member countries. At the most they have produced very general forecasts. This is due to the fact that the practical use of such models depends on the unity of purpose and policy within the regional group of countries concerned. Such models are fully viable only when the States concerned belong to a federation. Any other form of association or community with looser political ties makes for regional agricultural models of a purely academic significance.

The above considerations show that it is difficult to assess the effectiveness of a regional agricultural model except in the general context in which it was drawn up and specific objectives laid down for it. The general context and the objectives vary from one model to another even if the economic function of such models is similar as a rule (minimization of costs and maximization of profits).

The close reasoning which regional agricultural models imply, the internal consistency which they presuppose, the co-ordination of information they require, the retrospective factual verifications to which they lend themselves are factors which make regional agricultural model-making an interesting exercise for agricultural economists and one which is likely to be developed in the future in all economic systems. Agriculture would thereby become better integrated in the economy, the management of agricultural policy better balanced and the work of agricultural economists more effective. Although there may be no doubt as to the merit and usefulness of such models, the fact that more use is not made of them would seem to be due to lack of information and to the constraints involved in establishing and exploiting them. It would be interesting to hear Dr Sebestyen's opinion on the matter.

RAPPORTEUR'S REPORT – T. Matzugi, *Japan*

The discussion began by focusing on a further examination of the practical results of model-building, involving the question of their effectiveness and validity, whereas the paper deals more with the technical aspects of model-building. The need to create a framework of interregional models with horizontal and vertical differentiation with respect to different conditions in planned and market economies, small and large countries, etc., was explored. In response to the call for more participation of lower stages an example was given from Ireland, where no satisfactory solution had been found so far to the question whether top-down or bottom-up planning is to be preferred.

The view was expressed that the plurality of models built so far is not the result of theoretical exercises but corresponds to the variety of situations such as, for example, state of development, different political systems, different availability of statistics on regional levels. The lack of regional statistics, which also trace the interregional flow of factors and goods, and data on prices, are, so far, the main bottlenecks to achieving validity and efficacy of regional models.

The principles of socialist planning were explained in further discussion. The core of this explanation was that, in socialist planning, science and computer technology are applied to build a unified, integrated, management system in the field of food in order to transform inputs (that is, regional resources) into outputs in line with regional and national demand functions.

Further discussion developed the range of issues further. It seemed to be generally agreed that the high cost of collecting data, and its specialised manpower needs, call for a less sophisticated approach in regional model-building; that so far there has been an unnecessarily theoretical approach to the question due to the intellectual fascination of models; the different chances of participation by different levels of the population in goal determination under conditions of disaggregation and devolution of the society in question; the existence of different criteria for determination of regions; the role and the influence of the model-builder in the process. He is often requested to introduce other than profit maximisation functions, and preferably multiple objective functions related to the structure of the economy and its level of development, in his regional planning models.

In his final comment on the issues raised during the whole session the author extensively dealt with the practicability of existing models, which he thinks is a *sine qua non* of the success of socialist planning. This is the reason why much emphasis has been placed on the further adaptation of models to reality, for example, by the incorporation of several variables into the system and the development of both systematic and partial models. However, the possibility of utilizing the models developed for socialist planning are not restricted to that political environment, but rather depend on the objective functions and the content of variables involved. Consequently different planning targets – for example, even distribution of labour, meeting the demand of food in all regions equally, etc., – which determine the allocation of activities within the country, as well as different levels of planning, determine the structure of the model to be applied.

In order to overcome the lack of data needed to feed the models which imply new systems of planning, the author recommended carrying out sample surveys in order to limit the cost of providing new sets of data. While determining the size of such samples one should keep in mind that a substantial degree of aggregation precludes the tracing of micro phenomena.

One crucial factor in planning is the difficulty of price forecasting. Many price reaction models have been developed, such as the Cobweb theorem and simulation models, but their application is still unsatisfactory. Price forecasts are equally relevant for socialist planning; however, the price determinants are different, since prices have to be fixed centrally. Such price fixation has to anticipate price reactions and the cost of subsequent transformation. As such, the projection of price-induced changes is limited by the scarcity of funds available for computation. Moreover, in a politically indivisible plan all consequences of substitution have to be forecast and weighed in a system of multiple objectives represented by individuals and political bodies, which involves a maximization problem.

As to the incorporation of agriculture in an overall plan, there are possibilities of taking regional balances as a link – for example, balances of the labour force, financial requirements of industries, social and cultural facilities – and integrating them into national balances. However, in spite of the relevance of finding solutions for such technical problems, it remains a fact that social and political aspects deserve much more attention if planning is to be successful.

HUGH V. WALKER

*Models for Decision-making in Agricultural Marketing**

This is an age when scientific methodology and technology have achieved a number of towering successes after centuries of uneven yet persistent development. For example, man's safe journey to and from the moon bears valid testimony to his conquest of outer space. Few would disagree that this remarkable achievement epitomizes the totality of science that has so far been evolved by man. Already, in many branches of science, concepts and tools such as the mathematical model, information feed back, simulation, decision theory, stochastic process and the digital computer have come into common usage. The computer is now standard equipment in many institutions, where masses of data are to be processed. Agricultural economists have provided leadership in the application and use of these concepts and tools in quantitative research.

This session of the Conference focuses on the function of models for decision-making in agriculture. This paper attempts an appraisal of models for decision-making in agricultural marketing. It will offer some reflections on the results of research using these models, and will evaluate the models used at the different levels (i.e., micro, macro, regional and interregional interdependencies). This paper will also consider the need for coordination of the different model approaches, and the role of these models in the formulation of agricultural marketing policy.

1. CONCEPT OF AND PROBLEMS INVOLVED IN THE DECISION-MAKING PROCESS

A pre-requisite to appraising the models for decision-making in agricultural marketing would involve formulating some concept of the decision-making

† Executive Director, National Farm Products Marketing Council, Ottawa, Ontario, Canada.

* The author wishes to acknowledge the useful comments and suggestions of Dr. W. J. Anderson, Associate Director General, Agriculture Canada. These were helpful in the final preparation of this paper.

process, and an appreciation of the specific kinds of problems which the decision-maker needs to resolve in agricultural marketing. The literature on this subject abounds neither with clues about the "hard kernels" of the details of the decision-making process, nor with criteria used in balancing, comparing or choosing among alternative courses of action. Yet, these constitute the very core of the decision-making process. More attention needs to be focused on this key link in the chain of knowledge. In-depth analytical studies of the decision-making process either by participants actually involved in the process, and/or by adequately trained perceptive outsiders would make a definite contribution to knowledge. Much of the published material tends to gloss over, or to ignore the difficult decisions that have to be made in the development of programs and policies for agricultural marketing. Difficulties arise because decisions are normally made in an environment that is shrouded in uncertainties.

Nowhere does this problem of uncertainty seem to be more true than within the complex world of agricultural marketing. Few, who are intimately involved in this discipline to the extent that I am, would seriously question the assertion that the need for improved techniques of agricultural decision-making is already great, and shows signs of increasing. In fact, the growth of the marketing research and planning function attests to the concern for more systematic and better informed approaches to the problems of agricultural marketing.

No longer does the decision-making process merely involve a "collection of projects", or an investment program for the agricultural sector. It now involves the development of more integrated policies and programs of resource allocation to achieve clearly defined goals. This paper will indicate that over the past three decades, a body of tools, techniques and analytical models, that have been developed outside the field of agricultural marketing, shows promise of providing a framework within the functions of market planning and research can more closely be linked to the central decision-making process.

The development of effective public policies and programs to achieve certain strategic goals is a complex process involving a number of difficult decisions. For example, improving farm incomes and prices, reducing economic instability in agriculture, providing farmers with a greater measure of economic security, improving the bargaining position of farmers in the market place, providing the farm sector with a greater relative share of the national prosperity, and shifting more in farmers' favour the terms at which agricultural products are traded in the market place – are all typical issues facing most countries today. The decision-making process adopted by public and private planners to develop programs and policies to have a significant effect on these areas is complex.

What is the decision-making process? What are the fundamental steps in this process? What criteria are used in making a rational choice among alternatives? Petit, in his paper presented earlier, has suggested that "any decision process is a link between thinking and doing, between reflection and action".[1] Simon has also indicated that the decision-making process involves the following fundamental steps, viz: firstly, *problem finding*, which entails searching the environment for conditions calling for decisions to be made and new policies developed; secondly, *design*, that is inventing, developing and analyzing alternative courses of action; and thirdly, *choice*, which entails the selection of a particular course of action from among available alternatives.[2] In practice, a number of criteria may be used in the selection of a choice. The economic criteria may involve "cost-benefit" analyses of the alternative courses of action. This decision-making process presupposes that there is a clear statement of strategic objectives, which serve as criteria for balancing and evaluating alternative courses of action. However, the development of effective policies and programs to achieve these clearly defined strategic objectives is a complex and difficult task, which involves making critical decisions on problems as they emerge in the process.

Decision-makers are faced with an array of problems on which decisions must be made in the process of formulating and implementing policies and programs designed to affect agricultural marketing. The problems differ as between commodities, programs, sectors, institutions, regions and countries. For example, Canada has adopted a supply management approach to the problems of economic instability and economic insecurity in the poultry industry. In the case of eggs, decision-makers, who are responsible for the administration of the egg program, are confronted with a number of difficult problems on both the demand and supply sides of the market.

On the demand side, eggs have revealed three important characteristics, viz: firstly, a declining per capita consumption for table eggs, secondly, inelastic demand for table eggs, but elastic demand for processing eggs; and decline in consumption as disposable incomes rise. In full awareness of these characteristics, decision-makers are confronted with the problems of establishing the national quota, its apportionment among the provinces, and then its distribution among Canada's 5000 egg producers who are registered under the National Supply Management Program. In setting the national quota, decision-makers have to resolve the problem of determining the requirements of the components of the national market, viz: requirements for table stock, processing, hatching and export.

On the supply side, decision-makers have to resolve the problem of stipulating the number of layers, based on some acceptable rate of lay, to produce the national quota. In addition thereto, decision-makers must take into specific account imports and the production of the large number of unregistered producers who are exempt from the National Supply Management Program in order to avoid unmanageable surpluses. Moreover, decision-makers are faced with the additional problem of establishing producer prices for eggs at a level that would allow producers to recover their production costs, take into account the interests of consumers, and at the same time, not induce larger than normal in-flows of eggs from abroad.

2. CONCEPT AND ROLE OF ANALYTICAL MODELS

The literature on economics abounds with references about reality. Yet, there exists no universally acceptable proposition of what constitutes reality. Is it a mythical "will-o-the-wisp" concept that man forever pursues, yet never fully attains? Is there any other reality than man's concept, his perceptions of reality? Models are man's perceptions, his representations of reality. In a sense, they are his perceptions of himself in relation to his environment. They do not constitute a retreat from reality, but "stepping-stones" to reality. Hence, models – formally or informally – are an important part of, and play a crucial role in all decision-making processes. In fact, formally constructed models perform the function of refining these processes, thereby making them more systematic, and more scientific. Models enable man to think systematically and logically about reality. Therefore, the practice of building abstract models sharpens identification and analysis of problems. In themselves, such models aid in improving the logic of the decision-making process. The "acid test" arises when the model-builder attempts to convert an abstract model into an operational form by fitting it to the real world, thereby removing the model-builder from his "Ivory Tower". The link in this regard is to be found in observation and measurement. Models serve to apply the measuring instruments to those dimensions that are relevant in this exercise. They force both the builder and the decision-maker to set out their objectives more precisely, and so sharpen the analytical tools.

Models have often been evaluated either in terms of the reasonableness of their approximations to reality, or in terms of their predictive capability. However, the real purpose in building models goes well beyond these attributes. In fact, "the reasonableness of a model's approximation to reality may indeed be merely an academic quality, that is of little relevance to the pragmatic decision-maker". Furthermore, prediction is only a means to an end. The ultimate purpose of models is to aid decision-makers in answering the complex problems that emerge in the planning or policy formulation processes, thereby influencing the policies and plans of decision-makers in a manner that makes the outcome of decisions fall more in tune with the overall goals of society. Hence, models and model-building are not ends in themselves, and should not be viewed as such. They are means to achieve other ends. That is to say, they are tools or aids to decision-makers. As such, they provide inputs or aids to decision-makers in dealing with the complex problems encountered in the decision-making process. Models should therefore be problem oriented rather than be a mere academic exercise. Their real test relates neither to the reasonableness or their approximation to reality nor to their predictive capability, but to their appropriateness, relevance and capability in aiding human judgement in the decision-making process. Models are no substitutes for human judgement. They should, therefore, solve tomorrow's problems today, rather than resolve yesterday's problems tomorrow.

Thus, models assume a particularly crucial role in the decision-making process. It would seem appropriate, therefore, that the interrelations between the model-builder, the model and the decision-maker should be reviewed. The model-builder should ensure that the decision-maker is involved as intimately as possible in the model-building process, if the latter's confidence in the capability of the model is to be gained. The decision-maker should be involved in the conceptualization, formalization and specifications of the model, in the identification of the assumptions upon which the model rests, as well as in the review of the limitations of the model. There needs to be a meeting of minds in these areas. This may be accomplished in a variety of ways, the most effective of which could be a matter of mutual arrangement between the model-builder and the decision-maker. Moreover, the model-builder must be credible, that is to say, he must be able to communicate with the decision-maker.

However, there are many reasons, particularly from government's point of view, as to why models of systems such as the food system or the marketing system are needed. Such models aid comprehension of how the system operates, and assist in the identification, behaviour and interrelationships of variables within the system. They assist in sorting out significant variables in the system, thereby facilitating greater efficiency in studying and managing the entire system or any component thereof. In addition, models are useful in projecting the behaviour of the system, and in the identification and evaluation of alternative solutions. They are useful in the continuous monitoring of the system so as to identify changes that are necessary or that are likely to occur.

It does seem that some of the skepticism which decision-makers have revealed with respect to models arise from the fact, not that decision-makers are dull of comprehension, but that model-builders have isolated themselves from decision-makers, who ultimately bear responsibility for the decisions made. It is in the interest of acceptability of their models, that model-builders should ensure that decision-makers are more intimately involved in the model-building process. Decision-makers are shrewd and selective in their choice and judgement. Why should they not be? It is they who ultimately bear responsibility for the decision made.

3. PROBLEMS OF MODEL CONSTRUCTION

The practice of building models is an integral part of the scientific method in agricultural economics. Scientific classifications in the natural sciences have been found to be particularly useful, because they provide the basis for predicting the essential elements of behaviour. Human society, however, is the exception, yet it is with this behaviour that marketing is concerned. The needs and responses of human behaviour are neither uniform nor constant. In fact, if invariance is characteristic of the physical world, certainly variance is characteristic of the human world. The variability of human behaviour has often been used to defend inaction in the mathematical treatment of economic and social problems. The usual defence has been that human behaviour is not predictable.

The "new physics" have shown that the behaviour of atomic sub-structures are also individually unpredictable, but that the behaviour of masses of such particles are predictable by means of stochastic process methods. Does this notion not have some relevance to human behaviour? Certainly, what a single individual will do with his disposable income is not predictable. However, the prospects of predicting mass spending or mass savings seem pretty good. In any valid comparison of the problems encountered in the natural sciences and the social sciences, three important characteristics of the social sciences should be borne in mind. These are: the relatively many factors to be dealt with, their high variability, and their statistical instability.

Methodological developments in the natural sciences have had important implications for the social sciences in that they offer a methodology to smooth the path of the science of human affairs. They do nothing, however, to change the "state of turbulence" and variability of these affairs. Moreover, in the perspective of time, the physical world is a static world. This simplifies the accumulation of knowledge – a factor that is important in the development of a science. The variability and instability of the parameters of the human world are restraining factors that interfere with the development of empirical social science in that measurements become obsolete and inaccurate even though relevant for the analysis of the dynamic system. Furthermore, in the natural sciences the degree of accuracy of measurement is affordable, since the investment necessary for obtaining it could be spread as a social cost over all of perpetuity. The fact that measurement in the social sciences is perishable inversely effects the investment that can economically be made to obtain precision. Even if affordable, there could be no absolute precision of measurement.

However, agricultural economists need not be unduly overwhelmed by the towering successes achieved in the physical scientists' deterministic world. Current successes in the physical sciences have been achieved after centuries of uneven but persistent forward development. The foundations for this development were first laid by Claudius Ptolemy[3] in the second (2nd) century, and also by Nicholas Copernicus[4] in the sixteenth (16th) century. It may be noted that there were fourteen (14) centuries between Ptolemy and Copernicus, following whom new developments came in relatively rapid succession through the contributions of Johannes Kepler, Galileo, Isaac Newton, and more recently Albert Einstein.[5] It was through the cumulative efforts of these scientists that we have today a general model of the universe, which explains more satisfactorily the behaviour of cosmic phenomena. Surely, it is only time that will tell whether Einstein is to have the final say in the development of a grand model of the physical universe.

In contrast, economics had its birth as a science with the publication in 1776 of Adam Smith's "The Wealth of Nations". In the development of the science of economics, many lessons could be learnt from the experiences in the physical sciences. An important one is that the "coupling of models and measurement is a continuing process in science. Observation leads to the improvement or rejection of models and the progressive leap-frogging of theory and measurement goes on".[6] Moreover, although immediately inappropriate the methodology, a significant proportion of the technology and the attitudes of the physical sciences are transferable to the science of economics.

Marketing economists now need to focus on the uniqueness of economic and social phenomena, and so re-orient their models as to take into account the multi-dimensional aspect of human behaviour. A reason that some economic models have not received the accolades expected lies in the fact that their builders attempt to solve yesterday's problems tomorrow rather than resolve tomorrow's problems today. To be relevant, economic models should anticipate rather than follow problem development.[7] They should be more prescriptive and less descriptive.

4. PROSPECTS IN AGRICULTURAL MARKETING

The outlook for further progress in agricultural marketing must, therefore, be seen in the light of the model-builder's appreciation of the complex problems encountered in the decision-making process, the availability of tools for measurement, the complexity of the system that model-builders attempt to construct, and the availability of data to be used in making measurements. The promise of science hovering over agricultural marketing has certainly been illuminated in recent times by the advent of the electronic computer during the past decade. Concurrent with this invention, changes have also taken place in the physical structure of its marketing management, in the theories of economic activity, as well as the method of accumulating, organizing, processing, and utilizing information. The electronic computer has the incredible capability to extend and enlarge man's mental processes. Its future greatest contribution is likely to be in simulating systems or processes for scientific and experimental purposes.

In the construction and development of models intended to serve as decision-making aids in agricultural marketing, economists are still confronted with a number of conceptual, analytical and operational problems. Some of the more difficult of these involve those of perception, specification and measurement of reality; validation of models, procurement of reliable and up-to-date information to be used in measurements; perception and construction of relationships that depict the real world; the versatility of models as aids to the decision-making process, and the more intimate involvement of decision-makers in the model-building process.

Agricultural economists have used several classifications to categorize the various types of models which have been developed and used in agricultural marketing problems. Some authorities have distinguished between positive and normative models. Another classification is based on criteria relating to the mathematical model, or to the source of data used in estimating their parameters. However, in this paper three classifications will be used so as to facilitate presentation. These are: the analytical models based on statistical inference, mathematical programming models and simulation models. In this connection, I should like to point out that the research studies already completed, or underway in many world institutions and countries, are impressive. They bear unquestionable testimony to the dedication of agricultural economists in resolving a wide array of man's economic problems. Unfortunately, in view of the time constraint, it will only be possible to touch on a limited number of studies. This is, in no way, an attempt to attach labels of importance to particular studies, or to scale down the relative importance of others. Studies are cited only insofar as they illustrate a point being made.

5. ANALYTICAL MODELS BASED ON STATISTICAL INFERENCE

Many problems in agricultural economics deal with universes that are subject to human influence, and so change at least gradually over time. Reference has already been made to the particularly difficult problem that confronts decision-makers in Canada's supply management program in their attempt to maintain market balance in the poultry industry. A number of these problems involves estimation of the demand for and the supply of poultry products. An analytical tool, that is useful in providing an input into resolving such problems, is regression analysis.

Regression analysis, which provides a tool for testing empirically some hypotheses of classical economic theory, was the first type of alternative analytical technique that appeared in quantitative research in agricultural economics (1920's). Firmly based upon the relevant variables affecting market supply and demand as stipulated in the classical economic model, the application of the technique has focused, among other things, on problems of developing supply response functions, and of estimating demand relationships for a number of farm commodities using time series or cross-sectional data. The technique involves the development of an economic model, which is the theoretical stage; acquisition of adequate data; and the marriage of economic theory to the available data through the use of appropriate mathematical and statistical tools. This synthesis of economic theory, data, mathematical and statistical tools provides the analytical model, which is the operational version of the economic model.

In cases where the concern is with linear phenomena, adequately specified regression models are fairly realistic and computations are also easily handled, especially with the aid of the electronic computer. The performance of the statistical tests associated with the least-squares method are valid on the assumption that the samples are drawn from a normal population. It also gives good results even when the parent population is not normal.

5.1. *Time series models*

Historically, supply analysis has been based on single equation regression models using time series data. There have been a number of minor refinements over time in the application of such models. For example, Murray[8] used a lagged deflated price in estimating pig supply. Together with Cohen,[9] he used acreage planted in the United Kingdom to indicate the intended supply of wheat. Johnson[10] showed that in some circumstances, supply might be determined by external factors such as weather, rather than farmers' response to price. He also applied a number of statistical techniques to supply analysis: e.g., confluence analysis and first differences.

Some advancements have also been achieved over the simple correlation of supply with price (lagged). More varied developments in the use of time series data have been made in recent years, particularly in the United States. These have focused on the various deficiencies of simple time series models. However, they have been sporadic rather than cumulative in nature. First was the important development of simultaneous equation estimation model, whose importance lies in the fact that with a single equation model, in which one or more of the explanatory variables cannot be considered endogenous, the parameter estimates will be biased.

A single equation model is not appropriate when one hypothesizes that current supply of a product is a function of the current price of that product. In such a case, price is an endogenous variable, and a two equation model is required in order to obtain biased-free estimates. Girschick and Haavelmo[11] used this technique in determining the structural parameters of a system including a demand and supply equation for food in aggregate. This was followed by Foote[12] who developed in 1953 a more detailed model of the U.S. feed-livestock economy; and later by Cromarty's comprehensive model[13] of the agricultural sector in 1959.

The usefulness of simultaneous equation estimation models should be considered in the light of the fact that (1) parameter estimates from least-squares single equation models have generally been found to be very similar to estimates from simultaneous systems, and (2) within the agricultural sector, supply cannot readily be adjusted in the short-run period. Consequently, demand and supply responses are readily identifiable separately.

A more recent development in time series analysis may be ascribed to Nerlove[14] whose innovation, based on the much earlier work of Fisher and Koyck, is that of distributed lag models. This type of model focuses on the dynamic problems of supply. Other innovations include Griliches' attempt to derive supply relations from estimated demand for inputs – given a technical relationship, the inputs indicate intended supply in response to relevant stimuli; Suits and Kiozumi's[15] use of "dummy variables" in regression models for handling non-quantifiable factors such as policy, and the explicit inclusion of weather variables by Candler[16] and Johnson.[17]

Projection of the demand for food is one area of investigation where there is an immense and competent volume of theory and empirical research that is relevant to the planning and decision-making process in agriculture, particularly in the low-income countries. Indication of its usefulness can be gained from a review of the Food and Agricultural Organization bibliography.[18] The extent of interest is evident from the number of surveys cited by Houthakker,[19] and by the number of agencies in different countries that have been working on this problem as documented in the FAO list.[20]

In spite of the extensive theory and empirical research on demand projection, the scanty information available in case studies and plan documents indicates that to date agricultural planning in many low-income countries is still based on rather simple projections of recent trends or on simple expansions based on population growth. Although a significant attempt to use demand projections was made in the Economic Commission for Latin America (ECLA) Studies, the projections, however, have had little real impact on national plans.[21]

Projection of export demand is a critical and difficult problem for agricultural planning. It is one for which there is only an inadequate solution. This is certainly one area where international agencies and institutions have taken a great interest with substantial results, and can continue to make a special contribution. The most significant result to date from the point of view of agricultural planning and decision-making are the FAO commodity projections.

These represent major steps in the right direction. However, a serious limitation is that the projections are simply for high, medium or low estimates with no particular attention being given to price trends. This limits the direct usefulness to agricultural planning since, in projecting the contributions of the export sector to real national income and foreign exchange, exports should be measured in the prices expected to prevail during the projection period.

5.2. *Cross-section models*

Analysis of market supply need not be confined exclusively to time series data. Their combination with cross-section data has been reviewed by Hildreth.[22] Holme[23] developed a supply function for milk using survey data. However, the bulk of research on analysis of supply has, at least until recent years, been based on time series data at the market level.

Schaller and Dean[24] tested the recursive programming model against conventional regression analysis of time series data. They observed that regression analysis would provide better estimates during periods of smaller changes and relatively stable structure, and that the recursive programming would be more effective for prediction purposes under situations of sharp structural changes in technological, managerial skills, etc. However, conclusions based on predictive tests indicated that neither model does a particularly outstanding job of real prediction as opposed to explanation, though the regression analysis would still appear slightly superior.

A regression model may primarily be useful for short-run forecasting. The use of such models in forecasting calls for a statement of desired results in terms of a range of probabilities rather than in terms of exact predictions. Thus, the concept of statistical control makes the decision-maker more conscious of the limitations of his value judgement. An important aspect of the application of such models is concerned with the level of aggregation at which they may be used. For example, most regression models of supply response have so far been developed for individual commodities, thus providing little basis for inference about aggregate agricultural output.

In many respects, regression analysis would appear more directly useful for predicting farmers' response than conventional linear programming. Because regression results are based on actual past changes in production, they are more likely to take into account farmers' likes and dislikes and other considerations which are omitted in the usual programming model. Further advantages of regression are: (a) the relative accessibility of aggregate data compared with the difficulty of obtaining more detailed input, output and resource data required in linear programming; (b) relatively low cost and quick aggregative results; (c) the ability, given the satisfaction of certain statistical assumptions underlying the model, to make probability and confidence statements around the results.

6. MATHEMATICAL PROGRAMMING MODELS USED IN AGRICULTURE

Decision-makers are often confronted with problems of making decisions on questions of the most rational organization and use of resources in agriculture. The problems may also involve: forecasting the optimal organization of agriculture in some future time period; forecasting the effects of technological change on the optimum production pattern for agriculture; determining the effect on agriculture of changes in the domestic and export requirements for agricultural products; and determining the effect of different agricultural policies and programs on the optimal organization of agricultural production. Linear programming models of agriculture are useful aids to decision-makers in resolving these problems.

Following World War II, new methods of analysis have been developed that depend in large measure on the linear characteristics of economic problems, and which focus on the optimizing problem that is familiar to the science of economics. The most familiar of these are: game theory, input-output analysis and linear programming. Although John von Neumann enunciated the main theorem of game theory in 1928,[25] it was not until 1944 that its implications for economics were revealed.

Von Neumann's achievement lies in his demonstrating that something definite can be said about the array of cross-purposes and psychological interactions characteristic of human affairs. He demonstrated that under certain assumptions, each participant can act so as to be guaranteed at least a certain minimum gain (or maximum loss). Thus, each participant prevents his opponents from attaining any more than their minimum guaranteeable gains. The implications of this theory for economics are that it holds out hope of banishing oligopolistic indeterminacy where such economic situations exist.

Input–output analysis was the second of the three branches of linear economics to appear. Leontief[26] published the final clear statement of the method in 1936. The technique is based on the idea that a considerable proportion of the effort of a modern economy is devoted to the production of intermediate goods, and that the output of intermediate goods is closely linked to the output of any final products. A change in the output of any final product implies a change in the outputs of intermediate goods used in producing that final product and indeed, in producing goods used in producing that final product and so on.

In its original version, input–output analysis dealt with an entirely closed economic system – one in which all good were intermediate goods. Equilibrium in a system exists when output of the various products are in balance in the sense that just enough of each is produced to meet input requirements of all the others. The specifications of this balance and its pricing implications was Leontiff's first objective. Beginning with World War II, interest has shifted to a different view of the Leontief model. In this view, final demand is regarded as exogenously determined, and input–output analysis is used to find levels of activity in the various sectors of the economy consistent with the specified pattern of final demand. For example, Cornfield, Evans and

Hoffenbay[27] have computed employment levels in the various sectors and hence total employment consequent upon a presumed pattern of final demand. Leontief has also estimated the extent to which fluctuations in foreign trade influenced activity of the various domestic sectors.

Difficulties with studies based on input–output models relate to linearity in the technical coefficients. Gittinger[28] has suggested that although these models can be modified to minimize some of these difficulties, such as assuming a lag linear function, such devices do not really overcome the difficulties. He has pointed out that this type of planning model does not cope well with such problems as: the form that technological changes will assume, how these are to be incorporated in the models, and the institutional constraints on agricultural development, etc. Yet, these are precisely the kinds of considerations with which decision-makers are concerned.

In this regard, Bishop[29] has contended that the fixed coefficient assumption is untenable in agriculture because of enterprise combination flexibility on farms, the new technology available to farmers for producing the same products, and the problem of capital formation, where input-output theory assumes capital formation to occur in a sector other than the one where it is used. He concluded, however, that "input–output work is unrealistic concerning the conditions of growth in agriculture itself, but it may be helpful in estimating the availability of commodities entering into agricultural production and for commodities produced by agriculture which are generated in other sectors of the economy".[30] In addition, Seers has suggested that he has found input–output models useful as a "working sheet for gathering together statistical information", and as a device for helping "the economist gain insight into the working of the economy, especially its future potentialities".[31]

7. LINEAR PROGRAMMING MODELS

The last of the three branches of linear economics to originate was linear programming. It was developed by George B. Dantzig in 1947 as a technique for planning the diversified activities for the U.S. Air Force. Application of the linear programming model to economic problems in agriculture appeared in the 1950's. The model involves a computational method to determine the best plan or course of action from among several possible alternatives when a specific or numerical objective exists for it and the resources available are limited.

Linear programming is only one of several techniques available among the general set known as mathematical programming. The other techniques in this family are: non-linear programming, and dynamic programming. In special situations and under given conditions, techniques of recursive, stochastic and parametric programming or other variants thereof have also been developed. The technique has experienced vigorous growth since the conclusion of World War II, and has widely been used in agriculture, transportation and other industries. In agriculture, it has been used as aids in problems relating to: shipping of a product from several points of supply to various destinations; optimal marketing facilities; solution of profit-minimizing product flows;

interregional and national planning and optimal organization of agriculture; and models for the economic development of agriculture and other economic studies.

Pioneering work with the application of these models under the direction of Professor E. O. Heady at Iowa State has been underway on a progressive basis over a period of some twenty (20) years.[32] The initial models were simplified representations of U.S. agriculture, in which some rather gross assumptions were adopted primarily because of the limited computational facilities available in the mid-1950's, and the associated small size of the analytical models adopted. The progressive development of the Iowa State models could be traced from: (1) the Egbert–Heady model, (2) the Egbert–Heady–Brokken model, (3) the Heady–Whittlessey–Skold model, (4) the Brokken–Heady model and the Eyvindson–Heady model.

As is normal in the development of scientific methodology, the realism of these models grew and matured with time. Their development was enhanced by the electronic computer, more refined data and greater experience and facility in model-building by research workers.

These linear models used in forecasting, in policy development and evaluation relate to: (1) forecasting the optimal organization of agriculture in some future time period; (2) forecasting the effects of technological change on the optimal production pattern for agriculture; (3) determining the effect on agriculture of changes in the domestic and export requirements for agricultural products; and (4) determining the effect of different agricultural policies on the optimal organization of agricultural production.

The real test for these models lies in the extent to which the production patterns generated by the model approximate those that have occurred in reality. Although, no model can be expected to duplicate reality, it should at least be possible to explain most of the differences between what has, in fact, occurred and the results generated by the model. In general, the Iowa State models, particularly those that were earlier and less detailed, yielded production patterns that were fairly reasonable. The results generated by the earlier models indicated sufficient reasonableness to justify some confidence being placed in the results obtained when the models were used to project the effects of changes in economic conditions. Three main reasons explain the differences between the actual production and that generated in the optimal solution, viz: (a) the fact that the pattern generated by the model is an optimal solution which cannot be duplicated in reality because of the various restrictions which are not, and in some cases could not, have been incorporated in the model; (b) other limitations or errors in the model and data; (c) differences between actual conditions and those postulated in the model.

The models contain two types of limitations, viz: firstly, those that are inherent in all models of this type; and secondly, those that are specific to a particular model. The latter is usually the result of compromises (or errors) that were made in formulating the model and/or in collecting the data. Since most factors of production are considered variable, the optimal production patterns computed in the Iowa State models are valid only for the long-run. This limitation was especially true in the earlier models in which land is the only restraint on production. However, even in the latest model which includes labour and capital restraints, there was no differentiation between different types of capital and labour so that, in effect, the model assumes that one type of labour can readily be substituted for another. This may be possible in the long-run, but such is not possible in the short. As such, the Iowa models can say little about short-run changes resulting from changes in economic conditions or policies. This limitation may be remedied by including more realistic restraints in the model; but it may be more advantageous to use an alternative approach when short-run implications are of more interest than those of a longer-run nature.

Another limitation in these models is the timelessness of the analysis. That is to say, the models can identify changes in the optimal production pattern associated with changes in economic conditions, but they tell us nothing about the time required for this adjustment, or about the path that this adjustment might assume. In organizational planning in agriculture, information on the time requirement for adjustment and the path of such adjustment may at least be as valuable as information about the resulting optimal production pattern. A pitfall in these models, as is the case with all linear programming models, is the inherent assumption of constant input–output coefficients over the entire range of output possible for a producing unit. It is unlikely that this assumption is exactly satisfied in reality.

Another difficulty with these models relates to the problem of "aggregation bias"[33] which arises from the fact that all production units within either a given size group or a region are treated as a single production unit.

A final major limitation in the models is that requirement levels must be taken as given. That is to say, the models determine the production pattern that will satisfy a given level of product requirements at minimum costs or maximum returns. However, they do not allow for any inter-action between supply and demand. In this connection, Heady, Hall and Plessner[34] have developed "quadratic programming" models similar to linear programming models that incorporate demand functions thereby allowing demand and supply to be determined simultaneously.

A special case of linear programming that is of particular interest to agricultural marketing problems involves the application of the transportation model. This is appropriate to the determination of optimum shipping routes in the distribution of a product from several sources of supply to numerous destinations. This model has been found to be very useful in determining the least-cost routes in shipping feed grains from some sixteen (16) sources of supplies to eleven hundred (1100) destinations in Eastern Canada following the close of the St. Lawrence Seaway to shipping during the winter.[35] This model has also been found to be useful in several other economic and business management problems. Its appeal lies in its economic implications and its computational simplicity. The transportation model has obvious relevance to problems relating to the principle of comparative advantage and interregional trade.

In summary, the application of linear programming models to economic problems in agriculture have proven to be very useful tools of analysis to industry problems at the micro level, macro level, interregional interdependencies and even at the national level. The principal benefit of these models is also shared by other approaches to national problems in that it allows many segments of the industry and their inter-action to be considered simultaneously, thereby avoiding problems of approaches that consider each sector separately and so ignore the inter-actions between sectors.

An important aspect of the appeal that these models have is that they facilitate testing the effects of changes in conditions on the sector of the economy under review. Moreover, these models have some robustness and flexibility. The latest one developed at Iowa is quite large and contains considerable detail. It seems likely that even larger and more detailed models are possible in the future. It is also evident that smaller models could be developed that may focus on analysis of a specific region, a specific product or a specific problem. For example, Wilson and Wood[36] used a linear programming model in analyzing the effects of Canada's feed freight assistance on the regional distribution of livestock production. Linear programming models have proved to be flexible and powerful tools of economic analysis; and as such, they provide one of the valid approaches to micro problems and in developing regional or national models of agriculture.

8. RECURSIVE PROGRAMMING MODELS

In the attempt to develop rational policies and programs, decision-makers often require short-term quantitative estimates of production and resource adjustment by regions and for the country under alternative prices, subsidies, quotas, exports and resource inputs. Problems to be resolved in the decision-making process involve such considerations as: How will production change in the following year? What changes will occur in the pattern of land use as a result of program or policy changes? How will a reduction in subsidies affect supply of particular commodities? What effect will a change in prices paid to producers have on a supply management program? Recursive programming models are useful aids in dealing with problems involving considerations of supply response.

Recursive programming models are useful in analyzing a variety of problems in agriculture. In terms of solution procedures, recursive programming is similar to linear programming in that both are mathematical techniques employed to optimize a linear objective function subject to linear constraints. However, recursive programming is useful in predicting the actual behaviour of firms, whereas linear programming is designed to estimate optimal behaviour patterns

This special characteristic of the recursive programming model is possible through the use of "flexibility constraints" in addition to the conventional linear programming constraints. The recursive programming model may be used to make either short-run or intermediate-run predictions. In making short-run predictions for only one year ahead, a programming problem may be developed using the preceding year's prices and costs and the current year's flexibility constraints and resource supply constraints.

Research work done in Canada and the United States has demonstrated the flexibility of the recursive programming models. These models have some merits over linear programming in estimating supply response. Linear programming models provide optimal solutions which may be good predictions in the long-run. However, they do not estimate short-run or intermediate-run supply response, nor do they map out the actual process of adjustment. With recursive programming models, risk and uncertainty, norms other than short-run optimal positions and non-economic considerations are recognized through the use of both "flexibility" and "capacity" constraints.

Another important feature of recursive programming is that confidence intervals around forecasts of supply response may be estimated from the standard errors of estimates computed in the regression analysis of the flexibility coefficients. By contrast, in linear programming confidence intervals cannot be so estimated. An additional merit of recursive programming, as compared to linear programming, is that is provides a framework for analyzing the rates of adoption of new technologies and the abandonment of old ones.

Regression models provide positive tools of analysis. However, they have four major limitations relative to recursive programming, viz: (1) in the light of any recent or unexpected sharp structural changes, regression models cannot be used to predict production response because equations are fitted to historical data. However, by changing net income coefficients and/or restraints, recursive programming can be made to estimate the impact of new structures on production response; (2) in regression analysis supply functions are estimated without explicit reference to the technical structure of production, whereas in recursive programming the technical structure is first estimated and the supply function is then derived from it; (3) a simple time trend is introduced in regression models as a proxy for technological changes. In recursive programming technological change is analyzed in a much more precise way; (4) with regression models, statistical problems occur when the number of time series observations are limited, the independent variables are highly correlated. These problems do not occur in recursive programming.

There are two major limitations of recursive programming, viz: (1) research resources required to estimate a supply function are considerably greater for the recursive model than the regression approach; (2) when structural changes occur, the adequacy of the flexibility restraints to product supply response is greatly reduced because they are usually estimates from historical data.

9. SIMULATION MODELS

An alternative type of analytical model which has been acquiring substantial appeal among model-builders and decision-makers is simulation models or management information systems. These models have proven to be useful aids in facilitating integration of our knowledge of an economic system, thereby assisting in resolving complex problems such as some of those that confront decision-makers in Canada's supply management program for eggs as outlined earlier in this paper. Simulation is really the operation of an abstract model or the proto-type of a real system designed to trace out the dynamic inter-actions in order to answer specific questions about the system. Computer simulation is the operation of an abstract model or a real system on a computer. The simulation approach organizes the inter-actions between the parts of an economic system and the dynamic processes into a computer model. The model incorporates the theory of the system's structure and internal relationships. In turn, computer simulation reveals the dynamic characteristics of the system that is described within the computer program. Thus, by changing the guiding policies within the system, one can readily show how the behaviour of the actual system might be modified.

Based on this view of simulation, the methodology for the study of complex systems involves the formulation of a conceptual framework of the system to be followed by the simulation processes. The development of the conceptual framework involves the following: firstly, delineation of the problem, establishment of objectives, hypothesis, etc.; secondly, analyzing the interrelationships of the system and sub-systems; thirdly, the analytical and empirical modelling of the real system in mathematical format, specifying parameters to be estimated and the classification of variables; fourthly, translation of the model of the real system into a computerized system which involves many sub-systems and corresponding linkages between the systems. This also involves the collection of data and their organization for initial input into the system; fifthly, testing the validity of the sub-systems and the entire system in terms of simulators of the real system.

Simulation of the system involves firstly, performing experiments which provide insight into the validity of the hypotheses generated earlier; and secondly, performing additional experiments as inferences from previous results are used to develop new hypotheses to be tested.

Construction of a mathematical model of the real world forces the model-builder to put together the knowledge that he has about the separate components and their interactions. By simulating the model on a computer, the model-builder may reveal the consequences, the contradictions and/or internal inconsistencies and the assumptions of fragmentary knowledge about the system. From such an estimation, a vastly improved understanding of the real system can be gained. If such a model is representative of reality, then the output of the simulation should also indicate the kinds of consequences that are likely to be forthcoming from the real system, if it is operated, in fact, like a simulated model.

Research efforts have been undertaken at a number of institutions and countries, in developing systems simulation to deal with a wide array of problems. For example, a joint effort to generate more information on South Korea's agricultural sector was undertaken between Michigan State University in the United States, and the National Agricultural Economics Research Institute in South Korea. The problem, as envisaged in the Korean Agricultural Sector Model (KASM), involved the major concerns of Korean decision-makers such as: increasing farm income, achieving self-sufficiency in the important food grains; improving agriculture's trade balance and decreasing the contributions made by food prices to the country's inflation. A major problem, which the model-builders face in the development of this model, as well as in the Nigerian and the Venezuelan models, relate to the procurement of data of sufficiently good quality to measure the economic and behavioral variables associated with agriculture. Other limitations of the systems simulation approach are the high initial costs, a long period of model development, and the availability of qualified personnel with adequate background in the economic, technical and social features of the agricultural sector under study.

An important merit of the systems simulation approach is that it is comprehensive in scope, and versatile in its use of other conventional analytical tools and data, thereby providing model-builders and decision-makers with the flexibility of accommodating the diverse components of the system. Hence, the approach provides greater insights into the complexities and functioning of the economic system. Another positive feature of this approach is the scope that it offers for a reasonable degree of cooperation between economists, agronomists, extension workers, soil scientists, animal husbandrymen, farmers and policy-makers in generating data for the model. Certainly, the development of these models has benefited substantially from this interdisciplinary approach to model simulation.

This particular type of model is considered appropriate to the commodity systems approach to Canadian agriculture. Canada is among those countries with current experience in the use of a supply management program as a marketing approach to the low farm-income problem. In fact, Canada has gone the furthest along the road in giving farmers legislative authority to implement supply management programs.

A focus of Canada's national supply management programs is on the poultry sector, where consumption is more seasonal than most agricultural commodities, and where production is cyclical. These factors, coupled with change in consumption patterns, have led to difficulties in balancing supply with demand. In addition, the production of an egg, a broiler or a turkey involves several separate stages. Consequently, they are several key points in the decision-making process relating to the production and disposition of these commodities. The success of this process is dependent upon the level of managerial capability within the industry to coordinate the activities of each stage, particularly in regard to geographical movement and timing. The logical sequence of events in production and distribution may best be understood using a systems' approach.

Current information relating to the production and dispostion of eggs and egg products is acknowledged to be imprecise by most industry participants. Demand characteristics derived are at best indicative. There is a lack of information about demand forces and, to a lesser extent, product flows. The major pre-requisite of a responsive supply management program is a sound understanding of market forces. This understanding provides the means to forecast demand more accurately, and to predict the quantity required at a given location in a given time period. This can be accomplished by determining end-product use and the factors which influence this use.

Once future market needs have been established, and this has proven to be a difficult problem, it is then possible to schedule production to fill these needs. The technical coefficients of egg production are fairly standard, and consequently, hatchery egg and pullet placements can be directed to ensure adequate supply. For the optimum coordination of production activities, market requirements must be established a minimum of six months in advance. This approach also allows for monitoring of production. Since certain placement numbers are necessary to meet demand, high placements, or low fowl slaughterings, would signal a possible over-supply, whereas low placements or high fowl slaughterings, would signal a possible shortage. These indicators would become more precise as better information becomes available on hen numbers, placements, and fowl slaughterings.

Problems have been encountered in the broiler industry since early 1974 in producing and distributing broilers and broiler meat. Shortages and over-supplies have developed, exports have virtually disappeared, imports are reaching unprecedented amounts, and consumption is relatively stagnant. The trade situation in broilers is largely the result of sustained high price differentials between Canadian and world markets, particularly the U.S. market. The imbalances of supply and demand appear to be due to incorrect demand determination, while the reasons for depressed consumption have not been fully identified. There also seems to be some concern over the willingness of producers to respond to processing requirements.

In light of this situation, it is necessary that some research efforts focus on determining:

(1) Those factors influencing supply and demand for broiler meat.
(2) How to optimize use of resources within the industry to restimulate growth and competitiveness.
(3) How to best coordinate production and marketing to ensure adequate response to market requirements.
(4) The most useful method of communicating relevant market information among the various sectors involved in production, processing and distribution.

The use of a simulation model may be appropriate to determine optimum use of resources, the coordination of production and marketing, and to simulate product flow, financial flows, and the input-output relationships of each major decision-making unit. This model would not be based on expected theoretical relationships, but on actual, observable operating characteristics. The purpose of a representative working model of the industry would be to study realistically the inter-actions of each sector with the aim of improving over-all performance.

A simulation model would have the following characteristics: location of production, and flow of product through processing and marketing channels would be detailed; product flow would be outlined by kind of product, i.e., light or heavy birds, eviscerated whole or cut-up meat, further processed products, etc.; financial transactions would be categorized and their location indicated; critical decision-making points and the information required to evaluate alternative decisions would also be included; performance criterion for each point would be developed to assist in planning; the relevance and impact of imports and exports would be shown to provide a means of assessing alternative trade practices.

The success of the production and marketing process in poultry is closely related to the coordination of each activity in this process. Having a model which incorporates future market demand, productive and processing capabilities on a geographical basis, would provide decision-makers with a useful technique to assist in planning and scheduling. It would also help to identify those areas where improvement is needed in order to increase efficiency and competitiveness.

The technical coefficients for production and processing are fairly standard for broilers, and once market requirements are determined, supply can be geared to meet these demands. The results of this process can be optimized using the model outlined above. The success of this approach would depend on the willingness of decision-makers in each sector of the industry to co-operate and to carry out their respective responsibilities in relation to other industry participants.

A simulation model would be of benefit to decision-makers and others with an interest in the planning and coordination of the system. Regional capabilities and constraints would be better pinpointed and ways to capitalize on the capabilities and surmount the constraints could be more effectively evaluated. The effect of different policy alternatives could more efficiently be measured also due to the input–output relationships which would be incorporated in the model.

Present domestic disappearance statistics for poultry products are too general to provide the information necessary for proper scheduling of production and product movement. This is a particularly acute problem in the turkey sector, where demand is highly seasonal, product type diverse and length of production greater than for broilers and eggs. Due to the small market size for turkeys, fluctuations in demand have a greater impact on the industry than in other sectors.

One of the major concerns currently facing decision-makers is what amount of each bird size is needed to satisfy market requirements in a given location and at a given time. Little information exists as to regional consumption patterns, apart from holiday influences, and to the factors which affect this consumption. Because of this, market shortages and over-supplies have recently occurred, and concerns over imports have been expressed.

The development of a relevant demand system for turkey meat would assist greatly in alleviating the short-term problems of the industry. In order to develop a more effective long-term strategy, it is proposed that the demand system be incorporated into a larger sectoral model, similar to that considered for broilers.

Preliminary investigation of the market for turkeys indicates that most product is consumed over a relatively short time-span. Most of this consumption can be attributed to holidays. However, there appears to be a trend towards increased further processing of turkey products. The determination of regional consumption patterns and trends, and the factors responsible for these, would assist in stabilizing production and fluctuations in returns. This approach would allow for more efficient use to be made of production facilities through improved timing and routing of product flow.

Establishment of a model simulating the industry would aid in policy and decision-making. The model would include product flows, the resultant financial flows, and input-output relationships for each major transformation unit. Such a model would provide information relating to regional capabilities, flow capacities, and alternative possibilities. Performance criterion would be developed to assist in evaluation and planning. Major decision-making points would be outlined and the type of information required for assessment of alternatives.

Isolating decision-making points and determination of the information required by each, provide the foundation for the establishment of a modern market information system. Some progress along these lines has occurred in Canada, but many proven techniques have not made significant in-roads at the present time. As the complexities of the Canadian market for poultry products increases, and as competition between various products increases, a modern simulation computerized information system will become vital.

10. SUMMARY AND CONCLUSION

Models play an important role in the decision-making process, in the sense that they provide useful aids to human judgement. However, they are no substitutes for human judgement. Decision-makers are normally faced with an array of current economic problems in respect of which decisions must be taken. These decisions are intended to influence the performance of the system, such that its performance will be more in tune with the over-all economic and social objectives of society. If models are to make meaningful contributions in this regard, they must focus on the important economic problems of the day. Models should solve tomorrow's problems today, rather than resolve yesterday's problems tomorrow. Hence, they should be more prescriptive, and less descriptive.

If model-builders are to obtain decision-makers' confidence in these models, they must be credible. The decision-makers ought to be involved more actively in the model-building process. Moreover, in the specification and description of the models, and in the interpretation of the results generated by them, model-builders must speak less to themselves, and more to their clients, whose decisions they hope to influence.

An impressive array of analytical hardware has already been developed to deal with a wide variety of problems in agricultural marketing. It would seem that concerted efforts should now be made to coordinate the different model approaches. The merit in this is that such coordination will facilitate assessment of the relevance of the various techniques to the complex of problems in agricultural marketing. The several analytical techniques available for model-building may be combined in various ways when developing econometric models. For example, other techniques such as recursive linear programming, or other recursive models might be incorporated, as needed, within an over-all systems simulation approach to optimize certain components of the model.

Input–output analysis may constitute one phase of a model, which otherwise has the feature of a simulation model. What determines the analytic techniques appropriate for a particular model is the particular problem under study, and the over-all objectives to which the model is directed. I find myself at variance with those who would suggest that national economic models should be capable of solving a wide variety of problems, without first specifying precisely what those problems are. It would seem that one cannot expect a particular model to be appropriate for solving any more than a few of the major problems of agricultural marketing. Specification of a problem should precede the development of a model. The model is then developed to answer these questions. Some issues which would require the development of regional or national agricultural models are of a normative nature, while others may be of a predictive nature. For this reason alone, it cannot be expected that one model will be appropriate for analyzing a wide array of problems. This does not mean, however, that the research input to build two types of research models should be built in isolation.

The major research activity in working with regional and the national economic models is the synthesis of appropriate data. Model conceptualization may not be the major ingredient. However simple a model may appear to be, there is no question that considerable research resources are required in its development. I would expect that a large part of the research is for the generation of appropriate quality data. National economic models are highly dependent upon basic data which are consistent, and uniformly developed for all components of the model. The large data requirements for national and regional model development, and limited research resources make it essential for some over-all coordination of this type of research, as well as for the developed model approaches.

The diffusion of research results into the decision-making process suggests that models be problem oriented. The reason for this is that it is often necessary for decision-makers to be involved in the research process from its inception, if they are to make use of its findings. This presents a rather difficult problem, when one considers that considerable lead time is necessary in such projects, so that the research results may be timely. Part of the researcher's job is to develop an awareness on the part of the decision-maker as to the need for answers to forthcoming problems. There are various ways to involve the decision-maker in the research process, so that the findings do have credibility, and are understood when they become available.

One method being used in Manitoba, Canada, is that of an advisory committee of staff from the various provincial government departments, which might utilize the results of the study when completed. The purpose of this committee is to create an awareness on the part of the decision-maker as to the usefulness of the study to their supportive decision-making role; and to provide continuing insight into the type of problems which the model is capable of answering. In that way researchers are kept up-to-date with the dynamic and changing nature of the research problem. This also provides a means whereby greater realism is built into the model.[37]

There are important implications from the analysis of the various analytical models that have been reviewed in the previous section of this paper. In addition, it is quite evident that the application of a particular model must be related to the problem to be resolved. The problems which confront those concerned with agricultural marketing are complex and diverse. Consequently, there is a particular role that each model can assume, but this role must be consistent with the problem under review.

Each analytic model is constructed so as to test or to project the effects or outcome of different economic conditions and/or policies on agriculture, as well as other sectors of the economy. Linear programming models attempt to project the long-term effects of changes in domestic and foreign demand, in technological conditions, and in supply management programs on the optimal organization of agriculture. Recursive programming models test the short-run and intermediate-run effects of changes in prices, subsidies, production quotas, exports, and resource adjustments.

Analytical models based on statistical inference attempt to explain the interrelationships between independent variables, and also attempt to explain the association between independent and dependent variables in the short-run. Input–output models attempt to measure the effects of changes in domestic and export demand, productivity, prices and taxes on output, employment and income in agriculture as well as the rest of the economy. Simulation models attempt to test the effects of changes and prices, commodity systems production research extension efforts, private and public investment on farm income, national income, export earnings, the level of demand, and the level of employment.

With respect to the role of linear programming models in the decision-making process in agricultural marketing, it is evident that the models used by Professor Heady and his associates at Iowa State represent a continuous research input over a period of over 20 years or more. The initial models are very simple representations of American agriculture adopting very gross assumptions because of the limited computational facilities available in the mid-1950's, and considerable research input to generate data was also a contributing factor. However, the refinement of the Iowa models grew with time, associated development of the computer facilities, more refined data and even greater facility in model-building by research workers. In the Eyvindson–Heady model which was completed in 1970 there were 6800 rows and more than 40 000 columns, yet I am sure that these research workers would agree that considerable aggregation and some lack of realism still exist in the model. Overall, the Iowa State models are limited primarily to normative applications because of their reliance on the ordinary linear programming techniques. There is no predictive element to these models although there can be time based in the future.[38]

Recursive linear programming is identical with ordinary linear programming in terms of solution procedures, the difference being primarily of a conceptual nature with respect to the Manitoba model developed by Dr. Sahi.

The inclusion of flexibility constraints in the recursive model gives it some predictive capability than rather being strictly normative. The United States Department of Agriculture (USDA) production supply response model and the Sahi model for prairie agriculture are both capable of predicting acreage for the future years. However, like Iowa State models, these models consist of regional components. In the ordinary linear programming models, the solution for one region is dependent upon the solution for any other region.[39]

The simulation models developed at the Michigan State University particularly the Korean Grain Market Management Model, the Nigerian Model, and the Venezuelan Model could be identified as taking a more systems approach. These types of models do not necessarily start off with any given structure, but are rather based on what the research workers know about the system. The structure of such models becomes evident through branches and loops as included in the computer program. It may not be possible to describe the total structure in terms of a complete set of mathematical equations.

Simulation models may also be based strictly on econometric analysis. A computer program, however, will still be used to solve such models but the entire structure of the model can be described by mathematical equations. Most simulation models either follow a systems approach or are econometric in nature and are predictive rather than normative in terms of the specific type of solutions derived.

Input–output analysis in its most limited form is a means of describing an economic system. The various inter-dependencies in the economy are reflected through coefficients which indicate the sector requirements of every sector on every other one.

The experience in terms of analytical model-building thus far has tended to be rather institutionalized in terms of a particular model, for example, the Iowa State linear models and the Michigan State simulation models. This is not necessarily bad in that the research input into developing and perfecting these models has saved other research workers considerable time by way of not duplicating, but drawing upon the knowledge and experience gained in experimentation with these models.

It would seem that research effort should now focus on the coordination of the different model approaches, in order that the whole spectrum of analytical models could be brought together thereby facilitating assessment of their appropriateness to the diverse and complex problems confronting agricultural marketing. What is being suggested here is that research workers in the field of agricultural marketing should now take an inventory of the various analytic models available with a view to coordinating the various approaches. I say this because this has not yet been done with the same rigour that has applied to the development of these models.

Obviously, analytic models have a very distinct role to play in the formation and the development of agricultural marketing policy. In fact, the development of effective government policy programs in any complex is at best a difficult task. The decision-making process by which planners, model-builders, policy builders, and elected officials develop processes and programs to deal with these arrays of problems is in itself a complex process involving three fundamental steps: problem finding (searching the environment for conditions calling for decisions to be made and new policies developed), design (inventing, developing, and analyzing alternative courses of action), and choice (selecting a particular course of action from alternatives available).

The decision-making process presupposes that the planner has been given a clear statement of strategic objectives. When considered, these would serve as the criteria for developing alternative courses of action. The several economic techniques available for model-building can be combined in various ways when developing econometric models. For example, linear programming might be embedded within an overall simulation model to optimize certain facets of the model.

A major research input in the development of analytic economic models is obtaining appropriate data. However, data suitable for one economic model can often be used in the construction of other types of models. For example, the synthesis of enterprise budgets is a major research activity in the development of national economic models. Yet, it is quite conceivable that these same budgets could comprise appropriate data for some parts of a supply response model. Therefore, rather than initiating a major research activity to estimate these costs quite independently of other research, it is possible to use the same basic data as developed for an interregional model.[40]

Simulation models either with an econometric base or systems approach are probably much less costly. However, it must be recognized that they are not necessarily applicable to all types of problems, such as those that may better be resolved by linear programming models, or by mathematical models based on statistical inference. Familiarity with economic systems on the part of the research workers is probably required more in the case of mathematical programming models.

In closing, I would like to repeat the caution given by W. Craddock in his summary statement at a Canadian seminar on "national and regional economic models of agriculture". He cautioned that "Model-building can be a very fascinating and challenging type of research activity. As one becomes involved in the beauty of his model, he must not lose sight of the fact that: firstly, the model is being developed to deal with real world problems and secondly, no matter how much research input extended, the model contains many limiting assumptions and is based upon historical data, and as such it is, therefore, only a crude approximation of the economic system that it attempts to portray. Models can sometimes give the wrong answer due to the inability to describe adequately the various aspects of the social, political, and economic behaviour of individuals".[41]

NOTES

[1] Michel Petit (1976), *The Role of Models in the Decision-Making Process in Agriculture*, a paper presented at the XVI International Conference of the International Association of Agricultural Economists, Nairobi, Kenya, July 26–August 4, Page 1.

[2] Herbert A. Simon, *The New Science of Management Decisions*, Harper and Row, Page 2.

[3] Ptolemy was among the ancients who sought to unravel the mysteries of the physical

universe by constructing models to describe the movement of the planets. In the 2nd century Alexandria, Claudius Ptolemy formulated a model of the solar system which recognized that the world was round or spherical and that the planets had circular motion. See Peter Langhoff (1965), 'The Setting: Some Non-metric Observations', in *Models, Measurements and Marketing*, Englewood Cliff, N.J., Prentice Hall Inc., Pages 9–12.

[4] Copernicus was a Polish model maker, who near the beginning of the 16th century, stuck upon the radical idea that the earth and the other planets revolved around the sun. This was such a novel notion that nobody, including dignitaries of the church, took him too seriously. *Ibid.*

[5] *Johannes Kepler* improved on the graphic model of Copernicus by describing it in terms of a series of mathematical expressions. *Galileo* pushed Copernicus' theory so hard that he found himself in trouble with the Roman Catholic Church. When some Dutch lens makers developed the principles of the telescope, Galileo seized the opportunity to develop it and to use it in astronomy. *Issac Newton* was born in the year of Galileo's death. Newton built a new model of the solar system and removed some of the slight defects noted in Kepler's model. Later, *Albert Einstein* improved upon the Newtonian model which had served so well for such a long time.

[6] Peter Langhoff (1965), 'The Setting: Some Non-metric Observations', in *Models, Measurements and Marketing*, Englewood Cliff, N.J., Prentice Hall Inc., Page 11.

[7] See Hosan Ozbekham (1971), 'Planning and Human Action', in Weiss, P. A., *Hierarchically Organized Systems*, New York, Hafner Publishing Co., Pages 123–230.

[8] Murray, K. A. H. (1933), 'The Future Development of the Pig Industry in Great Britain'. *Empire Journal of Experimental Agriculture*, Volume 1, Number 3, September.

[9] Murray, K. A. H. and Cohen, R. (1935), 'Wheat Prices and the Acreage of Wheat in Great Britain', *Scottish Journal of Agriculture*, Volume XVIII, Number 4, October.

[10] Johnson, R. W. M. (1955), 'The Aggregate Supply of New Zealand Farm Products', *Economic Record*, Volume XXXI, Numbers 60–61.

[11] Girschick, M. A. and Haavelmo, T. (1953), 'Statistical Analysis of the Demand for Food: Examples of Simultaneous Estimation of Structural Relations', In *Studies in Econometric Methods* (Edition Hood, W. C. and Koopmans, T. C.) Wiley, New York, Pages 92–111.

[12] Foote, R. J. (1953), 'A Four Equation Model of the Feed and Livestock Economy and its Endegeneous Mechanism', *Journal of Farm Economics*, Volume XXXV, Number 1, February, Pages 44–61.

[13] Cromarty, W. A. (1959), 'An Econometric Model for U.S. Agriculture', *Journal of American Statistical Association*, Volume 54, Pages 556–574.

[14] Nerlove, M. (1958), 'The Dynamics of Supply: An Estimation of Farmers' Response to Price'. *John Hopkins Press*, Baltimore.

[15] Suits, D. B. and Kiozumi, S. (1956), 'The Dynamics of the Onion Market', *Journal of Farm Economics*, Volume XXXVIII, Pages 457–484.

[16] Candler, W., 'An Aggregate Supply Function for New Zealand Wheat', *Journal of Farm Economics*, Volume XXXIX, 1957, Pages 1732–1741.

[17] Johnson, R. W. M. (1955), 'The Aggregate Supply of New Zealand Farm Products', *Economic Record*, Volume XXXI, Numbers 60–61, May.

[18] Food and Agricultural Organization (FAO) of the United Nations, Bibliography on the Analysis and Projection of Demand and Production, 1963 (Commodity Reference Series Number 2) Rome: FAO, 1963, Page 273.

[19] Houthakker, H. S. (1957), 'The International Comparison and Household Expenditure Patterns, commemorating the Century of Engel's Law', *Econometrica* **25**: 532–551, October.

[20] Goreux, L. M. (1960), Institutions Dealing with Demand Analysis and Projections. Rome: FAO, Page 30 (mimeo).

[21] United Nations (UN), Economic Commission for Latin America (ECLA). Analysis and Projections of Economic Developments.

[22] Hildreth, C. (1960), 'Combining Cross-Section Data and Time Series, Cowles Commission Discussion Paper', Statistics Number 347, May 15.

[23] Holme, F. (1958), 'Estimating a Supply Curve', *Journal of Agricultural Economics*, Volume XIII, Number 1, June, Pages 67–71.

[24] Schaller and Dean, *Predicting Regional Crop Production: An Application of Recursive Programming.*

[25] Game theory is based on the proposition that there is a close relationship between parlor games of skill and conflict situations in economics. The theory suggests that in each situation, there are a number of participants with incompatible objectives. The extent to which each participant attains his particular objective depends upon what all other participants do. Each participant has, therefore, to formulate his plans so as to take into account the actions of his opponents, each of whom is so laying his plans as to take specific account of the first participant's actions. Hence, each participant must surmise what each of his opponents will expect him to do, and how they will react to these expectations.

[26] Leontief, W. W. (1936), 'Quantitative Input and Output Relations in the Economic System of the United States', *Review of Economic Study* **18**: 105–25, August. And a full exposition in 1941, 'The Structure of the American Economy, 1919–1929', *Harvard University Press*, Cambridge, Mass., 1941.

[27] Cornfield, Evans and Hoffenbay, 'Full Employment Patterns, 1950', *Monthly Labour Review* **64**: 163–190, February, 1948.

[28] Gittinger, J. Price (1966), *The Literature of Agricultural Planning*, Planning Method Series Number 4, National Planning Association, Page 63.

[29] Bishop, R. A. (1956), 'Input–Output Work as a Basis for Development Planning', *Monthly Bulletin of Agricultural Economics and Statistics* **5**: 1–10, May.

[30] *Ibid*, Page 10.

[31] Seers, D., 'The Use of Simple Input–Output Model for Drawing Up a Development Program', In I. Adelmen and E. Thorbecke, editors, *Theory and Design of Economic Development*, Baltimore, John Hopkins Press, Pages 9.f.

[32] Cited below are some of the linear programming models developed under the direction of Professor E. O. Heady at Iowa State:

Brokken, Ray F. (1965), 'Interregional Competition in Livestock and Crop Production in the United States: An Application of Spatial Linear Programming', unpublished Ph.D. thesis, Iowa State University.

Brokken, Ray F. and Heady, Earl O. (1968), 'Interregional Adjustments in Crop and Livestock Production, a Linear Programming Analysis', USDA, ERS Technical Bulletin 1936.

Day, Richard H. (1963), 'On Aggregating Linear Programming Models of Production', *Journal of Farm Economics* **45**: 797–813.

Egbert, Alvin C. and Heady, Earl O. (1961), 'Regional Adjustments in Grain Production: A Linear Programming Analysis', USDA, Ag. Res. Service Technical Bulletin 1241 and supplement.

Egbert, Alvin C. and Heady, Earl O. (1963), 'Regional Analysis of Production Adjustments in the Major Field Crops: Historical and Prospective', USDA, ERS Technical Bulletin 1294.

Egbert, Alvin C., Heady, Earl O. and Brokken, Ray F. (1964), 'Regional Changes in Grain Production, an Application of Spatial Linear Programming', *Iowa Agric. and Home Econ. Expt. Stat. Research Bulletin* **521**.

Eyvindson, Roger K. (1970), 'A Model of Interregional Competition in Agriculture Incorporating Consuming Regions, Producing Areas, Farm Size Groups and Land Classes", unpublished Ph.D. thesis, Iowa State University.

Heady, Earl O. and Skold, Melvin (1965), 'Projections of U.S. Agricultural Capacity and Interregional Adjustments in Production and Land Use with Spatial Programming Models', *Iowa Ag. and Home Econ. Expt. Station Research Bulletin* **539**.

Heady, Earl O. and Whittlesay, Norman K. (1965), 'A Programming Analysis of Interregional Competition and Surplus Capacity of American Agriculture', *Iowa Ag. and Home Econ. Expt. Station Research Bulletin* **538**.

Madsen, Howard C., Heady, Earl O., and Nicol, Kenneth J. (1970), 'Trade-offs in Farm Policy: Prices, Incomes, Costs, Interregional Payment Distribution and Patterns of

Land Retirement', Center for Agricultural and Economic Development, Iowa State University, CAED Report 36.

Mayer, Leo V. and Heady, Earl O. (1969), 'Projected State and Regional Resource Requirements for Agriculture in the United States in 1980', *Iowa Ag. and Home Econ. Expt. Station Research Bulletin* **568**.

Miller, Thomas A. (1966), 'Sufficient Conditions for Exact Aggregation in Linear Programming Models', *Agricultural Economic Research* **XVIII**: 52–57.

[33] Such a procedure is satisfactory only if it is true that the homogeneity of producing units in the group is such that the results so obtained will be identical with those that would be obtained if each production unit were treated individually and the results summed. If this is not the case, then the difference between the results so obtained with two procedures is referred to as "aggregation bias". It does seem unlikely that this problem could satisfactorily be resolved without a very detailed model.

[34] Hall, Harry H., Heady, Earl O., and Plessner, Yakir (1968), 'Quadratic Programming Solution of Competitive Equilibrium for U.S. Agriculture', *American Journal of Agricultural Economics* **50**: 536–555.

[35] Walker, Hugh V. (1967), *The Transportation of Feed Grains in Eastern Canada*, Ottawa, Canada. The Agricultural Economics Research Council of Canada, Number 8.

[36] Wilson, A. G. and Wood, A. W. (1969), 'Regional Livestock Production and Feed Freight Assistance', *Canadian Journal of Agricultural Economics*, Number 17, Pages 77–90.

[37] *Loc. cit.*

[38] See W. Craddock (1972), 'Summary on National and Regional Models of Agriculture" in Agriculture Canada – *National and Regional Models of Agriculture*, Ottawa: Economics Branch, Agriculture Canada, Number 7219.

[39] *Op. cit.*

[40] *Loc. cit.*

[41] *Loc. cit.*

BIBLIOGRAPHY

Agrawal, R. and Heady, E. O. (1972), *Operations Research Methods for Agricultural Decisions*, The Iowa State University Press/Ames, Iowa, U.S.A.

Anderson, J. R. (1972), 'An Overview of Modelling in Agricultural Management', Sydney, Australia, *Review of Marketing and Agricultural Economics* **40**(3), 111–122.

Baritelle, J. L. and Price, D. W. (1974), 'Supply Response and Marketing Strategies for Deciduous Crops', *American Journal of Agricultural Economics* **56**(2).

Bass, Frank M. King, Charles W., and Pessemier, Edgar A. (1968), *Application of the Sciences in Marketing Management*, J. Wiley, New York.

Beneke, R. R. and Winterboer, R. (1973), *Linear Programming Applications to Agriculture*, Iowa State University Press, Ames.

Bishay, F. K. (1974), *Model for Spatial Agricultural Development Planning*, Rotterdam University Press.

Bishop, Charles E. and Toussaint, W. D. (1958), *Introduction to Agricultural Economic Analysis*, J. Wiley, New York.

Black, J. D. (1933), *Research in Agricultural Policy Scope and Methods*, J. D. Black editor, New York, Bulletin No. 21.

Bromely, David W. and Beatlie, Bruce R. (1973), 'A Short-Run Price Prediction Model for Eggs', *American Journal of Agricultural Economics* **55**(3).

Bullock, J. B. and Logan, S. H. (1972), 'Cattle Feedlot Marketing Decisions Under Uncertainty'. Berkley, California, U.S.A. Giannini Foundation Monograph; California Agricultural Experimental Station, No. 28.

Burt, Oscar R. (1971), 'A Dynamic Model of Pasture and Range Investments', *American Journal of Agricultural Economics* **53**(2).

Byerbee, Dereck and Halter, A. N. (1974) 'A Macroeconomic Model for Agricultural Sector Analysis', *American Journal of Agricultural Economics* **56**(3).

Clark, G. B. and Simpson, I. G. (1959), 'A Theoretical Approach to Profit Maximisation Problems in Farm Management', *Journal of Agricultural Economics*, Volume 13, No. 3.

Cromarty, William A and Myers, Walter M. (1975), 'Needed Improvements in Application of Models for Agricultural Commodity Price Forecasting', *American Journal of Agricultural Economics* 57(2).

Crowder, R. T. (1972), 'Statistical Versus Judgment and Audience Considerations in the Formulaton and Use of Econometric Models, *American Journal of Agricultural Economics* **54**(5).

Review of Marketing and Agricultural Economics, Volume 40, No. 3, September 1972. Department of Agriculture, New South Wales, Australia.

Eyvindson, R. K. (1972), 'National and Regional Economic Models of Agriculture', Regina, Saskatchewan. Processed Report, Economic Branch, Canada Department of Agriculture, No. 72/9.

Foote, R. J., Craven, J. A., and Williams, R. R. (1972), 'Quarterly Models to Predict Cash Prices of Pork Bellies', *American Journal of Agricultural Economics* **54**(4).

Fuchs, H. W., Farrish, R. O. P., and Bohall, R. N. (1974), 'A Model of the U.S. Apple Industry: A Quadratic Interregional Intertemporal Activity Analysis Formulation', *American Journal of Agricultural Economics* **56**(4).

Gadner, Bruce L., 'The Farm Retail Price in a Competitive Food Industry, *American Journal of Agricultural Economics* 57(3).

Halter, A. N. and Dean, G. N. (1965), 'Simulation of a California Range-Feedlot Operation', Giannini Foundation Research. Report No. 282, University of California.

Handy, C. R. and Padberg, D. I. 'A Model of Competitive Behaviour in Food Industries', *American Journal of Agricultural Economics* **53**(2).

Hayenga, M. L., Manetsch, T. J., Halter, A. N., Carroll, T. W., Byerlee, D. R., Abkin, M. H., Chong, K. Y., and Johnson, G. L., Michigan State University Simulation Team (1971). 'Comparing Agricultural Adjustments and Policies via an Agricultural Sector Simulation Model'. Proceedings of the A.A.E.A. Annual Meeting. Southern Illinois University, Barbondale, Illinois. August 15–18.

Heady, Earl O. (edit) (1971), 'Economic Models and Quantitative Methods for Decisions and Planning in Agriculture', Proceeding of an East-West Seminar, edited by Earl O. Heady. The Iowa State University Press/Ames, Iowa, U.S.A.

Heady, Earl O. (1961), *Agricultural Supply Functions*, Iowa State University Press, Ames, Iowa, U.S.A.

Hockman, E, Regev, U., and Ward, R. W. (1974), 'Optimal Advertising Signals in the Florida Citrus Industry: A Research Application', *American Journal of Agricultural Economics* **56**(4).

Hutton, Robert F. (1961), 'Operations Research Techniques in Farm Management: Survey and Appraisal', *Journal of Farm Economics*, Volume 43, No. 2.

'Farmers in the Market Economy', Iowa State University Center for Agricultural and Economic Development. Iowa State University Press. Iowa, U.S.A., 1964.

King, Gordon A. (1975), 'Econometric Models of the Agricultural Sector', *American Journal of Agricultural Economics* 57(2).

Kost, W. E., 'Trade Flow in the Grain-Livestock Economy of the European Economic Community', Ann Arbor, Michigan, U.S.A. Dissertation Abstracts International (1972), 32(12) 6639. Michigan State University, East Lansing, U.S.A.

Kottke, Marvin W. (1961), 'Budgeting and Linear Programming can Give Identical Solutions', *Journal of Farm Economics*, Volume 43, No. 2.

Kulshreshtha, S. N. and Wilson, A. G. 'An Open Econometric Model of the Canadian Beef Cattle Sector', *American Journal of Agricultural Economics* **54**(1).

Langhoff, Peter (edit) (1965), *Models, Measurement and Marketing* Marketing Research Council, Prentice Hall, Englewood Cliffs, N.J.

Martin, Larry and Zwart, Anthony C. (1975), 'A Spatial and Temporal Model of the North American Pork Sector for the Evaluation of Policy Alternatives', *American Journal of Agricultural Economics* 57(1).

Martin, William E. (1972), 'More Sophisticated Tools for Less Important Problems: The

History of Range Improvement Research: Comment', *American Journal of Agricultural Economics* **54**(1).
'Systems simulation of Agricultural Development. Some Nigerian Policy Comparisons', Michigan State University Agricultural Sector Simulation Team. *American Journal of Agricultural Economics* **55**(3), 1973.
'A generalized Simulation Approach to Agricultural Sector Analysis with Special Reference to Nigeria', Michigan State University, East Lansing, Michigan, U.S.A. (1971).
Mighell, Ronald L. and Black, John D. (1951), 'Interregional Competition in Agriculture with Special Reference to Dairy Farming in the Lake States and New England', Cambridge, Harvard University Press.
Miller, S. F. and Halter, A. N. (1973), 'Systems Simulation in a Practical Policy-Making Setting: The Venezuelan Cattle Industry', *American Journal of Agricultural Economics* **55**(3).
Myers, W. M. (1972), 'Combining Statistical Techniques with Economy Theory for Commodity Forecasting', *American Journal of Agricultural Economics* **54**(5).
Naylor, T. H. (1971), *Computer Simulation Experiments with Models of Economic Systems*, J. Wiley, New York.
'Agricultural Production Systems Simulation', Proceedings of a Workshop, May 4–8, 1970. Oklahoma State University, Stillwater, Oklahoma.
'Marketing and Purchasing Strategies', West Lafayette, Ind.; Purdue University. Department of Agricultural Economics 59–63, January 1974.
Rosine, J. and Helmberger, P. (1974), 'A Neoclassical Analysis of the U.S. Farm Sector, 1948–1970', *American Journal of Agricultural Economics* **56**(4).
Roy, J. K. and Johnson, Phillip N. (1973), 'Econometric Models for Quarterly Shell Eggs Prices', *American Journal of Agricultural Economics* **55**(2).
Scoville, Orlin J. (1973), 'The Agri-business Sector: An Important Link in Economic Growth Models', *American Journal of Agricultural Economics* **55**(3).
Shepherd, Geoffrey S. (1969), *Marketing Farm Products; Economic Analysis*, 6th Edition. Iowa State University Press, Ames, Iowa, U.S.A.
Smith, Eldon D. (1972), 'Agricultural Marketing Research for Less Developed Areas', *American Journal of Agricultural Economics* **54**(4).

DISCUSSION OPENING – A. Weber, *Kenya/Fed. Rep. of Germany*

The increasing maturity of agricultural marketing as a subdiscipline of agricultural economics has been subject to several systematic reviews in Germany by G. Schmitt[1] 1968, in the United States of America by H. Breimyer[2] with a very useful extended international coverage and just recently by Bateman[3] 1976 for the U.K. Reviewing the existing literature was also done in 1973 at the Sao Paolo Conference for Farm Management.[4]

The report presented by Dr. Walker gives an immediate access to the most recent development of models in agricultural marketing. The concept of marketing used by the speaker is necessarily a little more narrowly defined than that used by the above quoted general reviews. The presentation gains, however, by concentrating on newly emerging types of models.

(a) *Type of decision-making and decision-makers*

Dr. Walker is not analyzing the various elements of the decision-making process itself, but investigates which models are available for specific purposes of policy decisions in marketing. He conceives, therefore, the task of model-builders as a contribution to marketing policy related to the public or to supply managment schemes directed to an industry or a commodity market.

As a consequence of this limitation, the decision-making on farms or of large firms in the food distribution system is not dealt with specifically, because public policy-makers normally use only a small part of the whole marketing mix: namely quantities and prices. Another element which has to be kept in mind is that the models of consumer behaviour are strictly based on the lines of neo-classical theory. There might be some new insights in the policy variables at hand if consumers reactions to changing conditions were not excluded.

b. *Class of Models*

Dr. Walker considers among verbal, graphical and mathematical models only the last ones and distinguishes three types:

(i) econometric models,
(ii) mathematical programming models,
(iii) simulation models.

His view of the three types of models follows the general professional discussion, but he observes that a certain sequence of econometric, programming and simulation models were the consequence of increasing computing facilities. This would imply that the progress made has not been dictated by the problem orientation of agricultural marketing economists as such. The progress has rather been the consequence of handling more data and testing more complex theories by applying computer programmes from other fields of science. In other words: Learning by doing paved the way to progress.

Another matter for remark is related to the interaction of increasing computer facilities and the intellectual advancement of policy-makers and agricultural economists. Does the increasing use of computer facilities bring better policies or do they inspire policy-makers to invent new areas of intervention with probably doubtful consequences?

It would be easy to argue that certain subtypes of models are not covered by Dr. Walker's review; lack of space obviously limited his selection. However, which type of models would Dr. Walker have excluded if he had to deal with decision-making in production economics. Let us assume the result of reviewing models of production economics would not have been very different from those in a marketing review. In that case we have to ask: are the models from production economics merely superimposed on marketing? Certainly very few marketing economists would agree that the same kind of problems occur in production economics and in marketing. Elaborating this question brings us probably a little nearer to the empirical relevance of the various types of models. If strict maximizing and minimizing rules in marketing cannot be followed the simulation approach seems to be more appealing than the strict linear programming type of model.

(c) *Further developments*

Since developing countries have to put a high emphasis on earning foreign exchange by exporting agricultrual products, two other types of models might usefully be investigated further in agricultural marketing to suit the following problem areas:

(i) Commodity modelling and supply management schemes at an international level have to be pursued in future. However, difficult aggregation and disaggregation procedures are a practical barrier to move faster ahead. They have to be overcome by marketing economists who involve themselves in this, much needed, research area.

(ii) Models are needed which describe the framework of international marketing of agricultural products and processed food in a more comprehensive way. These models have to stress a little more deeply traditional functions and institutions in an international setting without excluding the whole marketing mix.

REFERENCES

[1] Schmitt, G. (1968) *Landwirtschaftliche Marktforschung in Deutschland* Munchen. Bayerischer Landwirtschaftsverlag.

[2] Breimyer, H. F. (1973) The economics of agricultural marketing: a survey. *Review of Marketing and Agricultural Economics* **41**, (4) 115–165.

[3] Bateman, D. I. (1976) Agricultural marketing: a review of the literature of marketing theory and of selected applications. *Journal of Agricultural Economics* **27** (2) 171–225.

[4] Throsby, C. D. (1974) New methodologies in agricultural production economics in *The future of agriculture. Technology, Policies and Adjustment*. Oxford: Agricultural Economics Institute.

RAPPORTEUR'S REPORT – M. Hawkins, *Canada*

The nature of this session was to share experiences. The opening paper by Walker did an outstanding job of presenting specific economic models with their strengths and limitations. Therefore, the consensus of the group was to proceed from the paper into the broader issues which exist in market analysis.

The discussion reflected the varying philosophies and approaches to models utilized within marketing theory, yet it remained tolerant of ideological differences among the discussants. It appears that marketing is becoming increasingly concerned with group arrangements. This is in direct contrast to production economics which is primarily concerned with individual farm units. Vertically integrated systems, marketing boards and co-operatives are of increasing importance. Marketing issues have become more involved with equity and marketing has become a more integral part of development plans.

Models must involve many non-mathematical concerns. The conceptual framework of a market must be enlarged to consider institutional factors that is, government, agribusiness, human behavior, etc. Agricultural economists must become more adventurous in exploring other models and approaches which should appear as a result of problem-solving in the marketplace.

As we develop our models and concepts, marketing economists must prepare the strategy for policy-makers. We must deal with market imperfections. A consideration of the triangle involving commercial farms, government and a concentrated market place entails careful consideration of market structures, conduct and performance, while preparing our models and policy strategies. Here the theories of effective or workable competition deserve considerable attention as they blend with tenure systems, welfare considerations and social structures.

The challenge is pragmatically to structure a market system in order to get performance. In this regard there are many applied experiments in new systems throughout the world from which there are now applied research results which should be seriously examined.

The history and evolution of marketing events is largely ignored by model-builders as they work and thus they continue to lose contact with the core of basic economic thought. The construction of models that lack sensitivity to basic economic and policy criteria often leads to a large waste of resources.

As a final comment, the discussion showed an interest and maturity that was not apparent at previous conferences. Perhaps this illustrates the growth, maturity and concern of the profession of agricultural economists, in the discipline of marketing and its important role in economic and social development.

KAZUSHI OHKAWA*

The Relationship between Agriculture and the Economy: Issues for National Level Policy

1. INTRODUCTION

Economic policy or a system of policies is composed of aims, objectives and instruments or means. These terms are not used with exactly the same meaning through different economic systems, but broadly we have a concensus. Policy means government action in pursuit of certain national *aims,* which have an overall nature extending to a greater or lesser extent beyond the bounds of the economic aspect, and may differ with the political ideologies. The scope of the "economic" aspect can be defined more broadly or narrowly depending on whether a "social" aspect is included. Within this range we can recognise *objectives* of economic policy which specifically concern economic affairs and can be handled in economic terms. To achieve its objective a government takes certain actions, which we call *instruments* or means.

In applying such conventional concepts and terms to the problems of agricultural policy, the assumptions I would like to make are as follows:

(i) "The relationship between agriculture and the economy" is not the same as "The relationship of agriculture with the rest of the economy". What is intended by the theme selected is to treat agriculture *in* the overall context of the national economy. In this context, macro-economic policy of any kind, in terms of objectives and instruments, is all relevant to agriculture, which is one of the industrial sectors of the economy. Each sectoral component of the economy has its own characteristics, just as agriculture does. Despite these varied characteristics, macro-economic policy is designed to work based on its own logic (for example, fiscal and monetary policy). Within this framework, sectoral (or partial) objectives and their instruments of implementation in relation to the overall ones are taken up.

The relation of agriculture *with* the rest of the economy, on the other hand, is a theme of an approach of sectoral dichotomy. It has often been presented in terms of agriculture versus industry or the non-agricultural sector – well-known in a two-sector model approach. Intellectual knowledge drawn from research results of this kind no doubt actually contributes a great deal to our understanding of agricultural policy problems. However, conceptually this approach should be distinguished from the previous one

* International Development Center of Japan, Tokyo, Japan.

particularly for policy discussion. The problem here is the relation between objectives and instruments at the sectoral level (for example, competitive allocation of scarce resources between agriculture and industry). Actually, the need to make such distinction between the two differs for different stages of economic development: it is stronger for DCs, weaker for LDCs. I presume that the real issues for us are raised by the programme-makers in a varied complex combination of these two dimensions.

(ii) "The characteristics of agricultural policy common to all nations" needs qualification and specification. First, "common" does not mean a sweeping generalization. Instead, dissimilarities as well as similarities will be the point at issue. With respect to different economic systems, my knowledge is limited in respect of experience of the non-market economies and a bias thus produced will hopefully be remedied by other papers so as to arrive at a "balanced" discussion. However, so far as the objectives, if not the aims, are concerned, possible difference from the market-based economies may not be essential. The problem rather lies at the level of implementing procedures and machinery – instruments in a broad sense. Among varying stages of economic development, surely the problems vary and the task of clarifying the differences is important. However, I believe, that, rather than merely emphasizing the dissimilarities, an integrated understanding is useful and may be possible.

(iii) The time dimension is related to longer-term policies, in the sense that short-term (in particular counter-cyclical) policies will not be discussed. In the DCs based on the market-economic system, there has been rapid progress in policy-making during the postwar period. Most of them, however, pertain to aggregate demand adjustments of a short-run nature. Economic planning of an indicative nature in various degrees has also been introduced into a number of DCs as well as LDCs, bringing forth the structural, and hence longer-term, aspect of policy-making. Yet, in my view, most of them appear to be weak in the implementation process: between objectives (and targets in their quantitative indicators) and instruments less cohesion is recognised than is generally expected. Integration of agricultural policy into the process of economic planning in most cases seems to occur only in a loose way. Therefore, systematic empirical evidence for the longer-term dimension is limited except for a few cases – in contrast to non-market economies. Despite its importance, this aspect of the problem is not specifically discussed here.*

2. HOW AND WHAT HAS BEEN THE EXPERIENCE?

Let us first look briefly at what has taken place in the postwar period. The major objective of agricultural policy in DCs of the market economic system

* I want to acknowledge the encouraging comments received from Professor Dr. Schmitt and Professor Britton on this part of my earlier draft. It is a regret that because of the limits on length, special care could not be taken to introduce "concrete illustrations drawn from actual events to bring the paper to life", to meet Britton's suggestion. Schmitt's emphasis on the importance of different instruments in achieving even the same objective, among others, is a good point. Yet I could not well cover this point, hoping for fuller discussion in his paper, in particular regarding European experience.

has been towards giving farmers "income parity" with people in the rest of the economy – although the means and the scope have varied from one country to another. Such an equity objective for a particular sector of the economy has never before been so systematic and conspicuous throughout the long history of modern economic growth. This has recently also been the case for some socialist countries though to a lesser extent. An income policy for agriculture with an equity objective in DCs is a historical development of prewar price stabilization policy. It is to be noted that during the postware phase of high growth rate, this objective has been more or less accepted in most countries in association with realizing near-full employment in the economy. Conflicts did take place, however, with respect to instruments of implementation. Apart from detailed differences in the means used to effect income transfers, the major conflicts at the political and administrative level have concerned (i) internally, the costs to be paid by consumers and tax-payers and (ii) externally, the inconsistency with free trade policy of the protection and/or export subsidies benefitting agriculture. The agricultural surplus produced as a consequence of this policy has been, of course, of great significance to subsequent policy-making, although its magnitude has differed from one country to another.

There is a large literature on this notorious topic, most of it being critical as we well know. For the purpose of this paper, the following points are suggested:

(i) The basis of this policy is a recognition that the mechanism of both output and factor markets cannot operate in favour of farmers in achieving income parity, taking the urban level as the standard of comparison – which is an acceptable basis in highly industrialized countries.

(ii) Transfer payments of various kinds became feasible basically because farmers had become numerically – though not necessarily politically – a minority group in the entire economy.

(iii) Alternative means were less feasible and attractive to farmers.

A well-known alternative is a "structural" policy: a serious issue for Europe and Japan. May I add a few words about Japan to a fuller statement on Europe given below in Schmitt's paper. The so-called "structural policy" pursued under the terms of the Basic Law of Agriculture could not replace price support policy despite government efforts. This is due to the continuing lower productivity in agriculture despite its rapid pace of technological progress and labor outflow to the rest of the economy during the postwar upswing phase. This suggests the political importance of a time dimension in selecting and evaluating the instruments and their operation for meeting farmers' needs, though, of course, the policy aimed at avoiding the distorting economic effects of price policy operation and the like.

On the other hand, in most of the LDCs, after their political independence, economic objectives have been focussed on development – on transformation and modernizing the traditional economy. Agricultural policy is a major part of this, aiming at output and productivity growth. Income parity has not been the major issue. Such a dichotomy, DCs and LDCs, is of course too simple. A fairly wide diversity in varying stages of economic development and, accordingly, in agricultural policy should be kept in mind. For example, in a few LDCs with a socialist system like Sri Lanka, an equity objective has been planned for introduction to the agriculture-related field. In a market economy like Thailand, an income transfer policy has recently been tried for raising the farm price of rice. Furthermore, a number of LDCs has a minimum price support policy for the major farm product. Implementation in these cases seems to have met with difficulties. The objective between DCs and LDCs appears to be similar but we have to identify certain limitations in choosing objectives due to the practical effectiveness of the available instruments in LDCs.

In most of the LDCs during their initial phase of modernization, there have been serious conflicts in formulating the national plan with respect to the allocation of scarce resources – in particular, over allocation of investment between industry and agriculture, related infrastructures included. As a result of adopting for example, the "import substitution" policy, agricultural development has been discouraged in a number of LDCs. Sectoral terms of trade became favorable to industry (a historical counterpart of such a policy can be seen in the early industrialization phase of DCs of both economic systems). Even after these nations shifted to an "export promotion (substitution)" policy, mostly during the latter part of the 1960s, there were indications of policy bias towards industrialization and growth particularly in view of the previously discouraged state of agriculture and of society's tolerance for inequity and poverty. This is particularly so for the most seriously affected countries (MSA).

By now there is a large literature on these problems. For the purpose of this paper, the following points are suggested as being central.

(i) A balanced growth of agriculture with industry as the major objective and, except for a few nations, this objective has not been achieved.

(ii) One of the major reasons is failure in allocating scarce resources – the grounds of sectoral conflict.

(iii) Since the majority of people are engaged in agriculture, the problem of policy decision-making for agriculture is far more important to them for overall development than for DCs.

A few words may be added to qualify the objectives for LDCs. Recently, as suggested before, equity and social problems have increasingly become important in LDCs in forming development strategy, as exemplified by proposed "integrated rural development" (IRD, to be discussed later). This is a reaction to the diverse effects of "Green Revolution" for various classes within agriculture, but more than that, to the conventional pattern of development policies.* The problem for us here is an *ex post* recognition of the new issues arising as a result of carrying out certain policies – like the farm product surplus in the case of DCs.

What has been described above is merely an illustrative sketch to induce

* On this point I am indebted to Prof. S. Sawada's comment on my earlier draft.

comprehensive discussions by experts here. And yet it suggests that in discussions of agricultural and macro-economic policy conflicting objectives and instruments are involved and that these have been "solved" through policy-makers choosing one of the possible alternatives – for example, for DCs, alternative instruments such as promoting education and migration of farmers internally and free trade externally, and for LDCs alternative objectives such as more balanced growth between agriculture and industry (in particular in view of the recent policy of the Peoples Republic of China) and/or more equity than growth. I am not suggesting these alternatives were all feasible but merely intend to point out the fact of "choice" of certain objectives and instruments in the light of various conflicts. The system and machinery of carrying out these choices differ a great deal between countries of market economies and the centrally-planned economies. However, in both systems the final decision is made at the political level. The preferences of a political party or parties are decisive. The preference of the administration cannot be ignored in influencing decision-making in respect of objectives and in some countries administration is powerful in selecting policy instruments. The influence of interest groups is great in respect of objectives which concern particular groups, regions and industrial sectors of the economy. It is widely known that agricultural policy is most influenced by political preferences, and decisions are most likely to be made by judgement and compromise.

With regard to the instruments and their consistency with objectives, a few words may be in order particularly in comparison between DCs and LDCs. Technical precision, in conventional economic terms, cannot be expected for LDCs, unlike DCs, because of their different requirement for inducing technological progress in agriculture. For example, institutional innovations required for LDCs are one of the major instruments, but quantification of their effects, even apart from the difficulty of implementing them, is almost beyond our capability. Too-sophisticated methods may not help much as tools for decision-making in practice.

Such a brief decription, I am afraid, may lead to misunderstanding a complex process of decision-making related to the formation of agricultural policy. It is true that in both economic systems technical knowledge has increasingly become influential in the national plan and policy-making and such machinery as councils and committees has been organized to provide opportunities for economists to serve policy-decision-making. It is intended, however, to suggest that the professional functions and actual influences which economists could achieve have been rather limited.

3. THE CHARACTER OF AGRICULTURAL POLICY: SOURCES FOR DISAGREEMENTS

Now I would like to take up a subject of "the character of agricultural policy" which specifically relates to the last point. The reason for discussing such a seemingly obvious subject is that there is not necessarily a consensus between economists and policy-makers, and this has important influences on forming views, determining objectives and selecting instruments for agriculture. In largely abstract theory, for example, an optimum resource allocation principle can be applied to agriculture; a component of the economy, as a basis for planning sectoral policy. Because of the characteristics of this particular sector, however, the real problem is far too complex. How and to what extent? Answers to this question seem to differ widely.

First, natural endowment – land and climate in particular in relation to population – is the basic factor of agricultural production and it differs widely among nations. Despite types of technological progress adapted to the given conditions, this difference places varying restraints on output and productivity growth in agriculture at the national level (compare, for example, the U.S.A. and Japan). To what extent should this restraint be taken into consideration in forming national policy? Answers vary widely. If we can assume near competitive markets for both output and factors internally and near free trade externally, the natural endowment restraint would essentially disappear in forming national policy. On the contrary, if world-wide food supply appears to be unfavourable or its instability substantial, the restraint would be a serious matter for policy-making, particularly for food-importing nations. Security as an objective, rather than productivity growth, would be dominant and instruments for achieving self-sufficiency in food would be pursued. During the postwar upswing phase of near-full employment with surplus supply of food product, touched upon in 2, the latter assumption has been irrelevant. Quite recently, however, the situation has seemed to change and conflicts began to take place due to different judgements on the coming situation in food-importing nations. This suggests that the general optimum resource allocation principle would be modified to a considerable extent, depending upon the forecast one may make. Yet demand–supply forecast cannot be precise enough for this at the present level of technique.

Second, the unit of production is predominantly family enterprise, implying self-employed, and mixed income in most DCs as well as LDCs of the market economic system – except for modern enterprises in a few DCs and plantations in LDCs. This characteristic is not excluded in relation to some of the socialist countries where collective production units are dominant. In the case of mixed income of the family farm in particular, its relation with the factor markets of the economy is blurred in association with the fact that farmers' income formation depends upon the prices of products they sell. This not only characterizes agricultural policies but also makes their nature differ between countries depending upon the varying situation of agriculture in the national factor markets. For example, in evaluating the objective of an income policy for farmers, one analyst may assume competitive factor markets for both capital and labor between agriculture and the rest of the economy in asserting that the income support policy for agriculture will automatically result in raising land prices which will benefit land-owners (a modern version of Ricardian doctrine), whereas another may assume a fundamental disequilibrium of the labor market in order to defend income support policy for farmers whose major income comes from compensation for their labor. Between these two extremes we can think of intermediate cases of varying degrees, and whether the real situation of the country at issue is close to one of the assumptions thus made is a matter of debate. And yet, this is the basis for appraising the effects of income support policy: one asserts that it failed while the other defends it.

Third, as has previously been discussed, the characteristics of agriculture in the economy differ at varying stages of economic development. In DCs in pursuit of the equity objective, the income level in the non-agricultural sector is taken as a standard. In this sense it is really a partial objective, whereas in LDCs agricultural policy is a major integral part of the national development plan. The two characteristics mentioned above, therefore, imply greater significance than for DCs and accordingly different choices of its objective means of implementation would have more serious consequences for LDCs. Yet, with respect to the possible effects of agricultural policy measures, our professional knowledge based on reaearch results is more limited than it is in the DCs due, not only to data paucity, but also – and more important – to the different nature of the agricultural problem: structural transformation, institutional-organizational innovations, as touched upon before, in addition to the farmers' behaviour in respect of production response to output–input prices in association with possible technological progress. Coupled with the powerful influences of political preference, optimism versus pessimism, opinions about the rational behaviour of farmers, this situation would allow a large room for disagreement in formulating objectives and instruments consistently.*

Differing natural endowments, traditional production units represented by the family farm system, wider coverage of related areas – all these discussed above may be enough to illustrate the problems we face. Viewed from overall national policy, agricultural policy is characterized by these three features which are unique to this particular sector. The statement above may appear to emphasize dissimilarity of the problem between DCs and LDCs, but this is not the writer's intention. Historically, agriculture is the greatest traditional sector. During a long process of modern economic growth, irrespective of different economic systems, it has been modernized through its decline in relative shares in output and labor in the national economy. It is quite natural, therefore, for the objectives and instruments of agricultural policy to tend to differ correspondingly to the various stages of development. Technological and

* With respect to the last point, I would like to draw your attention again to the IRD – a "renewed" approach of rural development, characterized by its "integrated strategy and programme-making". This seeks to cover not only agricultural development but also all other aspects and factors relevant to improving rural life. As a reaction to a failure of the past development strategy focussed on output and productivity growth. The planners assert that the overall objectives thus aimed at should be acceptable. The critical view, however, is that the issues are found in the means of implementation to be designed in an "integrated" way. They appear difficult to make operational enough. The IRD pertains specifically to community and regional level, where it has problems closely related with other sectors of the economy. As seen from the standpoint of agricultural policy at the national level, however, it presents a widened coverage of both objectives and instruments, including interdisciplinary fields. One may say this is a new challenge to both policy-makers and researchers. Can we minimize disagreements?

institutional adaptation to the given natural resources tends to be underdeveloped in LDCs. The effects of changing supply–demand situation for food on forming policy at the national level may be even more serious, being aggravated by the pressure of foreign payments deficits (a problem outside the scope of this paper). The dominance of family farms calls for a "renewed" attention as the core of rural development.

Finally, in relation to the last point, I would like to touch briefly on the controversial issue of "trade-offs between growth and equity" – a central problem to-day, not in general but in particular regarding agriculture. On the one hand, the trade-off concept has often been applied to this particular sector, emphasizing the importance of equity both within the sector and to the rest of the economy. On the other hand, application of this concept to agricultural problems is rejected, beyond merely asserting that output growth is the primary objective. As far as the choice of objectives is not assumed to be "given" for us, we have to know the possible feasibility of the instruments relevant to the choice we may have to make. As has previously been suggested in discussing rural development strategy this aspect of the controversy may require much more discussion in order to arrive at its solution. In this respect, I wish to draw your attention again to the family farm. Output growth and income (Welfare) equity are, in principle, complementary in this case, both within agriculture and in relation to the rest of the economy. Fuller utilization of land and labor towards a higher output level can be expected and this brings forth a higher level of their mixed income towards parity with urban people. Furthermore, regarding the procedures and instruments for implementation, our accumulated knowledge is relatively rich and can be used once the programme is set in an operational way.

Of course conventional instruments may have certain limits, unless they are combined with institutional innovations, including land reforms – in particular in the case of heterogenous agrarian structure. Yet with this illustration I am suggesting the possibility of finding the area of least disagreement.

4. CONCLUDING REMARKS

By way of concluding remarks, I would like to take up two points; one concerns the pattern of policy and plan-making and the other a way of making professional contribution to it.

First, as has been touched upon earlier, the postwar rapid progress in economic policy with respect to instruments has mostly been realized in fiscal, monetary and trade aspects of the market-orientated economy. In theoretical terms they are mostly characterized as post Keynesian type. As compared to the past, this policy has recently become much more difficult to implement because of the emergence of the conflicts between the two objectives – full employment and price stability. Agricultural policy appears to have a new problem in being consistent with anti-inflation policies. However, what is much more concerned here is another aspect of the macro policy. With respect to the supply–production side, macro-economic policy has made less progress, despite experience of indicative medium and long-term planning by the use of a system of social accounts with econometric model approaches. Application of a similar approach to LDCs' development plans has also built up our knowledge on experience of both success and failure. What has been observed regarding the character of agriculture may lead us to state that a more systematic approach, in particular in designing policy instruments, is desirable in the context of overall policy in long-term planning. Integration of this request to the macro-economic policy cannot be effectively done, however, in the light of the present pattern of policy-making. From the agricultural side, therefore, it is suggested that macro-economic policy should be strengthened towards improving the long-term, production-structure, aspect of the economy. My knowledge is limited regarding countries in the non-market economic system, but in view of the recent performance of agriculture in Soviet Russia, for example, they may have a similar problem.

Second, what has been stated in Section 3, though illustrative, may contribute to indicating a high possibility that our insufficient knowledge – and/or disagreements on interpretation – of the facts and their future nature might be one of the important causes for providing wide scope for greater influence of political preference on decision-making. It is thus important to improve our professional knowledge in the direction of minimizing our disagreement so as to rationalize the process of policy-making through limiting the influence of political and administrative preferences. Progress in our research on the characteristics of agriculture, I believe, is an indispensable need for deeper understanding of the inter-relationship between this particular sector and the rest of the economy. I dare say this is reviewing the past experience of both the government planning touched upon before and the academic model analysis. In the latter, for example, in using a two-sector (agriculture and industry or non-agriculture) model approach, both theoretical and econometric, disagreements have tended to relate particularly to the performance of agriculture and farmers. To the various reasons mentioned in Section 3, perhaps our limited knowledge about evaluating dynamic processes of resource allocation should be added. To be more realistic in interpreting the relationship of agriculture and the economy towards improving policy-making, we need much more knowledge about the real process of the dynamic interaction mechanism in a two-sector approach, not only in terms of demand–supply relationships in the product market but also in terms of the choice of technologies which directly pertain to the factor market. I believe that much more complementary, instead of competing, relations between the two sectors can be created by choosing more appropriate strategies.

In this regard, it is desirable to have more communications through exchange of research results and analytical views, not only among agricultural economists, but also with general economists. Let them know about the real situation of agriculture. And we have to ask that industrial and/or macro-economic policies to be formulated with more regard to the viewpoint of agriculture. On the other hand preferably such macro-sectoral knowledge should be linked more closely with micro-analysis. For example, the conventional cost–benefit calculation used in evaluating and selecting individual investment projects, may not be without problems in taking fuller consideration of agricultural characteristics combined with the problems of income distribution and employment. This implies a desirability of much more co-operative study in related areas.

These efforts on the professional level may, I hope, lead us to making contributions towards solving the problems we face.

DISCUSSION OPENING – D. Bergmann, *France*

Being on the whole in extensive agreement with Professor Okhawa I intend to limit my comments to a series of remarks on three particular aspects.

In developed countries, agricultural and general economic policy conflicts with regard to employment are usually solved in favour of the dominant non-agricultural sectors of the economy. The task of agricultural policy is to regulate the decline of the farm population. When the economy is buoyant, industry draws heavily on farm labour. In a recession, agriculture is expected to take care of a larger number of its children and keep them fed and partly busy at no cost to the Unemployment Compensation Fund. This buffer-stock role of agriculture, which is the servant of industrial growth, appeared in the French Plans of the Fifties and Sixties and is visible in Polish planning according to the special group paper prepared on that theme. Such are the facts. Their consequences are not entirely negative. Partial employment on an overpopulated farm may be better than unemployment in a city. The consequences for the average farm income per active person in farming should, however, not be underestimated. Agriculture's role in serving as a labour reservoir should not be forgotten when budget funds are allocated for agriculture.

Conflicts between agricultural and non-agricultural interests are often acute with regard to land use and economic policies have often failed to recognise the agricultural interests in spatial planning. In Japan, for instance, the rate at which concrete has been poured over the paddy fields is appalling. In France, second homes are very numerous and their expansion is often detrimental to agriculture. It is important to note, for the benefit of those who believe in the advantages of free markets, that price mechanisms seem quite inadequate to ensure the protection of agricultural interests in land use. The only solution – and it is far from easy to manage – seems to be strict regulation such as is practised in a growing number of countries.

Lastly, a few remarks with a political science leaning. What are the factors determining the place given to agriculture in economic policy decisions?

In rich countries with a small and relatively declining agriculture, the importance still retained by farm policy is remarkable. This can be attributed to reasons of pure politics; the farm vote is, in several countries, able to tip the balance in favour of one of the two roughly equal parties or groups of parties. Also, in a country like France, farmers have invented highly effective methods of exerting pressure on the government – for instance by disrupting road traffic with tractor barrages, or by more violent methods. Lastly, there is growing recognition of certain external advantages to society of many farm activities.

In most poor countries, on the other hand, the striking fact is the neglect of agriculture, or of most segments of agriculture, in the general economic policies and development strategies. A recent FAO study on the Sahel countries which was conducted under the authority of our colleague Dr. J. P. Bhattacharjee, will be used to illustrate this. In those countries, farm income per farm worker is only about 10 per cent of non-farm income per active person. What is worse, "inequality, already strong in 1960, has increased during the next ten years". In Sénégal "the value of the groundnut crop fell from 22 billion CFA francs in 1967/68 to 20,5 billions in 1972/73 but the share received by farmers fell from 66 per cent to 43 per cent of this total and government's share increased from 7 to 41 per cent".

This disregard by city technocrats for the peasants, these anti-agricultural policies, will lead to explosive situations and/or dramatic retreat of abandoned communities in autarchy and revolt. This is the result of excessively industrial development strategies which are a sequel of "Stalinist" anti-peasant doctrines. China is one of the few countries refusing this type of reasoning.

Happily, there has been, in recent years, a trend towards a more balanced approach. Let us hope that this Conference may help reinforce this tenancy.

DISCUSSION OPENING – J.S. Marsh, *U.K.*

1. An important and recurrent theme in Professor Ohkawa's paper is the relationship between the policy-maker and the professional advisor. In discussing the choice of objectives and instruments, he reminds us that "in both systems" – market and centrally planned – "the final decision is made politically". Again at the end of the paper he suggests that the professional's contribution to policy-making may be more effective through improving scientific knowledge than by direct participation.

The relationship between the economist (professional) and the politician (decision-maker) represents one of the more fragile aspects of modern economic systems. In mixed market economies at least we seem to have drifted into a situation in which the policy-maker is dependent upon expert advice about the options open to him and the instruments at his disposal. Having abandoned the traditional role of policy, responding to and correcting specific features of the outcome of market operation, modern governments in most economies offer positive policies which it is claimed will secure goals such as equity in incomes, repaid growth in per caput income, stable prices and full employment – even though these may be incompatible. Since each decision clearly does interact in complex ways with other elements in the political/economic system, the politican turns hopefully to his "experts" for guidance. Sadly he is likely to be disappointed or misled. Not only is there no full understanding of the implications of the various actions open to him but also some of his mentors seem capable of thinking only along the tram lines of the particular set of jargon in which they have become expert. Within such a situation those professional advisors who wish to be heard must exercise considerable political skills. Not unnaturally one finds among the bureaucrats some of the consummate politicians.

For those who perceive that simple solutions of this difficulty such as a "return to free market economy" or the application of "socialist principles" are professionally inadequate and unsatisfactory, two lessons need to be learnt. First there must be a willingness to accept that problems come as a whole. There are not a group of special agricultural economic problems, which can be solved by our techniques alone, instead problems come complete with sociological, political, technological as well as economic aspects. Progress if it is to be made at all, requires attention to such interdisciplinary links. For instance the target of economic growth, if it is defined only in terms of our jargon and ignores other links, may prove disastrous for communities in both developed and less developed countries. It could be the case that some of the agricultural policies practised in developed countries in defiance of economic logic, have been attempting just such a balance.

Second, if we believe we have a contribution to make, it is our duty to make ourselves understood not only to one another, but also to other experts, to politicians and to those upon whom the politician depends for power. Many issues are of technical complexity but that significance must be explained in terms which can enable those who have to take decisions to really appreciate the options open to them. If we fail to do this then, in an important sense, we either become the decision taker or our advice will be ignored. Neither situation should be acceptable to the professional advisor.

2. Professor Ohkawa directs our attention to the attempts of agricultural policy-makers to achieve income parity between farmers and other people in developed countries. As one point for discussion he suggests that the "basis of this policy is a recognition that the mechanism of markets, both output and factor, cannot operate in favour of farmers in achieving income parity, taking the urban level as standard – an acceptable request made in highly industrialised countries".

It seems at least possible that the almost universal failure of agricultural policies to attain this objective indicates that the policy-makers may not have well understood the economic mechanisms at work. Market mechanisms, in general, do not lead to equality of returns except amongst those factors which have equal marginal productivity. Where unequal returns exist for factors which are in all other respects similar, the market sets up pressures towards equality. Income gaps perform precisely this function. Two types of attack may, within the general framework of a market economy, be sustained. First, that the gaps in income are unduly large because the market does not give adequate information or because factors are prevented from movement by institutional elements in the situation. Second, that the ultimate pattern of income distribution or the costs to the community of mobility implicit in the market mechanism, are too high. In the first situation we have too little mobility, in the second too much.

It is a source of confusion in agricultural policy that both criticisms are held to be simultaneously valid. Thus we have structural programmes designed to promote mobility and price policies which protect existing producers from the full blast of market processes. Certainly the overall effects of much of this "policy activity" is that markets for factors and outputs do not respond freely to changing supply and demand. On the other hand it is also true that the gap in incomes remains persistently high. At the same time massive dislocations occur in both domestic and world trade whilst exporting countries find that the income problems of farmers in prosperous countries deny them access to some of the largest markets in the world.

The possibility must be considered that, if we had decided which line of approach – more mobility or less mobility – was to be the route of attack on income problems, policies would have been more successful and less costly. Had we pinned our hopes more steadfastly to mobility, to attaining a better resource allocation, then market mechanisms would have provided higher incomes both for those who left and those who remained at the price of considerable, if short lived, discomfort. The greater total wealth at our disposal would ultimately have enabled us to deal generously with those who, through age, remoteness, etc., were unable to adjust. On the other hand, if we reject the shape of a modern society brought about by mobility, it would be more sensible to tax rather than to subsidise technical advance, to freeze the pattern of trade and to try to force prices up to keep pace with economic growth in general. Such a policy would, of course involve heavy and growing costs in the sense of lost opportunities for economic growth but it would tend to stabilise rural society and to maintain the incomes of the unprogressive farmers. Since in our society it is often the unprogressive who are the poorest, this, like the other approach might meet income goals reasonably well. It is a sad feature of much of our muddled approach to these problems that although we do transfer substantial amounts of real income, those who are poorest receive relatively little. Professor Ohkawa rightly cautions us about the complexity of the agricultural sector (p. 197) and the need in applying optimum resource allocation principles to take into account expectations about future market conditions as well as evidence about the recent past. Such caution is proper but should not obscure the need to work with, rather than against market forces; a neglect which has been the undoing of many agricultural policies.

3. Professor Ohkawa, in discussing the role of the family farm, towards the end of his paper, makes the thought-provoking suggestion that the "objectives and instruments of agricultural policy are modified corresponding to the varied stages of development".

For the moment, it must be left to others to spell out the details of a "layman's guide" to appropriate policies for different stages of development.

One guesses that the acceptable permutations are sufficiently numerous to be interesting. More important, perhaps, is the need to recognise that over time policies must change.

There is not inconsiderable risk, because of the politically determined character of much policy, that specific objectives or instruments will become sacred. Should this be the case, and Professor Ohkawa prove right – as I feel sure he is – then the development of policy may tend persistently to lag behind current realities. This seems to happen with the CAP of the EEC. It may well be true of our response to higher energy costs and to growth in world food demand. One important function for the professional agricultural policy-maker is to prompt responsiveness to change, to be sceptical about existing arrangements and to be constructive in offering alternatives.

In his far reaching survey Professor Ohkawa demonstrates just such an approach and encourages it in others.

Report of the general discussion

The breadth and depth of thought in Professor Okhawa's paper both occasioned comment to this effect from a number of speakers and led to the discussion taking the form of a number of separate lines of thought.

Several speakers, especially from the centrally planned countries, emphasised a need to treat agricultural integrally with associated sectors. Thus it could be usefully treated as a sub-sector of the whole food supply system. Equally, it was argued that there should be an integrated approach to rural areas, guided by broad social and political aims. Others took a more generalised view, seeing society as a system of interlocking and interacting groups. Any decision which affects the structure of one unavoidably influences that of others. The relationship of the agricultural to the non-agricultural sectors can only be rightly understood by some such approach. Furthermore, decision-making is not solely concerned with ways to achieve previously established objectives, the selection of objectives was also seen as a role of the decision-maker.

The influece of power structure – including the balance between the agricultural and non-agricultural sectors – on the tenancy of economic policy was another aspect which attracted attention. In consequence, an important need was to make agriculture more powerful and harmonious with itself, so that the benefits of agricultural policy were spread more widely.

Some of those who spoke with developing countries primarily in mind seemed pessimistic about the scope for agricultural economists to be influential. Partly this seemed to be because they felt that agricultural economists had come on the scene rather late. "Pure" economists are probably nearer decision centres than are agricultural economists and one might hope that they would read Professor Ohkawa's papers. However, the whole agricultural sector was seen as getting reasonable attention only when industrial growth led to a shortgage of food. Sometimes parts of it were the victims of expediency, as, for example, where funds accumulated in one fund – e.g. for groundnuts – have been used to buy imported foods, or export taxes or export taxes on groundnut oil has substituted for direct taxation which it was impracticable to administer: Such tendencies are reflected in an increasing share of value added assigned to the state compared to the farming sector.

The view was expressed that in respect to agricultural policy most European socialist countries lay somewhere between the developed and the developing countries. They resemble the LDC's in seeking development to DC levels but they are short of resources. They have achievements to show but have not solved their problems. The rate of development depends on resources available at different periods and the varying views of those in control about their allocation.

References to developed country experience underlined for the U.S.A. the two parts to the farm sector, namely the small farm sector where the key problems are low incomes and inadequacy of social services and the large farm sector where income stability is the main problem. It had been recognised that price policy was not the solution for both and the small farm problems have been attacked by way of farm job creation, vocational training, health services, etc. This seemed now the settled policy.

Output-reducing programmes with the aim of increasing prices had a variety of unintended results on consumer prices and the approach has changed fundamentally in the past three years to give a clearer run for farm production expansion. Farmers appear well satisfied with it but whether it was firmly established was unclear.

Participants in the discussion included: J. A. Akinwumi, *Nigeria*; P. C. Baillet, *France*; R. Benalcazar, *Ecuador*; A. Burger, *Hugary*; K. Jojima, *Japan*; V. I. Nazarenko, *U.S.S.R.*; D. Paarlberg, *U.S.A.*; G. Rocheteau, *France*.

K. Ohkawa (in reply)

I am grateful to all discussants, especially the openers, for their contribution to amplifying my paper either by presenting concrete examples of conflicts between general economic policy and agricultural policy or by extending in a much more thoughtful way what I stated with respect to the "problems of political preferences". There seems to be no serious disagreement with me. I share the view that in the developed countries with market economies industrialization creates a number of unfavourable effects on agriculture, particularly in relation to employment and land use.

Power structure is, of course, relevant to what I called the effect of political preferences, but with final decision-making at the political level this does not mean that power structure is the only determining factor. I recognize the "destructive" aspect of modern economic development, sharing Kuznet's view, but here we are concerned with the relation between agriculture and the economy and not the agricultural policy as such.

I have, myself, drawn attention to a possible shift of the relationship between agriculture and industry from the early stage to the developed stage, so that I fully recognize the importance of the essential nature of the problems which may arise at the "intermediate" stage, as for instance, in Hungary. The comments should be extended in a more general way. What I said about "income parity" does not necessarily express my own policy preference. Instead, I mentioned it as an illustration of the conflicts and the different views in interpreting them. To set the objective "socially" in a more comprehensive way, as asserted by a Soviet friend, is quite acceptable and desirable, if it can be fully implemented. In this regard "integrated rural development" touched upon in my paper would deserve attention.

I understand the two distinct policies – one for small farmers and the other for commercial farmers, and clarified with respect to the U.S. case. Japan's case may be similar to the former. I am happy to learn that in developing countries there is no distinction or separation between agricultural economists and general economists. Perhaps in developed countries we may be at a disadvantage from having too much specialization among us.

GÜNTHER SCHMITT*

The Relationship Between Agricultural Policy, the Economy and Economic Policy on the National Level in Different Economic Systems and at Varying Stages of Economic Development: Concepts, Frictions and Solutions: Western Industrialized Countries with Market Economic Systems

1. INTRODUCTION

The subject of this paper is the description, analysis and evaluation of interrelationships between agricultural and economic policy in industrialized countries with market economies under the specific economic and agricultural conditions of those countries. Its central hypothesis is that, subject to similar policy objectives, the dimension and structure of farm policy is determined by the specific economic and agricultural conditions prevailing in a country. In this context, "by dimension and structure of farm policy" is understood the volume and the composition of government aid to the farm sector (budget outlays and farm price support by import restriction etc.).[1] If this proposition is accepted, that is that the shape of the farm policy in the countries under review[2] is not so much subject to manipulation of political decision-makers but a function of actual economic conditions, etc., it follows automatically that inconsistencies between farm and economic policies are a systematic (structural) concomitance of market economies. It follows furthermore that the application of specific instruments by decision-makers in order to achieve the overall economic policy objectives – such as full employment, price stability, growth, etc. does have impacts on the specific economic conditions for the farm sector. Those impacts do influence in reverse the shape of farm policy and, consequently, the type and degree of inconsistencies mentioned will affect, of course, the performance of the total economy and can theoretically be measured in terms of welfare losses which the economy has to sustain in comparison with an optimal policy resulting in a Pareto-like optimum.[3]

In fact, unlike "economists", policy-makers do not intend to achieve Pareto optimality – even if they have all the information and instruments at hand necessary to achieve it. Such a political action would be in contrast to the basic principle of *political* decision-making: The policy-makers possibility function is not determined solely by the economic mechanisms. The policy-makers have to take into account a number of constraints, they cannot use any instrument that may be at hand,[4] and they are often also limited in the extent to which they can use any particular instrument. Furthermore, their

* University of Göttingen, Federal Republic of Germany.

decisions are very often the result of compromises between strongly divergent views of various members of the decision-making institution, for instance with respect to long run against short run. In general, policy-makers do not intend to maximise social welfare but to maximise the sympathy of voters in order to be re-elected.[5] This implies a specific choice of economic policy objectives not necessarily in line with the set of policy objectives of the economist. One reason is that the preference function of the policy-makers includes aims which are not directly related to the economy, such as individual freedom, international peace, etc. The resulting policy decisions therefore include some sacrifices of economic efficiency and will result in a possibility curve which will be lower than that of the economist. The actual policies will yield a lower level of collective well-being[6] compared with an optimal policy.

Those actual divergencies between the actual performance of the economy and its welfare maximising state can in a more systematic view be described (a) in terms of *incompatible* objectives of economic (and farm) policy, (b) an *inconsistent* application of instruments, and (c) a lack of *conformity* between expected (planned) performance and the forthcoming results of policy measures. Nonconformity may be due to insufficient information about the future course of events inside and outside the economy, incomplete knowledge of the (side-)effects of instruments, insufficient harmonization of decisions due to decentralized institutions, etc.

After having defined some basic propositions and terms to be used here, we will systematize the objectives of economic and farm policies in the countries to be reviewed. In the same section we will further include a definition and description of the main policy instruments to be used in those countries in order to achieve the policy goals mentioned. Resulting inconsistencies between these instruments will be discussed next. In the following chapter we have to deduce factors affecting the specific types and structures of farm policies in various industrialized countries with market economic systems in order to detect the reasons for existing inconsistencies. Then we will identify some sources of nonconformities of policy measures. Some remarks will be made in this context concerning inconsistencies of farm and economic policies as a result of the foundation of the European Common Market. In the final section we will present some ideas concerning possible means of resolving prevailing nonconformities, and inconsistencies as well as incompatibilities of farm and economic policies.

2. A SYSTEM OF DEFINITIONS AND CLASSIFICATIONS

First we have to define the terms "Industrialized Countries" as well as "Market Economic Systems" because the specific definition will affect the type and number of countries to be included in this analysis. With the term "*industrialized country*" we denote a country *showing an absolute decline in the size of farm population*[7] instead of using a specific level of gross national product per capita as a border line between industrialized (developed) countries and developing countries.[8] Using this turning point in the trend of agricultural population (and/or labour force) towards an absolute decrease in the size of farm population as the demarcation of industrialized countries, it will be demonstrated that *farm policies* of those countries are basically different from policies in countries beyond this line:[9] Passing this turning point will, generally speaking, reshape farm policy towards a definite redistribution of income in favour of the farm sector instead of a redistribution of limited resources (land) between an increasing number of farmers (land reform).[10] Consequently, classical land reform measures are not reviewed in our analysis. The term "*structural policy*" is limited to measures increasing productivity of resources within the farm sector by increasing factor (land) endowment of (decreasing) farm labour input.[11]

Economic policy in our context describes the process by which the decision-maker – executive (government) as well as legislative power (parliament – decides on the relative importance of certain objectives, and, if necessary, uses instruments or institutional changes in the attempt to obtain those objectives.[12] According to this definition, *farm policy* must be regarded as a subsystem of economic policy (which itself has to be regarded as a subsystem of overall policy) pursuing specific targets and applying specific instruments which idealisticly are compatible and consistent with economic policy. Both, economic as well as farm policy, will be executed within a certain framework of institutional arrangements (economic order) which will determine the character of the mechanism by which the decisions of economic agents will be coordinated as well as those decisions which will be subordinated to the objectives of the political decision-making body.[13] Within a market economy, the coordinating process will be fulfilled by the price mechanism; subordination will be approached by incentives and disincentives (gains and losses) due to changes in policy instruments used by decision-makers such as fiscal and monetary policy, etc. By market economic system in our context we refer only to economic systems of the latter type, sometimes called social market economies or organized capitalism, contrary to free (liberal) market economies (pure capitalistic systems) excluding subordinating mechanism on one side and centrally planned economies using imperative measure of subordination[14] on the other side – although the borderline between various economic systems in reality is blurred (mixed economies).

By the definitions given above, countries included in our analysis are North America (United States of America, Canada), Australia, New Zealand, Western, Northern and Southern Europe (excluding Albania) and Japan.[15] A further distinction between North America and Australia and the other countries mentioned has to be made according to the prevailing land-man-ratio:[16] The first mentioned New World-countries show an extremely wide land-man-ratio compared with the second group of Old World-countries. Within this group of countries we have to distinguish between Southern European countries (except Italy) and Western European countries, because of fundamental differences with respect to the stage of economic development and the institutional organization of farm policy (development plans). As is pretty evident, different land-man-ratios will determine the structure of farm policies, especially the magnitude and shape of the structural policy as well as the magnitude of price policy. This will be discussed in detail in Section 3.

Economic as well as farm policy has further to be defined as a system of policy objectives (aims, goals), policy instruments of various types (means) which are used or changed by policy-makers within the given institutional and/or functional constraints in order to achieve economic (farm) policy goals if the prevailing performance (situation) of the economy diverges (or is expected to diverge) from the desired one. This definition of policy as a systematic interrelationship between goals, instruments and performance of the economy holds for varying periods of time (short, medium as well as long run) as well as for various subsectors of the economy such as farm policy etc.:[17] The specific meaning of those ingredients of the policy system is changing with respect to the time period under review as well as to the specific subsector. Because our analysis is limited on the one hand to medium and, to a limited extent, to a short run period, and to economic policy in general and farm policy as a subsystem thereof, we need some further classifications. We use the term pure objectives – as economic proxies of political aims – in the sense that the existing economic and farm policy objectives can be translated in quantitative terms at least to a certain extent, although they are not quantified by decision-makers in any case. They can be divided in short-term and long-term objectives. *Quasi-objectives* are objectives which cannot be quantified (security of supply, promotion of international division of labour, regional distribution of income and resource allocation, conservation of national resources and environment, etc.). *Achievement* is the term used to describe the degree of attainment of an objective. Objectives can be achieved by policy-makers using *instruments* of *economic policy* which might be classified as fiscal, monetary, exchange rate and direct control instruments. Within farm policy, instruments are classified in this context as price (income) policy instruments, steering the level or the stability of input or output prices; structural policy instruments aimed to affect the allocation of resources within the farm sector and social policy instruments used to secure farm population against the risks of age, illness etc.[18]

Using the terms defined above, economic (farm) policy can be defined more precisely as an interdependent system of instruments used by decision-makers in order to achieve given objectives and quasi-objectives "at their best" by taking into account available informations concerning the specific economic situation, its likely future development, the main effects of possible instruments or objectives as well as the given institutional and political (ideological) constraints.[19] Within this system, the given objectives will be achieved "best", if the objectives of various subsystems of policy are totally compatible. This is normally not the case. *Incompatibility* may be the consequence of a specific economic situation (short run incompatibility), it can be structurally determined (medium run incompatibility) and may be due to the specific economic system (long run incompatibility). Each source of incompatibility, of course, requires a specific set of instruments. Consequently, the performance of an economy (or sector) with respect to the desired objectives will be further affected by the "constraints of the decision-making body": their ability to use an optimal set of instruments, optimal in terms of timing, combination and volume. The actual choice of instruments refers to the *consistency or inconsistency* of the instruments used. And finally, because in many cases the future change in a given situation as well as the future achievement of instruments in market economy can never be predicted as exactly as necessary, *nonconformities* between expected achievements and the actual outcome will occur. These nonconformities will again result in an inferior accomplishment of desired objectives.[20]

3. THE OBJECTIVES OF ECONOMIC AND AGRICULTURAL POLICY: THE SOURCES OF INCOMPATIBILITIES

In this chapter we first try to enumerate, to define, and to evaluate the actual objectives (and quasi-objectives) of economic and agricultural policy in the countries under review. Next, we will describe the relationships between the economic and farm policy objectives and the sources of incompatibilities of these objectives. In the following chapter we have to discuss *measures* of economic and farm policies resulting in inconsistencies; then we have to debate instruments which might minimize inconsistencies and welfare losses involved. In the then following chapter we will systematize the interrelationships between various instruments and their combination, taking into account the specific economic as well as farm "situation" in order to detect the actual sources of prevailing inconsistencies and nonconformities of objectives and instruments as well.

Reviewing the economic policy of Western industrialized countries after World War II, its performances,[21] the declared *price policy objectives*[22] and actual use of instruments,[23] main efforts of governments seem to be directed towards the accomplishment of full employment and price stability in the short run, economic growth and redistribution of income in the long run. Of course, various countries under review differ as far as the priority of short and long run objectives, the specific numerical size of short and long run objectives, the volume, structure and nature of instruments, and the ratio between actual achievements and targets[24] – the achievement ratio – is concerned. Furthermore, the countries also differ in objectives, instruments and performance of economic policy with respect to various time periods (business cycles). In fact, the central issues in the economic policies of Western countries all over the time period after World War II have been the never ending struggle between full employment and price stability, these fundamental incompatibilities of economic policy. Each country under review has been more or less successful in steering the economy between the Scylla of inflation and the Charybdis of underemployment.[25] In the long run experience it must be accepted that the achievement ratio of price stability in all countries has deteriorated,[26] due to rather liberal application of restrictive measures in fiscal and monetary policy, finally resulting in unemployment, devaluation and, sometimes, monetary crisis for the time being.

Contrary to price stability and full employment, *economic growth* and income redistribution between persons, regions and sectors of the economies have never been subject to systematic applications of proper instruments nor subject to definite quantifications of objectives.[27] In Western industrialized countries, especially the most developed countries, economic growth has been significantly lower since the Sixties compared with the Fifties. Factors explaining the lower growth rates are:

(a) the fact that the relatively unproductive agricultural sector serves no longer as an important source of economic growth,
(b) the rate of technological progress seems to decline,
(c) growth is heavily criticized as an indicator of welfare,
(d) institutional constraints hinder allocative efficiency.

It is possible that the increasing importance of the redistribution of income in favour of poor persons, regions and sectors affects the process of structural adjustment of the economy and, consequently, influences economic growth negatively. However, the actual redistribution of income in various countries of the Western world is difficult to measure and to evaluate, in most cases due to a lack of information.[28]

All countries under review protect a number of industries to some degree. In the West, *protection* is used especially for old declining industries, such as coal and textiles. They are protected in the one case from the competition of other forms of fuel, in the other from foreign competition. The basic method of protection is by tariffs, though quotas still play some rôle especially in the United States. Another important source of protection is the various non-tariff barriers. Agriculture is, of course, the classical example for protection. Protection here is justified partly by the increasing desire to improve the security of the food supply as a quasi-objective of economic policy.[29]

With respect to the subject of our analysis, as *quasi-objectives* of economic and farm policy relevant to agriculture and farm policy we would mention

(a) the equilibrium of balance of payments (as an intermediate variable affecting price stability, employment and growth),
(b) security of supply,
(c) promotion of international division of labour (with its repercussion on stability and growth), and
(d) protection of resources and environment.

In one or the other way, an achievement in favour of those objectives does affect the farm sector and/or farm policy. Nearly all countries mentioned have shown a fairly substantial gap between their balance-of-payments objectives and their actual achievements. Some countries had almost traditionally a deficit in their balance showing a worsening recently, whereas others still have a surplus, although a decreasing one. Others having had in earlier periods a positive balance, meanwhile experience a change towards a deficit.[30] The policy-makers endeavoured to influence the balance of payments by means of economic policy instruments especially by changes in the exchange rate. However, their efforts were very frequently inefficient due to various reasons. In some respect, farm policy directed towards export promotion for farm products somewhere did show a new strategic rôle within this context.[31]

Such a course of policy did have, or is expected to show, repercussion on the quasi-objective of *promotion of international trade*. Throughout the last three decades, the pendulum has swung in favour of international division of labour after a protectionist period between the World Wars. Until the recent energy crisis with the consequences on inflation, employment and the balance of payments, the trend in industrialized countries was clearly in favour of an intensified division of labour. Protection of individual countries (with the important exception of agriculture) was gradually rolled back and world trade, especially that between industrialized countries, has risen rapidly for over twenty years.[32] However, for the time being and very probably for the near future, the trend has changed towards more protectionism due to rising problems with the balance of payments in important industrialized countries, an import substitution policy in favour of more security of supply (of energy) and due to a policy for regaining full employment at the national level.[33]

In summing up, economic policies of Western industrialized countries with market economies do show great similarities as far as the nature and, to a certian extent, the priority of their objectives and quasi-objectives are concerned. Characteristic differences during the last decades can be observed with respect to the quantitative dimension of those objectives and the achievement rates of the policy instruments used. This is especially relevant to price stability and full employment, the most important policy objectives leading to almost permanent and, currently, to the most serious incompatibilities within economic policy. Furthermore, realized rates of economic growth differed widely between various countries.

3.1. *Farm policy*

As is the case in general economic policy, Western industrialized economies show quite similar objectives in their agricultural policy, at least as far as the nature of the goals is concerned.[34] *Income policy* concerning redistribution of income in favour of the farm sector has had the highest priority compared with other agricultural policy aims. This is true although the objective of income policy in almost all agricultural laws or official proclamations concerning farm policy is not expressed in quantitative terms with respect to the desired level of income, the specific measurement of income, the farmers or farms to be included and/or the size of non farm income as a standard for income policy.[35] Consequently, it is rather difficult to find suitable quantifications of the prevailing income objective nor is it easy to measure the rate of achievement of the relevant policy.[36] Even if the income goals were expressed in quantitative terms it does not follow automatically that an official guarantee of a certain income level for farmers means incompatibility with other economic policy objectives, such as promotion of economic growth. Whether farm income policy is consistent with economic policy aiming towards maximizing growth depends on

(a) the specific instruments used to achieve a satisfactory level of income in the farm sector,
(b) the relevant effects of these instruments,
(c) the prevailing conditions.

In the case of instruments increasing productivity, such as structural policy, this may result in a remuneration of the production factors within the farm sector similar to the remuneration of factors in other sectors. But traditionally when we speak about income policy we do not mean structural policy. With income policy we traditionally mean instruments such as price policy, subsidies due to tariffs, input or output restricting quotas, transfer payments etc. These instruments result in a price level for the producer above the equilibrium. Although tariff levels, the size of import quotas or the amount of subsidies are not appropriate measures of income policy objectives – because these are instruments rather than objectives – the degree or amount of agricultural protection and its probable effects on misallocation of resources represents a measure of the degree of incompatibility of the objective income policy versus economic growth – to be quantified by cost-benefit analysis.[37] Of course, the reasoning behind cost-benefit analysis is based on the (neoclassical) assumption that equilibrium prices lead to an optimal allocation of resources, an assumption which sometimes is subject to doubts as far as the allocative function of the price mechanism in agriculture is concerned.[38] Cost-benefit analysis clearly shows that welfare losses due to protective measures differ strongly according to the specific instruments being used by various countries. Therefore the level of support quantified by nominal protection rates is a measurement inadequate for the quantification of misallocation of resources by farm price policy.[39]

Income redistribution in favour of the farm sector by price policy measures, such as import restriction, is very often blamed as contrary to the objective of *price stability* within the economy. This problem arises specifically at times of inflationary processes in the economy as a whole if government tries to compensate the impact of inflation on farm income. However, the rate of increase of farm prices relative to the general price level does not indicate whether the agricultural income position really is affected by inflation, positively or negatively or, by the same token, is pushing inflation or contributing to more stability of the price level. This can only be judged with respect to agriculture's gains of productivity relative to those outside the farm sector. Provided that no additional income redistribution is attempted, price increases for farm products may well be consistent with the objective of price stability if the rises in prices are only compensating productivity gains smaller than those in the nonfarm sector and vice versa.[40] In fact, countries to be reviewed here, did differ to a large extent as far as the relation between inflationary price changes, productivity gains of agriculture and changes in the level of farm prices are concerned.[41] Obviously, the contribution of agriculture to an anti-inflationary policy of government depends on a number of factors, such as the prevailing income objective, the specific supply-demand situation and the state of the balance of payments. For the Common Agriculture Policy of the European Economic Community it must be added that a common price policy for countries subject to widely differing rates of inflation the interdependence between general price level, productivity gains and farm price level is disturbed for various member countries.[42]

This reasoning makes clear that increasing *productivity* by policy measures – such as structural policy in a rather broad sense[43] – is a rather important objective of farm policy of most countries under review, even if this objective is not mentioned in farm legislation of some countries. This objective is compatible with the growth objective of economic policy and, as has been said already, with the objective concerning income redistribution towards the farm sector. Therefore, it is not surprising that various countries underline productivity gains as an important farm policy objective. They allocate a considerable fraction of the budget for instruments to change farm and input structure.[44] It has to be mentioned that the rates of increase of productivity given in national long term plans do not correspond very strongly with actual achievements.[45]

Improved structural adjustments of, and within, the farm sector due to higher productivity of inputs result in a more efficient *allocation of resources* between sectors of an economy as well as within the farm sector. With respect to labour input, transfer of man-power to other sectors of the economy (showing higher marginal productivities) will not only increase gross national product[46] but also contribute to a higher degree of *employment* – provided that hidden unemployment in the farm sector is diminishing and the employment is not measured in terms of official statistics but in terms of effective employment. However, no contribution with respect to full employment will be achieved if the outflow of labour is only due to increased retirement of farmers as a consequence of "discontinuation schemes" being introduced in different types by almost all countries under review.[47]

Most "official agricultural laws" of Western industrialized countries emphasize the farm policy (quasi-)objective of securing *domestic supply* of food "at reasonable prices for the consumer".[48] Although no explanation is given as far as the level of consumer prices is concerned, the quasi-objective of security of supply in most cases seems to be only a camouflage for price policy – because the domestic supply of basic products in many countries under review permanently exceeds the minimum needed for emergency situations and, in some cases, even the total domestic demand.[49] Whatever the term "minimum required for emergency cases" quantitatively means, as far as I can see, only in Sweden has the degree of self-sufficiency by domestic supply been quantified. But in Sweden as in other countries actual production exceeds this level.[50] This in fact means that the objective of a minimum domestic production is in contradiction with the objective of reasonable consumer prices as world market prices are normally considerably lower than domestic prices. Welfare losses are the consequence of the misallocation of the resources implied.[51]

A more fundamental and important source of incompatibility between relevant farm policy objectives and the quasi-objective of *promotion of international trade* is bound to the impact of price policy on domestic production as described above. Price policy in industrialized countries of the West, as an instrument to increase producers' income, stimulates production increases. These production increases tend to be higher than the increase in demand due to population growth and rising real income per caput requires. As a consequence, most countries under review show a rising rate of self sufficiency, resulting in a relative decline (with respect to domestic demand) in imports of competing farm products.[52] Thus the income objective of farm policy contradicts the quasi-objective of intensifying international trade with farm products. This negative effect of domestic farm policy on international trade is intensified by the repercussion of various import restricting measures being used by those countries. In general, those import tariffs, quotas, etc. very often result in relatively low prices on world markets being subject to aggravated price fluctuations due to irregular export subsidisation in addition.[53]

Price policy directed towards stimulation of agricultural production is sometimes justified by decision-makers (and also farmers' organizations) as a measure towards *balance-of-payments-equilibrium* (where the balance of payments is in deficit).[54] Of course, this may be true in a purely statistical sense but, with respect to national welfare, import substitution seems to be only a second best solution of balance of payments problems. Import substitution results in losses of welfare of the own economy as well as of the export countries.[55] Consequently, price policy in the farm sector intended to counteract a negative balance of payments is, of course, not qualified to cure the fundamental reason for a balance-of-payments-disequilibrium. This cure only can be achieved by devaluation (in the short run) and by attacking inflation with fiscal and monetary policy measures.

A modern argument in defending income policy in Western industrialized countries nowadays stresses the (positive) *external effects* of agriculture not reflected in market prices. Those external effects are seen in *conservation of national resources*, environment, a specific structure of landscape and in impacts or urbanization, etc. Because the treatment, evaluation and measurement of external effects of the agricultural sector is a rather complex matter, not very deeply analysed, the interrelationships between various policy objectives and specific instruments and effects cannot be discussed in detail. It is supposed that, in general, these positive external effects resulting in social benefits higher than market prices are overestimated, especially by neglecting the social costs of farm production in terms of negative external effects. Therefore, to a certain extent this argument seems to be overemphasized. It is again a camouflage for a policy of high farm prices resulting in incompatibilities and inconsistencies of farm and economic policy.[56]

In summarizing this section, it must be stated first, that the nature of given objecitves in the field of economic and farm policy by themselves does not lead automatically to incompatibilities. In most cases, they seem to be compatible with each other. Incompatibilities of objectives in Western industrialized countries can only be detected if those objectives are expressed in quantitative terms and the effects and, especially, side-effects of instruments applied to achieve those objectives are considered. It must be remembered further that the nature and degree of prevailing incompatibilities between objectives of farm and economic policy are not only determined by the given structure of an economy (long run incompatibilities) which can only be solved by restricting the economy by relevant instruments or by adjusting the nature, and/or structure, of the preference function (objectives). Incompatibilities also can and do result from medium and short run conditions of the economy or the farm sector. These external conditions of the specific state of the economy and the agricultural sector are changing over time and differ between the various countries under review. Consequently, a detailed and a more precise definition, description and quantification of incompatibilities of given policy objectives in these countries cannot be presented within our analysis.

4. THE FARM AND ECONOMIC POLICY INSTRUMENTS: THE SOURCES OF INCONSISTENCIES

As stated above, inconsistencies of a policy follow out of the effects a specific instrument has on another rather than the objective to which the specific instrument is aimed (side-effect). Inconsistencies can be the consequence of:

(a) objectives which are unrealistic in a sense that they never can be achieved by any instruments available to decision-makers, that

(b) instruments cannot be applied due to institutional and other constraints, that

(c) instruments are insufficient with respect to their nature, timing, structure, intensity and combination and, finally, that

(d) insufficient coordination by the decision-making bodies prevents an optimal use of the instruments available.

Again, in our analysis a discussion of all instruments being used by governments in countries under review in different periods of time is almost impossible. Therefore we will try first to systematise farm policy instruments. Then we will identify the principal inconsistencies resulting from the use of these instruments. In the following section we will provide a systematic view of applied farm policy instruments in Western countries as far as their volume and structure is concerned. This consideration results in the hypothesis that some common factors, such as the level of economic development or the degree of structural adjustment, determine the nature, volume and structure of the mix of policy instruments used in the farm sector. Inconsistencies of policies therefore can be more systematically deduced from these determinants of farm policy.

Traditionally, farm policy instruments are classified according to the purpose (objective) they are supposed to achieve. We distinguish between instruments with a prime effect on *income* (income policy instruments), on *productivity* (structural policy), on *social security* (social policy instruments aimed to secure farm population against specific risks of such as age, illness, accidents etc.) and *non-sector specific policies* which may affect the farm sector although they are not specific to the farm sector (regional policy, measures to improve factor mobility, labour market regulations to enhance technological progress, etc.).[57] We will, of course, restrict our analysis to the sector specific instruments, mentioned above, although it seems to be clear that non sector specific instruments may be as effective in promoting adjustment of agriculture as sector specific instruments.

Price policy instruments are used to increase stability of farm (and consumer) prices and/or to improve producer incomes in the longer run.

Stabilization of prices in general may be consistent with the relevant economic and farm policy objectives, although stabilization of producers' income will not necessarily be the consequence of a stabilized price. The welfare gains are subject to some doubt – especially if the impact of stabilized prices on producers' risk is neglected.[58] However, the inconsistency of income policy via price increasing instruments with respect to optimality of resource allocation (as an economic quasi-objective), economic growth and promotion of international division of labour, as well as price stability of the whole economy, seems to be evident. Resource allocation within the economy, the farm sector and between nations is disturbed, growth rates are depressed because of a reduced transfer of resources from agriculture to more efficient uses outside this sector and an impact on the level of consumer prices, are the most striking consequences implying welfare losses for the economy as a whole. The size of welfare losses of course depends on a number of factors, not only on the protection rate (the difference between internal prices and equilibrium prices). Factors to be mentioned here are the degree of employment (affecting the opportunity costs of factors restrained in agriculture by price policy) and the specific instrument used in order to protect farmers' income.

The losses of welfare are less by transfer payments (subsidies) compared with border protection by tariffs, import quotas etc. simply because a distortion of consumption is prevented by the application of subsidies,[59] even taking into account positive balance of payments effects of protection.

The inconsistencies of price policy used as an instrument to protect farmers' income can be reduced to a certain extent by substitution of price policy by structural policy instruments. Although shape, organization and management of various instruments in the field of structural improvements are different in countries under review,[60] they are all directed towards increasing efficiency of factor input, either by reallocation of resources within the farm sector or between sectors and/or promotion of technological progress in production and marketing. Such a reallocation of financial resources in favour of structural policy may also unburden price policy with respect to its output increasing effect. This leads us to the hypothesis that countries are forced to substitute price policy to an increasing extent by structural (and social) policy measures because the domestic supply of agricultural products is exceeding domestic demand (at increasing rates due to decreased growth of population and income as well as due to decreasing income elasticity of demand). This refers especially to those Western European countries (including Japan) with an extemely narrow land-man-ratio equivalent to a land intensive production, low labour productivity[61] and a rather high protection rate of domestic production.[62] Of course, the achievement rate of structural policy instruments not only depends on the specific shape, volume and nature of policy instruments, but is also determined by the relevant economic situations during various time periods and in the countries concerned. This will be discussed in more detail in the next section.

In general, the achievements of structural policy measures depend on their effects on opportunity costs of labour inputs in agriculture. If opportunity costs are relatively low, even high government subsidies for adjustment in factor inputs – such as discontinuation schemes or retraining programs will result in rather low achievement rates.[63] As a consequence, structural effects will be highest in economically underdeveloped rural areas, worsening income distribution in agriculture. Therefore, a successful structural improvement only may be achieved by coordination of structural policy instruments and regional and infrastructural policy instruments, improving labour market conditions etc.[64] In many countries which are subject to great discrepancies in regional income and growth rates this link between structural policy for the farm sector and general economic and social policy is missing.[65]

Since World War II all industrialized countries of the Western world have introduced and extended their system of social security for farmers, especially with respect to age, illness and accidents.[66] However, the specific systems of social policy in various countries differ, even more than structural policy measures, with respect to persons included, fees, rents, etc. In some countries, retirement schemes are part of the general social system, in others specific systems are introduced for farmers, sometimes combined with discontinuation schemes for farmers, etc. Because of the prevailing wide range of differences between those social policy instruments it is, of course, difficult to evaluate them with respect to their consistency with various farm and economic policy measures. All systems, at least contribute to an intergenerational income transfer, most systems to an intersectoral one as well, because the number of active farmers and, very often, their individual fees are relatively small compared with total and individual old age pensions.[67] In this case social policy is – more or less – a vehicle for redistribution of income. The degree of consistency with the income objective of economic and farm policy, of course, depends on the specific scheme of old age pension regulations. Because in many countries old age pension for farmers is not related to the income of farmers, the impact on intrasectoral income distribution is the subject of debate, especially because in many countries under review retirement schemes are combined with one form or the other of discontinuation rewards for older and/or smaller farmers.[68] Sometimes the specific pensions schemes for farmers reduces the intersectoral mobility of farmers, thus resulting in reduced productivity growth.[69]

With respect to external effects of farm production on natural resources, environment and landscape, farm policy to an increasing extent is subject to specific instruments which subsidize agricultural production in certain areas (e.g. the hill farming program of the EEC). Such policies have an income redistributing effect. Again, the quantitative effects are difficult to estimate. They are dependent on size and structure of the instruments in use and the specific conditions prevailing. Consequently, the (negative) repercussion on resource allocation may be more or less strong.

It is the nature, size and specific shape of various farm policy instruments which determine extent and type of inconsistencies between farm policy and economic policy. The most important inconsistencies are due to income increasing instruments of price policy – although specific instruments are different with respect to various countries, time period and nature of the instrument – and may exist as far as domestic and international allocation of resources, economic growth and intrasectoral income distribution are concerned. Welfare losses involved seem to be extensive, although they are determined to a certain extent by the conditions of the economy as a whole. Structural policy instruments, if well suited (optimal) to the existing conditions of the economy and the farm sector, may reduce the welfare losses of income policy. This may also be the case with respect to social policy measures in agriculture, although welfare gains may be modest in most of the countries under review. As said already, the effects and side effects of various measures depend to a great extent on the specific (external) economic and agricultural conditions prevailing in the specific country and time period under review. This interrelationship between the state of the economy, policy instruments and given objectives, resulting sometimes in low or negative achievement rates due to nonconformities of instruments, will be discussed next.

5. THE REPERCUSSIONS OF THE STATE OF THE ECONOMY ON ACHIEVEMENT RATES: THE SOURCES OF NONCONFORMITIES OF POLICY INSTRUMENTS

By 'state of economy' we mean in this context

(a) the rate of inflation,
(b) the degree of employment,
(c) the extent of interregional disequilibrium with respect to economic activities and growth,
(d) the rate of economic growth,
(e) the balance of payments.

It is well known that industrialized countries in the West have achieved the relevant objectives to widely different degrees and that their achievements have been different at different time periods.[70] Therefore, only a systematic view on various types of nonconformities can be given here. The fact must also be stressed that the specific state of the economy described and measured by the relevant performances of the economy is the consequence of internal and external policy instruments, such as fiscal, monetary policy, etc. applied by national and international decision-making bodies. These effects, of course, cannot be discussed here.

Inflation, a permanent and, for the time being, accelerated attendant of economic growth of all countries – although at different strength – affects the farm sector especially with respect to the rise of most input prices. Rising input prices will be transferred to producers and consumers either by the (free) price mechanism or by political decisions on administered prices, thus resulting in additional pushes applied to the general price level. In fact, the impact of inflation on the farm sector is rather complicated and it depends on prevailing conditions (such as the decision on farm price level) whether agriculture is a loser or a winner with respect to its position in the struggle for income distribution within inflationary processes.[71] At least, inflation affects income distribution within the sector.[72] If inflation results in a worsening of the balance of payment, sometimes governments are likely to stimulate farm production by raising farm prices in order to substitute for imports of food. On the other side, policy measures to counteract inflation have tended towards rather small increases if farm prices are determined by administrative decision. However, no general and definite answer to the question concerning the impact of inflation on agriculture and the consequence on farm policy seems to be possible. Further analysis is necessary.

In contrast, the degree of *unemployment* in the economy directly and indirectly effects agriculture. Unemployment will result in a smaller increase of demand of farm products due to a depressed rise of incomes, and, even more important, a decline in off-farm jobs available for farmers.[73] This will lead to a smaller outflow of labour, resulting in a relative decline of income in the farm sector. This is especially relevant for the farm sector because overall unemployment in the economy means extremely high unemployment in rural areas distant from urban-industrialized centers of activity.[74] In the longer run, structural adjustment of, and within, the farm sector is reduced. This can be observed in countries showing permanently high rates of unemployment. Again, regional disequilibrium in adjustment is extended and the effectiveness of instruments for structural policy are hampered because of lack of job opportunities outside agriculture. This interdependence between unemployment and the nature, direction and speed of adjustment of the farm sector is increasing in its importance because decreasing growth rates of demand for farm products and the rapid rise of labour productivity within agriculture require a relative increase in labour transfer from the farm to other sectors of the economy. The rate at which labour leaves the agricultural sector, either through migration or though part-time employment, is highly responsive to the level and changes in unemployment. If labour input adjustment is restricted by unemployment in the non-farm labour force, pressure for higher producer prices will occur in order to achieve an income level comparable with non-farm income. The possible consequences have been discussed earlier.

Similar interrelationships between employment and structural adjustment as discussed above can be observed with respect to *regional discrepancies* in income, employment and growth. Because such regional imbalances do exist within various industrialized countries, although to different degrees,[75] the speed and direction of the adjustment process of agriculture in various regions differs to a large extent, even if the overall economy is sometimes growing at impressive rates. Low income, low productivity farms are concentrated in underdeveloped rural areas in contrast to urban industrial regions being locations of efficient, big and fastly growing farms. Rapid growth in the non-farm sector of a nation is a necessary but not a sufficient condition to overcome the regional, occupational, age (and racial) dimensions of poverty in the farm sector because interregional mobility is much too restricted.[76] This problem of interregional discrepancies in income and growth of the farm sector can only be solved by increasing efforts with respect to regional policy. Most of the countries under review have recognized this imbalances resulting in inconsistencies within farm as well as economic policies, and have increased financial support for regional policy instruments. However, the decline in employment and economic growth since the energy crisis has affected negatively the achievement rates in creating new job opportunities in underdeveloped rural areas.

The reasoning with respect to regional disequilibrium in labour markets above seems to be a further demonstration for the fact that rapid *economic growth* is a prerequisite for an adjustment of the farm sector without social unrest and tensions. Rather constant and high growth rates generate increasing demand for farm products, favourable conditions for transfer of labour to outside agriculture, high mobility on the land market, a stimulation for introduction of productivity increasing technological progress, low effectiveness of price and income policy measures on misallocation of resources (because of a strong competition from outside of agriculture and wide variety of individual choices between many alternative activities, etc.).[77] In fact, actual growth achieved by various countries did differ widely with respect to the size of growth rates, and the strength and length of business cycles disturbing the adjustment process of agriculture.[78] The present economic crisis demonstrates dramatically the interdependencies of agricultural adjustment process and economic growth and stability. These interrelationships also could be observed with respect to differences in the size and stability of economic growth in different countries of the Western world – although a number of additional factors have influenced this relationship. And there is still another aspect of economic growth. Because all countries to be reviewed did subsidize their farm sector by fiscal policy measures of one or the other type, growing economies were able to support increasing financial burdens of farm support because of rising tax resources and fast growing volumes of state budgets. As a consequence of this, almost in all industrialized countries of the Western world the government budget outlays for agriculture have increased absolutely (in real terms) throughout the period since World War II (and very often relative to total government budget and agriculture's net national product, etc.) although the share of agriculture in the national income, labour force, population etc. has rapidly declined. The reason for this astonishing phenomenon will be discussed in the following section.

Negative economic growth or even economic stagnation have (and will) effect agriculture negatively and probably to a greater degree than non agricultural sectors. Under such conditions, inconsistencies and nonconformities will consequently be increased, resulting in economic instability, social tensions and negative impact on international trade and, consequently, welfare. Those indicators represent the present situation in Western industrialized countries, of course, with varying magnitude. The solution of such difficulties and problems must be seen in regaining economic growth and stability, even the instruments to achieve these objectives seem to be less efficient than they were in earlier periods because of institutional constraints, a lack of confidence in the future, increased dependence on international division of labour etc. With respect to the latter, increasing imbalance of payments of some economies have resulted in measures to stimulate domestic farm production in order to substitute imports and to promote export of farm products. This, of course, may contribute to increases of farm income and therefore to the achievement of the income objective of policy. However, it must be doubted that those (short run) positive effects will be greater than implied welfare losses because of declining international trade leading to declining rate of economic growth, again sharpening inconsistencies of farm policy. At least in the medium run, such a policy will increase nonconformities of policy measures. Policy has to be changed later or earlier because implicit welfare losses cannot be borne any longer by the economies.

6. THE PROCESS OF SHAPING FARM POLICY IN WESTERN INDUSTRIALIZED COUNTRIES: TOWARDS AN ECONOMIC THEORY OF AGRICULTURAL POLICY

Until now, we have tried to systematize the main sources of incompatibilities, inconsistencies and nonconformities determining the interrelationships between objectives, instruments and the prevailing economic situation with respect to economic and farm policies in various industrialized countries of the Western world. However, this review may stimulate the impression that each country may be subject to inconsistencies, incompatibilities and nonconformities of the types described above without any systematic relationships between the type of conflicts mentioned and the state of economic development, the degree of structural adjustment of the farm sector, and the supply of farm products respectively to the demand of various countries. This remark leads immediately to the question whether farm policy may be determined systematically by specific conditions of agriculture determined by the levels of economic development and other relevant factors. The reasoning may be as follows: We do know that adjustment of the farm sector to changing economic and social conditions is determined by specific factors such as population growth, income and price elasticity of demand, productivity growth of factor inputs and demand for parity income for factors used in agriculture. These factors which effect in a projectable course the adjustment of the farm sector are indicated by the share of agriculture in national product, total population and labour force being a function of the absolute level of income per caput etc.[79] It may be further hypothesized also that policy and its changes over time are determined by the degree of adjustment of the farm sector to the prevailing state of the economy – provided that pure and quasi-objectives of farm and economic policy have not been changed fundamentally.

Of course, such a theory postulating that policy objectives and instruments are determined mainly by economic conditions is contradicting the hypothesis according to which policy decisions in a democracy are dictated by the political power of various groups in the society. Although we admit that the political theory of economic policy has been elaborated and refined in a number of aspects, especially within the context of the socalled "New Political Economy"[80] since Schumpeter's and Down's fundamental contributions,[81] with respect to farm policy formulation and implementation we have basic reservations. These reservations are mainly based on the paradox which obviously exists between the rise in the volume of farm support and the rapidly declining farm population.[82]

6.1. *The basic hypothesis*

We will attempt to solve this contradiction by showing that economic progress requires specific structural adjustment of a farm sector of an economy being subject to intensification due to advancements in economic growth. Assuming, as stated above, no radical change in policy objectives, farm policy measures used to achieve given goals have to be (a) intensified and (b) structurally adjusted according to the prevailing state of economic growth. It follows immediately that inconsistencies and incompatibilities between means and ends of farm versus economic policy are determined by the extent, nature and structure of farm policy instruments, being a function of the state of adjustment of the farm sector. Of course, this hypothesis has to be modified according to specific conditions of the countries under review due to differences in farm structure determined by historical development, etc. Such an economically based theory of farm policy enables us to analyze our issue more systematically as far as policy conflicts are concerned.

In this context, we will first discuss the problem involved by measuring shape, extent and structure of farm policy measures. Second, we will describe the underlying theory of adjustment to economic growth of (a) the farm sector and (b) the farm policy mix and its repercussion on concepts, conflicts, frictions and solutions of the relationships between agricultural and economic policy on the national level at varying stages of economic development in Western industrialized countries.

6.2. *Measuring the shape, volume and structure of farm policy measures*

The farm sector of the countries to be reviewed is supported by measures of two fundamentally different types.

First, agricultural support by measures influencing directly the supply of farm products such as import restriction, domestic supply control, etc. These instruments act as vehicles for transferring income from consumers to producers directly via higher market prices.

Second, farm support by a great number of fiscal policy measures such as subsidies for inputs, farm products, direct transfer payments, structural reform and social policy measures, etc. These income transfers are direct government expenditures for agriculture, financed by internal revenue, etc. Total farm support therefore can be measured by aggregation of both basic types of support measures, taking into consideration double counting[83] and tax concessions for taxing at preferential rates and similar privileges of the farm sector not included in budget outlays. It is quite obvious that to quantify total farm support a number of problems have to be solved:

- lack of statistical data,
- general aggregation problems,
- quantification of protection rates,
- incidence of subsidies.

These problems are discussed below. With respect to government expenditures (budget outlays) for agriculture the most striking difficulties are,[84] first, government aid to the farm sector is represented by budget outlays of the national Department of Agriculture. But agriculture is benefiting also from expenditures by other departments (such as the departments of commerce, traffic, social affairs, etc.) so that budget outlays of the Department of Agriculture do not represent the total fiscal expenditures for the farm sector. On the other side, budget expenditures of the Department of Agriculture are favouring other sectors of the economy – forestry and fishing, etc. Further, some governments are decentralized (federal states). Thus expenditures of communities, states and the central government have to be aggregated. Next, in various countries, agriculture is subject to special tax regulations, so that different types of tax exemptions should be included. However, in only a few countries, such as West Germany, official or private estimates of the volume of the relevant tax exemption are available. Very often tax exemptions are nothing else than income subsidies to farmers. Furthermore, it must be stressed that various types of budget outlays for agriculture favour the farm sector to a quite different degree. This varying incidence effect of transfer payments has the consequence that expenditures only represent gross instead of net transfers to agriculture.

In summarizing, it must be said that aggregated government expenditures for agriculture as shown by official statistics give only a rough picture of fiscal aid to the farm sector. Of course, more scientific investigations and analysis of this type of farm policy represented by fiscal instruments is urgently needed – especially because those budget outlays are of increasing importance, absolutely as well as relative to transfers of income by price policy, in most countries under review. That fiscal aid to agriculture has been a rather neglected area of research by agricultural economists probably can be explained by the fact that in the United States this specific type of government support does play a rather minor, however, rising rôle. Because research in agricultural economies is to a large extent dominated, or at least strongly influenced by US economists, this line of research is neglected also by economists in other parts of the world.

As has been said already, fiscal measures favouring the farm sector can be seen as a complement to or a substitute for income generating farm price policy measures mentioned first. Because price policy measurements vary to a considerable degree relative to the importance of fiscal policy measures between the countries and time periods being reviewed, the volume of price policy should be quantified, so that a comparison (and aggregation) for intercountry and intertemporal comparative analysis would be possible. It is obvious that an exact quantification of border protection by import quotas, tariffs, levies etc. resulting in higher producer prices is, theoretically as well as empirically, a difficult task.[85] The best and most objective measurement is, of course, given by the concept of the "Effective Rate of Protection" referring to the impact of protection on the relative value added of supported goods.[86] However, there are many difficult problems in measuring the effective rate of protection, so that the available data are rather limited. For our purposes we need an indicator which represents the level and change of price support relative to financial transfer payments. For this purpose it is sufficient to quantify the effects of protection on producer prices and income by calculating

"Nominal Gross Protection" of agriculture. Nominal gross protection is calculated in this analysis by multiplying the difference between the relevant index of domestic farm producer prices and the index of import prices (cif) of farm products which are substitutes for domestic goods by the volume of domestic farm production. The relative difference between domestic and import prices is based on calculations of absolute differences (nominal protection rate) at representative time periods. For international comparisons, the absolute figures of nominal gross protection expressed in national currency units have been converted into US-Dollars of 1970 in order to prevent disturbances due to revaluation of national currencies. As far as possible, double counting has been avoided.

6.3. *The results*

Some statistical results are given in the Appendix for a number of Western industrialized countries with market economies. The econometric analysis cannot be repeated here because of lack of space. The results have been published elsewhere.[87] Furthermore, only a rough summary of results will be given as follows, without discussion of details.

For Western industrialized countries, as far as data were available and manageable, we can observe that since World War II government expenditures supporting the farm sector increased absolutely, even if inflationary rises of the price level are taken into account.[88] The rates of increases are different with respect to country and time period reviewed. Rising government supports to agriculture cannot be explained by the displacement effect of Peacock and Wiseman,[89] because the rise of budget expenditures was not limited to the immediate post war period. For most countries expenditures increased also with respect to:

(a) agriculture's contribution to national income,
(b) gross revenue of agriculture, and
(c) total government budget expenditure.

Bearing in mind that agriculture in all countries was subject to a tremendous process of structural adjustment all over the post-war-period, budget expenditures for agriculture also increased, of course, per unit of farm output, per unit of labour and land. However, great differences between various countries can be observed with respect to all indexes of budget outlays relative to all indicators mentioned here. In principle, the analysis shows that

(a) in countries of the New World relative expenditures are less than in the countries of the Old World,

(b) within the latter group of countries government expenditures are relatively small in countries economically less developed than in countries showing high income per capita,

(c) expenditures increase more over time where economic growth is stronger than in countries with a relative modest rise in per-caput-income.

Concerning the level of nominal gross protection of the farm sector and its change over time, as measured according to the method described above, the following summarized results can be given. For almost all countries, the volume of gross protection has increased since World War II. The rise of protection was due mainly to increases of domestic farm production with respect to domestic consumption not so much due to increases in domestic prices relative to world market prices.[90] Nominal gross protection has increased, similar to government expenditures, with respect to (a) domestic product of agriculture, (b) total gross product of agriculture as well as (c) per unit of gross product, land and labour input in the long run. However, the *rates* of growth of gross protection measured in absolute as well as relative terms are less than those of government expenditures of almost all countries analysed. Price and income supports by border protection are less important than government expenditures in the more developed countries of the Old World, compared with less developed countries in the Old World and countries in the New World. Over time, nominal protection is declining relative to government expenditures in all countries. Border protection is generally more excessive in less developed countries of the Old World and for countries in the New World, although gross protection per unit of output and, especially, input is higher in the latter group of countries compared with the industrialized countries of the New World and the more developed countries of the Old World. In other words, economic growth results in a structural change of farm support towards fiscal policy substituting, as well as complementing, border protection. Both measures lead to an expansion of the total volume of farm support, although agriculture is a declining sector.

Increasing financial support of agriculture by budget expenditures is also subject to structural changes. The most striking changes occurred in those countries where agriculture is subject to a specific social security system (South and West Europe).[91] For these countries the share of government expenditure for social policy measures in the field of agriculture has increased rather rapidly. Because instruments aimed to improve the farm structure are very often also linked with social policy measures, to a rising degree, budget expenditures for structural improvements as a share of total budget outlay is in many countries decreasing, although absolutely it is rising. A large share of expenditures is still for structural improvements. No definite trend with respect to measures to increase farm income directly by subsidies for farm products, farm producers and/or inputs could be observed. Obviously, the share (and level) of government expenditures for structural policy depends on the degree of structural adjustment of the farm sector with respect to the state of economic development. The New World-countries are better off than the countries of the Old World, as demonstrated by figures of land-man-ratio, concentration of income and farm sizes etc. and their changes over time.

6.4. *Interpretation of results: the hypothesis of farm policy formation*

Changes over time in the volume and structure of agricultural support by border protection and by fiscal policy within the process of economic growth and, consequently, differences between Western industrialized countries with market economic order, are to be explained as follows: Economic growth is synonymous with changes in the structure of the relevant economy. Agriculture is, as we all know, the case in point as far as its relative declining position within an economy is concerned; low income elasticities of demand for food, low rates of population growth, changing relative factor prices result in strong pressures on the farm sector's relative income, a need for adjustment (reallocation of resources) and, consequently, a squeeze on farm labour to leave agriculture. What is even more important in our context is that these pressures on income, for structural adjustment and on labour outflow are the stronger (a) the higher are the growth rates of the economy and/or (b) the higher the level of economic developement already reached. This is due to a falling income elasticity of demand and declining population growth as well to increasing opportunity costs for agriculture. With respect to farm policy directed towards stability or improvements in personal income distribution between agriculture and non farm sectors, the volume of income transfer has to be increased and/or the adjustment of the farm sector has to be accelerated. Assuming that public expenditures for restructuring the farm sector are rising because costs for transferring most (labour) input are increasing over time, total government expenditures are rising more than proportional. Consequently, countries which are relatively less developed are characterized by a smaller volume of farm support compared with more developed countries. Furthermore, the structure of farm support in countries mentioned first is dominated by income transfer via measures directly affecting producer prices; in these developed countries government expenditures for agriculture are more important than income generating instruments of the type mentioned. Of course, further modifications of this simple hypothesis must be made in order to explain the most striking differences between agricultural support by countries in the New World compared with those of the Old one.

However, the observed rise in aggregated volume of support of the declining farm sector (as well as difference between countries) can, at least partly, be explained by different levels of economic growth. However, this hypothesis is unable to explain discrepancies in the volume and structure of farm support between (highly developed) countries of the New World and those of the Old World (including Japan) and, to a certain extent, differences existing between countries belonging to the latter group. Therefore, factors besides the level of economic development must be taken into consideration, such as (a) the relative degree of structural adjustment and changes thereof over time, (b) the prevailing degree of self-sufficiency and its changes over time and (c) the specific character of the prevailing market economy.

Regarding (a), the degree of structural adjustment in relation to the prevailing economic conditions may be measured by income per unit of input, excluding income transfers (net national income at factor cost.[92] This, of course, is a function of factor endowment of farms, factor productivity and (equilibrium) price relationship. A substitutive measurement for structural adjustment may be the land-man-ratio in various countries with respect to international comparisons.[93] The differences between New and Old World-countries are well known and they are reflected in the relative volume of farm support. The rate of changes of the indexes representing adjustment relative to income outside agriculture may determine growth-rates of the volume of farm support.

Regarding (b), the degree of self sufficiency as a numerical expression of the relation between domestic food production and consumption of home produced food determines further the volume and the structure of farm support measures. Surpluses of farm products, especially in countries having low degrees of structural adjustment, exert a restrictive pressure on farm price policies due to increasing financial burdens (export subsidies, etc.). This, of course, refers to changes in farm prices over time. Substitutive instruments, such as structural and social policy measures and direct income transfer payments, combined with output restricting provisions, do replace partly – or, at least complement – price policy.

Regarding (c), countries under review have been characterized as market economies being subject to fiscal and monetary policy instruments in order to subordinate individual economic activity to overall policy objectives. However, various countries are more or less tolerant as far as the demarcation line between individual and collective responsibility for actions to counteract unsatisfying situations.[94] In other words, some governments are less sensitive with respect to the use of policy instruments to cure undesired developments than others,[95] the latter trusting more to individual capacity for adaption. This means in our context that the extent to which farm policy measures are used is partly determined by the philosophy behind government interventions. In general, Old-World-countries seem to be more liberal at the present time than the other countries being reviewed. Of course, such differences are difficult to quantify.

6.5. *The policy implications involved*

If our analysis is correct that the shape, structure and changes of farm policy are mainly determined by:

(a) the state of economic development,
(b) the degree of structural adjustment of the farm sector,
(c) the degree of self sufficiency of domestic food supply,
(d) changes of these factors over time.

the certain implications seem to be evident.

First, given a great similarity of preference functions of decision-makers in Western industrialized countries, the degrees of freedom for manipulating the farm policy mix are rather restricted. This may be, among others, an explanation for the fact that the stress put by economists on the welfare losses of prevailing farm policies is rather unsuccessful in convincing governments to modify basically present policies.[96] This, of course, is a reformulation of Lindbloms[97] theory of "a disjointed incrementalism" which is pursued by decision-makers, only approaching "a piecemeal social engineering" (Popper).[98] Second, if our proposition holds, then the dimension, type and structure of policy conflicts with regard to inconsistencies, incompatibilities, and nonconformities described above can be located systematically in relation to the basic factors underlying relevant situations. Finally, possible solutions of those conflicts can be developed and transformed more systematically and in advance into relevant political actions. Those actions, of course, should not only consider solution of conflicts within and between national farm and economic policy objectives and achievements but also possible and foreseeable conflicts at the international level. Although, for the time being, various countries of the Western World are seeking those solutions by some returns to mercantilistic types of economic policy – the Common Agricultural Policy being a case in point – international coordination of farm policy of various countries is needed more than before. In any case the best conditions to achieve those goals will be provided if economic growth, full employment and price stability are regained. Such a course of future economic development will be the best presupposition for a more efficient, internationally coordinated, farm policy.

APPENDIX

TABLE 1. *Farm support by government expenditures for agriculture, selected industrialized countries with market economies, 1950–1974*

Period	Gross Domestic Product (GDP)			Government Expenditures (GE)				
	Total	Agriculture		Total		For Agriculture		
		Absolute	In % of GDP	Absolute	In % of GDP	Absolute	In % of	
							GE	GDP agriculture
France (Bill. of Francs)								
1955/59	271·54	25·45	9·4	59·30	10·4	1·83	3·1	7·2
1960/64	372·71	31·57	8·5	80·10	12·1	4·52	5·6	14·3
1965/69	590·91	40·01	6·8	128·50	17·0	10·04	7·8	25·1
1970/74	875·81	55·12	6·3	209·50	21·1	18·56	8·8	33·6
Federal Republic of Germany (Bill. of German Marks)								
1950/54	132·36	12·34	9·4	39·82	30·0	0·55	1·4	4·5
1955/59	218·27	15·36	7·1	65·11	29·7	2·21	3·3	14·2
1960/64	360·76	18·37	5·1	111·94	29·8	4·24	3·9	22·9
1964/69	519·44	20·92	4·0	155·43	30·0	6·53	4·1	30·4
1970/74	841·67	24·45	2·9	252·56	29·9	8·61	3·4	35·2
United Kingdom (Mill. of Pounds Sterling)								
1955/59	19211	848	4·4	5259	27·3	238	4·5	28·0
1960/64	22212	969	4·4	11034	49·6	287	2·6	29·5
1965/69	40006	1121	2·8	17141	42·7	242	1·4	21·6
1970/74	59811	1522	2·5	26398	44·1	357	1·4	23·4

TABLE 1.

Period	Gross Domestic Product (GDP)			Government Expenditures (GE)				
	Total	Agriculture		Total		For Agriculture		
		Absolute	In % of GDP	Absolute	In % of GDP	Absolute	In % of	
							GE	GDP agriculture
United States of America (Bill. of Dollar)								
1950/54	283·7	18·5	6·5	58·23	20·5	1·99	3·4	10·8
1955/59	363·1	16·6	4·6	78·08	21·5	3·36	6·9	20·2
1960/64	459·9	18·1	3·9	105·35	22·9	6·73	6·4	37·1
1965/69	663·1	22·5	3·4	154·94	23·4	7·64	4·9	33·4
1970/74	961·9	32·0	3·3	230·96	24·0	9·48	4·1	29·6
Japan (Bill. of Yen)								
1960	15873	2105	13·3	1765	11·1	139	7·9	6·6
1965	31480	3165	10·1	3745	11·9	346	9·2	10·9
1970	69268	4689	6·8	8213	11·9	883	10·8	18·8
1973	111004	6048	5·5	14284	12·9	1314	9·2	21·7
Australia (Mill. of Australian Dollar)								
1970	32702	2039	6·2	6121	18·7	357	5·8	17·4
Canada (Mill. of Canadian Dollar)								
1970	86151	3203	3·7	15500	18·0	348·6	2·2	10·9

TABLE 1.

Period	Gross Domestic Product (GDP)			Government Expenditures (GE)				
	Total	Agriculture		Total		For Agriculture		
		Absolute	In % of GDP	Absolute	In % of GDP	Absolute	In % of	
							GE	GDP agriculture
Belgium (Mill. of Francs)								
1970	1283·1	49·5	3·9	311·4	24·3	8·1	2·6	16·2
Finland (Mill. of Markas)								
1970	43592	5408	12·4	14800	34·0	953	6·4	17·6
Italy (Bill. of Lires)								
1970	57903	5072	8·7	18926	32·7	2279	12·0	44·9
Ireland (Mill. of Pounds Sterling)								
1970	1621·3	232·5	14·3	610·0	37·6	94·2	15·7	40·1
Netherlands (Mill. of Guilder)								
1971	129650	6990	5·3	30300	23·3	1135	3·7	16·2
Spain (Bill. of Pesetas)								
1970	2264·1	280·3	12·4	429·7	19·0	42·8	10·0	15·2
Sweden (Bill. of Kronor)								
1970	170112	6353	3·7	69745	41·0	8060	11·6	12·7

Sources: National Statistics

TABLE 2. *Distribution of farm support according to policy instruments, Federal Republic of Germany, France and United Kingdom, 1955–1974*

Period	Aggregated farm support[1]	Thereof in percent					
		Income Transfer by		Social policy[4]	Structural policy[5]	Education and training	Tax exemption[6]
		Border protection[2]	Direct transfer[3]				
Federal Republic of Germany (Mill. of German Marks)							
1955/59	6464	61·7	17·9	—	15·1	1·1	5·1
1960/64	10956	64·6	15·6	1·5	20·6	1·0	6·7
1965/69	15824	53·1	15·2	4·5	18·9	0·9	7·3
1970/74	22331	52·5	16·6	6·5	15·1	0·3	9·1
France (Mill. of Francs)							
1955/59	6010	69·5	16·0	3·3	10·7	0·5	n.a.
1960/64	11990	62·2	18·4	7·8	10·7	1·0	n.a.
1965/69	22501	55·4	16·0	15·5	11·3	1·7	n.a.
1970/74	33733	45·0	20·0	23·1	10·8	1·1	n.a.
United Kingdom (Mill. of Pounds Sterling)							
1955/59	414·4	42·7	52·9	n.a.	4·4	n.a.	n.a.
1960/64	504·8	43·2	50·1	n.a.	5·7	n.a.	n.a.
1965/69	474·2	48·9	45·4	n.a.	5·7	n.a.	n.a.
1970/74	635·5	43·9	43·5	n.a.	12·6	n.a.	n.a.

[1] Government expenditures and gross nominal protection.
[2] By import quotas, levies and import duties.
[3] By subsidies of output (deficiency payments) and input.

[4] By social security systems specified to agriculture; this is not the case as far as the U.K. is concerned.
[5] Factor productivity increasing government programs.
[6] Official estimates for West Germany. – For detailed information see text.

Sources. Statistisches Jahrbuch für Ernährung, Landwirtschaftund Forsten, var. issues. – Projet de Loi de Finances pour 1976, Presentation du Budget de l'Agriculture sous Forme de Budget de Programmes. Paris 1975. – Annual Review and Determination of Guarantees, var. issues. – Statistical yearbooks of countries reviewed.

NOTES

[1] For a more detailed system of farm support measures see: R. W. Howarth, *Agricultural Support in Western Europe*. The Institute of Economic Affairs (Research Monograph 25, London 1971, p. 11, with respect to structural policy: OECD, *Structural Reform Measures in Agriculture* (Agricultural Policy Reports). Paris 1972.

[2] These countries are defined in Chapter 2.

[3] See O. Gulbrandsen and A. Lindbeck, *The Economics of the Agricultural Sector*. Stockholm 1973, and U. Koester und S. Tangermann, *Alternativen der Agrarpolitik*. (Landwirtschaft-Angewandte Wissenschaft, H. 182). Münster-Hiltrup 1976. T. E. Josling, Agricultural Policies in Developed Countries: A Review. *Journal of Agricultural Economics*, Vol. 25 (1974), p. 237.

[4] See E. S. Kirschen (Ed.), *Economic Policies Compared, West and East. (Vol. 1, General Theory)*. Amsterdam and Oxford 1974, p. 31. – E. S. Kirschen et al., *Economic Policy in our Time (Vol. 1, General Theory)*. Amsterdam 1964, passim. – For a theoretical treatment see A. E. Bryson and Yu-Chi-Ho, *Applied Optimal Control*. Waltham, Toronto, London 1969.

[5] A. Downs, *An Economic Theory of Democracy*. New York 1957. For more modern development in the field of the New Political Economy see B. S. Frey, Eine Einschätzung der Neuen Politischen Ökonomie der 7oer Jahre. *Zeitschrift für die gesamte Staatswissenschaft*, Bd. 131 (1975), pp. 697–718, and literature quoted.

[6] See T. E. Josling *op. cit.* and references quoted.

[7] See J. Kumar, *Population and Land ın World Agriculture. Recent Trends and Relationships* (Population Monograph Series, No. 12). Berkeley 1973. – F. Dovring, *Land and Labor in Europe in the Twentieth Century*. The Hague 1965. – F. Dovring, The Share of Agriculture in a Growing Population. In: C. Eicher and L. Witt (Eds.), *Agriculture in Economic Development*. New York 1974.

[8] D. Cashley and L. Naiken, Note on the Turning Point in Trends of Agricultural Population and Labour Force in Developing Countries. *Monthly Bulletin of Agricultural Economics and Statistics*, 1974, pp. 14–16.

[9] See F. Dovring, *op. cit.*

[10] See among others: P. Dorner, *Land Reform and Economic Development*. London 1972.

[11] R. Hedtke, *Möglichkeiten und Grenzen einer Anwendung der Kosten-Etrags-Analyse in der Strukturpolitik*. (Schriftenreihe zur Industrie- und Entwicklungspolitik). Berlin 1973.

[12] See E. S. Kirschen (Ed.), *op. cit.* – T. E. Josling, *op. cit.*

[13] Th. Pütz, *Grundlagen der theoretischen Wirtschaftspolitik*. Stuttgart 1971, p. 17.

[14] G. N. Halm, *Economic Systems. A Comparitive Analysis*. London, New York, Sidney, Toronto 1968. – G. Hedtkamp, *Wirtschaftssysteme. Theorie und Vergleich*. München 1974.

[15] Compare: FAO, *Production Yearbook*. Rome.

[16] See Dovring, *op. cit.* and Y. Hayami and V. W. Ruttan, *Agricultural Development. An International Perspective*. Baltimore and London 1971.

[17] K. Müller-Heine, *Agrarpolitische Ziele und ihre Einordnung in den gesamtwirtschaftlichen Zielkomplex*. Göttingen 1972. – F. Mehler, *Ziel-Mittel-Konflikte als Problem der Wirtschaftspolitik* (Volkswirtschaftliche Schriften, H. 149). Berlin 1970.

[18] G. Schmitt und H. v.Witzke, *Ziel- und Mittelkonflikte sektorspezifischer Systeme sozialer Sicherung. Das Beispiel der landwirtschaftlichen Sozialpolitik in der Bundesrepublik Deutschland* (Sozialpolitische Schriften, H. 37). Berlin 1975.

[19] E. S. Kirschen (Ed.), *op. cit.*

[20] For a more detailed discussion of definitions see: P. Baumgarten and W. Mückl, *Wirtschaftspolitische Zielkonflikte in der Bundesrepublik Deutschland. Eine theoretische und empirische Untersuchung*. Tübingen 1969. – G. Schmitt, Planung, Durchführung und Kontrolle koordinierter Makro-, Struktur- und Mikrosteuerung in der Landwirtschaft. *Agrarwirtschaft*, Jg. 22 (1973), pp. 95–107. – J. Tinbergen, *Economic Policy: Principles and Design*. Amsterdam 1964.

[21] See for a general view: E. S. Kirschen (Ed.), *op. cit.*, and references quoted.

[22] E. S. Kirschen (Ed.), *op. cit.*, and E. S. Kirschen *et al.*, *op. cit.*

[23] P. Baumgarten and W. Mückl, *op. cit.*

[24] E. S. Kirschen (Ed), *op. cit.*

[25] E. S. Kirschen (Ed.), *op. cit.*

[26] E. S. Kirschen (Ed.), *op. cit.*

[27] E. S. Kirschen (Ed.), *op. cit.*

[28] E. S. Kirschen (Ed.), *op. cit.* – A. -H. Boyne, Changes in Income Distribution in Agriculture. *Journal of Farm Economics*, Vol. 47 (1965).

[29] D. G. Johnson, *World Agriculture in Disarray*. London 1973. – OECD, *Agricultural Policies in 1966. – Europe, North America and Japan*. Paris 1966. – W. M. Corden, *The Theory of Protection*. Oxford 1971. – J. P. Mackintosh, The Problems of Agricultural Policies. *Journal of Agricultural Economics*, Vol. 21 (1970), Agricultural Protection in US Agriculture. *American Journal of Agricultural Economics*, Vol. 53 (1971).

[30] E. S. Kirschen (Ed.), *op. cit.*

[31] This will be discussed in the following chapter.

[32] GATT, International Trade. var. issues.

[33] GATT, *op. cit.*

[34] This chapter is mainly based on a number of publications of OECD and FAO on agricultural policy of member-states, especially: OECD, *Trends in Agricultural Policies since 1955*. Paris 1961. – OECD, *op. cit.* – OECD, *Agricultural Policy Reports of various member states*, Paris 1973, 1974, 1975.

[35] For more detailed information see the systematization of countries given by OECD according to (a) defined income objectives comparable to given levels of non farm income (West Germany, Sweden), (b) rather well defined groups of farms subject to income redistribution without exact standards (Austria, Benelux, France, United Kingdom), (c) to restricting relevant definitions to instruments to be used in order to increase income (price policy: USA; improvement of efficiency: EEC, Canada, Denmark, Ireland) and (d) income objectives are not mentioned but implicit achievements of instruments (Mediterranean countries).

[36] Few countries (Germany, France, Netherlands, Sweden and EEC) regularly publish records on the income position of agriculture, sometimes as a consequence of prevailing legislation.

[37] See references quoted in note 3.

[38] Although there are quite a number of "Theories Explaining the Persistence of low Resource Returns" (by L. G. Tweeten, *American Journal of Agricultural Economics*, Vol. 51 (1969), the hypotheses that opportunity costs of main production factors in agriculture are less than their "equilbrium prices" has been accepted by most farm economists. This theory therefore may explain disequilibrium in production factor inputs and factor income prevailing even under conditions of polypolistic market structure. D. E. Hathaway, *Government and Agriculture*. New York and Toronto 1968. — G. Schmitt, Landwirtschaft in der Marktwirtschaft: Das Dilemma der Agrarpolitik. In: D. Cassel *et al.* (Hrsg.), *25 Jahre Marktwirtschaft in der Bundesrepublik Deutschland*. Stuttgart 1972. – A. Hanau, Landwirtschaft in der sozialen Marktwirtschaft. *Agrarwirtschaft*, Jg. 7 (1958), pp. 1–27.

[39] For more details see the following chapter.

[40] U. Koester, *Sektorale Preisentwicklung und Geldwertstabilität*. Meisenheim (1974).

[41] D. Manegold, Analyse der landwirtschaftlichen Gesamtrechnung des EUROSTAT – 1974. *Agrarwirtschaft*, Jg. 25 (1976), pp. 74–84.

[42] T. Heidhues und S. Tangermann, Der Einfluß von wirtschaftlichem Wachstum, Inflation und Währungspolitik auf die Landwirtschaft unter EWG-Bedingungen. *Agrarwirtschaft*, Jg. 21 (1972), pp. 173–182. – S. Tangermann, *Gemeinsame Agrarpolitik und nationale Wirtschaftspolitik in der EWG*. Tübingen 1974. – L. Kersten und D. Manegold, Aktuelle Fragen der Gemeinsamen Agrarpolitik. *Agrarwirtschaft*, Jg. 25 (1976), pp. 130–139.

[43] R. Hedtke, *op. cit.*

[44] OECD, Structural Reform . . . , *op. cit.*

[45] E. S. Kirschen (Ed.), *op. cit.*

[46] Scientific attempts to measure the contribution of structural adjustment to national income are numerous. See among others: OECD, *Agriculture and Economic Growth*. Paris 1965. – E. F. Denison, *The Sources of Economic Growth in the United States and the Alternatives before us*. New York 1962.

[47] OECD, *Structural Reform . . .*, *op. cit.* B. Felderer, Der Zusammenhang zwischen Allokation von Produktionsfaktoren und wirtschaftlichem Wachstum. Eine empirische Untersuchung. *Zeitschrift für die gesamte Staatwissenschaft*, Bd. 129 (1973), pp. 589–600.

[48] OECD, Structural Reform . . . , *op. cit.*

[49] See OECD-publication quoted in note 29.

[50] OECD, *Agricultural Policy in Sweden* (Agricultural Policy Reports). Paris 1975.

[51] A. Henze, Die Sicherstellung der nationalen Güterversorgung (*Sonderheft 48 der Agrarwirtschaft*). Hannover 1972.

[52] See OECD-*Country Reports and relevant statistics of the EEC.*

[53] FAO, *An Inquiry into the Problems of Agricultural Price Stabilization and Support Prices*. Rome 1960.

[54] T. Josling, Agriculture and Import Saving: A Cautionary Note. In: *Agriculture and Import Saving* (H. Samuel Occasional Paper No. 5). London 1970.

[55] T. Josling, *Agriculture . . .*, *op. cit.*

[56] See E. J. Mishan, Welfare Criteria for External Effects. *American Economic Review*, Vol. 51 (1961).

[57] See for a detailed discussion: V. W. Ruttan, Agricultural Policy in an Affluent Society. *Journal of Farm Economics*, Vol. 48 (1966), pp. 1100–1120, and references made.

[58] J. Bieri and A. Schmitz, Market Intermediaries and Price Instability: Some Welfare Implications. *American Journal of Agricultural Economics*, Vol. 56 (1974), pp. 280–285, and references quoted.

[59] For a systematic treatment of this subject see: W. M. Corden, *Trade Policy and Economic Welfare*. Oxford 1974.

[60] For an almost complete description of relevant measures see: OECD, *Structural Reform . . .*, *op. cit.*

[61] Y. Hayami and V. W. Ruttan, *op. cit.*

[62] R. W. Howarth, *op. cit.*, and references given. – G. Neidlinger, Versuch einer Quantifizierung der Wirkung unmittelbar preiswirksamer Agrarschutzmaßnahmen (*Sonderheft 23 der Agrarwirtschaft*). Hannover 1967.

[63] R. Hedtke, *op. cit.*

[64] V. W. Ruttan, *op. cit.*

[65] OECD, *Low Incomes in Agriculture. Problems and Policies*. Paris 1964.

[66] Kommission der Europäischen Gemeinschaften, *Vergleichende Darstellung der Systeme der sozialen Sicherheit in den Mitgliedsstaaten der Europäischen Gemeinschaften*. Brüssel 1974.

[67] Kommision der Europäischen Gemeinschaften, *Die Finanzierung der sozialen Sicherheit in der Landwirtschaft* (Reihe Sozialpolitik 20). Brüssel 1970.

[68] OECD, Structural Reform . . . , *op. cit.*

[69] G. Schmitt und H. v.Witzke, *op. cit.*

[70] E. S. Kirschen (Ed.), *op. cit.*

[71] See S. Tangermann, *op. cit.*, and references made.

[72] A. Henze, Inflationsursachen und gesamtwirtschaftlich-strukturelle sowie agrarstrukturelle Inflationswirkungen. *Berichte über Landwirtschaft*, Bd. 52 (1972), pp. 81–102. – C. Thoroe, *Inflation und sektorale Einkommensverteilung unter besonderer Berücksichtingung der Landwirtschaft*. Ph.D. Thesis, Kiel 1974.

[73] E. Guth, Analyse des Marktes für landwirtschaftliche Arbeitskräfte (*Sonderheft 52 der Agrarwirtschaft*). Hannover 1973.

[74] V. W. Ruttan, *op. cit.*, and references quoted.

[75] See D. Biehl *et al.*, *Bestimmungsgründe des regionalen Entwicklunspotentials* (Kieler Studien 133). Tübingen 1975.

[76] V. W. Ruttan, *op. cit.*

[77] V. W. Ruttan, *op. cit.*

[78] E. S. Kirschen (Ed.), *op. cit.*

[79] This theory of agriculture adjustment is generally accepted since A. G. B. Fisher's, C. Clark's, J. Fourastié's and T. W. Schulze's fundamental research.

[80] See especially: B. S. Frey, Entwicklung und Stand der Ökonomischen Theorie der Politik. In: H. -P. Widmeier (Ed.), *Politische Ökonomie im Wohlfahrtsstaat*. Frankfurt 1974.

[81] J. A. Schumpeter, *Capitalism, Socialism and Democracy*. New York 1942. – A. Downs, *op. cit.*

[82] P. Bernholz, Die Machtkonkurrenz der Verbände im Rahmen des politischen Entscheidungssystems. In: *Macht und ökonomisches Gesetz* (Schriften des Vereins für Socialpolitik, Bd. 74). Berlin 1973. – With respect to the controversy on determination of government policy by mainly political theories versus economic factors, in the field of fiscal policy. There is a strong support of the latter position. See: H. C. Recktenwald, Staatswirtschaft in säkularer Sicht (*Hamburger Jahrbuch für Wirtschafts und Gesellschaftspolitik, Bd. 15*), Tübingen 1970, pp. 119–138, and references quoted.

[83] In many countries minimum prices for farm products are guaranteed to farmers by a system of buffer stocks, export subsidies etc. These measures automatically result in an increase of domestic prices relative to world market prices. Adding relevant government expenditures for buffer stocks to border protection measured in terms of nominal protection would mean, of course, to count the extent of farm support price twice.

[84] For a more detailed discussion see: H. -G. Schlotter, *Die finanz und außenhandelspolitische Landwirtschaftsförderung in der Bundesrepublik Deutschland. Ausmaß, Struktur und künftige Möglichkeiten*. Hannover 1964.

[85] See R. W. Howarth, *op. cit.*, R. Dardis and E. W. Learn, *Measures of the Degree and Cost of Economic Protection of Agriculture in Selected Countries* USDA, ERS, Bull. No. 1384, Washington D.C. 1967. – G. Neidlinger, *op. cit.*

[86] W. M. Corden, *op. cit.*

[87] P. Schmidt und G. Schmitt, Auf dem Wege zu einer positiven ökonomischen Theorie der Agrarpolitik. *Berichte über Landwirtschaft*, Bd. 52 (1974), pp. 521–546. – P. Schmidt, *Bestimmungsfaktoren staatlicher Agrarpolitik in Marktwirtschaften*, Ph.D. Thesis, Göttingen 1975.

[88] For many countries an explicit price index of government expenditures is not available, most often consumer price index has to be used for deflation of nominal budget outlays.

[89] See A. Peacock and J. Wiseman, *The Growth of Public Expenditures in the United Kingdom*. Princeton 1961.

[90] See References quoted in note 85.

[91] Very likely, in countries where farmers are integrated in the general social security system, similar intersectoral transfers on income could be observed. This is simply due to the decline of the number of still active farmers relative to already retired farmers. However, no statistics are available to prove this hypothesis. Of course, the volume of income transfer probably is smaller than in countries with special systems for farmers. In the latter cases, farmers very often are priviliged as against non farm population.

[92] Difficulties in measuring national income at factor cost per unit of labor input (in international comparisons) are due to statistical differences in labour input figures.

[93] See F. Dovring, *op. cit.*

[94] In this line: P. Mosley, Towards a "Satisficing" Theory of Economic Policy. *The Economic Journal*, Vol. 86 (1976), pp. 59–72.

[95] Cases in point are the tolerance with respect to inflation and the extent of social security systems in various countries. A crude measurement may be the share of government expenditures on national income (see Table 1 of Appendix).

[96] See G. D. Hathaway, *op. cit.*; B. Talbott and D. F. Hadwiger, *The Policy Process in American Agriculture*. San Francisco 1968. R. S. Kirkendall, *Social Scientists and Farm Politics in the Age of Roosevelt*. Columbia 1966. – G. Schmitt, Zum Verhältnis der Agrarökonomik zur Agrarpolitik. *Agrarwirtschaft*, Jg. 21 (1972), pp. 213–220.

[97] D. Braybrooke and Ch. F. Linblom, *A Strategy of Decision. Policy Evaluation as a Social Process*. London 1963.

[98] K. Popper, *The Poverty of Historicism*. New York 1961.

DISCUSSION OPENING – R. W. M. Johnson, *New Zealand*

I would like to re-state certain aspects of the argument and draw some conclusions which I think are warranted from the paper in front of us.

The political objective is stated as getting re-elected – we read the phrase "maximise the sympathy of voters".

Economic growth is considered desirable as it meets political objectives – this can be roughtly paraphrased as "giving something to everybody".

The problem arises because of unequal growth in the different sectors. Different sectors grow at different rates and so do different parts of the same sector such as mountain regions and so on. Backward growth is not unique to the agricultural sector but is typical of it.

Governments take action to correct the resulting imbalances as political objectives set a value on the quantity of votes and not maximum growth of income.

This seems to be a summary of the main features of the argument presented. Now the most important manifestation of these economic forces is the movement of mobile resources out of agriculture – especially the labour resource. We see policies designed to aid movement, such as retirement schemes and structural programmes, and income programmes to alleviate its effects!

As I see it, one result of these policies, and not a cause of them, is a general tendency to increase national self-sufficiency in agricultural products beyond levels which comparative advantage would indicate as appropriate. This has great implications for trade in agricultural products, especially in third countries such as New Zealand.

The question that should be asked in my opinion is whether the adaptation of the agricultural sector to modern needs could have been achieved by other methods? Have the measures adopted slowed down or increased the rate of migration out of agriculture? Would this have happened anyway? And in the end would the social costs of letting it happen be any greater than those brought about by the present policies? For the countries concerned I believe the policies actually adopted have probably minimised social costs – but for the community at large I believe we are worse off.

DISCUSSION OPENING – John J. Scully, *Commission of the European Communities.*

I would like to make a number of comments, some rather general, some more specific, on some of the issues raised by Prof. Schmitt in his very exhaustive treatment of his subject.

In discussing the implications of farm policy for agricultural development it is necessary not to lose sight of the fact that agriculture is an extremely diverse economic sector. It is diverse

- in terms of the economic climate in which farmers operate
- in terms of topography, soil fertility, geography and climate
- in terms of farm units of varying shapes and sizes
- in terms of farmers and farm workers of differing physical and mental abilities.

The diversity of the sector explains in a very real way why farmers as a group, in any country or region, can never respond in a uniform manner to the needs of economic growth and development. In these circumstances, it is not too difficult to explain the apparent lack of conformity between the desired and the realised results of policy measures.

Insofar as incompatible objectives of economic and farm policy are con-concerned, one is tempted to ask to what extent is this incompatibility due to the lack of the necessary integration of agricultural planning and general economic planning. There is of course a political dimension, in the broader sense of the term, to this problem. To some extent one could say that a basic reason for the incompatibility lies in the failure of the relevant Government Ministries or other planning agencies to co-ordinate their respective measures of development policy within the framework of one national programme embracing the various sectors of the economy.

In the narrower sense of the political dimension, there is the apparent reluctance of decision-makers to look to the longer term rather than to the shorter term in seeking solutions to fundamental problems of development, greater political kudos being realised through the use of temporary short-term palliatives. In the long term they often tend to compound problems still further rather than to provide worthwhile solutions to them.

Within agricultural policy itself this political dimension is often apparent in the conflict existing between price policy and structural policy insofar as attempts to improve the income situation of farmers through progressive increases in farm prices in the short term tend to restrict the possibilities available for structural reform in the long term.

Even more fundamentally, in a number of Western European Countries certain aspects of land tenure, which are supported by existing national legislative arrangements, are a direct impediment to the implementation of realistic policies of structural reform. There are, for example, certain inheritance laws which permit the division of arms among heirs during the process of farm transfer within generations; there are estate duties which represent a substantial financial burden for farm heirs and which in extreme cases may lead to the disintegration of highly capitalised farms at the end of the family life cycle; there is the undue dependence on the occupier-ownership of land as the basic tenure system which is encouraged by the land laws of certain countries, and which is decidedly "anti-structure" in character, particularly in the current era of rapidly inflating land prices. There are mahy other such inconsistencies which could be mentioned in addition.

I am not quite as optimistic as Professor Schmitt regarding what he describes as "the tremendous process of structural adjustment" which has taken place in West European agriculture since World War II. Granted there has been a significant level of structural reform during this period. However, the available statistics show that currently some 50 per cent of farms in the European Communities are still less than 20 hectares in area. Between 1960 and 1970 the average increase in farm size was less than 3 hectares. During the same period, the average annual rate of land mobility for farm enlargement was no more than 0,7 per cent, and it is largely the farms over 20 hectares which have increased in size. This structural change has taken place mainly through the normal operation of the land market rather than through the implementation of official reform policies. Furthermore, in the original Community of the Six, young people left the land at an average rate of 9 per cent per year between 1968 and 1973. As a result the farm population is getting progressively older, accentuating further the basic structural problem of agriculture.

In conclusion, I would subscribe to Professor Schmitt's central hypothesis that, subject to quite similar policy objectives, the dimension and structure of farm policy is determined by the specific economic and agricultural conditions prevailing in the country. Obviously, in a period of sustained economic growth the agricultural problem becomes easier to solve in countries where the level of development is such as to facilitate agricultural adjustment. For countries lower down in the development scale, however – and there are still some of these in Western Europe – the solution to the agricultural problem is not all that easy. In such cases, where rural regions still represent important seats of political power, agricultural policy will, at least in the short term, be influenced to some extent by political as opposed to economic considerations. Unfortunately, this is one of the realities of the situation which cannot easily be wished away.

RAPPORTEUR'S REPORT – Hans van Miltenburg, *Netherlands*

The discussions that followed Johnson and Scully's opening statements could be summarised along the following lines. Until recently much emphasis in the analysis of agricultural policy has been given to categorising the goals and the way of financing this policy, rather than on stressing the volume of the expenditure involved. There was a wide-ranging discussion, and numerous valid points were made. For instance, concerning the costs of agricultural policy, attention was given to the relation between price policy and structural reform measures. One discussant argued that the reconstruction of agriculture should be realised on the basis of prices which have to be deduced from the level of the production costs of farm products in the new situation. He also recommended a transitional system of direct income payments. Another considered that support to agricultural policy should be related to social policy. The need of a structural approach to agricultural policy was also stressed and the view expressed that in the EEC market too much attention is given to the short term.

In the longer run the major problems are not price and income policy, but the problems involved in the competitive demands for land and energy. With high unemployment the real costs of agricultural policy are only slightly lower with direct payments than with price support.

As a consequence of the source of incompatibilities, inconsistencies and non-conformities, which Professor Schmitt summarised in his paper, perhaps more attention should be paid to the process of model-building itself.

The fact that the conceptualisation of models is often done without the involvement of the most important decision-makers and the more established economists frustrates the application of models, even though existing techniques have the potential for progress in agricultural policy. This might be solved to a certain degree by using relatively simple techniques and by paying more attention to the process of model-building as such.

The development of the so-called dynamic approach, as applied to several different fields by the MIT team of J. Forrester and the experience of the Michigan team in Korea and Nigeria, could underline this.

AUGUSTYN WOŚ AND ZDZISLAW GROCHOWSKI*

Relationship between Agricultural Policies and General Economic Policy

I

The economic policy of a socialist state finds its best interpretation in the long-term and operational development plans, because a plan formulates both the goals and means for their implementation, thus ensuring a dynamic equilibrium for the economic system as a whole.

Economic planning may only be effective if it functions as a system in which plans of different range, field and time horizon exist side by side. A planning system consists of:

(a) plans for the expansion of the national economy as a whole and of its branches and sectors (macro-economic planning),
(b) plans for town and country development (country-scale, regional and local planning),
(c) internal plans of enterprises (factory planning also referred to as operational planning).

In the practical management of the national economy of a basic importance are the relationships between the aforementioned kinds of plans so as to avoid conflicts between the basic economic goals and the methods and means of their implementation. To attain compatibility between the objectives and conditions of the general economic expansion and the tasks of individual sectors and branches is one of the more difficult problems of the theory and practice of economic planning. One should, however, emphasize that sector-and-branch planning is not autonomous and that it constitutes an integral part of central planning. The fundamental decisions concerning the national economy as a whole and its individual parts are made in a uniform system of economic planning. And although – as is the case in Poland – the individual peasant sector plays a dominant role in agriculture, embracing 80% of the whole area of arable land and giving 77% of the end production and 75% of the total commodity production in agriculture, it is wholly incorporated into the general plan of the socio-economic development of the country. Otherwise

* Institute of Agricultural Economics, Warsaw, Poland.

it would not be possible to develop a plan for the expansion of the national economy; peasant farming is linked, through many channels, with other branches of our national economy, and participates in their final products. At the same time, it supplies them with raw materials, manpower and various services. Due to the deepening relationships and a growth of the scale of interbranch flows, the points of contact between agriculture and the non-agricultural sectors and branches are gaining in importance.

Polish agriculture is characterised by a complicated multi-sector structure. State, co-operative, group and individual enterprises exist here side by side. The state farms are administered directly within the framework of the central plan. In relation to the co-operative and private sectors the methods of indirect influence by means of a set of economic instruments are adequate. Here, planning is not of an imperative nature. This does not mean that the production of the private sector is not a subject of central planning, but this plan is obligatory not for individual farmers but for the state administration and the enterprises concerned with supplying services to agriculture. Of a directive character is the part of the agricultural development plan which concerns (a) the supplies of production inputs of industrial origin for agriculture and the implementation of the basic investments in this sector of the national economy and (b) the tasks of the economic apparatus in the sphere of purchase and processing of farm products.

For agriculture, central planning is combined with the influence of the controlled market. This should be understood as follows: the market mechanisms (prices, credits, rates of interest, etc.) are made use of for the implementation of the tasks preset in the plan. The market with its mechanisms does not replace the plan; it does not occupy its place; but the market supports the plan in the course of the implementation of the planned tasks which have been determined at an earlier time. On the other hand, information of various kinds coming from the market is essential for the construction of an economic plan. As has already been said, to include agriculture in the plan for the national economy as a whole means going outside the state sector. This fact and the enormous diversity in conditions of production in farming are responsible for the fact that it is practically impossible at present to collect and process in a computer centre the information about the existing alternatives. As a result, it becomes necessary to use parametric control methods and to make wide use of the market mechanism.

The necessity of using parametric planning methods also results from the co-operative form of the ownership of the means of production in agriculture. Also the co-operative farming enterprises are not liable to imperative planning. The Centre can only shape, in a purposeful way, the conditions of economic choice in production co-operatives. Contracts for the delivery of farm products and credits are the two main instruments of planned management by the state of the production in the agricultural production farms.

In conclusion one can state that a socialist state can shape directly the production in the state agricultural sector and that planned influence methods are used in relation to the co-operative and private sectors. Having at its disposal all the more essential instruments of economic policy, the state can control agricultural production so as to meet the needs of the central plan and the community as a whole. In this way the conformability of the goals of the general socio-economic policy with the tasks of the individual sectors and branches, including agriculture, is secured.

II

The basic problem of economic policy is to determine an appropriate rate of economic growth. In a country which – like Poland – is on an average development level and where agriculture still plays a great part in supplying the gross product (17·5%) and the national income (11·8%), the ratio of the rate of growth of the national income to the rate of growth of agricultural production is of basic importance. In agriculture one has to do with some specific barriers which restrict the possibility of accelerating the rate of growth of farm production. With the present advance of technology, the rate of growth of agricultural production over longer periods of time, in the European conditions of the lack of non-cultivated land, generally does not exceed 2·5%–3·0% per year. During the past 25 years, the average annual growth of agricultural production in Poland amounted to 2·9%. Although this is quite a fast rate, in certain fields, especially that of animal protein, the rapidly growing demand is still higher than supply. Possibilities of increasing both the absolute level and the rate of growth of agricultural production determine both the general rate of economic growth and the rate of improvement of the national standard of nutrition. The implications of the feasible rate of growth of the agricultural production are the greater for the national economy as a whole the higher is the share of agriculture in the production of national income and in meeting the consumptive needs of the society.

The diminishing share of agriculture in the production of the national income – in addition to many other effects – makes it possible to greatly increase the ratio of investments in agriculture to the national income produced by this sector of the national economy. In a highly industrialised country where a relatively small part of the national income is produced by agriculture and where the share of agricultural investments is relatively low, it is possible to speed up the rate of growth of national production by increasing the share of agricultural investments in the total sum allotted for the investment projects. In a country of a low level of industrialization, agriculture has at its disposal practically unlimited resources of manpower; this restricts the scope of investments meant for the replacement of live labour and leads to keeping the capital absorption in production on a low level.

Economic growth is accompanied by a decline of the labour reserves in agriculture. In these circumstances, the selected rate of growth of agricultural production is, at the same time, the selection of an alternative of where the labour should be employed: in agriculture, or outside it. This being so, the social cost of labour in agriculture should be measured by the additional growth of the national income which might be attained should labour be shifted from the agricultural sector to the non-agricultural one. The necessity of attaining an increase in agricultural production with the decreasing resources of manpower calls for substantial outlays on the substitution of labour outflowing from agriculture. This is accompanied by the final capital absorption in farm production. These processes act – on the principle of feedback – on the level of investments which must be allotted by the national economy on the expansion of agriculture in the conditions of increased capital absorption in farm production. Thus, making a decision in the matter of the desired rate of growth of the national income one also decides about the division of investment funds between agricultural and non-agricultural sectors and – taking into account the high capital absorption of farm production – about the migration of rural people to jobs outside agriculture.

Agriculture affects directly and indirectly the rate of growth of the national income and the expansion of the national economy as a whole. Direct influence depends on the branch structure of the national economy, especially on the share of agriculture in supplying the national product. Indirect influence is revealed in several planes. First of all, agriculture influences the rate of growth of employment outside agriculture. As long as there exist differences in the level of consumption of agricultural products per one person employed in agriculture and outside it – the growth in employment outside agriculture increases the demand for farm products. If a country is not able to supply the additional amount of food required, the barriers to economic expansion begin to appear. In this way productivity in agriculture determines the possibility of increasing employment outside agriculture and the expansion of the economy as a whole, provided that the level of income and food consumption per capita does not drop. To maintain a dynamic equilibrium in the growth of the national economy it is necessary that the surplus of agricultural market production grows faster than the outflow of agricultural population to the non-agricultural sector.

If the labour released from agriculture is shifted to the non-agricultural sectors without improving the level of investments, the technical armament of work and, also, productivity will be correspondingly lower. The problem will be different if the investment fund is increased. Then, the technical armament in working tools of the additional labour, the migrants from agriculture, and their productivity will display a tendency towards growth. The problem will become much more complicated when the outflow of labour from agriculture would compel us to increase the stream of investments and capital goods meant to substitute for this outflow of manpower. Under such circumstances the national economy may face difficulties with maintaining the upward tentency in the technical armament in agriculture and in the non-agricultural sectors.

In conclusion, it may be said that in the period of accelerated economic expansion, agriculture may become a barrier to the growth of the national economy as a whole in that the rate of growth of agricultural production dictates the feasible rate of growth of the national income; the lack of flexibility in the supply of farm products constitutes an important factor which restricts the possibility of speeding up the rate of economic growth. This statement is confirmed by Polish experience from the period of fast industrialization.

In the socialist countries the central plan adapts the rate of growth of planned employment outside agriculture to the expected rate of growth of agricultural production. This is one of the basic principles of socio-economic planning.

One should, however, emphasize that with the expansion of the economy the influence of factors referred to as the "agricultural barrier of growth" diminishes. The reasons for it are as follows: the flexibility of demand for farm products is dropping; the differences in the consumption of farm products in the agricultural and non-agricultural sector become less pronounced; the reserves of labour in agriculture get slowly depleted; the expansion of industry becomes, to a growing extent, based on intensive methods and the increase in employment outside agriculture is more and more slowly hampering the general rate of growth of the national income. The experience of the past few years in Poland clearly confirms this opinion.

The change in the strategy of economic expansion introduced in Poland after the year 1970 has created a situation in which agriculture has become a factor stimulating general economic growth. This has occurred under the following circumstances:

Firstly, according to the new strategy, the growth of consumption has been considered to be a separate factor of economic growth. This view is based – quite correctly – on the feedback between the volume of consumption and the living standard of the population on the one hand and productivity and production – on the other hand. This strategy is based on the thesis that growth of income gives rise to factors accelerating the rate of growth due to its very strong effect on the motivation system. In the general program of improving the living standard of the population in Poland, agricultural products have been given a particularly exposed position. This called for the activation of factors stimulating the growth of agricultural production. Agricultural policy after 1970 in Poland has been clearly oriented towards production.

Secondly, criteria for the inter-branch allocation of investments have been changed so as to give preferences to agriculture. This is linked directly with the planned switching-over of the whole processing apparatus in agriculture to modern technology – the process of technical reconstruction of agriculture. This policy has created quite a different situation in the national economy expressed, for example, by a great increase in outlay on the branches of industry which produce modern technical facilities for agriculture, including the import of licences, machinery and equipment. A characteristic feature of the new policy in purchases abroad is that we concentrate on importing complete technological lines instead of single pieces of machinery and equipment.

Thirdly, the new development strategy has put forward with the whole acuteness the problem of structural changes, i.e., the process of passing from individual, disintegrated property in agriculture to large, modern agricultural enterprises. This is both the problem of increasing the scale of production and of system transformations (forms of ownership, human relations, organization of production, administration of farming enterprises). The new demographic situation of peasant farming (intensified migration aimed at wage-earning outside agriculture, biprofessionalism of more than one third of peasant farms, ageing of a great group of farm owners, lack of successors, feminisation of the profession of farmer, etc.) speeds up now the maturing of conditions for the socialist reconstruction of peasant farming. One of the most important forms is the taking over, in exchange for a pension, of the land of farmers who, arriving at the pensioner's age, have no successors. The processes of the socio-economic transformations in Polish agriculture are quite complicated and it would be impossible to discuss them in full in this paper.

The strategy of the dynamic growth of consumption is dependent the more strongly on the development of agriculture the larger is its share in covering the consumption needs of the population. Poland is now at such stage of general development when the share of foodstuffs in total expenditure on consumer goods on the part of the population is relatively high.

Agricultural foodstuffs account for about 34% of the total value of consumption. In these circumstances the "pressure" of the general economic growth and dynamics of wages and incomings on agriculture and foodstuffs is still relatively high. In the years 1971–75 the real income of the non-agricultural population has increased by 41% per one person employed. A great part (approximately 27%) went for the purchase of farm products and agricultural foodstuffs. In the last 5-year period the total agricultural production increased by 20·5% against a 35% increase of the market production and a 71% increase in the supplies of foodstuffs to the retail market. These data indicate how much the implementation of the general strategy of economic development is dependent on agriculture and on the dynamics of its development.

The Polish economy undergoes continuous dynamic transformations. As a result of a rapid expansion of the industries producing consumer goods the structure of consumption changes and the pressure of demand for foodstuffs and – indirectly – the pressure on agriculture decreases. This can be illustrated, e.g., by the fact that if in 1965, as much as 42·3% of the total amount of money spent on consumption was earmarked for food, in 1975 this index dropped to 34%, which is still high compared with the highest developed countries. The possibility of relieving the pressure of social demand for food and its pressure on agriculture depends at present on the non-agricultural sectors producing manufacturing consumer goods and rendering consumer services rather than on the agriculture and foodstuffs industry alone.

The considerations made in this paper and generalising first of all Polish experience were to indicate that agriculture strongly affects the rate of growth of the national income. A high share of agriculture in the national income and in its growth may hamper the general rate of growth of the national income and bring about increased consumption of capital and a decrease of the total effectiveness of outlays. In the countries which are at the medium and weak development stage any increase in non-agricultural employment must be accompanied by an appropriate increase in agricultural production. In this sense, the rate of growth of agricultural production dictates the feasible rate of growth of production and employment in the non-agricultural sector thus influencing the rate of the economic development of the country. On the other hand, the rate of development of the non-agricultural sectors strongly affects the expansion of agriculture not only in the sphere of the growth of farm production but also in the line of its technical and social reconstruction. Thus, there is a clear and strong feedback between the rates of growth of the two sectors of the national economy. The existence of these relationships has a vital significance when it comes to the choice of the strategy of economic development of the country. The high outlay-to-effect ratio in agriculture makes those responsible for central planning increase agricultural investments in order to overcome, in this way, the low efficiency threshold. Otherwise it would be impossible to implement the program for improved nutrition of the nation. The occurrence of the phenomenon of a high absorption of outlays and capital in agriculture induces the countries on a medium development level to free themselves from the barriers produced by a high share of agriculture in the growth of production. The actions in this case should aim at the optimisation of the structure of consumption expressed by replacing the consumption of goods of agricultural origin by that of the non-agricultural ones.

III

As has already been said, planning in agriculture in relation to the co-operative and private sectors affects only indirectly the level and structure of agricultural production by means of economic policy instruments.

The first indispensable instrument for influencing the development of agricultural production is to ensure, through the central plan, the inflow to agriculture of an appropriate stream of production means for current agricultural production and for investments. The total volume of these means and their variety depend both on the possibilities of the national economy (domestic industry and imports) and on the needs of agriculture, the latter being, however, very flexible. The potential demand on the part of agriculture for the means of production in Poland is many times greater than the effective demand for them. The task of the economic policy in relation to agriculture is to shape the demand for the means of production so as to make it possible for agriculture to absorb all the production means put at its disposal by the national economy, i.e., to adjust the demand for the means of production to the supply of these means. This is why economic policy is another instrument of the central influencing of the development of agricultural production. This policy is expressed, most generally speaking, in ensuring such a level of income from agricultural production as would guarantee to the farmers not only the appropriate standard of living (level of consumption) but also means for further increasing outlays on production and investments. Here, the prices of farm products and of the agricultural means of production constitute the main instrument for exerting influence on the level of income of agricultural producers. Of similar importance are such instruments, not connected with prices, as the tax and credit system and the system of subsidies.

The income of people employed in peasant farms, both private and co-operative, has a different character and fulfils other functions than the income of people employed in the non-agricultural sectors of the national economy. Unlike the income of the workers of state farms and non-agricultural sectors, that of the people employed in peasant farms depend, at the given price level, directly and completely on the results of farming.

The income of the peasant farms, besides the function of shaping the level of consumption, as is the case in other sectors, fulfils a productive function, too. Some part of the income is earmarked for extended production i.e. both on increasing outlays on the production in the next year and on new investment projects of a productive nature.

According to the results of the systematic many-year studies of the Institute of Agricultural Economics there is a close relationship between the growth of income in peasant farms in a given year and the growth of the amount of money spent on production next year. The higher is the growth of income the more money is earmarked by the farmers for a further growth of production in the forthcoming years. On the other hand – what is worthy of emphasising – in the periods of smaller income the farmers do not cut expenditure on production; they rather restrict its further growth which, naturally, inhibits the rate of growth of agricultural production in the years immediately after the year of reduced income. In such cases it is necessary to follow an appropriate agricultural policy in order to stimulate the growth of outlays on agricultural production.

The planned influence of the state policy on the level of income of the agricultural population calls for the determination of criteria of the required level of this income. The absolute level of income per person employed in peasant economy and in other sectors (the parity of income) cannot be compared directly, i.e., because of a usually higher degree of professional qualifications and education in the non-agricultural sectors. The rate of growth of income per one person employed can, however, be compared. In the longer periods of time the rate of growth of income of agricultural population should not deviate from the rate of growth of income of the non-agricultural population. This is justified by social considerations and – even to a greater degree – by production reasons.

In conditions of incomplete coverage of demand for farm products whose growth is faster than the growth of the income of population outside agriculture, a lower rate of growth of the income of the agricultural population would not stimulate increased outlays on extended reproduction and would inhibit the rate of growth of agricultural production thus widening the gap between the demand and supply of farm products.

In Poland a rule has been adopted that the income of the agricultural population should grow proportionally to the increase of the income of the non-agricultural population. This is the only just and rational policy for stimulating the growth of agricultural production to the benefit of the community as a whole. The implementation of the policy of equalising the rate of growth of the income of agricultural population compared to the income of those employed outside agriculture is not a simple task due to the unavoidable fluctuations in agricultural production. As results from the experience collected during the past 25 years, in every 5-year period there was one year of a greater or smaller decrease of production and, consequently, of the income of agricultural population which – after the rise of prices for farm products being introduced in such situations – were usually compensated and exceeded only in the third year after the year of poor production.

The necessity of equalising – through economic policy – the income of the agricultural population by introducing corrections into the division of the national income for the benefit of agriculture (which is effected mainly by raising the prices of farm products) results, first of all, from the slower rate of growth of productivity in agriculture compared with the remaining sectors of the national economy. Thus, for example, in the years 1950–1975 the average rate of growth of the national income produced by every person employed in the non-agricultural sectors of material production amounted to 5% per year compared to only 2·3% in agriculture.

The reason for the relatively slow rate of growth of income per person employed in agriculture could be found in the too slow rate of decline in employment (0·5% in agriculture as a whole during the past 25 years against only 0·3% in peasant farming) per unit of agricultural land. The slow rate of decrease of employment in agriculture, under the conditions of a continuous migration of rural population to towns, was due to the high birth rate (the population of Poland increased from 25 million in 1950 to 34 million in 1975) and to the restricted possibilities for greater employment than was its real level in the non-agricultural sectors in spite of the relatively high rate of growth – by 3·8% per year (from 4·7 million people in 1970 to 12 million in 1975*). The relatively high level of employment in peasant farming was one of the main causes of the slow changes in the agrarian structure which, naturally, had a decisive influence on the rate of growth of the income generated per one person employed. Agricultural production from 1 hectare of farm land was growing by 3·2% on the average per year in the period under study, but the outlays on production were being increased by approximately 8% which means that, in fact, the income produced from 1 hectare was growing by only 2% per annum and 2·3% annually per person employed.

In such a situation corrections to the benefit of agriculture had to be made in order to equalize the rate of growth of the agricultural population in relation to people from outside agriculture. The range of these corrections is best illustrated by the index determining the share, in the growth of agricultural income, of the growth from the increased farm production and from the changes in agricultural prices to the benefit of farmers. The parts of farmers' income obtained from these two sources are different in various periods and are mainly dependent on the rate of economic growth of the country and, consequently, on the rate of growth of real income per person employed in the non-agricultural sectors. For example, in the decade from 1961 to 1970 – when the average rate of growth of the national income amounted to 6·0% per annum, the corrections of the division of national income to the benefit of agriculture were relatively small. The share of income from the growth of

* It is anticipated that already in the current 5-year period (1976–80) the decrease of employment in agriculture will be accelerated. But a more drastic decline is expected after the year 1980. If the increase of labour (i.e., people in the production age) in the national economy as a whole in the period 1971–80 is to amount to 3·1 million, the same figure for the years 1981–90 is estimated at only 1·1 million people.

agricultural production in the general growth of the real income of agricultural population amounted to 73% compared to 27% from the changes of agricultural prices to the benefit of agriculture.

In the last 5-year period (1971–75) these proportions were changed drastically. The dynamic expansion of our country illustrated by the average growth of national income by 9·6% per annum and a 7·1 annual growth of real income per one person employed outside agriculture called for increased redistribution of national income to the benefit of agriculture in order to guarantee such an increase of the income to the agricultural producers which would stimulate increasing expenditures by them on production and investments, thus speeding up the rate of growth of agricultural production. This was necessary because of the high growth of demand for food resulting from the quickly growing real income of the non-agricultural population.

Starting from the preset strategy of accelerated economic development of the country, the prices of animal products were raised markedly already in the beginning of the 5-year period; in the next years the prices of other agricultural products and – once again – of certain animal products were raised.* This has produced so strong economic stimuli that the farmers increased the rate of growth of expenditures from 6·6% per annum in the years 1966–70 to 10·3% in the period 1971–1974 (this, naturally, was possible thanks to the growth of supplies of the means of production to agriculture). As a result, the rate of growth of agricultural production from 1 hectare of farm land increased from 2·8% to 5·3% p.a. and the rate of growth of the income generated per one person employed – from 2·1% to 3·4% p.a. On the other hand, the rate of growth of the income materialised increased by 6·8% in the same period of time which means that 50% of the growth of farmers income resulted from the increased farm production, the other 50% being the result of the changes in agricultural product prices to the benefit of agriculture.

In the first four years of the past 5-year period the income of farmers in terms of one person employed increased at almost the same rate as the income of people outside agriculture. Since the financial charges to be borne by farms (mainly taxes), were kept on an almost unchanged level, the part of income remaining at the disposal of farmers increased at a higher rate of 7·6% p.a.

As has already been said, in every one of the past 5-year periods one year was adverse for agriculture. For example in 1975, the agricultural production in the peasant sector dropped by 3·5% from one hectare because of a poor harvest of cereals (yields from 1 hectare decreased from 28·3 to 24·9 quintals).

* One should stress that the rise of prices of the farm products purchased does not affect the retail prices of food. The retail prices for the basic foodstuffs (bread, flour, meat, milk, butter and sugar) have been maintained at a constant level for many years. The difference between the prices of farm products purchased and the retail prices of food is subsidized from the State budget in the sphere of turnover or processing. The aim of this solution is to counteract the inflation processes, etc. This results, however, in a stronger pressure on the market of farm products. For example, meat consumption per capita increased from 53 kg in 1970 to 71 kg in 1975. To meet the growing demand for meat its export has to be partly reduced.

Still, the income of farmers decreased by only 2·2% per one person employed due to increasing prices for certain farm products during that year.

These few empirical data have been quoted to illustrate the mechanism of the economic policy of a socialist state in relation to agriculture in the sector of co-operative and individual farming (the data refer to these two sectors together), especially how this policy affects the level of farmers' income and, consequently, the rate of growth of agricultural production. The two instruments of the state economic policy in this sphere are: the system of agricultural prices and contracts with farmers for the delivery of farm products.

Individual and co-operative farms in Poland sell 90% of their production on the socialised (state or co-operative) market, 10% only being sold on the free market (mainly vegetables, fruits, potatoes, eggs and milk), in the form of direct sale of farm products to the consumers. Within the socialised market, 70% farm products are being purchased on the basis of contracts concluded with farmers. In this form, 100% of "industrial" plants, 99% of pig livestock and cereals, 91% of poultry and potatoes, 80% of vegetables, 72% of beef livestock, 66% of fruits and only 5% of eggs are purchased.

The system of contracts gives to the farmers certain price preferences (slightly higher prices, bonuses for delivery within the agreed schedule, or the guaranteed price, the latter covering vegetables and fruits whose prices outside contracts depend on demand and supply) or organizational preferences (collecting the products directly from the farm, free-of-charge advice and veterinary services) or enables starting the production of, e.g., industrial plants covered by the contract (the contracting institution delivers seeds to the farmer).

Recently, the parties are switching more and more widely to long-term contracts. This form ensures to the farmer preferences in obtaining investment credits on advantageous terms, priority in buying goods whose production does not fully satisfy the demand for them, and a partial remission of payments of the investment credits. This latter is a form of state subsidy and covers only farms which start specialization in specific branches of animal or vegetal production on the basis of long-term contracts stipulating that products would be sold to state or co-operative enterprises.

In spite of the complicated multi-sector structure of Polish agriculture, especially the dominance of non-integrated individual farming, the system of instruments of agricultural policy developed in our country makes it possible to control effectively the expansion of agriculture in full compliance with the requirements of the centrally planned national economy. This system – as shown here in a brief outline – cannot be rigid. It must react in a flexible way to changing production, and to demographic and social conditions in agriculture and in the non-agricultural sectors. The efficiency of a planned system of controlling agricultural expansion depends, to a great extent, on the properly prepared system of information which would enable both the evaluation of the current production and economic situation in agriculture and short-term forecasts concerning changes in this situation. A correct evaluation of the situation makes it possible for the central organs of economic policy to make decisions which may counteract any disadvantageous phenomena. As has been confirmed by long experience, decisions counteracting such adverse phenomena are, from the social point of view, more effective than decisions made in order to offset the effects of these adverse phenomena.

H. SCHLICHT*

Rural Nutrition and Living Conditions in Relation to GDR Agricultural and Economic Policies

As in other socialist countries the economic policy of the GDR is focussed on doing everything to raise further the material and cultural levels of living of the people on the basis of a high rate of development of socialist production, increased efficiency, scientific and technological progress, and higher labour productivity.

This objective serves the welfare of all working people in our socialist society. It is through high individual work results, a sustained and systematic growth of production – making full use of all the potentials of the country, and increasingly close co-operation (especially with the Soviet Union and the other socialist countries) that we will be able to amply satisfy the steadily growing needs of all people.

The objective for the agricultural policy of the GDR derives from these considerations:

(a) We increase systematically both agricultural production and its efficiency in order to ensure stable, steadily improving and continuous supplies of high-quality food, to the population, meeting demand and increasing the degree of self-sufficiency. The same aim applies to the demand of the manufacturing industry for raw materials of agricultural origin. As experience shows, the demand for high-quality foodstuffs increases continuously.

(b) At the same time, we do our best to bring the working and living conditions in the villages up to those in towns, thus gradually overcoming the major differences between rural and urban life.

The laws governing social development imply, and the experience gained in the past years has demonstrated, that we will only be able to solve these problems in a rational way, and with efficiency for the whole society, if we continue to intensify agricultural production, passing over gradually to factory-like, large-scale production crop and livestock farming in specialized co-operative and State-owned enterprises.

* Bernburg, German Democratic Republic

The provision of modern, efficient means of production – particularly machinery systems and fertilizers, extensive experience in co-operation within and between socialist farms, farmers' high level of technical skills, and other factors enable us to make this step now.

Based on the theory of scientific Socialism, encouraged by many years experience of socialist development in the Soviet Union and supported through planning by the whole society, workers and farmers have successfully embarked upon this road.

From the very beginning the economic policy of the Party of the working class and of the socialist State has been directed towards promoting the alliance with the farmers, towards incorporating them in State management, continuously expanding their production basis, hastening the process of social development and improving the rural living conditions on a long-term and planned footing.

The democratic land reform under which almost 30 per cent of the nation's farmland was given to farm-workers and small-holders, or converted into national property, helped to strengthen the confidence of farmers in the working class.

Thanks to a systematic economic policy a powerful farm-machinery and fertilizer industry was set up. State-owned seed-growing and animal-breeding farms were major prerequisites for an overall boost of yields and performances in cropping and livestock farming, respectively. Tractors and implements of the Machinery-Hire Stations of the State gave priority to work on the small- and medium-sized farms at particularly favourable rates. Stable prices for agricultural products fixed by the State, the guaranteed sale of all produce, bonuses to stimulate a high production increase and allowances for improving production conditions (improvement schemes, purchase of livestock, construction of farm buildings, etc.) enhanced production and the sale to the State of a steadily growing share of the farmer's output. All this resulted in a continuous improvement of supplies and a strengthening of the position of the working class and its State in the countryside.

Able cadres of the working class, politically experienced and good organizers, were sent to rural areas. The conditions of schooling and general education in the countryside were gradually adapted to those in towns. The example of socialist State farms also helped to create favourable conditions for the transition to co-operative production.

From 1960, when all farmers had joined the agricultural production co-operatives (co-operative farms), our agrarian policy concentrated on the all-round implementation of co-operative work on the development of Socialist democracy by broadly incorporating the co-operative members into the management and planning of their farms and communities, and on the systematic in-service training of the farmers to enable them to apply scientific findings and to the conscious building of Socialism.

The general principles of Lenin's Co-operative Plan have always been applied in a creative way, taking account of our conditions. These principles include, among others, the central direction of the process, the consideration of differentiated stages of development, forms of organization corresponding to the practical experience of the co-operative farmers, convincing examples and comprehensive knowledge, voluntariness and, not least, the systematic promotion of the co-operative way by the socialist State, for example, when creating the material and technical foundation.

The continuous increase of production and the fundamental transformation of the farmers' social situation confirm the correctness of the road embarked upon. It is a way understandable to farmers, matching their interests, which are also the interests of the socialist society.

During this period fundamental social changes occurred in the villages. The former rivalries and economic dependence were replaced by common interests, co-operation and mutual assistance. The improvement of living conditions became a community issue. The local State bodies took responsibility for the establishment of a general polytechnical secondary school system providing the children of the co-operative farmers with the same chances of education as under urban conditions. A comprehensive vocational education and in-service training system was set up. So far it has given more than 80 per cent of all those engaged in farming a complete technical training. Some 6.3 per cent of those employed in agriculture have had higher education (university or medium level). In the majority of the villages, facilities for children of pre-school age have been created.

Over the whole country there is a dense network of retail shops, rural department stores, cultural facilities, cinemas and restaurants, rural outpatient clinics and dispensaries, accessible to the rural dwellers because they live near them or can easily go there using public means of transport.

Working conditions and social security have fundamentally changed.

The majority of operations have been mechanised. With the exception of temporary labour peaks, the agricultural enterprises have a working schedule comparable to that in manufacturing industry. Rural workers are granted almost the same social benefits (such as paid holidays, sickness allowance, temporary leave, free medical care, right of pension, temporary paid leave for women for one year after childbirth, shorter working hours for mothers with several children, etc.). The State grants stimulation funds, price additions and other allowances to enable farms working under unfavourable conditions of production – which must be made use of to ensure food supplies – to produce adequate returns so that they can pay equally high wages and may improve their production basis. Therefore, the average income of comparable jobs done under differing economic conditions does not vary much.

Granting credits on soft terms helps to ensure normal incomes for co-operative farmers even in years of poor harvest.

By way of financing policies, especially long-term stable prices for agricultural products and continuously rising labour productivity, the average income of a co-operative farm member has reached a level equal to that of comparable branches of industry. State-fixed agricultural prices have been arranged in such a way that all agricultural enterprises are able to obtain the necessary means for expanded reproduction. Price increases for inputs which might have a negative effect on financing are temporarily compensated by

State subsidies. This is done with a view to not endangering the stability in retail prices for staple foods which we have been observing for more than 20 years – an essential principle of social policy.

The GDR co-operative farmers earmark continuously rising sums of money to improve, through higher social consumption, the living conditions for all those living in rural communities. This is done in close collaboration with other enterprises and institutions. This year, for example, a State decision will be taken providing for an extensive housing programme. It will encourage enterprises and co-operative farmers to build houses on a collective or private basis. Apart from what has been achieved already, this programme is estimated to substantially improve, over the next 5 years, the housing conditions of 8 to 9 per cent of the people employed in agriculture. In addition, agricultural workers and co-operative farmers combine in efforts to set up social services, to improve village scenery, to create recreational facilities, etc. In this way, too, rural social conditions are gradually approaching those in towns.

In the meantime, co-operative farmers and workers have gained extensive experience in the socialist way of production. The material basis of production has been considerably enlarged, and there has been experience over several years of contractual co-operation going beyond the limits of one co-operative, or State farm, thus adding to the efficiency of production. All this made the co-operative farmers and workers embark upon the road to factory-like, large scale crop and livestock production.

Economic experiments conducted over several years regarding contractual inter-farm co-operation of several enterprises in the same area for producing crops along industrial lines, experience in the collaboration of specialized breeding farms and commercial farms specialized in one product, of handling and processing enterprises, marketing facilities and wholesale trade, of production increases going hand in hand with a higher degree of specialization and efficiency, more favourable conditions in the national economy as a whole for providing complex machine systems and up-to-date livestock installations as well as new findings in science and technology adding to higher yields and performances have been factors stimulating that process.

This corresponds fully with the objective interests of society as a whole. Already Lenin arrived at the conclusion: "Mechanized large industry and its transfer to agriculture is the only economic basis for Socialism".

Industry and agriculture have a uniform social basis in our society. The same economic laws of increasing socialization, division of labour, growing concentration of production and specialization are acting in it. The further development of the productive forces, new findings of science and technology – if these are to be used efficiently – imply new forms of organization of production and systematic planned co-operation of the enterprises engaged in the division of labour.

Co-operative and State farms with diversified production have taken account of this by farming their land jointly, pooling their machines in large complexes (for example 10 combine-harvesters with the appropriate transport capacities), by setting up plots of 200 hectares (500 acres) or more, by pooling funds to purchase efficient machinery, to employ stores, application and handling equipment for fertilizers and other agro-chemicals or to build processing units (such as dehydration plants, feed pelleting plants or easiling facilities). Often, self-contained enterprises have arisen by these facilities being linked by economic relations and through long-term contracts.

In livestock production, several enterprises pool their funds, animal stocks, feeds and labour force, to produce more efficiently and with less labour in up-to-date, fully mechanized units. These newly arising, inter-farm units are run on the basis of their own plan. Standardized projects with equipment and management systems which have stood the test of time, are available for building such units. These are dairy units housing some 2,000 cows, young-cattle raising units with more than 5,000 head, pig raising and fattening units with an annual output of 6,000 to 25,000 tons of pork.

The formation of such specialized units of production has proved successful as higher efficiency, high and stable production results and improved working conditions clearly show. However, it is a long-term process, requiring substantial inputs. It must be developed according to a central plan. It needs thorough discussion, the selection of the most appropriate variants and the participation of the participating enterprises in the democratic bodies set up by them for their management.

The vertical co-operation of autonomous enterprises of different types of property engaged in the production of one commodity has proved very useful for improving supplies. In these vertical co-operation unions an increasingly close agro-industrial integration is effected in a democratic way and to the benefit of all partners. On the basis of equal rights, enterprises doing breeding-work, reproduction, commercial production, handling, processing and marketing work together.

They co-ordinate their production, make joint efforts for applying scientific and technological advances, for the training of their staff, etc. Special endeavours are aimed at reaching a high product quality, cutting losses, completing the assortment and making the supplies more continuous. They co-ordinate their plans, build joint storage and processing capacities, marketing centres and sales facilities.

The enterprises are mutually linked through long-term contracts and stable prices and supply relationships.

All forms of co-operation are directed by democratic management bodies such as co-operation councils, union councils, assemblies of delegates and plenipotentiaries. The participating enterprises act within them on the basis of equal rights irrespective of their size and share in total output. The decisions taken by the democratic management bodies need to be confirmed by the member assemblies of the participating enterprises or the manager of the respective State enterprise.

Through its economic policy the socialist State supports and promotes this process of transition to the factory-like way of agricultural production because this is in the interest of society.

The State directs this development as a whole, planning the balance between the branches, both on a long-term basis and for the different territories. The State plans the contributions to be made by science and technology and takes charge of generalizing the experience of the best. Under the leadership of the working class Party new things are continuously analysed, the level reached being evaluated and the macro-economic possibilities considered as regards the further implementation of that process. Especially in State enterprises, long-term economic experiments are conducted to gain experience and judge results likely to be obtained under commercial conditions. At enterprise level, staff-members of State bodies and scientific institutions participate in organization and extension.

The transition to this way of agricultural production is also economically stimulated by granting credits at low rates of interest, by partly subsidising the application of new scientific and technological findings, by price additions for products from factory-like units, etc.

In particular, the State promotes socialist economic integration between the socialist States, for only in this way the growing requirements from agriculture can be met efficiently. A large part of the machinery, agro-chemicals and projects for units are jointly developed, designed and jointly (or by division of labour) produced following mutually agreed standards. The same applies to establishing new crop varieties and animals breeds. This kind of integration is feasible because the development problems of agriculture in the socialist countries are tackled fundamentally in the same way, on the basis of the Marxist–Leninist agrarian theory. This also contributes to the further drawing together of our Socialist countries.

The maintenance of peace is the basic condition for the systematic long-term implementation of the agrarian policy and the successful construction of our society. Through the peace policy of the Soviet Union favourable external and internal conditions are created for the undisturbed realization of our projects. Our efforts for increasingly closer economic and trade relations and scientific and technological co-operation with mutual benefit and advantage are also aimed at this. This co-operation is in the interest of the people. It fosters national independence, equality of rights in international relations, economic development and social progress to the benefit of our people.

As agricultural economists we should regard it our permanent commitment to take an active part in this fight for peace, social progress and the improvement of food production to the welfare of the working people.

V. R. BOYEV *

Problems of Agri–alimentary Complexes in Territories with Extreme Production Conditions

New data about natural resources in connection with the progress of science and technology and increased demands for food products make it necessary to bring into national economic activity land which is extensively used or not used at all. Such land is usually remote from established centres of culture and production. They have weakly developed transportation and show great differences in pedo-climatic conditions of production. During the industrial development of such regions providing the rapidly expanding population with food-stuffs of good quality presents a problem.

The USSR has built up a varied experience of industrial development of territories with extreme conditions of production. In general such territories are situated in the north of the country, in desert and semi-desert regions. A thorough study of this experience shows that attracting manpower into the new regions of industrial and transport construction by means of higher payment does not give the necessary results for ever. The economic development of such regions, especially when intended for long term functioning, should be based on the application of the complex of economic and practical measures which may provide an attraction and a necessary specialist consolidation. First of all it requires the provision of the cultural and welfare facilities appropriate to the local conditions and good quality food – a main condition for effective vital activity.

There is no need to emphasize that the problem to be solved is difficult even for the regions where we have favourable soil-climatic conditions for agricultural and food industry development. The difficulties of solving the problem increase greatly in regions with extreme conditions of production and particularly in the construction zone of Baikal–Amur railway in the Soviet Union. Almost throughout the whole territory which will be crossed by this railway there are low average annual and seasonal temperatures, with mountainous relief in one place and marsh-ridden areas in others. The length of the growing season is 50–116 days in some places where new towns and workmen's settlements are being built. Such a season is 2–4 times less than in active agricultural regions. The average annual precipitation is 180–530 mm, positive temperature sum is 850–1350°C.

* U.S.S.R.

The successful fulfilment of a large scale programme of industrial development on the lands poorly used before is only possible by a systematic approach to their development and realization. Programme-objective planning is a highly reliable methodological basis for the development and realization of such programmes. From the methodological point of view programme-objective planning is to be considered as the complex method of development of national economic plan and its parts which make it possible to realize the planning logic "from objectives". At the same time the achievement of one or two purposes is based on co-ordination with the complexes of organization-economic, and other, measures including financial and material-technical resources and manpower. Therefore, the most important feature of programme-purpose planning is a directed usage of resources in time and space for the achievement of specific purposes.

The co-ordination of purposes and resources creates a firm basis for selection of rational solutions, optimisation of industrial development proportions within one branch and between different branches, and the creation of specific features of social economic infrastructure. The infrastructure provides harmonious combination of public interests with interests of development of separate regions and each executor who takes part in the achievement of specific purposes.

Scale, structure, dates and manner of working out the programme for industrial and agricultural development of new regions depend on their concrete content and purposes and on the presence of production resources that can be used.

The elaboration of long-term forecasts and objectives of production forces development within the limits of definite territories is the conceptional (purpose) stage of programme-objective planning. So possibly exact determination of the following factors is of principal significance in the elaboration of food subprogramme at the forecasting stage of the development of new regions. There are three general factors: (a) intermediate and final objectives of economic development of the present territory; (b) volume and structure of necessary food products in near and far future; (c) the sources of food products at minimal total expenditure.

The basic feature of programme-objective planning for food product supply of specific regions consists in the fact that this problem solution is achieved by means of agri-food complex creation as a part of a complex programme for the solution of the main purposes of industrial assimilation of new regions and not by means of realization of separate episodical measures.

The main pecularity of the agri-alimentary complex problem consists in system creation of food products production and delivery to direct consumers by means of arrangement of food production in areas of consumption or its delivery from other districts in fresh or processed state. The complex includes the solution of such questions as ploughing up new earth, ensuring manpower and corresponding material-technical means, processing, storage and realization of production, creation of social infrastructure, organization and execution of research and design development, etc.

Realization of the whole programme of developing regions' supply with food products depends, to a considerable extent, on accurate matching of these problem solutions in time and space. Though the above-mentioned questions do not include its whole range, they reflect, to a considerable extent, the specificity and many-sided problem solution of agri-alimentary complex creation. The total expenditure is the most important optimum index of separate questions solution of agri-alimentary complex creation. However, it is impossible to proceed only from mechanical comparison of expenditures for these purposes during problem solution of supply of population with food products in new industrial regions situated in extreme conditions of production, and so it may be advisable to deliver it from other regions or to organize local production. It is necessary to take into account that relatively high expenses for milk, vegetables and a number of other products near areas of consumption, especially at the first stages of arrangement of agri-alimentary complex, are caused very often by unfavourable nature and practical economic factors, particularly by dissipation of production among small sections or by application of obsolete technology. It is impossible also to leave out of reckoning that during a whole year natural milk, potatoes – to say nothing of fresh vegetables – cannot be delivered to regions with low and very high temperatures without special and expensive transportation and packing. And these very products are the most important components of quality nourishment. The defect mentioned cannot be eliminated completely by means of storehouse construction, with regulated conditions of storage, and by means of product delivery in warm seasons because we must bear in mind the great volume of consumption of these products.

Our estimations of technical economic indexes of production and realization of general agricultural products obtained from different expenditure comparison and generalization of accumulated experience in analogical problem solution in similar production conditions of other regions are the basis for conclusion that we should solve the problem of meeting the requirements of the population in the zone of the Baikal–Amur railway for food products in three main directions:

(a) by means of organising the production of products which are perishable and travel badly in areas near the consumption point, for example, field vegetables, potatoes, milk, eggs;

(b) by means of intensification of agriculture in neighbouring regions having more favourable economic-natural conditions for potatoes and vegetable growing, and dairy-meat production;

(c) by means of delivery of heat-loving vegetables, fruit, grapes and melons from more distant regions.

The significance of these sources of food products for the separate parts of the railway under construction will be unequal because of considerable differences in soil-climatic conditions of production, location, transport communications, etc. And in future it will change in different ways. For instance, in the eastern part of the railway the demands for potatoes, field vegetables, milk and meat will be met in the near future by local arrangement for production. But the majority of food products will be delivered in the near future from other regions to central parts of the railway which have a less favourable soil-climatic conditions for agricultural development and where we assume there will be intensive road and industrial construction.

The above mentioned facts and others point out the impossibility of routine methods of approach to the problem of meeting the population demands of new regions in specific food products. We must speak not about difficulties but about rational combination of different variants of supply with the necessary products (in volume, assortment and quality) with minimal expenditure of labour and facilities taking into account a variety of assortments of the necessary food products and the physical possibility of their production in the required volume.

DUŠAN TOMIĆ *

Decision-making in the Self-management System in Yugoslav Agriculture at the Macro- and Micro-level

1. GENERAL FEATURES

Decision-making in agriculture is a very complex process including all aspects of the development of agricultural production and all agents of agricultural development. The promotion of the process and the methodology of decision-making in agriculture gains all the more in importance in contemporary circumstances.

Realistic and timely decision-making at the national, regional and international levels is the condition of primordial importance for the efficient development of agricultural production, for the solution of the world food problem and the balancing of the world food supply and demand in the long run.

The analysis of the decision-making process in agriculture requires a complex approach. It is essential to take into account both its methodological and its social aspect.

The methodological aspect of the decision-making consists in working out the most convenient methodology for defining and setting up the basis and background for making realistic and efficient decisions concerning agricultural development at all the levels where they are made.

The methodology of decision-making depends on the level of scientific theory and research, the application of scientific and professional methods and the availability of systematic statistical data and documentation.

The social aspect consists of the analysis of the position and the role of agricultural producers in the decision-making process in agriculture. It requires an answer to the question whether agricultural producers are independent in their decision-making and what is their influence on decision-making at all levels, from the individual farm right up to national agro-economic policy.

Both aspects of the decision-making process in agriculture are mutually connected and conditioned. There are, however, qualitative differences between them. The methodological aspect is extremely important for the

* Scientific Adviser, Economic Institute, Maršala Tita 16, P.O.B. 402, 11 001 Belgrade, Yugoslavia.

rational constitution of decisions, but its importance is primarily of a technical and technological nature. The focus lies in the social aspect of decision-making, which is determined by the particular features of the socio-political system and socio-economic relations.

The social position of the decision-makers in the process of social reproduction represents the basic element of rationality and efficiency of decision-making.

2. SELF-MANAGEMENT AND DECISION-MAKING IN YUGOSLAV AGRICULTURE

The social aspect of decision-making in Yugoslav agriculture is based on self-management, which represents the basic feature of the socio-political system in Yugoslavia. The ideological and socio-economic principles of the self-management system are elaborated in the Constitution of SFR Yugoslavia and the Law on Associated Labour.

The essence of the self-management socio-economic relations in the Yugoslav economy and society resides in the principle that the worker-producer is the only, and the essential, agent of the formation and distribution of income and the beneficiary of the surplus work.

The socio-economic position of the workers in the social reproduction process determines the mode of decision-making in Yugoslav agriculture as well as in the whole economy and society. Decision-making is effected through the essential forms of self-management:

(a) by way of personal declaration by the workers in the basic organisations of associated labour, which represents the basis for all other forms of direct decision-making;

(b) by way of decision-making by the elected delegates in the self-management bodies of the organisations of associated labour, associations and institutions of the producers and political bodies;

(c) by way of effecting the direct workers' control in the organisations of associated labour and other bodies.

The main sense of direct decision-making in the self-management system consists in the principle that every worker or agricultural producer makes decisions in his organisation of associated labour about the conditions and results of his work and the whole social reproduction.

Direct decision-making in Yugoslav agriculture takes place at different levels, beginning from the farms, through various institutions and authorities in municipalities, provinces, republics and at the national level.

The functioning of the system of direct decision-making is discussed here at the national level as well as at the level of the social agricultural farms and private farms since at these levels the most important decisions for the development of agricultural production are made.

We shall consider the basic questions of the domain, content and structure of direct decision-making at the national level and at the level of agricultural farms.

3. DECISION-MAKING AT THE NATIONAL LEVEL

The most important document on which the agricultural producers and their organisations and institutions make decision is the Social Agreement Among Republics and Provinces on the Development of the Agro-industrial Complex in Yugoslavia. In this document the following basic questions of the strategy of development of the agro-industrial complex are worked out:

(a) the position of agriculture in the policy of the socio-economic development of Yugoslav economy and society;
(b) the basic aims and directions of development of agricultural production and the possibilities and conditions of balancing the supply and demand for food;
(c) production policy, rate of growth of agricultural production and development of crop production and livestock breeding;
(d) development of social farms and the main lines and possibilities of the growth of the socially organised agricultural production;
(e) basic principles of policy concerning individual farmers;
(f) relations and frameworks of the linking of agricultural producers, food processing industry, commerce, industry producing fixed assets, working capital for the needs of agriculture and consumption of agricultural products;
(g) essential conditions and frameworks of the development of expanded production in agriculture, such as the volume and pattern of investment, price policy and the level of protective prices by categories of agricultural products, credit policy and credit terms, conditions and orientation of foreign trade, and commodity reserves.

The Social Agreement defines the long-term policy for development of the agro-industrial complex in Yugoslavia for the five year period. On the basis of this document are worked and implemented the annual development plans of Yugoslav agriculture.

This document is taken as the basis for the elaboration of the programme of development of the agro-industrial complex by republic, province and community, in accordance with their prevailing natural, climatic, land, technical and technological and socio-economic conditions. The whole system of agrarian and economic measures and institutional conditions of the agricultural development is worked out more broadly and concretely in these programmes.

Many scientific and research institutions as well as the socio-political organisations are included in the elaboration of the social agreement on the long-term development of the agro-industrial complex in Yugoslavia. The most responsible institutions for its preparation and elaboration are the Federal Committee for Agriculture and the Federal Office for Economic Planning, as well as the corresponding republican and provincial institutions.

The Social Agreement on the Development of Yugoslav Agriculture is the subject of wide discussion and argument at the national level among the agricultural producers and their institutions and organisations as well as in the framework of the widespread scientific and research and socio-political activity in the country. The opinions and tenets exposed in the public discussion are the decisive elements for its final adoption by the delegates in the Federal Parliament of SFR Yugoslavia and the republican and provincial assemblies.

4. DECISION-MAKING AT THE LEVEL OF AGRICULTURAL FARMS

In accordance with the basic aims of the long-term development of the agro-industrial complex in Yugoslavia agricultural farms work out their development policies. In doing this they are completely independent in defining and decision-making relating to the direction and the framework of their development.

The socio-economic structure of the Yugoslav economy is characterized by the existence of the socialist social farms and private individual farms. The way of decision-making in these farms is different.

4.1. *Social sector*

In the structure of Yugoslav agriculture the social farms are playing the more important role. They are the agents of technical and technological progress and intensification of agricultural production. The social farms have reached the productivity level of developed countries and in some agricultural farms it has even been surpassed.

Socially organized agriculture has already been yielding 46–47% of the production of commodities.

In the social farms the workers are the main agents of decision-making. By personal declaration as well as by referendum they make decisions on the most important questions of the building up and development of the social farms. These are the following questions: (a) statute of the organisation of associated labour; (b) guidelines of the development plan and programme of the establishment of the organisation of associated labour; (c) production policy in crop production and livestock breeding; (d) efficiency of business management and formation and distribution of the income obtained; (e) policy concerning personal incomes and reward according to work; (f) investment policy, particularly long term investments; (g) provision of financial resources of a large volume and the building up of the organisation of associated labour and (h) conditions for the association of labour and resources with other organisations of associated labour, food processing industry and commerce.

The direct decision-making by workers represents the basis of the work of the self-management bodies, above all of the workers' council and other bodies in the organisations of associated labour in the social sector. The self-management and professional bodies work out the generally adopted decisions, carry out the operational and long-term policy of development of social farms and direct the process of intensification and increase in the socially organised agricultural production on the basis of modern technical and technological progress.

The workers follow by way of direct self-management workers' control, the working out, and implementation of the decisions generally adopted by the self-management and other bodies in the social farms.

The self-management workers' control has the right to demand that the decisions are respected together with the rights and interests of workers and has to ensure their consistent implementation.

4.2. *Private sector*

The private sector comprises a larger part of productive capacity and agricultural production in Yugoslav agriculture. Small farms are, however, characteristic of the private sector since it consists of about 2·6 millions individual farms with an average size of only 3·8 ha of total land. The linking up and association with the social sector is the right way to promote, increase and to socialize agricultural production of the private sector. In this field considerable productive and socio-economic results have been obtained.

The individual producer is completely independent in his farm and free in decision-making concerning all the questions of development of his farm and especially in the following: (a) production policy; (b) promotion of agricultural production; (c) farm construction and (d) way of use of the income obtained.

The individual farmers influence the agro-economic policy measures and conditions for their activity through their cooperatives and association of cooperatives. Their influence is particularly focused on the following questions: (a) policy of supply of intermediary goods; (b) the use of services by the social sector; (c) the conditions to be fulfilled to obtain and repay credits; (d) way of buying agricultural products; (e) cultivation of land; (f) building up of infrastructure in agriculture, and (g) conditions for obtaining pension, health and social insurance for the farmers.

Individual farmers are completely independent in making decisions on the association of labour and resources with the social farms. When the development and socialization of the common agricultural production of the individual and social farms is in question the individual producers have the same rights as the workers in the social farms and they are completely equal in making all the important decisions on the development of joint production and the formation and distribution of income obtained through joint production in accordance with the work and invested resources.

5. THE MAIN LINES OF FURTHER DEVELOPMENT AND PROMOTION OF THE SYSTEM OF DIRECT SELF-MANAGED DECISION-MAKING

The system of direct self-management and decision-making has contributed to a great extent to the development of Yugoslav agriculture during the past two decades. This system incited the strong interest of agricultural producers in the promotion and growth of agricultural production and in obtaining income and improving the standard of living.

The agricultural production in Yugoslavia has more than doubled. The productivity of Yugoslav agriculture increased relatively greatly per active farmer and per hectare of arable land. The intensification and socialization of agricultural production in the coming period will enable Yugoslavia to satisfy the home needs for food in the near future, by its own production and provide a part of production for export. Yugoslavia will become a net exporter of food and will make a comparatively important contribution to the solution of the world food problem in relation to her available resources.

The long-term development of Yugoslav agriculture particularly emphasizes the need for further improvement of the system of direct self-managed decision-making, as the essential factor of promotion and increase in agricultural production. It is indispensable to ensure the systematic study, analysis and improvement of this system. In this field the investigations must be particularly directed to the following questions which are characteristic of all levels of decision-making:

(a) Promotion of the system of direct self-managed decision-making should enable the timely decision-making at all levels which is the essential condition for the efficient and stable development of agricultural production;

(b) It is necessary to make a continuous and complex study of the socio-economic relations in agriculture, which are the basis of the system of direct and rational decision-making;

(c) The more efficient and complete agreement of interests of various agents in agriculture in the process of understanding, agreement and decision-making must be ensured;

(d) The system of professional training and informing of the agricultural producers must be promoted so that they are in a condition to make timely and realistic decisions;

(e) It is necessary to examine and improve the system of collecting and filing the statistical data and documentation material in order to ensure the elaboration of trustworthy and complex grounds for making realistic and efficient decisions at all levels;

(f) Programmes of scientific investigations are needed which should be directed to those problems which are important from the point of view of more intensified development of all agriculture and making the corresponding decisions on the most important development problems;

(g) Greater attention should be paid to the training of professionals at all the levels, especially of agroeconomists, with the aim of preparing more efficiently the analytical basis for decision-making;

(h) Building up of the consistent system of direct self-managed decision-making should ensure its efficient functioning both in the short-term and the long-term development of Yugoslav agriculture.

DISCUSSION OPENING – Wiktor Herer, *Poland*

I would like to concentrate the participants' attention on two problems.

The first problem is the comparison of the two planning systems for agricultural development in Poland and Yugoslavia.

These systems show some similarities. In both, peasant farms, which produce an essential volume of produce in these countries, are free in their adjustment to the economic environment. This is shown, primarily in the price system; the indirect influence of the state on peasant farms through a price policy provides the basic state operating method in both countries.

This matter seems clear to me but what should, perhaps, be clarified in the discussion are the differences between these systems. It would be useful to have explained whether in Yugoslavia, to the same degree as in Poland, the central planner determines the basic elements of the economic environment, including the prices of agricultural products and those of input flowing from the non-agricultural sector to the agricultural one, and the volume and product structure of that input, as well as the demand for manpower flowing from the agricultural sector to the non-agricultural one.

The second problem, connected with the paper of Dr. Boyev, is that of obtaining the necessary agricultural production in the Soviet Union, and also in some other socialist countries, in regions marked by high marginal costs.

It seems that this phenomenon, which occurs both in the Soviet Union and in Poland and, perhaps, in some other socialist countries, results from the co-existence of the following phenomena:

- a very high income elasticity of demand for agricultural product which results, among other things, from a fairly egalitarian division of the growth of consumption fund,
- High marginal costs of obtaining agricultural products.

The co-existence of these two phenomena brings about specific difficulties in Poland. In my opinion, these difficulties must be solved not only by accelerating the rate of growth of agricultural production but also by gradual changes in the consumption increase structure, aiming at a growing share in consumption of industrial articles characterized by marginal costs not varying from the average costs, or even declining – as a result of economies of scale. Among other things, it should be realized that a country like Poland, characterized by poor soils and by climatic conditions rather unfavourable to agricultural production, cannot aim at imitating the consumption structure of countries, such as France and the U.S.A., characterized by abundance of fertile soils situated in regions of favourable conditions.

RAPPORTEUR'S REPORT – P. E. Temu, *Tanzania*

Considerations presented as forming a background to the discussion included the fact that not withstanding differences in conditions and levels of economic development among socialist countries, the accelerated development of productive forces and living standards remains the main characteristic of economic and social development in all the socialist countries. Agriculture had always played an important role in this process and can be expected to play an even bigger role in the future. The view was expressed that more attention should in future be paid to the close functional interdependence between agriculture and other sectors – the papers underlined this – and measures adopted which would give agriculture the same recognition as the rest of the economy.

Further, there was held to be need to raise the quality of the diet as opposed to the quantity of food produced, in order to raise nutritional levels, to take appropriate measures to modernize agriculture and increase the output of food and fibre sufficiently to cope with increasing demand, to foster the growth of large-scale agro-industry organised along socialist lines, and to improve marketing and distribution of farm products, the development of socialist production relations, and the process of decision-making, etc.

It was pointed out that peasant agriculture in Poland was linked to the rest of the economy in many diverse ways. State farms were subject to imperative planning, but co-operatives and private farms were more subject to controlled planning, influenced by various signals of the market mechanism, itself conditioned by the policy of the central planners. Considerable emphasis was put on the relation between the overall rate of growth of the economy and the rate of growth of agriculture, and circumstances were spelt out in which agriculture could either, on the one hand, become a fetter to general economic growth or, on the other hand, its key propellant.

With respect to Yugoslavia, it was generally observed that there is close connection between development of agricultural production and the system of self-management in decision-making. This system has been institutionalized, and permeates management and decision-making in all sectors and at all levels. It is a system through which peasants and workers at all levels, either by themselves or through their elected representatives, are enabled to move increasingly towards more effective forms of socialization of the economy, including the agricultural sector. As in Poland, the private farm sector has full freedon to reach its own decisions on farm production, investment, etc., reflecting market circumstances. As in other cases of centrally planned economies, this freedon is exercised within the general framework and guidance of the national plan.

The particular feature of the USSR situation discussed concerned the supply of adequate food and other provisions in the development of remote rural areas such as the northern parts of the USSR, in desert and semi-desert regions. The illustration chosen was the area to be traversed by the Baikal-Amur railway in the Soviet Union. It was stressed that the successful fulfilment of such a programme, under difficult conditions of climate, soils and topography, lies in proper programme planning and co-ordination of objectives and the means (or resources) for their implementation. The use of resources in time and space have to be directed to the accomplishment of the desired goal at minimum cost.

Thus, the principal problem of supplying food for the population within the railway construction zone would have to be solved by a judicious combination of measures to (i) transport perishables, e.g. dairy products, as cheaply as possible, (ii) grow on the spot whatever can be produced economically, and (iii) arrange the delivery of fruits and vegetables from more distant areas.

Further references were made to other situations, notably Czechoslovakia and Hungary, where it appeared that the central plan, as defined by the Party and Government, laid down the global and often specific goals, subject to which various organisations, including agricultural enterprises, planned their activities. The use of different types of mechanisms and incentives to achieve the desired goals was often cited.

It was clear that there were certain lessons from Eastern Europe for the developing socialist countries such as, for example, Sri Lanka and Tanzania. It seems that chief among these is the essential nature of interdependence between the agricultural and non-agricultural sectors in planned socialist growth. A second important consideration concerns the way in which the public (or socialized) sector of the economy, the co-operative sector and what remains of the private sector, can all be made to bend their energies in the direction desired by the national plan. For this to happen, however, careful planning and implementation mechanisms must be worked out in order to minimise or eliminate possible incompatibilities. The really unanswered question concerns the types of mechanisms or system of incentives that would effectively enable a developing country to achieve a full and rapid transition to socialism.

M. L. DANTWALA

Agricultural Policy In India Since Independence

My paper deals with agricultural policies adopted from time to time in India and the context in which they were adopted. My acquaintance with agricultural situations in other Most Seriously Affected Countries (MSA) is very limited, and I felt that one should refrain from expressing views on countries and people with whom one's emotional and intellectual involvement is, at best, remote.

The theme of this session covers a wide field – The Relationship Between Agricultural Policy, the Economy and Economic Policy, on the National Level in Different Economic Systems and at Varying Stages of Economics Development. I have taken the view that a detailed account of the agricultural policy of a country, describing the nature of conflicts and the choices available in the context of specific situations, and a critique of the decisions made by the policy-makers, will implicitly serve the purpose of the programme; and that it is not necessary to refer everytime to the varieties of relationships mentioned in the theme. For example, during the latter half of the 1960's, when the food situation was critical and the High Yielding Varieties of cereals became available, the policy-makers in India faced a conflict; The adoption of HYV's would augment food production but was likely to aggravate inter-class and inter-regional disparities. Were they right in the decision they took? If not, what were the available alternatives? Such are the issues discussed in this Paper.*

India's agricultural policy, and perhaps that of most LDC's, has often been criticised for its "neglect" of agriculture. The criticism acquires legitimacy because of what is generally characterised as the "failure" of agriculture. The alleged failure may have a reference to either the growth of agricultural production or the promotion of social justice, or both. It is, therefore, necessary to get a more precise idea of the performance of Indian agriculture in both these fields and identify policies related to this performance.

* I am grateful to Dr. (Mrs.) R. Thamarajakshi, Dr. V. M. Rao, Dr. L. S. Venkataramanan, Dr. C. H. Shah and Dr. Pravin Visaria for their many helpful comments on the earlier draft of this Paper.

A reference to agriculture's performance in the pre-Independence period (approx. 1901–1950) may not be considered quite relevant, but legacies do matter and the state of agriculture as inherited from the colonial days – whether it be in India or Taiwan – is not quite irrelevant for the assessment of the post-Independence performance. George Blyn's study has revealed that between 1891 and 1947, aggregate grain output in British India increased at an average rate of 0.11 per cent per year. In fact, in the latter half of the period, the growth rate was a negligible 0.03 per cent. Rice output, constituting half of the total output, actually declined over the 56 year period at an average annual rate of 0.09 per cent [1]. During this period, population increased at a mercifully low rate of 0.67 per cent per annum in British India. Even so, between 1911 and 1941, per capita availability of foodgrain – taking into account international trade flows – declined by as much as 26 per cent.

This was from where agriculture in Independent India took off. Its subsequent performance though not a shining example of success is not as dismal as is sometimes depicted. Between 1951 and 1971, foodgrains production increased from 55.0 million tonnes to 108.4 million, or at the annual rate of 2.7 per cent, keeping slightly ahead of the growth rate of population. The growth was, however, not smooth and there were quite a few years – particularly 1966, 1967, 1973 – when the country experienced a severe food crisis. Throughout the period – with the exception of the year 1972 – foodgrains had to be imported, the maximum being 10 million tonnes in 1966. After 1971, foodgrains production started declining again, dropped to 97 million tonnes in 1973, and rose to about 104 million tonnes in the next two years. In 1975–76, it regained it momentum and reached 115 million tonnes, The Economic Review 1974–75 of the Reserve Bank of India observes: "Notwithstanding the expansion of irrigation since 1951, the degree of vulnerability of the agricultural sector to vagaries of climate does not seem to have diminished significantly" [22].

Equally germane to the assessment of agriculture's production performance is the fact that in these two decades (1951–1971), India's population increased by 187 million, and by 1976 another 60 million will have been added. It is worth noting that in spite of this tremendous increase, this backward agriculture has been able to provide a per capita availability of foodgrains – with marginal imports – of about 450 grams per day.

The above should not be interpreted as reflecting a sense of complacency about India's agricultural production performance. Indeed, in years to come, India will have to do much better than its best performance in the past. According to the "medium" projection, by the end of the century India's population will be about 1,000 million. Making a few balanced assumptions regarding the growth of population, the growth rate of national income and its (more equitable) distribution, V. M. Rao has estimated that by the year 2001, India's requirements of foodgrains (assuming low population growth) would be 2.5 times its consumption in 1964–65; requirements of "other foods" would be as high as 4.35 times [20]. We can ignore at our peril, the warning sounded by David Hopper. Presumably reflecting the world opinion and employing "whom to save, whom to abandon" life boat analogy, he has warned: "India, along with some of its neighbours in South Asia, is *seldom considered a candidate for salvation*" [6] (emphasis added).

Let us revert to the post-Indendence period, and briefly review agricultural policies germane to agriculture's performance. We shall confine our review of agricultural policy in India to a few specific issues which have figured prominently in the literature on the subject. These may be listed as below:

(a) inadequacy of plan investment for agricultural development,
(b) price policy and terms of trade,
(c) urban bias,
(d) "Green Revolution" and inegalitarian growth, and
(e) failure of agricultural policy to make significant contribution to the reduction in rural poverty and unemployment.

I

Plan Expenditure on Agriculture

The "neglect" of agriculture for which the Indian policy-makers have often been criticised is generally identified with the failure to allocate an adequate share of public expenditure to agriculture. Every one was happy that agriculture was given pride of place in India's First Five Year Plan (1951–1956). The share of agriculture and community development in the Public Sector[1] Outlay in the First Five Year Plan was 15.1 per cent, as against 6.3 per cent for industries and minerals. The Second Five Year Plan reversed the ranking by allocating 14.4 per cent to "industries" and 11.8 per cent to "agriculture". Apart from this, the major sin of the Second Plan was alleged to be its preference for "rapid industrialization with particular emphasis on basic and heavy industries". We shall not discuss here whether for a country of India's size and geo-political situation it would have been wiser to ignore the establishment of basic industries. Apart from that, the accent on rapid industrialization does not *ipso facto* prove neglect of agriculture; modernization of agriculture is incompatible with such a sectoral view. In any case, the importance attached to a sector should not be judged by its share in the public sector outlay. The absolute quantum of public expenditure on agriculture in the Second Plan was raised to Rs 568 crores from Rs 357 crores in the First Plan. Besides, it may as well be argued that the First Plan "neglected" industrial development, as the planners were not yet ready with a plan of industrial development and allocated to it a meagre share of 6 per cent. Had the First Plan provided for a steel mill or better still a few power generation units and fertiliser factories, the allocation to "industries" in the First Plan would have been larger and the appearance of reversal of priorities would have been avoided. Besides, industry-agriculture linkages make it inappropriate to talk in term of "shares" of sectors in public expenditure. What is relevant is investment *for* agriculture, rather than investment *in* agriculture.

Our contention is not that investment on agriculture has been adequate. Our submission is that the charge on inadequacy needs a more substantial proof. There is no sector of Indian economy which has not – perhaps justifiably – complained about inadequacy of public investment, be it power, transport, family planning, education, social services, and even coal, cement and steel. Scarcity of investible resources is chronic in all developing countries and no sector of the economy should use inadequacy of funds to explain away its poor performance. In fact, it should look inward and examine whether it has used the resources made available to it efficiently. This imposes an unpalatable self-scrutiny and needs more rigorous analysis than a populist demand for more funds.

II

Agricultural Prices

One of the most persistent criticisms of agricultural policy in India and other poor countries is that they have been deliberately "forcing producers' prices down". In 1964, T. W. Schultz [23] asked: "Why are so many poor countries (including India) placing a low economic value on their farm outputs?" Edward Mason [12] wrote that the prices of foodgrains and some other farm outputs were held down by Government action. Michael Lipton [10] asserted "farm prices have been systematically kept down since 1960 in India". He reiterates the charge in his most recent article and quoting S. R. Levis avers that in Pakistan, in the early 1960's, producers' food prices were forced down by as much as two-thirds of their real value [10]. An exactly opposite view is expressed by Walter Falcon [3] : "With the new wheat-fertiliser technology and a government-guaranteed price in West Pakistan almost double the world market price at official exchange rate, wheat was extremely profitable . . . . The Government tied up more than $ 75 million in supporting the prices of wheat. These funds delayed, perhaps even precluded, other expenditures that were more productive". Writing about the same time as Lipton, Keith Griffin [5] complains that "in many cases the cost of innovation has been heavily subsidised by the government. The innovating farmers have not only *high* prices for their products but also low prices for their inputs". And more specifically, "at the moment, however, the governments of several countries, e.g. Pakistan and India, are supporting domestic grain prices at levels which exceed world prices by a considerable margin" [5]. Whom should one believe? In any case, it seems that both those who allege high prices and those who allege low prices are agreed that LDC's are following a wrong price policy.

As will be seen from the facts presented below, the relative prices of agriculture in India have remained consistently high and agricultural price policy in India has been on the whole favourable to the surplus producing farmers. Year after year, the Government of India gave higher procurement prices than those recommended by the Agricultural Prices Commission, appointed by the Government to advise it on price and precurement policies.

Table 1 – *Index Number of Wholesale Prices : Relative Prices of Manufacturers and Agricultural Commodities*

(Base : 1961-62 = 100)

	General index of wholesale prices	Index for manufacturers	Index for agricultural commodities	Prices of manufacturers as per cent of the prices of agricultural commodities
Weights	100.00	32.26	33.20	
(1)	(2)	(3)	(4)	(5)
Average of months				
1965-66	131.6	117.0	141.7	82.6
1966-67	149.9	125.3	166.6	75.2
1967-68	167.3	129.1	188.2	68.6
1968-69	165.4	132.8	179.4	74.0
1969-70	171.6	139.7	194.8	71.7
1970-71	181.1	149.7	201.4	74.3
1971-72	188.4	160.5	199.6	80.4
1972-73	207.1	168.8	219.7	76.8
1973-74	254.2	189.3	280.6	67.6
1974-75	313.0	240.7	350.8	69.5

Source: See ([9], Table 5.4, p. 97).

Tables 1 and 2 show that the terms of trade as judged by (a) Relative Prices of Manufactures and Agricultural Commodities and (b) Relative Prices of Agricultural and Non-agricultural Commodities have been, by and large, favourable to agriculture. A more sophisticated exercise by R. Thamarajakshi [26] pertaining to inter-sectoral terms of trade (all agricultural products purchased by non-agriculture: non-agricultural products purchased by agriculture) also clearly indicates favourable terms for agriculture. Thamarajakshi has also calculated the index of income terms of trade (by correcting the indices of net barter terms of trade with the value at constant (1960–61) prices of the actual marketed surplus) of the agricultural sector to the domestic non-agricultural sector for all uses. The index of income terms of trade has risen at a rate of 4.53 per cent per annum during 1951–52 and 1973–74 (see Table 3).

Information regarding the farmers' terms of trade, i.e. the ratio of prices paid and prices received is fragmentary. However, we find that while the weighted index of paddy input prices increased by 18.2 per cent between 1971–2 and 1973–4, the increase in the wholesale price index of rice was as much as 60 per cent between July 1972 and July 1974. The increase in the price of diesel oil in March 1974 and in the prices of fertilisers in June 1974 would lead to a further rise of 9.5 in the input index [7]. However, in 1975, prices of fertilisers were reduced in two successive instalments. A further substantial reduction has been made in March 1976[2].

Table 2 – *Relative Prices of Agricultural Commodities and Non-Agricultural Commodities*

Year (July-June)	General index of wholesale prices		Index for agricultural commodities		Index for non-agricultural commodities		Prices of agricultural commodities as percentage of non-agricultural commodities
Weights	1000		332		668		
	Index	Percentage increase	Index	Percentage increase	Index	Percentage increase	
(1)	(2)	(3)	(4)	(5)	(6)	(7)	(8)
Annual average							
1961-62	100.6	–	100.4	–	100.7	–	99.7
1962-63	105.2	4.6	102.9	2.5	106.4	5.7	96.7
1963-64	112.3	6.7	112.4	9.2	112.3	5.5	100.1
1964-65	124.6	11.0	134.0	19.2	119.9	6.8	111.8
1965-66	135.9	9.1	147.5	10.1	130.1	8.5	113.4
1966-67	155.2	14.2	174.0	18.0	145.9	12.1	119.3
1967-68	167.0	7.6	185.2	6.4	158.0	8.3	117.2
1968-69	166.6	(–) 0.2	183.1	(–) 1.1	158.4	0.3	115.6
1969-70	174.3	4.6	198.8	8.6	162.1	2.3	122.6
1970-71	182.1	4.5	198.9	0.1	173.8	7.2	114.4
1971-72	191.2	5.0	200.7	0.9	186.5	7.3	107.6
1972-73	216.2	13.1	234.3	16.7	207.2	11.1	113.1
1973-74	271.0	25.3	299.5	27.8	256.8	23.9	116.6
1974-75	316.8	16.9	352.6	17.7	299.0	16.4	117.9

Source: Derived on the basis of Index Numbers of Wholesale Prices (Base: 1961-62 = 100). See ([21] [22]).

Table 3 – *Composite Price Indices and Inter-Sectoral Terms of Trade : 1951-52 – 1974-75*

Years	Agricultural products purchased by non-agriculture for inter-mediate consumption	Agricultural products purchased by non-agriculture for final consumption	All agricultural products purchased by non-agriculture	Non-agricultural products purchased by agriculture for intermediate consumption	Non-agricultural products purchased by agriculture for final consumption	All non-agricultural products purchased by agriculture	Net barter terms of trade of all products Col. (4) over col. (7)
(1)	(2)	(3)	(4)	(5)	(6)	(7)	(8)
1951-52	99.13	93.23	95.44	81.65	96.50	94.76	100.72
1955-56	70.92	75.57	73.83	82.86	77.24	77.90	94.78
1960-61	100.00	100.00	100.00	100.00	100.00	100.00	100.00
1965-66	132.42	141.64	138.18	125.42	120.08	120.71	114.47
1970-71	195.34	200.53	198.58	159.89	155.45	155.97	127.32
1971-72	189.40	202.92	197.85	165.85	164.63	164.77	120.08
1972-73	200.72	222.96	214.61	179.99	180.57	180.50	118.90
1973-74	289.61	275.25	280.64	209.21	204.30	204.87	136.98
1974-75	324.86	355.16	343.79	268.25	255.17	256.71	133.92
Compound rate of growth	5.89	5.96	5.94	4.53	4.43	4.45	1.43

Notes: – (1) These composite price indices have been prepared by combining the relevant indices of wholesale prices (Government of India Economic Adviser's Index Numbers) of the individual items identified as being purchased from or sold to the non-agricultural sector by the agricultural sector for different uses, and using as weights the estimated value in 1960-61 of the actual purchases or sales as the case may be.

(2) Composite indices using the estimated value of the actual purchases or sales in 1968-69 as weights are being separately prepared by the author.

Source: See Thamarajakshi ([26] and [27]).

The impression that the Government of India has been deliberately keeping the prices of agricultural commodities low is perhaps due to the fact that in some years foodgrains procurement prices fixed by the Government were below the prevailing market prices, though in quite a few years of good harvest, the Government has also prevented foodgrain prices from falling below the same procurement price as was fixed in deficit years. (For all practical purposes, the distinction between procurement price and the minimum support price has been obliterated). Thus, when both good and bad years are reckoned together, the accruals to the farmers from levy and non-levy sales would not be generally less than under free market conditions. Further, it should be noted that the Government does not procure the entire marketable surplus of foodgrains. Even in the best years, foodgrains procurement has not exceeded 8.8. million tonnes or 10 per cent of the net foodgrains production. Till 1964, it did not exceed 2 per cent and varied between 5 to 8 per cent between 1965 and 1970. For particular crops and regions, the incidence would, however, be higher.

It is our contention that the rise in the post-levy free market price, consequent upon the withdrawal of a part of stocks from the market through procurement, more than compensates the farmer for the "loss" suffered by him from selling to the Government at below the market price. In other words, the weighted average price of levy and non-levy sales is likely to be higher, and certainly not less, than the price the farmer would have received in the absence of the levy. The magnitude of the difference between the two would depend on (a) the price flexibility coefficient, (b) the proportion of the marketable surplus procured by the Government and (c) the relative level of the open market prices before the procurement operations commence and the levy price.[3]

While considering the question of price policy it is necessary to examine carefully the price effect on (a) production and (b) income distribution. While it is true that a change in the relative prices of two substitutable crops is likely to have a favourable effect on the production of the crop in whose favour the price is changed, it will simultaneously have an adverse effect on the production of the competing crop. In other words, the aggregate supply elasticity for the agricultural sector as a whole is considerably lower than that for individual commodities. Thus in a situation such as that prevailing in India where almost all agricultural commodities are in short supply – and also the critical inputs, including land – price is not an appropriate instrument for augmenting agricultural output. Mahar Mangahas and his colleagues [11] also confirm that "there is little evidence to indicate that price changes are an effective device for influencing *aggregate* agricultural output". This is particularly so in the context of traditional technology.[4] As John Mellor [13] points out "even if increase in production takes place (as a result of higher prices), it would be a movement along the production function, hence at increasing real cost in resources".

More important for our present purpose is the income distribution effect of the increase in foodgrains prices. Mellor has shown that an increase in foodgrain prices actually reduces the income of small farmers belonging to the lowest three deciles of expenditure classes, as they are not purchasers of foodgrains. (Many Western writers probably do not know that the majority of rural households are net purchasers of food, otherwise they would not have confused consumer bias with urban bias). Income transfers resulting from increased prices of foodgrains cause the largest declines in the income of low income consumers and the largest increase in the income of high income producers. "A ten per cent increase in foodgrains prices compels the bottom two deciles to reduce their real expenditure on foodgrains by 5.9 per cent and consumption of milk and milk products by 18 per cent".[5] Mellor concludes: "an increase in foodgrains prices has a substantial income effect in reducing consumption of high nutritive value".

To sum up, the facts are that, far from "forcing down producers' food prices", the policy-makers in India have kept food prices high and displayed a big farmer and anti-urban and anti-poor bias and *thereby* "damaged" the nutrition of the poor. As such, those vicariously concerned with the LDC's "wretched on earth" would do well to advise policy-makers against the folly of high prices of food. The (big) farmers in India have enough incentives from the negligible agricultural taxation and heavily subsidised critical inputs like irrigation water and electricity for the pump sets.

To the extent that any bias can be deduced from price policy, one can as well discern an anti-industry bias on the part of Indian policy makers. For several years, the controlled prices of cement and steel were kept so low that the manufacturers were unable to plough back adequate funds for replacement and modernisation of their units.

III

Urban Bias

Before commenting on urban bias in India's (and all LDC's) agricultural policy, let us admit that such a bias does exist in several fields of Indian policy, particularly in health, education, and organised labour. In regard to agricultural policy, however, the allegation of urban bias seems to be based on misinformation. On the count of deliberate underpricing of foodgrains we have adduced enough evidence to dispel the impression of urban bias. We shall here deal with only one more misleading example of urban bias, namely "encouraging farmers to devote more resources – especially land – to rich men's food". Specific instances mentioned are shift from millets to rice (sic), maize to wheat and to milk production. Apart from the facts, which we shall presently cite, it may be mentioned that the most potent factor influencing changes in the cropping pattern in recent years has been the availability of cost reducing technology. In India, the highest increases in agricultural productivity have taken place in wheat and bajra (bulrush millet) – the latter being the most important millet. The rate of increase in the production of bajra has been markedly and consistently higher than that in rice. So much for the shift from "millets to rice". As for maize, the rate of increase in its acreage has been next only to wheat.

True, hybrid jowar (Sorghum) has not been a success, and in pulses there is complete stagnation. The failures in these crops are mainly attributable to the non-availability of suitable high yielding varieties. According to our information, however, neither funds nor scientific efforts have been lacking for evolving suitable varieties.

Milk no doubt is a rich man's food at present, but in areas where milk production has increased, consumption of milk in poor households has increased both in rural and urban households. In cities like Bombay, perhaps for the first time in recent history, cheaper milk has become available to the low income households, though not yet to the very poor. More importantly, many scholars and policy-makers consider production of milk, poultry and vegetables as the most promising source of additional income for the smaller farmers and the landless [8]. The following facts may be of interest in this connection. Several field investigations indicate that about 70–75 per cent of households owning cattle belong to the category of small farmers and agricultural labourers [24, 18]. A study by V. S. Vyas and his colleagues [29] reveals that in Nadiad Taluka (Kaira District, Gujarat), the share of income from dairying in total farm income was as high as 78 per cent in farms below 5 acres. In dryland agriculture. animal husbandry provides substantial additional employment to agricultural labourers, particularly to women in the form of self-employment [15]. In a recent study of milk production and marketing, it was found that the landless labourers and small farmers accounted for 57 per cent of milk producers in "dairy" villages, 48 per cent of total milk production, and more than 50 per cent of the marketed surplus of milk and milk products. This, however, did not deprive them of the requisite home consumption of milk of more than 200 milli litres per capita per day [16].

Milk consumption is not a mere urban luxury; it is an important source of income and employment to the poor households in rural India and a valuable source of animal protein in near future, if the discernible trend in lowering the cost of production and distribution of milk is maintained. There would be little hope for the small farmers if they were restricted to growing poor man's food. With state-sponsored irrigation, extension and marketing facilities, they should be encourged to grow what is most profitable for labour intensive small-scale farming. In India cattle are fed with fodder and oilcakes (and seldom with inferior cereals) and the encouragement of milk production does not involve any significant diversion of land capable of yielding more calories (or nutrition) per acre of land.

It may also be pertinent to mention that in Kerala whereas the per capita availability of rice from internal production remained most stationery, the production of tapioca (poor man's potato) increased from 1.6 million tonnes in 1961–62 to 5.4 million tonnes in 1971–72. An authentic report from Kerala states [2]: "The drop in the availability of cereals (mainly rice) would have produced under-nourishment among the low income families, say, even the middle class families, who could not afford to buy sufficient quantities of rice at the going price. The sharp increase in the output of tapioca has not only averted a deterioration of the situation, but even improved the average level of calorie intake in the State". It adds "It may be presumed that, by and large, the increase in the production of tapioca, has made a greater impact on the diet of the lower income households".

While Lipton criticises the policy which encourages shifts towards "inappropriate foods" such as meat and dairy products, Carl Gotsch [4] is unhappy over the fact that "relative prices skewed in the direction of cereals through Government support mechanisms have discouraged diversification and further expansion of acreage under vegetables, fodder for dairying, pulses, etc" [4]. Since cereals require a relatively low labour input, such policies reduce the aggregate demand for labour significantly. As pointed out earlier, a shift towards dairying, within reasonable limits, is likely to promote both nutrition and employment.

IV

The Green Revolution: A Bimodal Development

The two successive severe droughts in 1965–66 and 1966–67, gave rise to international apprehensions about India's capacity to feed her huge and growing population. The harshest critics [17] recommended the application of the "triage" formula to countries like India which were considered beyond redemption. Fortunately for the country, at this very time the High Yielding Varieties (HYV) of cereals became commercially available. India's policy-makers plumped for it with alacrity. Dr. Norman Borlaug complimented the then Minister for Agriculture as "the first high officer to recognise the significance of the new wheat strains and willing to take the risk involved in importing 18,000 tonnes of dwarf Mexican varieties". The Pearson Report [18] characterised the speedy adoption of HYV as "one of the authentic marvels of our time". Others described the process of agricultural transformation as "one of the most amazing stories of our time". While this was the general observation, the economists, who had neither anticipated the Green Revolution nor played any part in its adoption by way of even policy advice, did not take kindly to it. Their reaction varied from scepticism ("Cornucopia or Pandora's Box") to downright condemnation on the ground that it was leading towards the emergence of dualism. Let us accept that technological changes ushered through the application of HYVs "*as such* have contributed to the widening of the income disparities between (1) different regions, (2) small and large farms and (3) landowners on the one hand and tenants and agricultural labourers on the other". But the question is: situated as the country was in the mid-Sixties, when its capacity to feed its people was seriously being questioned, and some critics were advocating the application of "triage" and "life boats" formula to food aid, what was the choice before the policy-maker? Highest priority had to be assigned to augmenting food production and the HYVs offered an excellent means of doing so. The possibility of its inegalitarian effects – assuming that these could be clearly perceived at that time – had to be weighed against the obvious inegalitarian effects of food shortage and high prices, under which the poor suffer the most.

Did the adoption of HYV technology increase food production? It is contended that "the so-called Green Revolution has failed to raise the overall rate of growth of agricultural output in the country above the level achieved in the 15 years prior to 1965" [19]. It is also asserted that "despite technological changes, the growth of agricultural output in India slowed down in the 1960's compared to 1950's". Such statements are, at best, half truths based on selective time spans. Let us accept the suggestion that "the comparison of output between successive peaks (in production) would give an idea of output growth adjusted for weather". According to the data provided by the critic himself, the annual percentage difference between the three pairs of pre-green revolution peaks (1953–54 to 1958–59; 1958–59 to 1961–62 and 1961–62 to 1964–65) was 1.8, 1.7 and 2.7 respectively. As against this, the two post-Green Revolution pairs (1964–65 to 1967–68 and 1967–68 to 1970–71) gave the annual percentage change of 2.1 and 4.5 or an average of 3.4 ([19], Table 1.2). How then does one say that the Green Revolution has failed to raise the overall rate of growth? Let us look at the statistics differently. The impact of the "Green Revolution" was first felt in the year 1967–68, after the two successive drought years. The linear rates of growth – based on Index Numbers of production – in the five years preceding the year 1964–65 (the year 1964–65 was the best year till then) and the five subsequent years were 1.81 and 3.86 per cent per year respectively.

Nanumantha Tao maintains "there are reasons to believe that even without the Green Revolution, the growth rate would have been maintained at 2–2.5 per cent per annum". The reasons he adduces are: "The growth of population at about 2.2 per cent per annum has been exerting an upward pressure on prices of agricultural commodities. This would have provided incentives to the farmers for expanding output and would have induced the Government to invest in irrigation, fertilisers, etc". Apart from the fact that under static technology, high prices have little impact on aggregate production, it is surprising that one so deeply concerned with the poverty of the Indian masses should recommend, or prefer to rely on, high food price path of growth of production instead of welcoming the cost reducing technology for achieving increased production! Besides, his argument that "some of these inputs including fertilisers which were known before the onset of the Green Revolution would have been used at a certain rate even in its absence" is equally questionable. As is well known, application of higher doses of fertilisers to the traditional seeds was unremunerative, since it resulted mainly in vegetative growth and subsequent lodging and did not increase output. Thirdly, there is clear evidence to indicate that the growth in cropped area was slowing down, from 2.1 per cent per annum during 1949–50 – 1960–61 to 0.6 per cent during 1960–61 – 1970–71. Under the circumstances, adoption of the HYV's was the only solution to the food problem of the country.

Many studies of the distribution of gains of technological changes are vitiated by the fallacy of single factor analysis. There are at least two components which determine the additional gains of different classes of producers over time: (1) change in production and (2) change in prices. There could be a third, namely changes in the shares of different classes of growers in the total area cultivated. The second and the third have nothing to do with the technological change *per se*. They reflect the effects of (imperfect) market structure or market behaviour of different classes of growers and the land market.

There is substantial evidence which indicates that big farmers obtain much higher prices for their produce either because of their bargaining power or capacity to withhold stocks in a rising price situation. But even if the same price is obtained by all classes of producers, the gain from the price rise – which was substantial in the post-Green Revolution period and had nothing to do with it – would be much larger for the big farmers because of the higher percentage of their marketable surplus.

Geoffrey Swenson [25] has analysed the production and the price effects in the distribution of benefits in a situation of technological change. Analysing the sources of change in the total value of paddy production between 1965–66 and 1970–71 for a sample of farm operators in survey villages, by farm size, Swenson found that the small farmers (2.5–5.0 acres) and the very large farmers (20 acres +) had gained almost equally (21.7 and 22.3 per cent) from changes in production. The main difference in gains emanated from price changes. The smaller farmer gained from this source by only 17.9 per cent, while the very large farmers gained by as much as 47.6 per cent. Swenson sums up the position by observing: "Looking at the Gini ratios, it is evident that the change in the distribution of paddy income would have been in the direction of greater equality *with equal paddy price for all operators* in 1970–71" (emphasis added).

Input Revolution

The contribution of HYV technology should not be judged exclusively in terms of the increase in output which is often distorted by the vagaries of the weather. Its impact on the behavioural response of farmers judged by a sharp step-up in investments in irrigation and increased purchases of modern inputs is an equally relevant criterion for judging its contribution. The number of private tubewells increased from 0.1 million in 1965 to 0.47 million in 1971 and the number of pump sets – diesel and electric – from 0.88 million to 3.24 million during the same period. The net area irrigated by wells (mostly private) which had increased prior to the advent of HYV technology from 6.5 million hectares in 1951–52 to 8.6 million hectares in 1965–66, sharply increased in the next four years to 11.1 million hectares, and its share in the total net irrigated area increased from 32.8 per cent to 36.7 per cent. Similarly consumption of chemical fertilisers per cropped acre increased from 4 kg.–16 kg. or by 400 per cent. It is estimated that expenditure by agriculture on modern inputs in real terms (1960–61 prices) increased from Rs 207 million to Rs 734 million during the first decade ending 1960–61. In the second dacade ending 1970–71, it went up to Rs 4,355 million and has further gone up to Rs 6,181 million in 1972–73. The percentage of expenditure by agriculture on modern inputs to total expenditure on all inputs at constant (1960–61) prices, has increased sharply from 6.19–21 [27]. There is, therefore, no doubt that the Green Revolution has made a significant contribution to the modernisation of Indian agriculture.

Reverting to the issue of the inegalitarian effects of the strategy of agricultural development in India, let us accept that it has led to the acceleration of (a) inter-regional and (b) inter-class disparities. When one refers to the aggravation of inter-class or inter-regional disparities in the context of the Green Revolution, let it be clearly understood that the Green Revolution *per se* has not made any class or region poorer than what it was or would have been in its absence. As a matter of fact in several regions many small farmers have adopted the HYV's and improved their incomes. In many districts of Punjab, for example, the adoption rate has been as high as 90 per cent. Besides, the fact that HYV's are technologically – as distinct from economically – neutral to scale has lowered the threshold of non-viability. Similarly, there is hardly any region which has not benefited at least to some extent from the HYV's. All that the charge of inegalitarian distribution effect of technological change means is that the better endowed farms and regions have benefited relatively more than small farms (farmers) and regions with low irrigation and/or low rainfall.

It is in the very nature of all innovations that they tend to be adopted first by the more resourceful or, more simply, the rich. This by itself need not be a cause of alarm, *provided* the innovation is inherently capable of more universal adoption,if not autonomously through a deliberate public policy.

As for the removal of inter-class and inter-regional disparities, it may be said that while the latter needs a long-term programme involving large-scale investment, the former needs above all political will to undertake appropriate institutional reforms. Major policy measures relevant for preventing or reducing inter-class disparaties are (1) more equitable distribution of land ownership and land tenure reforms to abolish exploitative tenancy arrangements, (2) expanding institutional arrangements for provision of credit and other inputs, extension and marketing.

In the matter of equalising land ownership and protecting the share-croppers and tenants from exploitative arrangement, in spite of the plethora of legislation, the Government's performance to date is admittedly unimpressive. Yet a sombre thought keeps simmering in our mind. Given the heavy and constantly growing pressure of population on land, no restructuring short of collective or co-operative ownership of land can eradicate inequalities, and even then it would be well-nigh impossible to accommodate *productively* the entire labour force in agriculture, more so two or three decades later. Perhaps, the collectives can conceal redundancy from the public view.

The second line of action was to adopt countervailing measures in favour of the class of farmers who are incapable of benefiting from the technological change with their own resources. The policy-maker in India can claim to have taken several such measures, such as the establishment of the Small Farmers' Development Agency (SFDA) and the agency for Marginal Farmers and Agricultural Labourers (MFAL), orienting co-operative credit towards small farmers – it is mandatory to provide at least 20 per cent of the total credit to small farmers – and the more recent decision to establish a chain of Regional Rural Banks which would cater exclusively to the credit needs of the small farmers. It is beyond the scope of this paper even to attempt to evaluate the achievements of these programmes.

As against this, there is a widely shared view that most of the benefits under the scheme have been diverted and appropriated by better-off farmers with political influence. From this a general conclusion is drawn that in the absence of a radical change in property relations and socio-political power structure, such reformist measures as outlined above will make little impact on the conditions of the rural poor and in fact "tend to be unproductive and as such add to the inflationary pressure in the economy". We shall revert to these two questions later.

As regards the aggravation of inter-regional inequalities brought about by the Green Revolution it is obvious that the disparities arise from differences in natural endowments like soil, climate, underground water and river flows which could be harnessed for canal irrigation. It is admitted that there are limits to what public policy can do to reduce the inequality arising from natural endowments. "Public investment in irrigation including the exploitation of ground water potential" is suggested as having "the largest prospect" for equalising opportunities of growth. Subject to technical feasibility and a judicious view of social benefits and costs, we would fully support the plea for larger investments in flow and lift irrigation. But a few facts about irrigation in India may be noted. During the last 25 years, public investment of around Rs 35,000 million has been made in the construction of major irrigation projects. Apart from the fact that there is considerable under-utilisation of the irrigation potential, "costly irrigation waters, impounded at great expense to the community, have not only not yielded the benefits expected of them, but have in many cases been allowed to destroy or lower the fertility of large tracts of land". Approximately 7 million hectares of once fertile land have been affected by waterlogging and salinity or alkalinity. Besides, "the gap between the (irrigation) potential created and "utilised" is over 3.8 million acres. Considering the fact that the cost of creating irrigation potential works out on an average at Rs 900 per acre, the unutilised potential of 3.8 million acres represents an investment of nearly Rs 3,500 million, on which no return is being presently obtained". Our purpose in drawing attention to these facts is to emphasise better planning and management of investments.

V

Rural Poverty and Unemployment

In the preceding sections we have attempted to refute some of the charges against India's agricultural policy, such as neglect of agriculture, deliberate under-investment, under-pricing of agricultural commodities and urban bias. We have also contended that the Green Revolution, or more modestly the

HYV's, has helped to step-up cereal production, stimulated investment and substantially increased the use of modern inputs. While it is accepted that the technological change has led to a widening of the inter-regional and inter-class disparities, we are not sure that any region or class would have been in a better position, had the policy-maker decided to forgo it. Besides, its price restraining effect – more than negatived by monetary inflation – has relieved to some extent the burden of poverty. As against these positive aspects, agricultural policy has not contributed significantly to the removal of rural poverty and unemployment or to making the agrarian structure more egalitarian. Thus, while Indian agriculture has slightly improved the per capita consumption of foodgrains, in spite of the addition of 287 million people since 1951, it has failed to provide land or employment to a large segment of the additional labour force. Only one question may be asked: Was it the sole responsibility of Indian agriculture to provide employment to all and as many people born in rural India, or atone for the failures of population policy or for that matter industrial and monetary policy? Our dissent is mainly with this fragmented view which looks at agricultural policy, isolated from the totality of economic policy.

As pointed out earlier, efforts have been made to help the weaker sections of the rural population through programmes like SFDA, MFAL, DPAP, Employment Guarantee Schemes, and the earmarking of a percentage of co-operative advances to small farmers. Apart from a few exceptions, their overall impact has not been very perceptible. The recent 20-Point Programme enunciated by the Prime Minister gives pride of place to agriculture and especially to the problems of the rural poor. The programme includes items like bringing under irrigation at least 5 million more hectares of land, provision of drinking water especially in drought-prone areas, implementation of (land) ceiling laws and redistribution of surplus land among the landless with redoubled zeal, vastly expanded programme of providing house sites to the landless in rural areas, abolishing the practice of bonded labour, liquidation of rural indebtedness by stages, enhancement of minimum wages of agricultural labourers wherever necessary. It is too early to judge the performance of the programme.

The failure of agricultural strategy – and its economic policy content – to make any impact on rural poverty and unemployment or equitably distribute the gains from technological change has been variously attributed to socio-political factors such as lack of political will, the elitist composition of political leadership and bureaucracy – no less than that of its critics – structural inequalities in the ownership of land and other assets, a bias in favour of big farmers, etc. There is a strong element of truth in each of these criticisms. Yet perhaps this is not the whole explanation. Agricultural growth has not taken place in many countries where such socio-political factors as inequality and unemployment have been eliminated and more surprisingly agricultural growth has taken place in several countries where social inequality exists in varous degrees. While the economists know enough about stimulating growth, perhaps their knowledge and understanding are not adequate enough to suggest solutions to the problems of poverty and unemployment.

Economics is dubbed a dismal science. It would, however, be more appropriate to transfer the epithet to the practitioners of this science. They are a difficult lot to please. At the time when agricultural production in the LDC's was stagnating and dependence on food aid appeared unending, they prophesied doom and advised the USA to apply "triage". When the HYV's of seed held promise of a Green Revolution, they highlighted the consequential accentuation of inter-class and inter-regional disparities. Some even apprehended such abundance as would lead to adverse terms of trade for agriculture and ruin farmers. Some criticised food aid as detrimental to farmers' incentive others considered LDC's desire for self-sufficiency in food as nutritionally damaging to the poor. Low foodgrains prices – a rare phenomenon – were considered as damaging farmers' incentives and an indication of urban bias – though there are more poor consumers in rural areas. Others felt that food prices were being supported at too high a level, tying up funds and delaying, or even precluding other more productive expenditure. Subsidising inputs was denounced as inappropriate pricing, but since profitability should not be "squeezed", restraint on product prices was considered inadvisable. Encouragement of dairy production was exposed as catering to the rich men's food, but others considered concentration on cereals production as inhibiting a cropping pattern with higher employment potential. However, when the economists traverse (sometimes unknowingly) into other disciplines – sociology or politics – they become more cheerful and confidently recommend – the remoter the country, the greater the confidence – "deep structural change" and convincingly expose "reformism" as faint hearted and "modernization" as an elitist ploy.

NOTES

[1] "Public Sector Outlay" should be distinguished from Investment. The latter excludes "current outlay" but includes (estimated) private sector investment.

[2] "It has been estimated that farmers would benefit to the extent of Rs 1,050 million as a result of the cut in the prices of fertilisers" [28]. Since then there have been further reductions in fertiliser prices.

[3] Thamarajakshi has estimated the coefficient of flexibility of the price of rice with reference to changes in availability of the grain at – 1.10, given other things equal [27]. Mellor and Dhar estimate price flexibility coefficient for all cereals at – 2.0 [14].

[4] In spite of the persistently steeper increase in the prices of pulses relative to wheat since 1960–61, production of pulses has declined over the years.

[5] See Mellor ([13], pp. 28–35).

REFERENCES

[1] Blyn, George (1966) *Agricultural Trends in India, 1891–1947; Output, Availability and Productivity*, University of Pennsylvania Press, Philadelphia, U.S.A.

[2] The Centre for Development Studies: Poverty, Unemployment and Development Policy, Trivandrum.

[3] Falcon, Walter P. (1970) "The Green Revolution", *American Journal of Agricultural Economics*, Vol. 52, No. 5, December.

[4] Gotsch, Carl H. (1973) "Economics, Institutions and Employment Generation in Rural Areas". Paper prepared for Ford Foundation Seminar in Rural Development and Employment, Ibadan, April 9–12, (mimeo).

[5] Griffin, Keith (1974) *The Political Economy of Agrarian Change,* The Macmillan Press Ltd, p. 210.

[6] Hopper, W. David (1976) "Sixth Coromandel Lecture: A Perspective on India's Food Production", January 5, 1976; Reproduced in *Eastern Economist,* January 30.

[7] Government of India (1974) Report of the Agricultural Prices Commission on Price Policy for Kharif Cereals for the 1974–75 Season, Ministry of Agriculture and Irrigation (Department of Agriculture), New Delhi.

[8] Government of India (1971) Interim Report on Milk Production through Small and Marginal Farmers and Landless Labourers, National Commission on Agriculture, New Delhi, December.

[9] Government of India (1976) Economic Survey, 1975–76, Ministry of Finance, New Delhi.

[10] Lipton, Michael (1975) "Urban Bias and Food Policy in Poor Countries", *Food Policy,* Vol. I, November.

[11] Mangahas, Mahar, Recto, Aida E. and Ruttan, V. W. (1966) "Price and Market Relationship for Rice and Corn in the Philippines", *Journal of Farm Economics,* Vol. 48, No. 3, Part I, August.

[12] Mason, Edward (1966) *Economic Development of India and Pakistan,* Harvard University Press, Cambridge.

[13] Mellor, John W. (1974) *Agricultural Price Policy and Income Distribution,* Department of Agricultural Economics, Cornell University, Ithaca, New York, June 24 (mimeo).

[14] Mellor, J. W. and Dhar, A. (1967) "Changes in the Relative Prices of Agricultural Commodities in India, 1952–53 to 1964–65", *Agricultural Situation in India,* Vol. XXII, No. 7, October.

[15] Patel, H. F. A Study of Integrated Dryland Agricultural Development, Agro-Economic Research Centre, Sardar Patel University, Vallabh, Vidyanagar.

[16] Patel, S. M., Thakur, D. S. and Pandey, M. K. (1975) Impact of the Milk Co-operatives, Institute of Co-operative Management, Ahmedabad (unpublished).

[17] Paddock, William and Paul (1968) *Famine 1975!,* Wiedenfeld and Nicolson, London.

[18] Pearson, Lester B. (1969) *Partners in Development: Report of the Commission on International Development,* Praeger Publishers, Inc., Pall Mall Press, London.

[19] Hanumantha Rao, C. H. (1975) *Technological Change and Distribution of Gains in Indian Agriculture, Institute of Economic Growth,* The Macmillan Company of India Ltd., Delhi.

[20] Rao, V. M. (1975) *Second India Studies: Food,* The Macmillan Company of India Ltd., Bombay.

[21] Reserve Bank of India (1974) Report on Currency and Finance for the Year 1973–74, Bombay, Table III. **48,** p.70.

[22] Reserve Bank of India (1975) Report on Currency and Finance, Volume I: Economic Review, 1974–75 Bombay, Table III. **50,** p.65.

[23] Schultz, T. W. (1968) *Production Opportunities in Asian Agriculture: An Economist's Agenda,* Agricultural Economics Paper No. 68: 12, Department of Economics, University of Chicago (Revised July 12).

[24] Shrivastava, R. K. (1970) Impact of Cattle Development on Rural Economy of the Kaira District, Institute of Agricultural Research Statistics, New Delhi.

[25] Swenson, Clyde Geoffrey (1976) The Effect of Increases in Rice Production on Employment and Income Distribution in Thanjavur District, South India, unpublished Ph.D. Dissertation, Michigan State University, 1973. A summary of results of this research is published under the title, "The Distribution of Benefits from Increased Rice Production in Thanjavur District, South India", *Indian Journal of Agricultural Economics,* Vol. XXXI, No. 1, January–March.

[26] Thamarajakshi, R. (1968)"Intersectoral Terms of Trade and Marketed Surplus of Agricultural Produce, 1951–52 to 1965–66", *Economic and Political Weekly*, Vol. IV, No. 26, June 28, (Review of Agriculture), pp. A-91–A-102.

[27] Thamarajakshi, R. "Role of Price Incentives in Stimulating Agricultural Production in a Developing Country" (forthcoming in a book).

[28] *The Times of India*, March 18, 1976.

[29] Vyas, V. S., Tyag, D. S. and Mishra, V. N. (1969) *Significance of New Strategy of Agricultural Development for Small Farmers*, Agro-Economic Research Centre, Sardar Patel University, Vallabh Vidyanagar, 1969.

DISCUSSION OPENING – J. W. Mellor, *U.S.A.*

Professor Dantwala brings to us further opportunity to learn from the extraordinarily rich and varied development experience of India. One of the obvious, but often ignored, changes in the world is the accumulation of two or three decades of post-colonial experience with many approaches and many forms of development effort problems. There has also been a vast increase in experience that rises from a direct, continuous contact with those processes on the part of nationals of the developing countries, who include in their number persons highly involved in policy as well as those more removed from commitment to specific positions and views. One can only deplore that the view of these processes and observations from the industrialized nations is so often clouded by current political fashions that exaggerate general tendencies, one way or the other, and make it difficult to sort out divergent pieces of the development puzzle for answers to sub-questions. If we are to accelerate development we must increase our capacity to dissect, within a broad conceptualization, and to judge policies in a broad socio-political context.

Important lessons can be learned, relevant to other countries, by understanding why Indian grain production growth rates accelerated so radically from the negligible growth rates (0.11 per cent or less) in the colonial period, to the close to 3 per cent rates of the post-colonial period, the problems of the broad coverage rural development efforts of community development of the 1950's – so similar in concept and practice to the currently fashionable Integrated Rural Development Programs – and from the complex political economy of price and taxation policy vis-a-vis the agricultural sector.

Professor Dantwala demonstrates throughout his paper how the many aspects of agricultural policy interact one with another, how change in one part changes a general equilibrium which implies change in the others, and how all relate to the broader political context. It is probably failure to recognise these points, as well as the effect of divergent and unstated political perceptions and objectives, which explains the extraordinary inconsistency of so much of foreign analysis and comment on the Indian scene. Professor Dantwala notes the common Western criticism that India, and other low income countries as well, have set agricultural prices too low for adequate incentives to farmers and conversely that the government of India, and others, have favored "rich farmers" by subsidizing technology and inputs and supporting farm prices above international levels. Professor Dantwala feigns confusion at this conflicting criticism – although one suspects that he knows perfectly well what the correct positions are and what motivates the conflicting positions. In any case the detailed discussion of price policy is perceptive, broad and useful, despite the pressure to deal with what must eventually be considered the straw men of much currently conventional wisdom.

In addition to a thoughtful treatment of the "old" issues of agriculture-industry balance in development, Professor Dantwala provides a thoughtful treatment of the equity issues. He rightfully shows impatience with the extreme shift from criticism of lack of emphasis on agriculture to gross criticism of equity effects which resulted from the very production increasing policies essential to producing more food for the poor as well as others. He draws attention to the particularly difficult regional aspects of the problem, the complexity of specific programs for including the poor in production processes, and illustrates the overall complexity by the role of milk production as an intensive employer of labour, thereby distributing income and access to calories to the poor and as a product to be consumed largely by relatively higher income people.

From Professor Dantwala's edifying exposition one senses future possibilities for studying and learning more of the class conflicts relating to agricultural production and equity, the specifics of operational problems and solutions incident to pursuing now known policy directions, and the potentials for solving regional aspects of growth and equity. Analysis of the complex and diverse historical record with the benefit of the insights of the practitioners and close observers offers us opportunity for accelerated growth in the quality of understanding and, we may hope, in the quality of practice as well.

G. BARBERO*

Agricultural Development and Regional Economic Integration

1. INTRODUCTION

During the last twenty years economic integration at the regional and sub-regional level has gained a vast audience and commanded increased attention in both developed and developing countries.[1] Several schemes encompassing a variety of approaches have been implemented with varying degrees of success and, with or without modifications of the original structure, are still in existence. Although the performance is, to say the least, highly uneven, it cannot be denied that a vast body of practical knowledge has become available to the international community through these efforts. The literature has been enriched not only by an increasing number of case studies illustrating the objectives, approaches and problems of the integration schemes but also by significant advances in the theory of economic integration. Attention has gradually shifted from integration ventures between developed economies to the specific environment, solutions, and conflicts of similar initiatives among developing economies. At the same time theoretical speculations have moved away from the rather simplified propositions, concerning customs unions, put forward by the pioneering work of Viner in the early fifties and have lead to more elaborate formulations covering a wider set of objectives and variables than those postulated by the comparative static analysis of welfare gains, based on the balance between trade creation and trade diversion effects (Robson, 1972). A new branch of economic theory has thus emerged alongside and interacting with development economics and international trade theory (Andic, *et al.*, 1971; Belassa, 1961). Similarly, political scientists have devoted great attention to regional organizations and to the conditions and circumstances under which regional economic integration can flourish and contribute to political union and the preservation of peace. (Bussey, *et al.*, 1971; Hansen, 1968; Nye, 1971). (See fig. 1 for summary of selected integration schemes).

* **While preparing this paper I had the opportunity of consulting the preliminary report of a seminar on "Agriculture in Regional Integration", organized by FAO in Rome, September 1975. I am therefore grateful to Dr. James O'Hagan of FAO for kindly providing the report as well as other background material discussed at the Seminar.**

TABLE 1. *Summary Data of Selected Integration Schemes*

Scheme	Population (millions) (U.S. dollars)	GNP per capita	Member countries
EEC – European Economic Community	255	2,900	Belgium, France, Germany F.R., Italy, Luxemburg, Netherlands, Denmark, Ireland, Great Britain
CMEA – Council for Mutual Economic Assistance	360	1,500	Bulgaria, Czechoslovakia, German D.R., Hungary, Mongolia, Poland, Romania U.S.S.R., Cuba
LAFTA – Latin American Free Trade Association	250	670	Argentina, Bolivia, Brasil, Colombia, Chile, Ecuador, Mexico, Paraguay, Peru, Uruguay, Venezuela
Andean Group	70	600	Bolivia, Colombia, Chile, Ecuador, Peru, Venezuela
CACM – Central American Common Market	16	415	Costa Rica, El Salvador, Guatemal, Honduras, Nicaragua
CARICOM – Caribbean Community	5	740	Barbados, Guyana, Jamaica, Trinidad and Tobago*
EAC – East African Community	36	140	Kenya, Tanzania, Uganda
UDEAC – Custom and Economic Union of Central Africa	9	240	Cameroon, Central African, Republic, People's Republic of Congo, Gabon
CEAO – West African Economic Community	26	170	Ivory Coast, Mali, Mauritania, Niger, Senegal, Upper Volta
ECOWAS – Economic Community of West African States	120	155	Benin, Guinea, Guinea Bissau, Ivory Coast, Niger, Nigeria, Liberia, Mauritania, Senegal, Sierra Leone, Togo, Upper Volta, Gambia, Mali, Ghana

* The following countries have joined the Community in 1974: Antigua, Belize, Dominica, Grenada, Montserrat, St. Kitts–Nevis–Anguila, Santa Lucia and St. Vincent.
Source of data: World Bank Atlas, 1974 (data refer to 1972).

The existing integration schemes practically cover the entire spectrum of integration forms, from free trade areas to customs unions, common markets and communities (the latter involving a certain amount of common policies, policy harmonization or common services). Full economic union has not been achieved yet by any of the schemes. Beyond the official labels, which often embody a good deal of aspiration, one can observe considerable differences in the degree of involvement of participating countries, one with another, and at times a combination of ingredients which theoretically belong to separate levels of integration. For this reason it seems preferable to speak of forms rather than of levels of integration which imply the existence of a *continuum* from free trade areas to economic union, i.e., an increasing degree of reciprocal involvement (Belassa, 1961).

Important differences are also observable with regard to the degree of supra-national authority, namely the types of decisions which regional institutions can make and the procedures through which decisions are reached (whether by a judicial organ acting alone, or by representatives of states with or without veto power). In some instances there is practically no supra-national authority; regional institutions advance proposals which then need the approval of the partners concerned.

There are, finally, other types of arrangements – which some would consider as belonging to regional economic integration and others to simple forms of regional cooperation – deliberately restricted to specific development projects (river basins, interstate roads, etc. . . .). These arrangements are considered to fall outside the scope of the present paper.[2]

Another point is worth stressing at the outset. The declared or implicit fundamental reasons of any treaty or agreement concerning regional economic integration is the stepping up and diversification of the economic base of the group as a whole and of each participating country. Whatever the stage of development, promotion of industrial development is the main goal. Increased agricultural output, modernization of the sector or improvement of the well-being of the rural population are never high on the priority list; at best they are placed on equal footing or even considered as by-products of economic growth stimulated by the other sectors. Not infrequently the agricultural sector is the source of strong embarassment at the time of deciding if, to what extent and how, integration arrangements should extend to agricultural commodities and should incorporate something more than loose goals of agricultural development. The result is that when examining reports on achievments in the agricultural field one inevitably discovers that progress has been slow, although several commissions have been at work, problems studied and endless negotiations carried out. If in one particular instance, the EEC agriculture has had for a considerable time a role far greater than its sheer economic weight, it is not because the founders had envisaged that it should be so. The reasons were of an entirely different nature and almost certainly not forseen at the time of drafting the Rome Treaty.[3]

Starting from these observations, in designing the structure of the paper, I have had to make some important choices. Rather than describing the differences of integration approaches and the role assigned to the agricultural sector, I have attempted to explain the reasons behind observable differences. Furthermore, and in so doing, rather than confining political factors into the background as intruding elements in what would otherwise be a beautiful field of activity and study where order and rationality dominate, I have given much space to political factors and to their linkages with economic interests and objectives. I am aware that this choice may have unduly sacrificed the treatment of agricultural development problems. However, given the introductiory nature of the paper, to be followed by others discussing specific cases, I hope that its content will serve the useful purpose of helping to place agricultural problems in a broader setting.

Consequently, the paper is divided into two parts. The first deals at some length with the background factors and motivations leading to regional integration arrangements, with specific reference to experiences in Europe, Latin America and Africa. The second part purports to review critically the problematic areas of regional integration in the light of the main propositions of the theory of international economic integration and then tries to appraise how agriculture can contribute to the integration process or derive benefits from it.

PART I

Background Factors and Motivations for Regional Integration

Steps towards regional economic integration can be looked upon as a response to the challenges (in terms of constraints and possibilities) inherent in the world economic order issued from the international political divisions and alignments produced by the last world conflict. The changing relations between the two super-powers (from cold-war to bipolar stability or peaceful coexistence), the long period of economic growth coupled with the great expansion of international trade, the acceleration of the decolonization process jointly with the perpetuation of economic dependence of the developing world, the emergence of a great number of new nation-states and the increasing political weight of the Third World, provide the overall setting for tracing the origin, shapes and vicissitudes of economic integration at the regional level. Obviously these world events and related trends have had a different impact in the various continents due to their pecularities in terms of history, geographic features and role played in international affairs. Since it is in Europe, Latin America and Africa that economic integration has advanced most, the following general remarks mainly reflect conditions, economic and political, of these three continents.

Even if political union is rarely an explicit aspiration of the contracting parties of an economic integration scheme and even if, where pursued, any attempt in this direction has been rather disappointing, the fact remains that economic integration treaties or agreements are politically inspired acts. As such they involve both internal actors (political, economic and bureaucratic forces or elites) and external actors whose degree of cogency is a function of several variables reflecting the extent of economic dependence, type of political alignments, development aspirations of each nation and of the regional groups as a whole. Consequently, the formation of regional economic groupings is often founded on a multiplicity of motivations which may differ not only among participating countries but also between groups within the countries themselves.

Europe

To exemplify this multiplicity of motivations it may be useful to recall that the Rome Treaty establishing the European Economic Community among the Six was the end product of a complex set of forces and the final outcome of previous various forms of cooperation in the economic, defence and political field.[4] Without any ambition to cover all of these forces or to list them in order of priority I shall simply mention those which, in my judgement, were the most significant. Starting from the external actors, one must recall the interest of the United States, during the cold-war climate, to reinforce the economic and political power of Europe both as a market for American exports and as a cushion against the pressure from the East. In this contest the need to settle the German question, (rearmament, French–German balance of power) certainly had a considerable weight. Furthermore, the establishment of the EEC owes much to the impact of a federalist movement which even before the end of the war was already active in pursuing the idea of a united Europe and was later able to convince the political leaders of the need for supra-national authorities and strong community institutions. On the economic front, the stronger industrial firms, national and multinational, were pressing for an enlarged market and for the freer movement of goods, labour and capital, while some nations, with a weaker industrial base and therefore reluctant to rapidly move towards liberalized trade of manufactures, were nonetheless interested in expanding their agricultural exports to the potential partners.

The need to find a compromise among divergent expectations by ensuring, in perspective, a balanced distribution of opportunities and benefits, made it therefore necessary to include agriculture within the scope of the common market, but at the same time its very inclusion called for a substantial departure from the general philosophy of the Treaty. Industries, banks, commercial services were looking forward to a more liberal climate, away from the restrictions imposed during the thirties and the war period, while the expectations could hardly apply to the agricultural sectors of most of the countries; clearly the strict application to the agricultural sector of the general competition rules would have been incompatible with existing national farm prices and income support policies and with the internal balance of political power, and would have caused, politically and socially, unacceptable adjustment pressures.

If the approaches and the processes of integration are influenced not only by the consensus they gain but also by the forces opposed to it, then it is appropriate to mention among the background conditions of the EEC the existence in two countries of large and voiceful leftist parties and trade unions strongly opposed at that time to the strengthening of the European economy along neo-capitalist lines. The formation of a large common market and the expected benefits in terms of increased welfare were probably envisaged by the common market proponents as a means to dilute the influence of these organizations and to undermine the causes of their strength.

The interplay of these forces contributes to explain not only some of the special features of the EEC, namely why only six countries joined in the initial effort, why a relative high degree of supra-nationality was introduced, but also why the basic propositions of the Treaty and its instruments are inspired by a *laissez-faire* philosophy with great reliance on the liberation of market forces under competitive conditions and their capability to bring about increased efficiency, faster economic growth and widespread distribution of benefits. It also explains the degree of protection accorded to agriculture and the adoption of a common agricultural policy, resulting in common market organizations and common prices.

It is worth stressing, in this connection, that the formulation and the implementation of a common agricultural policy, especially during the early sixties, accentuated the political role of the Commission, while the agreements reached by the Council of Ministers on agricultural issues, in the absence of other relevant common policies, were often regarded as strategic steps for asserting the identity and the liveliness of the Community as an institution.

As it is well known, the background conditions or structural factors of the initial period have changed through time and not only from the enlargement of the Community from six to nine countries. The European economies have become increasingly integrated, mainly because of the great expansion of intra-regional trade, and even if little progress has been achieved in the harmonization of economic and social policies. The integration process itself has given rise to a number of conflicts both within the Community and with the rest of the world. The increasing competitiveness of European manufactures and the effects of the agricultural policy have created a conflicting situation with the U.S., of which the monetary crisis is an important outcome. And with the monetary crisis the working of the CAP has become increasingly complex introducing further grievances over the distribution of costs and benefits. Also the compounding difficulties of the seventies have made it evident how much the EEC is a case of interdependence among unequal partners. The polarization effects of an integration process dominated by market forces (widening rather than reducing disparities between and within nations) and the different impact on the balance of payments of the energy crisis and of inflationary trends have accentuated the assymmetry between partners: an asymmetry which up to second half of the sixties had been masked by a situation of relative stable prices and fixed exchange rates, high rates of economic growth in all countries and the expectation of a continued flow of benefits.

In this connection the modest adaptive capacity of the Community must be noted. The Community has been substantially unable to recognize with concrete actions the great internal, economic, social and political changes which have taken place since the time of the Rome Treaty and the new challenges which have come from the world scenery. If in some instances (the Lomé Convention is the outstanding example) it has been able to speak with one voice and to act with illuminated self-interest, on the whole its action, under the impact of the world economic crisis, has been dominated more by the revival of nationalistic interests than by a redefinition of objectives and a renewed sense of identity. (Amoa, 1974; Dams, 1975; Awad, 1974).

Regional economic integration is not, as is well-known, a prerogative of Western Europe. The Council of Mutual Economic Assistance (CMEA) among socialist countries was established as early as 1949, and can be interpreted substantially as a move towards the consolidation of the socialist block in the face of the rapid recovery and development of the Western European economies. The declared scope was the intensification of economic cooperation among partners, on the basis of the socialist international division of labour, in order to speed up the rate of economic development of each individual country. Primary attention has been given to industrialization, through the coordination of national economic plans and with regard to the most important industrial branches. Several international institutions have been established in order to promote the desired level of cooperation as well as to facilitate inter-state trade, but it is interesting to note that none of these institutions enjoys supra-national decision power, even after the adoption of the Complex Programme of 1971. The mechanism for the coordination of the five year economic plans provides a framework for determining, normally on a bilateral basis, the structure and volume of goods entering intra-regional trade. Agriculture has so far received only minor attention, on account of the priority accorded to industrialization and of the considerable differences among countries, in the size and structure of the agricultural sector and in national policies. An overall tendency towards greater self-sufficiency in food is however detectable and the cooperation at the regional level is mainly directed to stimulate scientific and technical progress, as a means to foster growth in output and productivity. Measures to minimize the effects of imported inflation as well as to increase the value added in agricultural exports to third countries have lately been considered (Lavigne, 1973; USDA, 1974/1975).

Latin America and Africa

(a) *Common features.* The formation of the EEC, on account of its considerable share in world production and trade, has in many ways influenced the development of regional integration initiatives elsewhere, especially in Latin America and Africa. In these continents, however, integration attempts and their realization have been nurtured and conditioned by quite a different set of historic and structural factors. Let us first of all look at the features which are common to both continents.

The two continents have a similar position with respect to the international division of labour inasmuch as they are still to a large extent suppliers of oil, minerals and primary agricultural products to the industrialized countries and buyers of investment and consumption goods, including food.[5] Their share of world exports is on the whole a modest one (from 4 to 5%); this is due to the limited number of commodities traded and to the heavy dependence of some countries on one commodity only. As to the destination of exports and sources of imports, the markets of the U.S. and Western

Europe account for approximately two thirds (but for Africa the European markest are by far the most important). The extent of intra-regional trade is also very small compared to that of Western Europe or CMEA countries; in Africa only 10% of total exports go to other countries of the continent and only 5% of total imports are of continental origin, while in Latin America the relevant percentages are 18 and 13 respectively.

Expansion of inter-country trade is hindered by the lack of communication and transportation facilities, the restricted pattern of production, high cost of manufactured products and international payment difficulties. By and large, development efforts have been directed, with nuances from country to country, towards the twofold objective of increasing exports of traditional commodities and of expanding the production of manufactured goods for internal consumption, through a policy of import substitution involving a high degree of protection from external producers. In so doing countries have had to withstand great fluctuations in export revenues, and to suffer from the deterioration in the terms of trade of primary commodities (except for the very recent period and certain primary products); at the same time for their industrial development they have had to rely on foreign investors, external sources of financing and imported technology and to face the limitations derived from the meagerness of internal markets. The rapid growth of population and the urbanization processes have increased food demand and stimulated the adoption of imported consumption models by the higher income strata with the result that the scarce foreign exchange resources have often had to be diverted to importing consumption goods rather than investment goods.[6] These trends have tended to accentuate rather than lessen their economic dependence on the developed world. Therefore, long before the Third World was able to challenge the righteousness of this economic order, regional economic integration among developing countries came to be looked upon as a means to surmount the limitations of the individual country approach to economic development; more specifically as a suitable framework for fulfilling a number of deeply felt aspirations or needs such as: to improve terms of trade for their exports, to place import substitution on a sounder basis thanks to wider markets (with greater potentialities for specialization in production and for savings of hard currencies for import operations) and to adopt other policies (concerning foreign investments, infrastructure, cooperation in production of services) appropriate to foster economic growth.

But beyond these common features which can easily be generalized to the whole of the developing world, to understand development strategies and approaches to regional economic groupings one must also look at those factors which are specific to each continent or large groups of countries.

(b) *Latin America.* The most distinguishing feature of Latin America derives from its special relations with the United States. Since the time of the Monroe Doctrine, the U.S. has claimed a protective role over the Latin American continent and manifested its tight control over the strategic aspects of economic and political life of most of the countries. The main consequence of this situation has been the maintenance of the *status quo* in the internal power structure, which originated out of the colonial period, and was based on oligarchies, large concentration in land ownership and, later on, a diffused connivance between foreign investors and the ruling class. As a reaction to this dependency relation, in some countries as early as the beginning of this century, there developed nationalistic movements attempting to assert the right of each nation over the control and utilization of its own resources and to affect a certain redistribution of power in line with ongoing changes in the social structure. These and other later similar movements gained strength due to the relaxation of restrictions imposed by the cold war and the increasing involvement of the U.S. in other parts of the world. In an apparent paradox, since the early sixties, nationalism and regional integration efforts have interacted, giving rise to a number of conflicts and dilemmas which have at their heart the choice of development strategies and of the relevant approaches to integration. Should priority go to national integration or to regional integration or should the two be consciously pursued concomitantly? For those asserting the latter position the real choice is not whether to integrate or not but whether integration should proceed from "inside and from below" and with the support of development planning, at both national and regional level, or from "outside and above" which is substantially the result of market forces, aided by trade liberation measures (Garcia, 1969). Under the influence of external forces and the internal dominant groups it is the latter approach which has been priviliged so far, although significant departures have lately found some room (with the creation of CARICOM and the formation of the Andean Group). Political and economic elites are also accused of having resorted to integration arrangements (mainly involving freer trade) as a means of "exporting" their internal problems; horizontal expansion of markets through regional integration is considered as an alternative to increased internal demand through fuller employment and better income distribution between areas and between social groups (Lizano and Willmore, 1975). These remarks help to explain, I believe, why the regional economic groups in existence have placed very little emphasis on supra-national institutions or have deliberately avoided any supra-nationality, and why rather elaborate procedures based on multilateral negotiations characterize the decision-making process.

It is also in the light of the previous remarks that one must appraise the little attention accorded to the agricultural sector, compared to industry, in the design of integration schemes and during the integration process. The main feature of the agricultural sector in most of the countries is the sharp dualism between the modern export-oriented sub-sector (whose products go mainly outside the region) and the traditional sub-sector producing for the internal markets but also incorporating large subsistence components, hardly integrated with the rest of the economy. Great differences can be observed among countries in price levels (for products and for inputs), in price policies and in the role and weight of state marketing institutions charged with the procurement of major staple foods. Under the circumstances, without tackling the problem of policy harmonization, trade liberalization directly or indirectly affecting agricultural commodities soon run up against objective limitations, as no country is willing to accept or able to cope with the disarray in the price system and the probable negative employment effects, on the great masses of small producers, that would be caused by a widespread elimination of barriers to trade.

A greater concern for agricultural matters has been recently displayed. (ECLA/FAO, 1975). The Caribbean Community has shown awareness of the special needs of small or less developed countries and envisaged an important role for regionally planned economic development projects, among other goals, to raise agricultural production and to affect a better distribution, in time and space of available foodstuffs. Considerable attention has also been given to the creation of Community institutions and the joint production of services.[7] Similar developments can be observed with respect to the Andean Group where, notwithstanding the priority accorded to industrial programmes, some steps have been taken towards the coordination of national agricultural policies and the design of joint programmes in various agricultural fields (BID, 1974).

(c) *Africa.* In Africa, in spite of the fact that the delocalization process is a recent phenomenon and still an unfinished task, regional economic integration already has firm roots and involves a large number of countries. This is partly due to the durability of colonial influence after the proclamation of political independence. In fact, the formation of economic groups has a long standing tradition thanks to colonial ruling. The classical case is provided by the common market among the countries now forming the East African Community which was set up during the twenties under British rule. Similarly, the Custom and Economic Union of Central Africa is an outgrowth of the Federation of French Equatorial Africa. The existence of common currencies, discontinued, however, by EAC after 1967, was an indication of the degree of actual integration among the participating countries. The continuation of pre-existing economic, political and cultural relations, even if under modified forms, conditioned not only by the pecularities of the production and trade structures set up by the colonial ruler but also by the coming into power, with independence, of political and bureaucratic elites favouring the maintenance of special relations with the mother country (see *African Digest Guide,* No 8/9). However, the old aggregation forces gradually lost their impact and have been replaced by new aggregation of forces, of both internal and external origin. Population increase, advances in economic development and urbanization have caused the emergence of new classes and social strata, of which often the military regimes are the expression; also the formation of new nation-states and their political structuring more than elsewhere have been influenced by the climate of "peaceful coexistence" and competition between the super-powers and as well by the role played by other emerging powers. The search for new forms of regional cooperation, in economic and political matters, and the explosion of antagonisms have proceeded side by side, aggregations or divisions being centered as much on ideologies and concrete forms of political organizations, as on control of resources, attitudes towards former imperial powers (and their new forms of dominance), meaning and shape of African unity, acceleration and total completion of decolonization. The establishment of ECOWAS (Economic Community of Western African States) in 1974 is the most outstanding example of the new aggregation trends, in that for the first time it brings together a large number of former French and British colonies (see, Africa Research Bulletin). It is also an indication that if African unity is a long-term objective and nationalism a present necessity, economic integration at the regional level is a vital instrument for economic growth and social development objectives.

To appreciate progress along these lines it is worth remembering the importance of nationalism within present African context.[8] The nation-state may well be an obsolete form of political organization in the old world but it is far from having exhausted its role in the developing countries and particularly in Africa. In the latter, "nationalist leaders are in a position more analogous to seventeenth century than to nineteenth century Europe in that they still need the double-edged weapon of sovereignty to cut both internal tribal ties and external ties to consolidate the state they have captured" (Nye, 1971). The national state in the hands of modernizing elites can be an important tool for the needed internal changes of their societies. The existence of national states and of rivalries among them may of course play to the advantage of the multinational businesses which tend to exploit location advantages, fiscal benefits, low cost manpower and ease of profit repatriation. But rather than denying the validity of national sovereignty this danger simply stresses one of the many contradictions developing countries must face in their relations with the developed world and among themselves. A partial way out of this problem, according to proposals already being considered, could be the creation of multinational enterprises owned by the partners of an integration scheme.

The present stage of development of most African countries must also be borne in mind when considering the types of integration arrangements which African countries have privileged and the special obstacles which regional activities must overcome. The small weight of intra-regional trade is not, or not only, the outcome of trade barriers but rather a direct consequence of the weakness of the economic base: simply, there is not much to trade among partners (Penouil, 1974). Furthermore, in many instances, products cannot or are not allowed to move freely even within the same country, and taxes on imports are often an important source of state revenues. Free intraregional trade cannot therefore rank very high in the list of development tools. Expanded trade, if it is to flourish, will have to derive from a more complex and sophisticated pattern of production specialization between partner states; it cannot rely solely on complementarity, unless, discarding the developed country pattern of production, more rigid planned arrangements are followed. Consequently, priority must go to other fundamental objectives; the building up of a physical and a social infrastructure, the choice of types and location of industries (including agricultural processing), comprehensive programmes of rural development, acquisition and allocation of foreign financing; in essence to the direction of investments which will establish the production and trade patterns of many years hence. The objectives of simultaneously pursuing internal and external integration is probably even more important here than in the Latin American case. Perhaps, this explains why integration arrangements now operating in Africa have taken the form of economic communities with emphasis on the joint running of services and joint development programmes rather than on the freer movements of goods.[9]

PART II

Benefits from Integration and Agricultural Development

We must now deal more specifically than in the preceeding sections with the relationship between agricultural development and economic integration processes. In order to do so some references to the theory of economic integration are necessary with the understanding, however, as already underlined at the outset, that what matters here is not theory *per se* nor the analytical tools which have been developed but, on the contrary, the problematic areas to which attention has been directed and the conclusions reached through theoretical speculations as well as applied work. The relevance of these conclusions to the specific problems of the developing countries and to their agricultural development objectives will be the main concern of this second part.

The principal sources of economic gains from arrangements for international economic integration, providing an incentive for countries to participate, include:

(1) Increased production due to specialization according to static comparative advantage.
(2) Increased output due to better exploitation of scale economies.
(3) Gains from improvements in the terms of trade of the group as a whole with the rest of the world.
(4) Forced changes in efficiency arising from increased competition.
(5) Integration-induced structural changes affecting the quantity or quality of factor inputs such as labour, capital (including improvements in the rate of technological progress) and enterprise (Robson, 1972).

The basic pure theory of integration as developed by Viner and others has been primarily concerned with the first of the sources listed above, with specific reference to customs unions and trade in final goods. The basic argument is that if internal tariffs are eliminated and a common external tariff set up, there will occur shifts in trade: if sources shift from a high-cost domestic production to a lower-cost production in a partner country, trade creation is the result, whereas trade diversion occurs when the shift takes place from the lowest-cost external producer to a higher cost partner. If trade creation is predominant, then overall efficiency is increased since more can be produced from given resources. A custom union in this case benefits at least one member and the world at large; if trade diversion predominates the union is injurious to at least one member and the world at large. Although later extensions by also explicitly considering consumption effects (due to union-induced changes in relative prices) have pointed out that trade diversion does not necessarily produce negative welfare effects, the limitations of the theory and particularly of its heuristic value derive from it static nature and from the rather rigid assumptions on which it is based (balanced trade, capacity of prices to reflect true opportunity costs, constant returns to scale). Furthermore, the theory assumes that countries joining a custom union are essentially pursuing the objective of increasing allocative efficiency within the existing economic structure and that the alterations in trade patterns are the unique source of benefits.

Further extensions of the theory have had therefore to recognize the existence of a multiplicity of objectives and of sources of benefits and the need to incorporate structural changes and to account for dynamic effects. It is interesting to note that this broader approach has become necessary not only to reflect more accurately the conditions and potential effects of integration arrangements among developing countries but, first of all, to provide a more appropriate framework for evaluating the effects of European integration. Obviously, if considerations of such factors as internal and external economies of production, changes in the supply of factors, efficiency in their application and rate of technological progress are important for advanced economies they are even more so for developing countries for which, on the one hand, the impact on intra-regional trade due to reduction of tariff and non tariff barriers is bound to be limited in the initial period and, on the other hand, economic growth consideration in the face of great internal distortions and balance of payments disequilibria are paramount.[10]

What is at stake here is a fundamental change in the structure of production and trade requiring considerable investments and time; to bring about this change an appropriate trade mechanism at the regional level is no doubt of some help but would hardly suffice without more direct approaches such as commonly agreed upon development policies and other mechanisms to ensure an acceptable distribution of benefits and costs, both between and within countries. The alternative path to regional integration, for most of the countries is trying to produce as much as possible of a greater number of products and the most likely consequence of unused capacity (due to limitation of internal demand), high costs, waste of foreign exchange and possible diversion of resources from production of export commodities to production of goods for internal consumption.

One question which has received increased attention by theorists and students of integration is that of harmonization of economic policies, a problem which is of particular relevance for integration arrangements which go beyond trade in products and extend to factor mobility. Policy harmonization may be necessary with regard to structural differences among partner countries in jurisdictional principles, taxation systems, social security, transports, marketing regulations, attitude towards foreign investors, and it applies also to macro-economic policy measures such as monetary and fiscal policy, exchange rate policy and balance of payments. Short of total harmonization which is equivalent to full economic union, an objective under present circumstances, as unrealistic as generalized free trade, the most important problem in this field is the identification of those areas in which the failure to harmonize policies may create the greatest difficulties for the process of integration.[11]

But the difficulties to harmonize must also be appraised at the light of possible and legitime differences in development objectives among participating countries since any progress in policy harmonization necessarily involves a certain surrender of national sovereignty. Especially in developing countries the problem of harmonization is further complicated by the growing importance of multinational corporations whose decisions may considerably affect the pattern of trade within an integrated area. Although they are in themselves a form of international integration, it does not necessarily follow that through their decisions concerning location of plants, allocation of markets, transfer of finance and technology, they will also bring about optimal specialization in the region concerned. Apart from the well-known initiatives in higher circles for partial control in the mode of operation of these corporations) it is obvious that also within integration arrangements some forms of legislative harmonization in various fields affecting their businesses may demand high priority.

But perhaps the most interesting area in which both analysts and policy makers (national, regional or international) have exercised intelligence is the one concerned with distributional aspects between the integrated group as a whole with respect to the rest of the world and among the individual countries of an economic group (UNCTAD, 1975). The importance of this problem is well exemplified by the fact that the greater part of the crisis occurred within the most advanced integration arrangements having arisen because of conflicts over the distribution of costs and benefits of integration. Considering that countries within an integration scheme are often characterized by different degrees of economic development and of efficiency in internal organization, when product and factor liberalization measures are enacted there will be a tendency for economic activities to concentrate on those countries and regions which are best equipped to take advantage of the new opportunities. Experience shows that these "backwash" effects are detrimental to the process of integration unless partners are able to persuade each other that integration brings, over time, each one of them more gains than losses; thus continued participation in the integration scheme may require compensatory measures to be adopted. Attempts to define the concept of "equitable distribution" and to analyse the factors promoting the unequal distribution of benefits have brought forth a number of proposals for alternative compensatory measures, such as payment of lump sum fiscal compensations, regional policies affecting the allocation of investment and other indirect means to prevent the polarization effects of economic development and to possibly correct for any such distortions already existing.

With reference to the list of principal sources of benefits (see Part II above, second paragraph) which may be derived from integration, it is necessary to discuss in a more detailed fashion, if and to what extent they may apply to the agricultural sector and to point out the specific problems which may arise through attempting to fully incorporate the agricultural sector into various integration arrangements. The subject can be approached from two different but interrelated viewpoints. The first is to see if and under what conditions the agricultural sector within an integration scheme affecting the entire economy can directly contribute to the production of those static and dynamic effects which lead to greater economic efficiency and growth. The second is to see if and under what conditions the agricultural sector can benefit from the integration-induced effects taking place in other sectors of the economy, especially in the industrial sector.

With respect to the first viewpoint, it may behove us to stress the conditions, both internal and external, in which the affected countries must function. As already pointed out, the conditions required in a comparative static analysis of welfare gains through increases in allocative efficiency of production barely exist. Most countries, on the contrary, have neither full employment, balanced trade, nor an efficient price mechanism in operation.[12]

In many instances the dualistic structure of agriculture displays not only a profound difference between the modern export-oriented sub-sector and the domestic market-oriented sub-sector but also the cleavage between a restricted number of large modern farms, supposedly competitive by international standards and the great number of small, traditional, peasant farms, providing only meager incomes for the largest part of the rural population. Any move to freer trade by reducing trade-barriers will stimulate the competitive sub-sector towards greater specialization and increased efficiency and will, however, also simultaneously and negatively affect the traditional sub-sector by displacing a great number of farmers. The phenomenon of disguised unemployment may be greatly transformed into productive employment, contributing to the growth of production, but most likely the largest part will be transformed into open unemployment.[13] The outcome is uncertain, but considering the relatively small contribution of the industrialized modern sector to generate employment, it is probable that the negative effects will outweigh the positive ones. Furthermore, since each country's balance of payments, even in the most advanced integration schemes, remains a national responsibility, it is also possible that freer movements of goods will lead, at least for some countries, to further structural disequilibria. Thus, the emphasis on bilateral negotiations and reciprocity in trade which has prevailed so far, particularly in Latin America, and especially within LAFTA, has a certain amount of justification.

Another aspect of integration which is becoming a permanent feature of many developing countries is the increasing role of the State in the organization and operation of agricultural markets and in the handling of foreign trade of agricultural commodities to ensure adequate supplies of major staple foods. Under these circumstances a simple alteration of the level and structure of tariffs is not a sufficient condition for expansion of trade, since the decision to trade ultimately rests on a centralized public organization and not on profit seeking operators. As a consequence, systematic patterns of trade are slow to develop and trade is most likely to occur only occasionally between countries with shortages and those with surpluses.

In appraising the potential of integration in the agricultural field, it must be realized that in the absence of integration the most likely solution is not a move towards freer trade, inside and outside regions, and increased specialization led by market forces, but rather a move towards a two-fold attempt on the part of each country to improve its standing in foreign exports and to increase its internal self-sufficiency. The limited impact of present integration arrangements on the intra-regional trade of agricultural commodities, as contrasted to expansion of trade in manufactured goods, is a clear sign that any expectation of production integration through market forces is ill-founded as far as agriculture is concerned. Regional cooperation in agricultural matters must necessarily involve a deep concern for common development objectives and jointly planned initiatives.

In conclusion, the probability of increased agricultural production due to regional specialization according to static comparative advantages is rather small. Nor can the agricultural sector, as traditionally defined, hope to exploit as well as industry the benefits associated with the economies of scale. As a result of the integration improved opportunities for economies of scale exist for certain agricultural processing industries which, to produce competitively, may require large dimensions and specific locations. This applies to an even greater extent in the agro-allied industries sector which are the suppliers of agricultural inputs.

Much more favorable to the agricultural sector are the prospects of gains from improvements in the terms of trade with the rest of the world by regional groups and in particular by those countries which rely heavily upon the production of only one or a few agricultural products. Several steps in this direction have already been taken, for example some producers of coffee, meat, bananas and sugar, heavily dependent upon the export revenues of these commodities have been stimulated to defend or improve their position against excessive fluctuations of export-revenues and the growing prices of imported goods through the forming of associations. No one expects, however, that this cartel-like operation in the agricultural commodities market will achieve the same degree of success as that of the oil-producing nations, OPEC. It must be added that actions aiming at the improvements of the terms of trade do not necessarily presuppose the existence of regional integration arrangements although integration-minded solidarity may add additional strength to initiatives of this type.

Changes in efficiency arising from increased competition among agricultural producers of partner countries is again not an immediate prospect for the reasons already outlined. Such changes will materialize over time only insofar as countries are able to diversify their commodity production for export purposes, both inside and outside the region, and to acquire or develop improved information, transportation and marketing systems.

The structural changes affecting the quantity and quality of factor inputs, as resulting from the integration process, will vary in extent and direction depending upon the type of integration arrangements, the structural peculiarities of the country before integration and the impact, positive or negative, of national policies directly or indirectly affecting agriculture. In this field more than in others, the potential benefits of integration must be regarded as opportunities which can only be translated into development if effective use is made of them through appropriate internal economic policies. The ensuing structural effects may well be perverse and in this respect there are undoubtedly many lessons to be derived from the EEC experience.

As pointed out by Yudelman with respect to Latin America, the major problems that affect agricultural development involve the relationship between industry and agriculture or arise from the fact that within the agricultural sector many national economies are not integrated in themselves (Yudelman and Howard, 1970).

The effect of an integration scheme on agricultural development must consider the impact of industrial efficiency on the internal terms of trade between agriculture and the rest of the economy. National policies creating import substitution industries which produce for insufficient markets, have often led to high-cost inputs for agriculture (machinery, fertilizers in particular). High prices for purchased inputs and relatively low prices for products sold (a policy dictated by other internal reasons) is still a widespread situation in many countries, a siutation which obviously penalizes the earning power of the agricultural sector and is contradictory with respect to agricultural development aspirations.

To fully appreciate the possible contribution of industry to the production of agricultural inputs it must be remembered however that most of the developing countries are confronted with a set of realities quite different from those which the European nations had to face in their earlier stages of economic development. The main difference may be found in the fact that the developing economies must find productive employment in agriculture for a growing labour force which can not be fully absorbed by the other sectors whatever may be their rate of economic growth. The experience of the developed world no matter how valuable cannot be exported as a package. In order to obtain increased farm output, greater agricultural employment, higher productivity and rising farm income (possibly with greater distributive equity), they must search for original policies, especially with regard to the development of technology and institutions.

Consequently, among other things, they must pay great attention to the promotion of agricultural research, tailored to their specific needs, to the dissemination of scientific information and to the translation of this information into new and more efficient techniques. Regional economic integration certainly offers a better framework for attempting a solution to these enormous problems than could be provided by each nation alone. To think however that integration through the functioning of market forces can provide an easy answer is probably the surest way to greater disappointments.

NOTES

[1]The term "regional" is used here as defined by Nye (1971) and others; it applies to selective organizations which restrict their membership, in principle and in practice, on the basis of geographical contiguity. The distinction between political and economic organizations should also be borne in mind. Whereas the former tend to be "macro-regions", at times extending across continents, the latter tend to be "micro-regions" because of the important role of physical contiguity and identity.

[2]Considering the difficulties encountered by more ambitions integration schemes and the opportunity for cooperative arrangements which are less demanding in terms of political will, longer-term commitments and surrender of national sovereignty, integration projects may in the future gain wider attention. In fact, proposals in this direction have been recently advanced (Belassa and Stoutjesdijk, 1975). Integration schemes and projects, however, are not necessarily alternative paths, since the two can well coexist.

[3]Elsewhere I have dealt at some length with the place of agriculture in the history of the European Economic Community (Barbero, 1974).

[4]An outstanding precedent was the establishment of the European Steel and Coal Community marked by a high degree of supra-nationality; on the contrary, the projects for an European Defence Community and for political union did not gain enough consensus and were eventually abandoned.

[5]In Latin America this pattern displays greater variations as some of the countries have been able to enlarge their industrial base and to become exporters of manufactured goods.

[6]The extent of external dependence is often aggravated by the burden of indebtedness for even countries which enjoy a positive trade balance (coming from exports of agricultural commodities, minerals or oils) find themselves with negative balance of payments because of debt servicing, payments of royalties, dividends on foreign capital, rising shipping costs, etc. . . . (Onitiri, H.M.A.).

[7]Furthermore, there is a scope for a country to be member of the Community and not of the Common Market.

[8]The preservation of national sovereignty is an explicit objective of the Organization of African Unity.

[9]The two, of course, are not incompatible. If in EAC interstate trade is relatively more developed than elsewhere it is probably for the long history of common services (railways, harbours, post and telecommunications, collections of custom duties). But even in this case special devices had to be introduced in 1967 to avoid the polarization of benefits to the advantage of the country having a more diversified economic base. (Hazelwood, 1975).

[10]The list of internal distortions is a long one; it suffices to mention: high rates of unemployment and underemployment, made more acute in perspective by the rapid population growth rates of the past; dualism among regions, sectors and within sectors, great personal inequalities in control of income distribution and of resources; deficiencies in physical integration; foreign exchange gap and growing import requirements; lack of qualified manpower and entrepreneurial experience.

[11]For instance the failure to arrive at a common monetary policy by the EEC among other effects has introduced great disturbances in the operation of the common market for agricultural products, a market which had been conceived and organized in time of stable exchange rates.

[12]An efficient price mechanism requires that prices reflect the true opportunity costs.

[13]The limited employment potential of the modern export sub-sector has been repeatedly emphasized.

REFERENCES

Amea, R. K. (1974) 'Euro-African Association and the Economic Development of Africa South of the Sahara', *Politica Internazionale,* No. 7–8.

Andic, F. *et al.* (1971) *A Theory of Economic Integration for Developing Countries,* London.

Awad, J. H. (1974) 'The impact of Association upon Industry and Agriculture in Sub-Saharian Africa', *Politica Internazionale,* No. 7–8.

Banco Interamericano de Desarrollo (BID) (1974) *Progreso economico y social en America Latina* (Cap. IV, Integracion economica), Informe anual.

Barbero, G. (1974) 'L'agricoltura della politica economico-sociale della CEE' (Agriculture in the EEC economic and social policy), *Rivista di Economia Agraria,* XXIX: 2.

Belassa, B. (1961) *The Theory of Economic Integration*, Homewood.

Belassa, B. and Stoutjesdijk, A. (1975) 'Integracion economica de paises en desarrollo', *El Trimestre Economics,* July–September.

Dams, T. J. (1975) 'Agriculture Cooperation in the new Convention', paper presented at the J. F. Kennedy Institute Colloquium on *The Lomé Convention and Development Cooperation, Old Links or a New Departure?,* December.

De Bussy, M. *et al.* (1971) 'Approches théoriques de l'intégration européene', *Revue Francaise de Science Politique,* XXI: 3.

ECAL/FAO, (1975) *The Agricultural Sector in the Economic Integration Systems of Latin America,* ECLA/FAO Joint Agriculture Division, Santiago, August.

'Economic Alignments in Sub-Saharan Africa', *Africa Digest Guide No. 9,* June 1973.

'Economic Cooperation in Africa', *Africa Research Bulletin,* various issues.

Garcia, A. (1969) 'El dialogo latinoamericano en al Seminario de Africa', *Estudios Internacionales.*

Hansen, R. D. (1968) 'Regional Integration, Reflections on a Decade of Theoretical Efforts', *World Politics*, XXI: 1.

Hazlewood, A. (1975) *Economic Integration: The East African Experience,* London.

Lavigne, M. (1973) *Le Comecon,* Paris.

Lizano, E. and Willmore, L. N. (1975) 'La integracion economica de Centroamérica y el informe Rosenthal', *El Trimestre Economico,* Jan.–March.

'Les mouvements d'integration économique en Afrique', *Bulletin hebdomadaire de la Krecietbank,* 8 Nov. 1974.

Nye, J. S., (1971) *Peace in Parts, Integration and Conflicts in Regional Organization,* Boston.

Onitiri, H. M. A. (1973) 'Self-reliance', *Ceres,* July–Aug.

Penouil, M. (1974) 'L'évolution du Commerce intrafricain et du commerce Cee-Afrique Noire', *Politica Internazionale,* No. 7–8.

'Political Alignments in Sub-Saharan Africa', *Africa Digest Guide No. 8,* April 1973.

Robson, P. (ed.), (1972) *International Economic Integration,* Penguin Books.

UNCTAD (1975) *Current Problems of Economic Integration,* The Problem of Distribution of Benefits and Costs and Selected Correction Measures, New York.

USDA, Economic Research Service, *The Agricultural Situation in Eastern Europe, Review of 1974 and Outlook for 1975.*

Yudelman, M. and Howard, F. (1970) *Agricultural Development and Economic Integration in Latin America.* London.

DISCUSSION OPENING – H. F. Breimyer, *U.S.A.*

Dr. Sen opened this conference by declaring that we are concerned for the political economy of agriculture. Dr. Schultz repeated the theme. In this bicentennial anniversary of the publication of *Wealth of Nations* it is worth remembering that economics as a field of knowledge was an offshoot of political economy. For my part I regret that a separation later took place.

Dr. Barbero's paper is clearly in the tradition of political economy. Few economic models are available to help in describing principles of regional integration. There is not even a good definition. Dr. Barbero says integration is more than co-operation and less than full economic union. Within that zone there is only diversity.

I make these points. First, although his style is subtle Dr. Barbero rejects several rather conventional economic propositions, primarily on grounds of welfare. I find myself in sympathy, as we should be wary of economics by epigram. As one example – expanded trade does not necessarily benefit all trading partners and definitely will seldom benefit all proportionately. In the same vein, removing restrictions on trade between countries that exercise substantial control over their internal economies does not constitute achieving free trade.

Second, Dr. Barbero warns, correctly, against understating the problems or overstating the potential in regional integration. Here the doctrine of the second best comes into play. Those of us who work in policy are keenly aware of that doctrine. Perhaps we can conceive of a best-of-all-worlds where neo-classical economic thought applies widely rather than selectively. We do not live in that kind of world. On the contrary, I fear we live in a world of fair weather liberal trade and foul weather protectionism. Nations co-operate when times are good but scurry towards nationalism and autarky when times turn bad. If we cannot have the "first-best" of a whole world of relatively liberal trade policies, regional integration may be a realistic possibility for a second-best.

My final comment relates to the relinquishing of national authority. Mere multilateral trade "understandings" under revocable terms are virtually worthless. Integration implicitly involves some transfer of sovereignty. Furthermore, in my opinion, some kind of supra-national authority is required. Perhaps this view of integration does not appeal; perhaps it is not even second best. It may be third best. But, if so, it is better than third worst, which is how I would classify a rampant worldwide trend towards national autarky. Regional integration is better than that!

DISCUSSION OPENING – A. Nussbaumer, *Austria*

Problems of integration are manyfold and may be very different from area to area, depending on historical circumstances, stages of economic development, cultural trends, and political values; thus only a multi-disciplinary study can do full justice to all the phenomena concerned. We should be very grateful therefore to Dr. Barbero for the wide perspective and the broad approach he has chosen in dealing with his topic. However useful the scientific method of isolation and abstraction may be, and hence the concentration on just one specific aspect, e.g. the economic one, it might also frequently prevent us from taking full account of the wide spectrum of social and economic problems of agricultural development and regional integration. Let me direct your special attention to a few economic points only nonetheless.

First, for all the broadness of his approach, Dr. Barbero obviously has had to confine himself by selecting some specific problems. Having decided to give much space to political factors and to their linkages with economic interests and objectives, he certainly could not have dealt in detail also with abstract models of economic integration. Economists having devoted much effort to designing ever more advanced models since J. Viner and J. E. Meade made their original contributions in the early 1950's. Such a report about the development of pure theory might even have absorbed all of the time and space available here.[1] Dr. Barbero being mainly concerned with the agricultural problems of developing countries, also chose to put only limited emphasis on Western European experience.[2] Yet we should not be disappointed because of this. He has made it quite clear already in the introduction to his paper that economic integration being a very complex phenomenon, and economic as well as political issues having to be considered, different conditions make for different challenges and hence for different policy aims and solutions. He had to deal, therefore, in his own paper with a special case viz. that of the prospects for agriculture in developing countries under conditions of economic integration. The many examples quoted from different parts of the globe might even raise the question of whether a single study could possible suffice to tell the story of all developing economies.

Secondly, speaking of economic integration in general, it might also be worth while to remember that already in 1958, the EEC treaty having just become effective, T. Scitovsky made a special study of the prospects for Western European economic integration[3] in which he examined the set of conditions which had to be fulfilled if economic integration was to be successful. Just to mention a few of them: a high share of the region in total world trade, countries integrating being each others natural trading partners, low transportation costs, a high level of technical development and similar political objectives. Scitovsky found that these conditions prevailed in Western Europe, and therefore concluded that economic integration there stood a good chance of success: history has proved him right. One might even say that the occasional failures of the EEC to reach declared goals can be explained, consistent with Scitovsky's review, by a lack of political agreement.

Applying this analysis to integration between developing countries, and following Dr. Barbero's presentation of conditions frequently prevailing there, one should not be surprised to find that integration in other parts of the world frequently has not been nearly as successful as in Western Europe.

Thirdly, Dr. Barbero is very doubtful as to whether free international trade and even worldwide economic integration really are beneficial to all participating countries, and if I understand him correctly, he is also very sceptical as to the validity of the standard theories of international trade. But should we not be less doubtful as to the validity of pure theory, yet the more sceptical as to its correct application? It may suffice to keep in mind the nine conditions established by Samuelson and Stolper[4] to be satisfied if international free trade and factor movements were to lead to complete factor price equalization and hence to optimum conditions for factor productivity. If one is to understand that in the real world these conditions not being met, there is ample room for theoretical arguments in favour of interventions to help the disprivileged.

These conditions are: full employment, perfect competition, neutrality of money, all countries produce all goods, identical production functions, their linearity to scale, absence of transportation costs, complete factor mobility, and the other usual conditions for Pareto-optimality. Whenever any of these conditions are not satisfied, free trade and compensating government interventions may be superior in their effects upon welfare to free international trade alone. There are many special theories proving this, e.g. the theory of optimal tariff protection, the theory of infant industry development, anti-dumping theory, terms of trade theory, etc . . . Therefore, a general and dynamic theory of international trade and integration applied under realistic conditions might lead us to conclusions substantially different from those following from the traditional model case. We may continue to believe in the core of the economic theories of international trade and integration and yet reach solutions on the basis of these theories which satisfy the needs of development policy.

Last, but not least, we should not forget about the importance of a good infrastructural base for every kind of economic activity. We may account for it in theory by accepting infrastructure as an additional factor of production available only in some locations. Or we may regard it as an immobile public good helping to save on private costs of production. It may also be viewed as a fact which makes for external economies available only in some locations, and thus attracting all mobile factors of production; it makes little difference whether advantages of location are due to natural resources, favourable climatic conditions, or man-made infrastructure.[5] We may conclude from this that, under conditions of free trade and integration, there are no equal chances for development as long as such differences exist; while natural advantages of location should be allowed to influence the distribution of economic activity according to spatial equilibrium theory, advantages due only to better man-made infrastructure certainly call for development policies. As far as infrastructural development in general is concerned, we should however be in favour of economic integration, since it may help to save costs for all participating governments. Regarding the future of agriculture, the necessity to provide for a good infrastructure in all developed economies if they are to be competitive, may place agriculture at a disadvantage to urban development because it is not concentrated in space and the creation of the infrastructure needed is therefore very expensive.

My fourth comment concerns the goals and aims of integration. Dr. Barbero certainly is right in preferring to speak of different forms rather than of levels of integration because speaking of levels always implies that some are higher and some lower, and that we should always strive for the higher ones. Yet depending on their general economic situation and on their political and economic goals it may not be desirable for some countries to go beyond a certain stage of integration. Or complete integration might only be considered if special disadvantages to a country's economy, or to some sectors of it can be compensated for by offsetting transfer payments.

Different goals and aims of integration also call for different methods and institutions. When deciding upon methods of integration, value judgements therefore can never be avoided.[6] This even applies to the fundamental decision of whether market methods of integration are to be used as in Western Europe today; these are certainly in agreement with the economic and political systems there, but they do not necessarily meet all of the needs of developing countries. The same applies to agriculture. Since agricultural production and the distribution of income with regard to the agricultural sector to a great extent do not follow from the rule of free competition alone in all national economies, we should not be surprised that the free market method of economic integration meets with special difficulties if applied to agriculture.

I fully agree therefore with Dr. Barbero that there must be a balanced distribution of opportunities and benefits if integration is to be achieved in agriculture. Such a distribution may even call for international transfer payments short of which no agreement about integration can be reached.[7] Yet even the experience of the European Common Market shows that there are limits to the willingness of national taxpayers to agree to international transfer payments for the benefit of farmers in other countries. If common prices are introduced in an international agricultural market, this problem will be still more important since the new and international prices have to be made compatible with at least some of the old income policies. This is not solely for the sake of avoiding immediate disturbing consequences for production because when accepting common international pricing, price subsidies paid by consumers have to be replaced by income subsidies paid by governments. National governments frequently are not willing to carry the burden alone.

My fifth comment is on economic theory again. We should face the fact that a dynamic theory of economy integration is needed today. Dr. Barbero has said that it was one of the aims of integration to step up and to diversify the economic base of a country, and he has listed some of the dynamic elements of integration in the introduction to part II of his paper. Yet he did not discuss the consequences for international trade theory. The classical *Viner–Meade* model of integration used operates under the limiting assumptions of resources not being mobile, and technical standards and all of the other conditions of production remaining unchanged. It also deals nearly exclusively with factor allocation, assuming perfect competition. Economies of scale are disregarded, and so are changes in income and consumption. It may be difficult to build and to solve a dynamic model of integration considering all of these influences; but it would certainly have been most interesting to hear more about it.[8]

Also increasing government intervention in national economic affairs should be accounted for by theoretical analysis. The classical theories of international trade and integration are founded upon the assumption of *laissez faire* on the national level. Should the international market alone be sufficient to provide for optimal results of economic integration there must therefore be no national intervention even on the home market, an assumption which is entirely unrealistic.[9] This is why economic union today calls for harmonization of national economic policies sufficient to exclude at least major disturbances.

Certainly many government interventions are not directly concerned with foreign trade but primarily with economic development, personal and regional distribution of income, and monetary and exchange stability. But indirectly all of these interventions have a considerable influence on national and international economic equilibrium. Optimum international allocation of factors of production need not even be the principal aim of economic policy, including trade and integration policies; it may be full employment or a specified rate of growth. The quality of the international allocation of factors may be considered as of only secondary importance, a second best solution in this respect therefore being preferred. Yet governments may try to reach such aims not necessarily by interfering directly with the working of the market, they may just change the "external" conditions under which markets operate.

This is the reason for my sixth point, that Pareto-optimality alone no longer serves as a sufficient guideline and measure of economic welfare. As soon as we cease to start from the assumptions of full employment, perfect competition, a "normal" rate of growth, a pre-determined distribution of income, and an established set of other, frequently institutional and legal, conditions surrounding the market, there is not simple Pareto-optimal solution. Instead there are as many Pareto-optimal solutions as there are sets of external conditions under which the markets operate.[10] Many of these conditions are controlled by the state and subject to government influence. For the neo-classical model of a self-adjusting economy these are external variables, but applied Political Economy has to internalize them. In a world where the best of many possible market optima has to be chosen, we cannot do so without a set of assumptions about external conditions, including economic policy aims.

Let me finally conclude with a statement on economic policy. Barbero in his paper juxtaposes the systems of the free market and of economic planning. In doing so he shows a certain preference for planning, at least as far as developing countries are concerned. Should we not consider yet another alternative, at least as far as economic systems in Western Europe are concerned? Governments identify problems, they set targets, they try to realize them at least by indirect action, changing market conditions. If this is done systematically, we may call it planification. In theoretical terms, governments have decided on the kind of Pareto-optimum they want to see realized.

If international economic integration is to be successful under conditions of government intervention, there has therefore to be agreement about common objectives of economic policy. Policy instruments have to be harmonized and applied considering the needs of partner countries. There has to be common understanding about the distribution of benefits as well as of disadvantages following from economic integration, if necessary about compensatory international transfer payments. Short of such agreements no durable union can be built. It is here that the most difficult problems of economic integration still remain to be solved, especially as far as agricultural development is concerned.

NOTES

[1] For a summary of recent developments in integration theory see e.g.: M. Kraus ed.: *The Economics of Integration*, London 1973.

[2] For the institutional and political aspects of Western European Economic Integration see: M. Meerhaeghe: *International Economic Institutions*, London 1966; C. J. Friedrich: *Europa-Nation im Werden*, Bonn 1972; G. Mally: *The European Community in Perspective*, Lexington, Mass. 1973.

[3] T. Scitovsky: *Economic Theory and Western European Economic Integration*, London 1958.

[4] P. A. Samuelson: International Trade and Equalization of Factor Prizes. *Economic Journal*, vol. 58 (1948); *idem*: International Factor Price Equalization once again. *Economic Journal*, vol. 59 (1949); J. Bhagwati: *The Pure Theory of International Trade: a Survey. Surveys of Economic Theory, voll. II*, London, New York 1969, pp. 156, p. 174.

[5] R. Jochimson: *Theorie der Infrastruktur. Grundlagen der marktwirtschaftlichen Entwicklung*, Tübingen 1966; H. W. Richardson: *Regional Growth Theory*, London 1973.

[6] M. Weber: *Die "Objektivität" sozialwissenschaftlicher und sozialpolitischer Erkenntnis. Gesammelte Werke zur Wissenschaftslehre*, 3rd. ed., Tübingen 1968; *idem*: *Der Sinn der "Wertfreiheit" der soziologischen und ökonomischen Wissenschaften*, op. cit.; F. Kaufmann: *Methodology of the Social Sciences*, London 1958.

[7] Inside the EC there are presently four major transfer mechanisms at work by which funds are being made available on a non-commercial basis for employment in other member countries: European Investment Bank, European Social Fund, European Development Fund, and European Agricultural Guidance and Guarantee Fund.

[8] See: B. Balassa: Trade Creation and Trade Diversion in the European Common Market: an Appraisal of the Evidence. *Manchester School Economic and Social Studies*, vol. 42 (1974), pp. 93; S. K. Ghosh: Towards a Theory of Multiple Customs Unions. *American Economic Review*, vol. 64 (1974), pp. 91; W. E. Prewo: Integration Effects in the EEC: An Attempt at Quantification in a General Equilibrium Framework. *European Economic Review*, vol. 5 (1974), pp. 379.

[9] See: A. Nussbaumer: Western Market Economies. *The Year Book of World Affairs*, vol. 30 (1976), pp. 223; *idem*: *Die Stellung des Staates in der Wirtschaft. Verhandlungen des Fünften Österreichischen Juristentags*, Wien 1973.

[10] E. J. Mishan: *A Survey of Welfare Economics, 1939–1959*. Surveys of Economic Theory, vol. I, London, New York 1968, pp. 154; J. H. Henderson, R. E. Quandt: *Microeconomic Theory, A Mathematical Approach*, 2nd. int. stud. ed., New York 1971, chapter 7, pp. 254.

Report of the general discussion

The general tone of the discussion lacked wholehearted enthusiasm for the EEC as it has been operated so far. The question was posed as to where regional economic integration fell along the scale from liberal to protected trade policy, since external barriers might be raised while those inside were removed. A speaker held that in the EEC effective policy-making was hampered when price policy became paramount. Structural policy would have led to positive moves in harmonisation of monetary, regional and other policies. Such success as the EEC had attained had been at the expense of Southern Hemisphere meat producers and of price stability for LDC procucers – effects which were firmly in line with price theory.

The nature of any cause/effect interrelationship between CAP and the monetary crisis was clearly debatable. Participants saw that research was potentially useful in this subject area but would have been more impressed had there been some indication of what changes in operational techniques might arise from research results.

The diversity of types of integration attracted some attention. The EEC was unusual since there was hope of political integration – that is, of a supra national organisation, whereas others were international in scope. Security considerations were mentioned in connection with the creation of both the EEC and the CMEA and the role of ideological orientation of the partners' efforts towards economic integration in LDC's was questioned. The influence of large differences in wage rates between possible partners in discouraging union and the view that the status of CAP stems from the political status of those who have headed the Commission, rather suggests that the reply to a discussant who sought for a statement of principles governing the process of economic integration was that any such model would be highly complex.

Participants in the discussion included: P. C. Baillet, *France*; P. C. Bansil, *Zambia*; L. Folkesson, *Sweden*; F. S. Masinde, *Kenya*; D. Paarlberg, *U.S.A.*; J. F. Rimsdijk, *Netherlands*; G. Schmitt, *Fed. Rep. of Germany*; T. W. Schultz, *U.S.A.*; A. Weber, *Kenya/Fed. Rep. of Germany*.

G. Barbero (in reply)

I am grateful to Dr. Nussbaumer for his contribution outlining the conditions which must be fulfilled for integration to be successful. I particularly agree with him that integration schemes require previous agreement on the distribution of benefits and costs and that this distribution is not static. Most of the conflict met in studying actual integration schemes arises from the fact the there are structural changes, induced by the integration process, which tend to cause deviation from an acceptable distribution of benefits. International transfers have been used, not only in the EEC, but in the East African Community as well as in other African communities.

Regarding the relations between the working of the market and planning, I am not saying that the market is not doing any good – it does not seem to be doing a good job in the west – but I am sure that the combination of planning and direction from government in determining the sphere in which the market is to work is important. Even the experience of the Comecon countries where some form of market is working is, I think, proof that there must be a combination of the two.

I confess that I gave little space to the relations between blocks in relation to integration schemes and this is the reason why I did not refer to the hardships which certain countries, especially the meat-producing countries, have had to face in recent years. In relation to the Common Agricultural Policy and the world monetary crisis, I was trying to say that the policies followed by the European Community may have created problems in relations with the United States, first in the industrial sector and also in trade in agricultural commodities. The changed attitude of the United States to the Economic

Community is to some extent explainable on the relations which have developed between the two blocks, which account for a very large share of total world trade. It is useless to evade the fact that what the EEC did had an impact on the rest of the world. I think that the policies of the EEC have had that effect on the rest of the world and that some of the developments of world trade and world relations have their roots precisely in the fact that the EEC block is an important one in respect of its share in production and in total trade.

This raises the problem of the relations between regional groups. Will the outcome be a more liberal climate in trade relations or more protection? I am not well qualified to answer but it seems to me that the development of the regional groups – the size of which is very important, and I overlooked this point – would probably lead ultimately not to a more liberal climate but to planned arrangements between groups. This seems to me to be the tendency.

My paper has been criticised for Western bias, and, being from the Western world, this may well be the case, but I have gone to great effort to isolate the differences, the background factors and motivations which exist in the various continents and I have tried to understand why regional integration in Africa has taken a certain shape in full cognizance of the changes which are taking place. I have mentioned the influence of the colonial period and neocolonialism as a starting point in explaining why regional integration is already so developed in Africa although national states are of recent origin. I think I have mentioned, too, that there are new aggregation forces operating and political structuring in the various countries is quite different. It is a subject which would require a great more elaboration than is possible here but Africa is a very important source of experience in this field.

I said that one of the motivations for the formation of the European Community was precisely the building up of a block as a reaction to the consolidation of the Socialist block. I think this was one of the reasons why the two interacted. It took place in the period of Cold War and this is an important point to remember.

I mentioned the different rate of expansion in industrial and agricultural products as an indirect measure of the ease with which the expansion of trade within the two sectors can take place but it is certainly not the only measure. If I mentioned that in order to expand trade there is a need for common development objectives it is because, in my opinion, in order to have greater trade in agricultural products we have to have a broader pattern of production of agricultural products and a greater quantity and this cannot be achieved in the short-term without co-operation among countries. It has been said that common development objectives in the EEC have not been so successful, and I fully agree, but I think the experience of the EEC is important not only for the positive aspects but also for the negative ones. One of the aims of regional groups may well be improvement of the terms of trade, especially for the developing countries relying on few commodities exported mainly to the developed world. This is the meaning I had in mind in this context. In my view the existing regional groups may enhance the solidarity among the members and hence make the association more effective in improving the terms of trade.

MICHAEL TRACY*

The Decision-making Process in the European Community with Reference to Agricultural Policy

The general Conference theme of "decision-making" suggests, in relation to the European Community's agricultural policy, an analysis of the processes by which major policy decisions are reached. Such an analysis should throw light on the complex institutional structure which is involved, and on the numerous forces which come into play before a result is obtained. The annual price-fixing for agricultural products is obtained. The annual price-fixing for agricultural products provides a major illustration of this process. It also seems appropriate in this context to discuss briefly some of the main areas of interaction in the making of agricultural policy as between the Community and the outside world.

1. THE POLICY-MAKING FRAMEWORK

The European Community represents a unique and novel attempt at integration between sovereign States: it is neither an intergovernmental organization, nor a Federation. The rules by which it operates are to be found in the basic Treaties and in a mass of derived legislation; over the years, moreover, the Community has evolved practices which do not appear in any legal text but which represent pragmatic responses to political realities. In the following pages, an attempt is made to explain as briefly as possible these various aspects.

1.1 *The legal basis of decision-making*

The basic authority for decision making in the European Economic Community[1] is provided, as regards the original six Member States, (Belgium, France, Germany, Italy, Luxembourg, Netherlands), by the treaty of Rome of 1957, and for the three new Members (Denmark, Ireland and the United Kingdom) by the Act of Accession of 1972.

The Treaty of Rome established the basic principles and features of the EEC (Articles 1–3). It instituted a common market, and aimed to harmonize

* General Secretariat of the Council of the European Communities. The author is indebted to numerous colleagues in the Council Secretariat for assistance with this paper. Any opionions expressed in it are his own responsibility.

the economic policies of Member States. Custom duties and quantitative restrictions on trade between Member States have been abolished, and a common tariff has been set up as regards imports from third countries;[2] freedom of movement of persons, services and capital as between Member States is to be established. The Treaty provided for a common agricultural policy and a common commercial policy in relation to third countries (also a common transport policy). It also laid down rules to prevent distortions of competition within the common market.

1.2 *The Institutions of the Community*

It may be said in general terms that policy is formed on the basis of proposals from the Commission, by decision of the Council following consultation of the European Parliament and in many cases also of the Economic and Social Committee; policy measures are subsequently implemented by the Commission. The Court of Justice ensures that decisions are consistent with Treaty provisions and subsequent Community acts.

The composition and tasks of the Institutions are to a large extent laid down by the Treaty, but in practice important developments have occurred over the years.

(a) *The Commission.* In the strict sense, the Commission is composed of thirteen members,[3] appointed by agreement between the member governments but, once appointed, independent of individual Member States.[4] The Commissioners have previously held high posts in national governments or administrations – many of them are former Ministers. In the wider sense, the term "Commission" is often used to include the services numbering about 7,000 officials.

Besides its role in proposing policy to the Council, further discussed below, the Commission's executive tasks are numerous and important. This is particularly the case as regards the management of markets under the common agricultural policy (the Directorate-General for Agriculture – D.G. VI – is one of the biggest and most active in the Commission). Another important responsibility for the Commission is the preparation and management of the Community's budget.

(b) *The Council.* The Council is composed of Ministers from each of the nine Member States. While in principle there is only one Council, in practice it meets in different forms: Foreign Affairs, Finance, Agriculture, etc. (The Agricultural Council meets about once a month.) Under the Ministerial Council are numerous Committees and Working Parties, also composed of national representatives (officials). The Committee of Permanent Representatives (COREPER) has a general responsibility for preparing Council sessions, and has over the years grown to be a very important element in the Community's decision-making process.[5] Sessions of the Agricultural Council are prepared mainly by the Special Committee on Agriculture. Both COREPER and the SCA meet weekly. The meetings of the Council and its dependent bodies are serviced by a relatively small Secretariat, including a Legal Service.

The Presidency of the Council, and of its dependent bodies, rotates between the Member States at six-monthly intervals.

"Summit meetings" of Heads of State or Government of the nine member countries are not, strictly speaking, part of the Council framework: such summits, however, are now referred to as the "European Council", and although the European Council does not itself take legislative decisions, there are increasingly close links between the proceedings on these occasions and the regular sessions of the Council proper.[6]

Decision-making in the Community involves a particularly close interplay between the Commission and the Council. Their respective roles are not easy to define in general terms: it would not be quite correct to say, for example, that the Commission has the executive and the Council the legislative authority. It is important to note that the Council normally cannot act without a proposal from the Commission (it could not for instance initiate, on its own, discussions of a sheep-meat policy). The Commission's right of proposal gives it a major influence over policy. It also has influence through its participation in all meetings of the Council and its dependent bodies.

The *voting procedure* of the Council is an important subject and has inevitably attracted much attention. Under the Treaty (Article 148), the Council can take decisions by a "qualified majority" vote[7] on many subjects (including agricultural policy); unanimity is however required to amend a Commission proposal (Article 149). Practices in the Council have in fact evolved differently: the Council avoids taking decisions by majority vote in cases where important national interests are involved. It is therefore necessary to seek compromises to which all Member States can agree. (In this process, the Commission may itself modify its proposal to reach an acceptable solution.)

It was opposition by France (under President De Gaulle) to the supranational aspects of majority voting which led to the so-called "Luxemburg compromise" of 1966 on this point. It is however questionable whether the Community could have progressed if Member States had frequently found themselves isolated and overruled on matters of particular importance to them. In practice the Council has to seek a balanced solution, which may not be wholly satisfactory to all delegations but at least is not completely unacceptable to any of them.

At the same time the decision-making machinery would be severely handicapped if *all* decisions had to be reached unanimously, and this was recognized at the Paris Summit of December 1974. In fact, majority voting has continued to be used for budgetary matters, and has recently been resorted to on several occasions by the Agricultural Council.

(c) *The European Parliament.*[8] The Parliament consists of delegates so far designated by national Parliaments from among their members. It has however been agreed that members of the European Parliament will in future be directly elected and need not be members of national Parliaments.[9]

Hitherto the Parliament's main function has been to debate Commission proposals before they are adopted by the Council, and to deliver an opinion (an Agricultural Commission examines in detail proposals related to agriculture before they are submitted to a plenary session). Members of the

Council and of the Commission participate in major debates. The Council is not bound to follow the Parliament's view. However, the Parliament's powers in the establishment of the annual budget have been substantially strengthened.

The Parliament has the right, under Article 144 of the Treaty, to censure the Commission (it has no corresponding right concerning the Council): if a motion of censure is carried by a two-thirds majority the Commission must resign. Censure motions have been tabled only twice: once in November 1972, on the issue of budgetary powers – the motion was however withdrawn without being voted upon; the second time in May 1976, by the British Conservative group, protesting at the scheme for compulsory incorporation of skim milk powder in livestock feed (this was rejected on 16 June by a large majority but provoked an extensive debate on the dairy surplus problem).

(d) *The Economic and Social Committee.*[10] The Economic and Social Committee consists of representatives of various economic and social categories, including, for example, farmers, workers and "the general public". It has advisory status: in some cases it must be consulted on a Commission proposal, more often consultation is optional. It is consulted on all important agricultural policy matters (it has an Agricultural Section which holds detailed discussions on agricultural questions). Its opinions tend to be more of an economic and technical and less of a political nature than those of the Parliament. They are taken into account by the Council, though the Council is not bound to follow them. In recent years the ESC has also taken the initiative in producing reports and opinions on its own.

(e) *The Court of Justice.* The Rome Treaty set up the Court of Justice to "ensure that in the interpretation and application of this Treaty the law is observed" (Article 164). The Court consists of nine Judges, one from each member country.

The Court may be called upon to give a ruling in disputes between the Commission and a Member State, or between two Member States: if the Court finds that a Member State has failed to fulfil an obligation under the Treaty, the State must take the necessary measures to comply with the Court's judgement. The Court also has the power to give to national jurisdiction binding "preliminary" rulings on the validity and interpretation of acts of the Council and the Commission, and may declare such an act to be void.

As regards agricultural policy, though the number of cases actually brought before the Court has not been great, its rulings have exerted considerable influence, and the possibility of recourse to the Court is an important factor particularly where possible distortions of competition are concerned.[11]

1.3 *The form and status of Community acts*

The Council and the Commission, within their respective fields of competence, can "make regulations, issue directives, take decisions, make recommendations or deliver opinions" (Article 189 of the Rome Treaty). Most acts of agricultural policy – in respect of market management – take the form of "*regulations*". These are applicable throughout the Community: they do not require further national legislation and take precedence over any existing national laws. A "*directive*" stipulates the result which is to be achieved but leaves to national authorities the choice of form and method: this device is often applied in relation to structural policy in agriculture since it permits national authorities to adjust the form of legislation in accordance with their own circumstances. "Decisions" are usually of a more limited and specific character: they may relate to a single Member State. "Recommendations" and "opinions" have no binding force, and are not common under the common agricultural policy; however, the Council sometimes adopts a "*Resolution*" which may be significant in laying down principles for future action.[12]

Community acts, as well as Commission proposals, Parliamentary opinions, etc., are published in the Official Journal.

1.4 *The Community budget*

The existence of a Community budget, provided for in Articles 199–209 of the Treaty, is essential for the implementation of common policies, particularly the common agricultural policy. Following certain transitional arrangements, the Decision of 21 April 1970 on "the replacement of financial contributions from Member States by the Communities own resources" provided that from 1 January 1975 the budget of the Communities would be financed entirely from "own resources", in particular duties under the common external tariff, receipts from levies, compensatory amounts, etc. on imports from third countries (agricultural products): in so far as these receipts do not suffice, they are to be completed, in principle, by a share of proceeds from value-added tax, but until the arrangements in this last respect have been agreed upon, Member States make contributions based on their relative Gross National Product.

Within the overall budget, the agricultural fund – European Agricultural Guidance and Guarantee Fund (EAGGF or FEOGA)[13] – accounts for by far the greater part of expenditure,[14] largely because the CAP has been developed well ahead of other common policies, so that expenditure formerly borne by national exchequers has been transferred to the Community, but also because of the rise in cost of support.

The FEOGA has two sections:

(a) *the Guarantee Section*, financing restitutions on exports to third countries and intervention to stabilize markets within the Community (5,160 million units of account in the 1976 budget);
(b) *the Guidance Section,* financing action to improve structures (currently limited to 325 million units of account per annum).[15]

It is worth noting that receipts from import levies on agricultural products do *not* appear in the FEOGA accounts, but – as already mentioned – go into the overall Community budget.

The annual Community budget is now established, for each calendar year, on the basis of proposals submitted by the Commission by June of each year, but detailed figures for agricultural credits are put forward only in September, so that harvest results can be taken into account. Following examination by the Council (Finance Ministers) and the Parliament, the budget is usually adopted in December. The estimates of FEOGA expenditure are based on current price guarantees: in its 1977 budget proposals, the Commission allows an extra 200 million units of account for increases in the subsequent price review. If in the course of the year actual FEOGA spending does exceed the budget appropriation, a supplementary budget may be required.

Every Commission proposal having budgetary implications must include an assessment of these implications, which is examined by the Budget Committee of the Council.

In view of the rise in FEOGA expenditure over recent years, due to market imbalances and monetary factors, the budgetary aspect has become very sensitive, and the Commission makes every effort to contain spending within the original annual estimates.

1.5 *Specific provisions concerning policy-making and management under the common agricultural policy*

In Title II of the Rome Treaty, which deals with agriculture, Article 43 lays down the relevant procedures: "The Council shall, on a proposal from the Commission and after consulting the Assembly, acting unanimously during the first two stages [of the transitional period] and by a qualified majority thereafter, make regulations, issue directives, or take decisions . . ."

This Article provides the basis upon which the common market organizations have been set up and developed. For the various commodity sectors (e.g., cereals), the Council has adopted a basic regulation, laying down the system of price determination, the methods of price support, the arrangements relating to trade with third countries, etc.

Authority for administering various features of the regulation is delegated to the Commission (e.g., the periodic fixing of import levies on the basis of world market prices). Certain matters must be referred to a *Management Committee,* consisting of officials from the Member States and chaired by a Commission official.[16] Voting in the Management Committees is by qualified majority (the same weighting system as that already described for the Council). If the Management Committee disagrees with the Commission (this does not happen often), the Commission can introduce measures contrary to the view of the Committee or may defer their application for a month, but must submit the matter to the Council: the Council can take a different decision from that of the Commission by a qualified majority (majority voting does in fact continue to be used on such occasions).[17]

The basic commodity regulations provide a "*safeguard clause*" under which "if by reason of imports or exports the Community market experiences or is threatened with serious disturbances which may endanger the objectives set out in Article 39 of the Treaty, appropriate measures may be applied in trade with third countries until such disturbance or threat of disturbance has ceased". If such a situation arises, the Commission can decide the necessary measures on its own initiative or at the request of a Member State. Measures decided by the Commission can be referred to the Council by a Member State, and the Council can amend or repeal the measures in question by qualified majority.[18]

The Commission has also set up *Advisory Committees* for various products, on which producers, food manufacturers, distributors, trade unions and consumers are represented. These Committees may be consulted by the Commission in the process of formulating policy; they do not have any powers.

2. THE ANNUAL DETERMINATION OF AGRICULTURAL PRICES

The annual price-fixing exercise is a crucial aspect of the common agricultural policy, and it brings into play all the institutional features which have so far been described. It begins with an analysis by the Commission of the various economic factors which should be taken into consideration. In recent years the Commission has developed an "objective method", which on the basis of trends in farm costs and non farm incomes seeks broadly to determine the price increase which would be necessary to give "reference farms"[19] an increase in net income comparable to the increase in non-farm incomes over the recent period. This leads to the calculation of an overall price "requirement" for all products and all Member States. In recent circumstances, the calculation has been complicated by changes in exchange rates, and ways have had to be devised of taking this into account.

The market situation for individual products is also taken into account, and this influences the price proposals made for each product. In recent years, the excess supply affecting the markets of certain commodities, in particular dairy products, has inevitably had to be taken into account. The price proposals for 1976/77 included various measures aiming to improve market balance for dairy products, beef and cereals in particular (the regulation of the wine market had already been under consideration for some time). The final set of proposals becomes as a result a complex document incorporating not only the proposed prices for each of a wide range of products but also various related measures. Adjustments to take account of monetary changes have also become an important feature.

The proposals are submitted to the Council around November. The Parliament and the Economic and Social Committee are then consulted: both hold committee meetings and plenary sessions before delivering their opinions, usually in the following February. The debate on the subject in the Parliament usually occupies a day or more, and the "Opinion" may make detailed comments as well as expressing a view on the overall price increase. Most of the Parliamentary speeches tend to emphasize producers' points of views, and the resulting opinion usually advocates a higher price increase than that proposed by the Commission.

An important though non-official influence is that of the COPA (Comité des organisations professionnelles agricoles), grouping the farmers' unions of the nine countries. The COPA transmits its views to the Council; its President and Bureau are normally granted an interview with the President of the Council. Pressure from the farmers has not always been limited to diplomatic channels: there may be demonstrations at the national level, and sometimes in Brussels, which have on occasion led to regrettable scenes of violence.

The Economic and Social Committe has the opportunity to take into account the views of different sectors of the population. Consumer influence is however weak relatively to that of producers; there exist certain consumer unions at the Community level, but they do not directly make a significant impact. (A more important factor is the extent to which some Member States take into account consumer interests in the positions they adopt in the Council).

Meanwhile, the Commission's proposals are being examined in detail by the various Committees and Working Parties of the Council. During the months of January and February, there is hardly a day without some meeting on the subject, sometimes more than one. These examinations take place on a product-by-product basis, but also overall aspects are considered, such as the "objective criteria", the "agri-monetary" features, and the budgetary implications.

The positions of the various Member States may be widely divergent at the outset of this process, depending on their respective interests. Countries in which agricultural production and exports are particularly important will press for relatively high price increases, though even as between such countries there may be different priorities as regards price relationships between products. Other countries may be concerned to minimize price increases, in the consumer interest and in the context of anti-inflationary policies. For some countries the budgetary implications may be an important consideration. Differences of opinion will also exist about the adjustments to the general prices increases which are necessary to take account of monetary changes. Proposed changes in the methods of support, as well as the price levels themselves, may give rise to particularly complicated discussions.

Each Minister of Agriculture is subject to strong pressures from various sides within his own country: from producers' organizations in particular, but there may also be differences of view within each Government which have to be reconciled.

Several Council sessions will be necessary to clarify the national attitudes to the various aspects of the proposals. The divergences tend to remain wide until deadlines for decisions approach: these consist of the beginning of the marketing year for certain products, usually some time in March for dairy products and beef. The dates in question, having been previously fixed by the Council, can be – and sometimes are – postponed, but it is difficult to go beyond the end of March.

The final "package" of decisions is usually adopted at the end of a "marathon" Council session lasting several days, often culminating in an all-night meeting. Such occasions attract considerable attention from the press and public opinion; the publicity in fact often tends to harden national or sectoral positions and to impede the move to compromise. Nevertheless, national divergences are gradually narrowed as concessions are made: it may still take more than one attempt at an overall solution (often a first attempt is made by the Chair, the final one by the Commission) before agreement is reached. The ability of the responsible member of the Commission to suggest compromise solutions to a variety of problems is an indespensable ingredient of success. Majority voting is out of the question on a matter of such importance: in the end, all Ministers must agree, and be prepared subsequently to support the "package" in the respective countries. It is inevitable that no Minister can be fully satisfied: all have to make concessions.

On these occasions, the Ministers of Agriculture have always been conscious of the importance, in the overall context of integrations in the Community, of reaching a solution. Failure to agree on such a vital matter as the fixing of common farm prices would have incalculable consequences for the Community as a whole.

Even when agreement is reached on the "price package", this relates to the content and not necessarily to the exact form of the acts which have to be adopted. A substantial number of Council regulations are required, as well as numerous subsequent implementing regulations of the Commission, to put the agreement into effect. The process of translating political decisions by the Council into precise legal texts is often difficult and may in some cases take several weeks or even months.

3. RELATIONS WITH THIRD COUNTRIES IN AGRICULTURAL POLICY ISSUES

The general issue of the Community's status in international law is a complex matter which goes well beyond the scope of the present paper. The Treaty of Rome (and the Treaties setting up the other two Communities) include certain provisions concerning the Community's relationship with international organisations (article 229),[20] the conduct of its trade policy (Article 113), relations with the overseas countries and territories (dependencies) (Articles 131–136), agreements with third countries in general (Article 228) and association agreements in particular (Article 238). These provisions lay down, among other things, the respective roles of the Council and the Commission. In practice, however, the Community's relationship with the outside world has evolved in a variety of ways over the years.

It is possible to distinguish the following general cases:

(a) *Informal bilateral or multilateral contacts.* Decisions taken under the common agricultural policy and relating primarily to the management of the Community's domestic market may have important repercussions on third countries. Problems are liable to arise particularly when the Community market is oversupplied: such a situation led to the suspension of beef imports from July 1974, and more recently to the scheme for compulsory incorporation of skim milk powder in animal feed – an arrangement which in the view of the U.S. in particular could adversely affect trade in soya.

Such situations may lead to bilateral or multilateral contacts between Community representatives and representatives of the country or countries concerned, often on a relatively informal basis. Representatives of third countries stationed in Brussels frequently make their views known to Commission representatives, sometimes also to the Council, by submission of written memoranda or by interview (for example with the Chairman of the

COREPER). The Council and the Commission take the views of the third countries into account as far as is consistent with the aims of the common agricultural policy.[21]

(b) *Participation in international organizations and conferences.* In most cases, the Member States of the Community, and not the Community itself, are members of such organizations, but the Community is present through the Commission, as an observer, often however with the right to speak. In conferences of a particularly political nature, the Council may also be represented through the Member State holding the Chair (as in the recent Nairobi session of UNCTAD IV). In advance of international conferences, and during the sessions, the Community seeks to coordinate its position and to establish a common brief for its spokesman.

There are some cases in which the Community participates as such, though Member States retain voting rights (e.g. the International Wheat and Sugar Councils). The Conference on International Economic Cooperation (the "North-South" dialogue) is an exceptional case in which the Community participates and not Member States.

(c) *Formal negotiations with a view to the conclusion of trade or association agreements.* The general principle is that, on the basis of a proposal from the Commision, the Council establishes negotiating directives. The task of conducting negotiations is attributed to the Commission (Articles 113 and 228), though when a political responsibility is at stake, the Council also is closely involved). It is for the Council to conclude any agreement (Articles 114 and 228) (except in cases falling within Commission competence): the final act may be signed on behalf of the Council by two plenipotentiaries, e.g., the President of the Commission and the President of the Council. If the agreement covers matters which are within the competence of Member States, then representatives of the latter will also sign the agreement.

The three examples set out below fall within this latter category.

(i) *GATT.* The individual Member States of the Community are contracting parties to the GATT; as part of a customs union, they are exempt from the general "most-favoured-nation" clause. In some Gatt matters (such as the admission of new Contracting Parties), the EC Member States continue to act individually. However, Article 113 of the Rome Treaty makes it clear that commercial policy is a Community-wide responsibility: on tariff and trade matters, therefore, the Community must act collectively. The Council has to define a mandate for negotiations, and to conclude any acts which may result. Article 113 entrusts the Commission with the responsibility for conducting negotiations, and sets up a special committee, known as the "113 Committee", to ensure constant coordination of the Community's position before and during negotiations (this Committee is composed of top-level officials responsible for trade matters from the Member States.

A Community mandate for the current round of multilateral tariff negotiations, formally opened in Tokyo in September 1973, was adopted by the Foreign Affairs Council in February 1975. As regards agriculture, the

Community insists that its common agricultural policy cannot be called into question during the MTN's: since virtually all governments intervene extensively in agricultural markets and trade, agriculture should be dealt with as a special subject, and in a special group in GATT. The mandate emphazies the role of commodity agreements to stabilize world trade in major agricultural products:these would involve minimum and maximum prices together with stock-piling arrangements for cereals, sugar and rice, and minimum and maximum price agreements for dairy products. In the negotiations conducted in Geneva so far, the Community representatives have explained these proposals in more detail. Progress has however been slow: U.S. position is different from that of the Community on both procedure and basic issues.

(ii) *The "overall Mediterranean approach".* Over the years, the Community of Six had concluded trade or association agreements with a number of Mediterranean countries. In 1972, when these agreements had to be adapted to the Community of Nine, it was decided to adopt an "overall approach", with the aim of gradually removing barriers to trade, taking account however of the degree of development and the competitive strength of the industries of the countries in question: as regards agriculture, the Community would grant preferential treatment for the exports of the Mediterranean countries, so far as consistent with satisfactory functioning of market organizations within the Community.

On this basis, negotiations were launched in 1973 with Israel, the three Maghreb countries (Algeria Morocco and Tunisia) and with Spain. In the first half of 1975, agreements – involving tariff concessions by the Community on agricultural products – were reached with Israel and were nearly completed with the Maghreb. However, the concessions on products which are important also for the Mediterranean regions of the Community were causing growing concern to producers in these regions. Already in June 1973 this had led to a number of decisions by the Agricultural Council aiming to safeguard Community production of wine and of certain fresh and processed fruits and vegetables. In May and June 1975 the implementation of the agricultural concessions to Israel, and the completion of the whole negotiations with the Maghreb countries, were held up while the Agricultural Council debated further measures to protect Community producers. The Council agreed in June on a number of measures to this effect. The case of wine was particularly difficult, because of the pressure of excess wine production within the Community itself: it was agreed that wine imported from the Maghreb countries should respect the Community's "reference price", otherwise the full duty would be applied. It was also agreed that if wine imports from the Maghreb were to disturb the Community market, there would be a special distillation of wine to relieve the pressure.

Subsequently (in April 1976), the agreements with the Maghreb countries have been signed; negotiations with the Mashrek countries (Egypt, Syria and Jordan) began early in 1976. The Council has authorized the opening of negotiations with Greece with a view to full membership of the Community. Discussions are taking place with Turkey concerning an improvement of the association agreement with that country (particularly in view of the agricultural concessions now made to other Mediterranean countries). Discussions are also under way with Portugal and Spain.

The prospect of increased competition however continues to cause concern to Community producers.[22]

(iii) *The Lomé Convention – Relations with ACP countries.* Economic relations between the original Community of Six and 17 African countries and Madagascar were governed by the Yaoundé Conventions of 1963 and 1969.[23] Following the enlargement of the Community, negotiations were undertaken on the basis of Protocol 22 of the Act of Accession, for the extension of the arrangements to a wider group finally totalling 46 ACP (African, Caribbean and Pacific) countries.

These negotiations were officially opened in July 1973: they culminated in the signature, on 28 February 1975, of the Lomé Convention which now links the 250 million people of the Community with 250 million people of the developing countries.

Under the Convention the ACP States grant the Community most-favoured-nation treatment and have undertaken not to discriminate between Member States of the Community. The Community grants unrestricted, duty-free access to the Community market for nearly all ACP products. In the case of products covered directly or indirectly by the common agricultural policy, the Community grants duty-free entry for those products which on entry are subject only to customs duties, and preferential treatment for those products subject to other import measures such as import levies. The Convention provides (Article 2) that the import arrangements may be modified if changes are made in Community's market organization for the products concerned, but only after consultation with the ACP countries.

The Convention sets up institutions through which such consultation can take place: it provides for a *Council of Ministers,* composed, on the one hand, of the members of the Council and members of the Commission of the Community, and, on the other hand, of a member of the government of each ACP State. This Council of Ministers has the power to take certain decisions, and may formulate resolutions, recommendations or opinions. It will normally meet once a year. There will be frequent consultation in the Committee of Ambassadors, and in various sub-Committees or working parties (e.g., for sugar). The Convention also sets up a *Consultative Assembly,* made up on a joint and equal basis of members of the European Parliament and of representatives (not necessarily parliamentarians) appointed by the ACP States. The Convention also provides for an arbitration procedure.[24]

Article 11 of the Convention specifies that consultation shall take place, at the request of the Community or of the ACP States, in relation to trade measures, preferential agreements, etc. Such consultation has already been significant in the preparation of the Community's offer on generalized preferences (which originated under UNCTAD II) and on tropical products in the context of the GATT multilateral trade negotiations.

The arrangements concerning *sugar,* to replace the Commonwealth Sugar

Agreement, played a vital role in the negotiations leading up to the Lomé Convention, because of the important interests at stake on both sides, and because of uncertainties on the world and Community markets. The Community's own sugar policy was under discussion in the Agricultural Council at the same time, and the Agricultural Council as well as the Foreign Affairs Council was responsible for the Community's position in this field. After last-minute difficulties, a Protocol on sugar was agreed upon, annexed to the Lomé Convention. It contains a Community undertaking to buy and import for an indefinite period (the Convention itself is limited to five years) up to 1,375,000 tons of sugar from the ACP at guaranteed prices. The guaranteed price is to be fixed each year by negotiation between the Community and the ACP representatives, "within the price range obtaining in the Community".[25]

In this connection it may be noted that, when negotiating the Convention of Lomé, the Community agreed that the Commission would each year consult the sugar-exporting States before submitting its proposals to the Council of the European Communities for the determination of sugar prices within the Community for the following marketing year.

The particularly favourable arrangements to be applied to imports of other agricultural products of the ACP States (e.g. beef, oilseeds, tobacco, cereals, fruits and vegetables) were the subject of a unilateral declaration by the Community annexed to the minutes of the signature of the Lomé Convention. The form of a unilateral declaration was used in order to maintain the principle that the common agricultural policy is not negotiable with third countries.

The provisions of this declaration have been applied in the case of *beef*, the main export of Botswana and an important source of revenue for three other ACP countries. When in the summer of 1974 the Community had suspended imports of beef from third countries, the Council agreed however to admit specified quantities of ACP beef (related to the highest exports in the preceding years) with a 90% abatement of the import levy. This concession has recently been extended to the end of 1976.[26]

4. FINAL REMARKS

It will be clear from the examples given in this paper that the processes by which the European Community reaches its decisions are complex and varied. This seems inevitable: the Community is a unique case of integration between sovereign States, and it is still in an early stage.

It is possible to argue that the development of the Community has not fulfilled the expectations of those who originally hoped for the creation of a supranational power. The virtual abandonment of majority voting in the Council, the development of the Council rather than the Commission as the major decision-making centre, the increasingly important role of "Summit" meetings, tend to support this view. These tendencies have had their consequences in the inevitable slowness with which compromise agreements are reached in the Council, and in the equally inevitable fact that compromise solutions may not be the best from the economic point of view. In agricultural policy, where interests are strongly divergent, these handicaps are obvious. One can also point to the difficulty which the Community often has in coordinating its position in international conferences.

At the same time, the Community is very far from being merely an intergovernmental organization. The commitments contained in the Treaties, particularly as regards the creation of a common market, the priority of Community over national legislation, the authority of the Court of Justice, the existence of a Community budget, and the achievement of a common agricultural policy, are significant features. In difficult negotiations, such as the agricultural price-fixing, the will to reach agreement has always been uppermost, even at the cost of painful concessions. In the agricultural area moreover, the executive powers of the Commission, and the practice of majority voting in the Management Committees, are not to be forgotten. And in the international arena, as the examples in the last part of this paper have shown, the Community exists in its own right and has made important contributions to the expansion of trade and to economic cooperation and development.

NOTES

[1]There are strictly speaking, *three* Communities: the European Coal and Steel Communtiy, the European Economic Community and the European Atomic Energy Community. The plural form is therefore legally correct, but it is convenient to refer to the "European Community". In 1967, under the "Merger Treaty" of 1965, the Councils and Commissions of the three Communities were merged (the Assembly and the Court of Justice were already common to the three Communities).

[2]Under Article 8 of the Treaty of Rome, the Common market was fully established after a transitional period of 12 years – i.e., after 31 December 1969. As regards the three new Member States, the transitional period will be completed by 1 July 1977 as regards the tariff on intra-Community trade and the common external tariff, and by 1 January 1978 as regards agricultural prices.

[3]Two each from each of the four biggest Member States, one from each of the others.

[4]The method of appointment is currently under discussion; it is suggested that the President of the Commission should have more scope to choose the other Commissioners.

[5]Its role was formally recognized in the "Merger Treaty" of 1965, establishing a single Council and Commission (Art. 4).

[6]For instance, an agreement in principle that the Community would continue to import certain quantities of New Zealand butter at least up to 1980 was reached, in the context of U.K. "renegotiation" of its terms of access to the Community, at the Dublin "Summit" of 10 March 1975; it was finally implemented by the Agricultural Council on 22 June 1976.

[7]A qualified majority consists of 41 votes out of 58, the weighting being as follows: 10 each for France, Germany, Italy, and the U.K.; 5 each for Belgium and the Netherlands; 3 each for Denmark and Ireland; 2 for Luxembourg.

[8]The Rome Treaty refers to the "Assembly", but it has become customary to call it the "Parliament".

[9]The date for the introduction of direct elections, and the numerical composition of the directly-elected Parliament, are under discussion at the time of writing.

[10]The Rome Treaty does not give the ESC quite the same status as the other Institutions (cf. Article 4, and Articles 193–198).

[11]The "Charmasson case" is particularly significant. On the basis of a plea by an importer of bananas against the application in France of import quotas under a national market organisation system, the Court ruled (on 10 December 1974) that the existence of a national market organisation could not justify the maintenance of quotas beyond the transitional period (cf. note 2). This ruling has reinforced the need for *common* market organisation for any products not already covered by such organisation and for which trade restrictions remain (in particular sheepmeat and potatoes).

[12]For example, the Resolution of 25th May 1971 on "new guidelines for the common agricultural policy" (Official Journal No. C 52 of 17.5.71), which was the outcome of the "Mansholt Plan" debate and laid down the basis for subsequent structural policy (implemented mainly in three directives of 17 April 1972 – O.J. No. L 96 of 23.4.72).

[13]The French initials "FEOGA" are more pronounceable.

[14]In the 1976 budget, 5,485 million UA out of a total of 7,449 million. (One UA currently equals US $1·24.).

[15]The basic text defining the activities of the FEOGA is Council Regulation No. 729/70, Official Journal No. L 94/13 of 28.4.70.

[16]Besides the Management Committees for each commodity sector subject to common market regulations, there are similar Committees for the FEOGA, for structural policy and certain other matters. Apart from formal consultation according to the procedures described above, the various Committees permit the exchange of information and views.

[17]This procedure was applied in May–June 1976 over the issue of export restitutions for malt. To avoid excessive payments, the Commission had proposed (in particular) to shorten the period of validity of the export certificates. This proposal was rejected by a majority in the Management Committee and referred to the Council, which finally approved by majority vote a somewhat different scheme with less stringent conditions.

[18]It was under this procedure that the suspension of beef imports was decided by the Council in July 1974.

[19]i.e., farms obtaining approximately an income comparable with other sectors.

[20]This Article refers to "appropriate" relations: in practice it permits a Community presence (through the Commission) at meetings of the UN and its specialized agencies, of the OECD, etc.

[21]The suspension of beef imports has been somewhat relaxed since 1974, though the state of the Community beef market has not so far made it possible to admit imports freely once again; new arrangements for beef imports are at present under discussion. In the case of the milk powder incorporation scheme (which is temporary and limited to 400,000 tons), the Council adopted on 23 June 1976 a complementary scheme by which protein products (i.e., mainly soya) can be stored, to reduce the risk of a falling-off of imports. The decision, also on 23 June last, on New Zealand butter imports is also a case in point.

[22]The problems involved were studied in detail in a report by the Economic and Social Committee in November 1975, "Current Prospects of the Agricultural Products of the EEC's Mediterranean Area". They were also discussed at a "Colloque Méditerranéen Franco–Italien" held in Paris in October 1975 by producers' organisations. One of the political groups in the European Parliament has in May 1976 proposed a Mediterranean Agricultural Conference.

[23]Mauritius subsequently became a party to the 1969 Convention.

[24]The Convention entered into force on 1 April 1976. The Consultative Assembly met for the first time on 2–4 June in Luxemburg; no date has been fixed for its next session, although it is due to meet at least once a year. The first meeting of the Council of Ministers is due to take place in Brussels on 14–15 July.

[25]The negotiations on the price for 1976/77 were finally concluded, after considerable difficulty, on 5 June 1976 and endorsed by the Agricultural Council on 23 June.

[26]Other important aspects of the Lomé Convention, such as the scheme for the stabilization of export earnings ("STABEX"), and the provisions for financial and technical cooperation, fall outside the scope of the present paper. The latter will be dealt with in the paper to be presented to the IAAE Conference by Mr. Van der Vaeren of the Commission.

DISCUSSION OPENING – P. Gajo, *Italy*

Mr. Tracy illustrates the dynamics of the EEC structure, influenced by its enlargement from six to nine countries and by the variations in background conditions, and we appreciate the synthesis he presents. He places particular importance on the principle of respecting the Sovereignty of member States and putting into action a policy which can modify the internal structures through step-by-step, and agreed, actions.

We are concerned with the activities of the EEC in the field of decision-making relating to agricultural policy and, in particular, price-fixing. This delicate aspect is emphasized by Mr. Tracy when he says, "The annual price-fixing exercise is a crucial aspect of the Common Agricultural Policy". The ups-and-downs of EEC life have put some constraints on the enthusiasm of the decision-makers and on public opinion. This has been particularly important in recent years, owing to the monetary disturbances.

A consequence of the varied economic and structural backgrounds of members has been the introduction of two tools of agricultural policy, namely, (a) the creation of the compensatory accounts and (b) the adoption of different rates of exchange with respect to the unity accounts. These two new elements have, in fact, weakened the spirit and the scope of the Treaty of Rome.

Thus it seems clear that in order not to have decisions made under political constraints it is, at present, necessary to set up a common monetary policy without EEC currency fluctuations, if we want the Community to return to its original line, for its own welfare and that of all countries of the world. Up to now efforts in this direction have shown unoptimistic results, but we hope that all those responsible at various levels will find the way to put the EEC on the right track.

Apart from general problems, other points of no less importance are under discussion in Brussels. For example, in the field of equalisation of income between agricultural and non-agricultural sectors, action on structure is becoming very important. In this respect we should note the recent directives in sectors of vital importance, such as mountain areas, farm reform, vocational training, etc. These directives have been achieved by the EEC members under their respective policies, at different times and by different measures, even if sometimes in conflict with reality. The first enthusiasm, as we have noted, is fading owing to certain facts, such as the "integrative premiums", which lead to some bottlenecks – even frauds. On the other hand, other facts lead to a certain optimism such as the intervention policy for sugar beet, characterised by guaranteed and profitable prices up to a certain point beyond which the price follows the market.

But apart from these general problems we think that, as IAAE is now meeting for the first time in Africa, we should devote some of our time to discussing the ACP association with EC especially from the point of view of decision-making, which is the theme of our conference.

Mr. Tracy has pointed out all the main aspects of the Lomé Convention and on his exposition I suggest that some topics could be put on the agenda for our conference, among them – the criteria for other associations to the Convention, which have not yet been decided in Brussels. We can envisage that, as other countries get independence, they will want to be associated with the EEC. Other issues might be the relationships between ACP countries and other developing countries not associated, the eventual impact of EEC enlargement on the Lomé Convention – as we know, Greece has asked to become a member of the Community. We understand that difficulties arise among ACP countries, and between them and Brussels, because of their association, but we have to remember that even among EEC members difficulties arise from the agreements which have been signed.

DISCUSSION OPENING – D. Hussey, *Australia*

My comments will strongly reflect the fact that I am an Australian. Not only are there deficiencies in my understanding of Community agriculture and agricultural policy, but there are also a number of differences between the policies and objectives of Europe and those of agricultural exporters elsewhere – such as Australia. The latter point is illustrated by the last phrase in Mr. Tracy's paper which reads that the "Community . . . has made important contributions to the expansion of trade and to economic co-operation and development". I would strongly assert that the extent to which you agree with that statement will be markedly influenced by where you sit.

I would first like to make a general point which I consider to be of relevance and importance to the discussion. Intervention in agricultural production, pricing and trade is not a sin confined to the European Community. It occurs world-wide. Every country, or group of countries, faces the need to strike a balance between national aspirations and objectives and international objectives and obligations. To do this some form of decision-making process is required. These processes will inevitably have to take into account sectional interests and conflicting pressures and will always involve subjectivity and a large portion of the *ad hoc*. They are decision processes which the economist and econometrician will find very difficult, if not impossible, to model. By their very nature they seem likely to ensure that the formulation of objectives and movements towards these objectives will be very slow. For these reasons my sympathies lie particularly with the European bureaucrats and economists who are responsible for designing and managing effective decision-making procedures. Not only is it necessary to resolve intra-country differences and objectives but it is additionally necessary to achieve consensus among all the countries which compose the Community. However, while you have my unqualified sympathy you do not necessarily have my support.

Turning now to the paper I congratulate the author on this detailed outline of the extremely complex decision-making machinery. My knowledge and understanding of the procedures have certainly been enhanced. So also have my doubts and suspicions about its effectiveness and efficiency.

However, in one or two respects the paper does not quite go far enough. At least some brief reference is needed to the major objectives underlying this decision-making machinery. I, personally, would also be interested in the more detailed aspects of how the fundamental and long-term agricultural objectives are formulated and how it is decided whether or not they are pursued. On a solely pragmatic basis it might be suggested that from the point of view of a country like Australia the actual decisions are of far more relevance and importance than the machinery that produces them. However, third countries have a legitimate right to attempt to influence European decisions and the exercise of this right is dependant on a good understanding of the procedures available for consultation with third countries.

My final observation requires some expression of disappointment at the lack of critical comment. Being a bureaucrat myself I most certainly appreciate the limitations and restrictions that often surround self-criticism and criticisms of our masters. However, I am presuming that the lack of critical comment is not implicitly reflecting the views that all is well with the system.

As you might expect, however, I am only too keen to be critical of the decision-making procedures and the decisions they produce. As an outsider from a country affected by these decisions my overall conclusion is that there must be a better way – a better way of achieving Europe's objectives at a lower cost to Europe and internationally.

I acknowledge that such a statement is largely subjective and judgemental and sweeps under the carpet all the complexities and rigidities which impinge on the decision-making procedures. I have already admitted to an incomplete understanding of these problems. However, I remain firm in my view until convinced otherwise. It seems to me that there are many changes and improvements which could be made to both policy objectives as they are practised, and the procedures for formulating and implementing the objectives, which would benefit both Europe and affected third countries.

From what I remember the original objectives of European agricultural policy were generally viewed as being sound and rational. The basic thesis of facilitating long-term structural adjustment supplemented by necessary income support has many similarities with the broad features of Australian agricultural policy. However, as time has proceeded it seems increasingly that something is being lost in the translation. I ask myself why?

It seems to me that the answer lies, at least in part, in the demands placed on the decision-making processes and the way these processes are forced to react as a result. Let me attempt to expand on this point.

The size and complexity of the European Community have meant that administrative and decision-making machinery have become correspondingly large and complex. This has obviously limited the ability of the machine to react rapidly and effectively to change, particularly in a manner consistent with longer term objectives. Unfortunately, the last decade has seen plenty of change in the world agricultural situation. In this situation, it would appear that the bulk of resources devoted to agricultural administration and decision-making in the Community have been involved in finding immediate solutions to current developments and problems. This has, it seems, been at the expense of ensuring that the train has remained on the long-term rails.

I acknowledge that this has not been totally the case. The recent past has seen a number of in-depth evaluations of long-term objectives and effective procedures for achieving them. However, I will need to look to other participants at this discussion for views on how effective these evaluations have been in term of concrete policies and instruments.

To conclude – Australia, like all agricultural exporters, is critically influenced by agricultural decisions and decision-making procedures in the Community. Third countries have a legitimate right to try and influence these decisions and the Community acknowledges this right in both theory and practice. However, in the final analysis, after the rest of the world has exerted what influence it can, the decisions which emerge from Europe must be taken as exogenous inputs into the planning and implementation of agricultural policies in third countries. This is so even if third countries consider that these decisions are, in the international context, unacceptably sub-optimum. It is, therefore, essential that the European Community not only formulates and clearly states long-term objectives and policy instruments, but that there is a reasonable degree of concurrence between these objectives and practice. I see that as essential to national policy development in third countries.

Have I departed from the real world in expressing this hope? I really do not know the answer to that question. I recognise the difficulties and complexities involved but I refuse to be a pessimist. If I took a particularly cynical stance I might conclude that there is little ground for optimism given that the Community is unable to agree on the relatively simple question of a single set of initials to denote the Guarantee Fund.

However, what I do believe firmly is that if these improvements I suggest do not occur, then, when we build our models in Australia for the purposes of formulating and evaluating rational agricultural policies, we will not only need to include stochastic elements for prices and climate, but also similar procedures for taking account of European agricultural developments.

RAPPORTEUR'S REPORT – H. Weindlmair, *Fed. Rep. of Germany*

Many of the discussants acknowledged the paper as giving a very comprehensive and detailed insight into the decision-making process relating to agriculture in the EEC and the institutions involved. However, as some speakers remarked, they would have expected more critical comments and an evaluation of the current state, and the future prospects, of the CAP.

The discussion centred around the following problem areas:

(1) *Long run perspectives of the CAP:*

Various discussants criticized the fact that very little is known about the long-run aims of the EEC concerning agriculture. As Hussey put it, outsiders get the impression that the large apparatus of the EEC institutions is only used for day-to-day decisions while the long-run problems are not sufficiently discussed.

Knowledge was stated to be lacking, especially in the following fields:

(a) the plans of the EEC with respect to price policy, self-sufficiency, and international trade,
(b) the prospects and aims with respect to the harmonization of different policy instruments (price-, structural-, and social policies),
(c) the problems which arise from the growing tendency in the EEC member states to transfer their structural and income problems to the EEC,
(d) the objectives for co-ordination of monetary policies,
(e) the ideas about the contribution of the EEC to the world food problem (price stabilization, storage), and
(f) the criteria for further associations with the EEC and their impact on other countries.

Answers to these questions are important not only for European agriculture but also for countries for which the EEC is an important export market.

In his answer Mr. Tracy stated that the objectives of the EEC for agriculture are set out in the Treaty of Rome and they are very similar to the aims of many other countries, e.g. to supply food to consumers at reasonable prices, to provide for sufficient incomes of producers, etc. However, during the last few years additional problems have arisen, e.g. the monetary crises, erratic world market prices, commitments of the EEC on international trade, problems with respect to structural differences, etc.

The CAP is under constant review and some recent reports on long-run policies had significant influence towards creating new measures concerning the markets for grains and milk. However, some of the problems, especially the ones with respect to structural policies, are so complex that they could not be solved so far.

With respect to the world food problem and to international trade, Mr. Tracy mentioned again that the EEC made specific proposals at the recent negotiations at Geneva. But there is still no agreement, especially with the U.S.A. Concerning the ACP countries Mr. Tracy pointed out that the Council of Ministers of the ACP–EEC countries in a meeting in July 1976 in Brussels agreed upon the association of six additional member states.

(2) Another point of discussion was the minor influence of consumers in the decision-making process of the CAP, though some discussants remarked that the consumer interests are, of course, taken into account by the national representatives in the Council. The view was expressed that in the future more emphasis might be put on consumer interests.

(3) The third main topic of discussion concerned the research activities performed as a basis for decision-making in the EEC, especially with respect to price fixing. The challenge which the problems of the CAP provide to model-builders cannot be adequately met because of the insufficient data base available.

Mr. Tracy had already pointed out that the research service of the EEC is much less well equipped than the ERS of the USDA.

In this relation speakers wanted to know:

(i) why the outlook work of the EEC is not conducted more intensively,

and why no economic advisory board has been established so far,
(ii) to what extent the implications of price decisions concerning trade, inflation, consumer prices, etc., are assessed in advance, and
(iii) which facts of development in productivity, exchange rates, etc., are taken into account when employing the "objective method" for price fixing, and why the details of this method are not published.

It was pointed out that the "objective method" is a very recent development, and it had been continuously improved through experience. Through the introduction of additional policy measures, besides price policy in recent years, a more adequate price setting activity should be possible. It was stated that the outlook work and the forecasting activities have increased considerably during the last few years. In a current research project for the Commission, forecasts of supply and demand for 50 commodities in 1982/83 are undertaken.

Participants in the discussion included: N. C. Westermarck, *Finland*; G. Schmidt, *Fed. Rep. of Germany*; P. van den Noort, *Netherlands*; R. G. F. Spitze, *U.S.A.*; D. K. Britton, *U.K.*; H. J. Block, *Fed. Rep. of Germany*; D. Paarlberg, *U.S.A.*; W. Frank, *Belgium.*

JERZY KLEER*

Economic Integration within the Framework of CMEA in Relation to Agriculture

1. BACKGROUND

Experiences of both socialist and capitalist countries prove that integration processes have objective and supra-institutional character. Processes of economic growth occuring in conditions of especially rapidly developing technical progress constitute the basis for international integration.

International economic integration is determined by internationalisation of economic and technical life in a sense that certain solutions in productive techniques may be applied in different countries. The facility of their transfer and popularisation caused a specific imitation effect in the field of production and technique, contemporary economic integration is a process taking place mainly in Sector II, that is in the manufacturing sector.

This fact assured the course of changes for integrating national economies and explains the rapid internationalisation of technical and organisational progress.

Development of modern technique and its efficient application in economic life overstep national boundaries, especially in the case of small and middle-sized countries. Contemporary development assumes and makes inevitable the use of large scale production, especially since technological development enlarges that scale significantly; contemporary technical progress is not only universal but also unusually expensive. This, in turn, makes common scientific research inevitable and requires the assurance of the possibility of those researches being applied in production. The further consequence of this is the development of co-operation in economic activities in different countries which contributes to specialisation and international division of labour. The result of these processes is an increase in economic and technical interrelations between different countries.

However, what relationships occurs in the contemporary integration process between Sector I and Sector II? The most general relationships between these sectors are as follows. The bigger the share of Sector I in the economy – with the level of economic development of the country or the integrated group thus relatively low – the bigger the significance of this Sector for the

* Warsaw, Poland.

economy. On the other hand, if the share of Sector I in the process of creation of national income and employment is relatively small, the interest in Sector I – which means mainly agriculture – is mainly of social and political character; the economic aspect is much less important. Thus the harmonious development of Sector I in the economy as a whole is in the first case a *sine qua non* for economic growth and in the second case is only an important social factor, protecting against the existence of underdeveloped sectors causing social tensions. These differences are of great significance for integrational processes. Contemporary international economic integration concerns first of all Sector II. Nevertheless, if the process starts in conditions of a rather low level of economic development of integrating countries, the integrational processes cannot exclude agriculture. They determine, on the one hand, the field for integrating process and, on the other, they define the use for integration purposes of different tools and mechanisms of an economic and administrative character.

These general remarks will be considered in detail below from the point of view of socialist integration with CMEA countries.

2. DIFFERENCES BETWEEN SOCIALIST AND CAPITALIST SITUATIONS

Though the general basis for economic integration in present conditions is the same for socialist and capitalist countries, there are differences which are significant enough to make the principles of socialist and capitalist integration differ. Because of the lack of time, I will concentrate on the different features of both kinds of integration so that their common features will not be discussed.

The feature that creates socialist integration is social ownership of the means of production. That fact that the means of production are social property has certain consequences. The most significant are the macrosocial and macroeconomic characteristics of the task of economic activity. That means that, independently of particular solutions in different CMEA countries, the macroeconomic and macrosocial approach is typical for development processes in these countries. These conditions have a direct influence on the harmonious development of the economic structure, both as a whole and seen in its different parts.

The problem of appropriate proportions between different sectors is an essential feature.

The macroeconomic and macrosocial approach to the processes of development is followed by a specific approach to the economic calculation. The macroeconomic approach is essential to the economic calculation.

Although the economic calculation should not be seen separately for the enterprise, branch and the whole country, the differences are significant. They also appear when we consider the national economy in conditions of isolation, that is, when we study economic activity and socialist economic growth of one country. They also appear when we analyse integrating groups of two or more countries. Thus it should be remembered that macroscale is not synonymous with the sum of microscales.

The next specific feature is connected with the time horizon in which we analyse the function of objective. In the socialist economy, and also in the integrated groups, the objective is maximised in a longer run than in the capitalist economy, because of two conditions. Firstly, under the conditions of social ownership of the means of production the economic surplus can be transferred from one field to another as well as from one branch to another relatively easily and without causing any social tensions. It also means that the effects of economic activity because of those redistribution mechanisms can be gained in a longer period.

Secondly, in the socialist economy the relations between social and economic activity are much more direct than in the capitalist economy. In the conditions of socialism the social aims are equal, if not superior, to the economic aims. Since the economic effects created by the fulfilment of social aims may be achieved in a longer period than the realisation of purely economic aims, so in estimating the results of economic activity in socialist countries and effects of an integrated group a longer period of time should be considered.

In the present state of economic development, economic integration of socialist countries assumes sovereignty of different national economies. So far, in the process of socialist integration, the dominating role of supranational institutions has not been accepted. Those institutions play a much less significant role than in the case of, for example, Common Market countries, and their main function is limited to branch co-operation. This aspect of economic sovereignty was explicitly accepted in the Complex Program of Socialist Integration during the conference in Bucarest in 1971. In this sense the concept of economic sovereignty and the principle according to which all basic problems have to be discussed and all solutions accepted by participating countries contrast with the so-called institutional or authoritative concept of integration, which assumes the existence of supranational organisations having the right to force certain solutions upon its members. What are the reasons for which the domination of national aims and sovereignty of those aims are specially firmly emphasised these days?

I believe this may be explained in two different ways: one concerns social ownership, which is untransferable, the second is related to the level of economic development. A developed formula of those rules may be expressed as follows:

All integrating activities are to serve further development of all participants of the integrating community, also none of the integrating countries should pay for the success of others. In this sense the transfer of national income between countries is not accepted.

Similarly, capital "means" may not be transferred in the case of the developed internal needs of the capital exporting country. This is a result of the above mentioned macrosocial motivations. Although in socialist countries the transfer of production factors is not excluded, the internal interest has been – so far – of a dominating character. Is the superior character of national interest going to be a permanent phenomenon in the process of socialist integration? It seems that it is a rather temporary rule. Assuming a temporary character for that rule it is difficult to estimate for how long it is going to be obligatory in the activity of an integrating group.

3. INTEGRATION IN SOCIALIST COUNTRIES

A great deal of literature on integration problems shows that the concept of integration is neither precise nor convertible. Generally we may distinguish three points of view.

Integration is defined as a state that is a certain final model of integrated national economies in which the integrated economies create a new value as far as the level of mutual adaptation and interrelation is concerned.

Integration as a state is at present only a hypothetical – although an important – model since it is the destination state for integrating national economies.

A modification of the concept of integration understood as a state is a concept which sees integration as an aim. This approach stresses mainly the quantitative aspects that are the basis for integration – whereas, in the case of integration treated as a state, it is mainly the character of structural relations between integrating national economies which is considered.

A different point of view, and most probably a much more significant one, is represented by a concept which treats integration as a process. In this case not only the final plans but also ways of achieving certain aims in the transition periods, leading to integration as a state or aim, can be seen. It is particularly important since in these stages a new type of co-operation between different national economies is created, new relationships appear, and gradually a new international division of works – mainly in Sector II – comes into being and, speaking more precisely, interbranch specialisation is developed. The adaptation processes are here important not only in the field of economy or technique in a broad meaning of this word, but also in the behaviour and value systems.

It may be clearly seen from experience, documents and programs of CMEA countries, that integration is understood first of all as a process, and aims in the different stages of development are being modified or even changed. Generally it may be said that in each case the tasks, on the one hand represent the economic policy of the member countries of the integrating group and, on the other hand, they result from the level of development and character of the national economies of the countries concerned.

Those tasks are each time formed by policy of the country and by the objective state of the national economy of the integrating group. Accordingly to the changes in these factors the tasks of integration are changed or modified.

In the history of CMEA different kinds of tasks have been performed and it is impossible to describe them precisely. I will concentrate on the tasks, accepted by the Complex Program of Socialist Integration, resolved in 1971 in Bucarest during XXV Session of CMEA countries. These tasks are as follows:

(1) Accelerated development of productive capacity of all member countries and achievement of the highest level of technical and scientific development;

(2) Improvement of the structure and enlargement of the scale of production according to the rules of scientific and technical revolution;

(3) Fulfilment of the increasing need of the national economies for fuel, energy and raw materials, modern equipment, agricultural products, foods, and other consumer products, mainly by increased production and national use of resources of the member countries;

(4) To increase the material and cultural level of living of the people in the member countries;

(5) Gradual unification of the levels of economic development of member countries;

(6) Development of absorptiveness and stability of the world socialist market and protection of the proper place of the CMEA countries on the world.

These tasks show, first of all, that there are many of them; secondly, they have different time horizons; and thirdly, they can be achieved with the help of different mechanisms, methods, and so on.

In this sense they have, or more precisely they may have, in the short run, a competitive character. For example, where there are difficulties or growth is limited, most important are tasks which protect the present functioning of the national economy and the longer run tasks must be postponed. In this context, special attention should be paid to the following fact: though all the tasks of Complex Program represent the interests of all countries in the community, some of them tend to favour the national interest, some the interest of the whole community. In the long run there is no contradiction, but in the short run, some differences may and, indeed, do appear.

4. METHODS IN SOCIALIST COUNTRIES

Mechanisms in a system of socialist integration vary but roughly may be divided into two groups: direct mechanism – that is planning, and indirect mechanism – that is goods/money mechanism.

The basic mechanism of socialist integration within CMEA is the planning mechanism which, however, should not be understood as one universal plan but, firstly, as a group of undertakings, in co-operation and co-ordinated with the planned development of different national economies; and, secondly, as a common planning undertaking, for example common international planning. This sort of activity has been recently undertaken on the wider scale. International planning must not contradict national plans of economic development, that is because of the previously mentioned principle of national economic sovereignty.

As direct mechanism we can classify: long term prognosis, economic consultations, co-ordination of economic development plans, international co-operative relations, bilateral and multilateral specialisation relations, and long term trade agreements. Direct mechanism has administrative and state sanctions.

The money-market mechanism in the integrated group of CMEA countries is inferior to the planning mechanism both in the sense that it cannot contradict the planning mechanism and also that its effects are mainly included within the relations of plan of social and economic development. However, this mechanism has its own independent functions. First of all some attention should be paid to the creation of prices.

The price system in the integrating group of CMEA countries has a dual character. There is a difference between home market prices and foreign market prices and there is a lack of relation between them. They are set according to different rules. We may neglect the home price system.

I would only like to stress that in different countries they are set in different ways. Thus prices for individual goods are not comparable.

On the other hand prices that are obligatory in foreign trade exchange between CMEA countries have, and had, an objective base in world market prices. World prices are the starting point for creation of prices in the CMEA market. However, they are not exact world market prices. Modification not only implements stabilisation over a longer period – usually five years – but also the smoothing of conjunctural fluctuations, and above all there is the fact that these prices are only a base for negotiations between individual countries.

In exchange between CMEA countries up to the end of 1974 so-called "stop prices" were obligatory. This means that for the five year period a price was set which was the average world price in the previous five year period. For example, in the years 1971–1975 only up to 1974 the bases were corrected contract prices from the 1965–69 period. Modification of the "stop price" system took place in March 1975 when all CMEA countries accepted the so-called "walking-price" system. The idea of this system is that for each year average world market prices from the last five year period are obligatory; for example: average prices from the period 1971–75, for the year 1975; average prices of 1972–76, for the year 1977; and so on. This system is, and will be, a base for price setting for all agricultural and food products exchanged between CMEA countries. In spite of some advantages – price stability – this system has also some faults, for example it is not very flexible. Some exporters of agricultural products think the home prices are a better base since world prices do not secure enough profit for their country – for example, due to poorer technical equipment. The formation of a proper price system still remains one of the most important tasks for socialist countries, integrating in CMEA.

One of the specific features of the present stage of socialist integration of CMEA countries is a lack of transferable currency. None of the internal currencies is transferable, that means they function only within national economies, and none of the currencies fulfil functions indispensable for an international currency. The base for the calculations between CMEA countries is the transfer rouble, established in 1964 by International Bank for Economic Co-operation between CMEA countries. This unit is used in calculations between CMEA member-countries and actually only in the case of articles exchanged within quantity or value quotas established by bilateral agreements. On the other hand, the surpluses exchanged are calculated on a different basis – sometimes even a transferable currency is used. The lack of transferable currency is a significant shortcoming in the integration processes. According to integration plans formulated in the basic document, transferable currency is going to be introduced in the future.

In the present stage of the socialist integration of CMEA countries the economic institutions and multinational enterprises do not play a really significant role; multinational enterprises especially have only a marginal role. In the process of integration multinational enterprises play only a subsidiary role because the basic decisions are made in the central decision centers at the whole country level. Multinational organisation, similarly to multinational enterprises, are controlled – either directly or indirectly – by central decision centers in individual countries. This, among others, also differentiates socialist and capitalist integration. For capitalist integration, from the very beginning, develops together with supranational organisation and enterprises. However, in the future a more significant role for those institutional forms is planned; though at present it is very limited.

5. AGRICULTURE IN SOCIALIST INTEGRATION

Having in mind the above conclusions I would like to discuss the place and role of agriculture in the socialist integration of CMEA countries. There are three factors determining the function of agriculture. The first is connected with the achieved development level of the integrated group; the second is a result of the minimal economic complex in the national economy – and similarly in the integrated group; and the third is formed by the present and long term tasks of the integrated group.

CMEA countries have already achieved a high level of development, measured by GNP per capita, but agriculture still plays an important role in the economy of those countries, and this fact has a significant meaning for our considerations. Let us take a few data that show the role of agriculture.

TABLE 1. *The share of agriculture and forestry in production of national income (material product) in CMEA countries*

Country	1960	1970	1974
		per cent	
Bulgaria	32·2	22·6	20·7
Czechoslovakia	14·7	10·1	9·8
German Democratic Republic	16·4	11·6	10·6
Poland	30·3	17·5	14·5
Romania	34·9	19·1	16·4
Hungary	30·8	17·7	16·6
Soviet Union	20·7	22·0	18·3

From: *Statisticzeskij jezegodnik stran czlenow SEW 1975*, Moscow 1975 page 42

Table 1 leads to the following conclusions. During the last 14 years the share of agriculture in the creation of national income decreased significantly. The strongest downward trend appeared in the less developed countries. The decrease in the share of agriculture in national income creation, however, was not proportional to the GNP per capita. The differences, measured in points, in 1974 were much smaller than in 1960. The dispersions decreased from 20·2 points to 10·9 points, but still were large. Actually the share of agriculture in the creation of national income was somewhat higher than the figures included in Table 1 suggest, for in CMEA countries prices of agricultural products are relatively low compared to industrial products. If we add some additional points to give an indication of the share of agriculture in the creation of national income, then the significant role of agriculture in the economy of CMEA countries becomes clear.

An additional argument for the above conclusion may be provided by the Table 2 showing the share of agriculture and forestry in employment.

TABLE 2. *Employment in agriculture and forestry as a proportion of employment in CMEA Countries*

Country	1960	1970	1974
		per cent	
Bulgaria	55·5	35·7	30·1
Czechoslovakia	25·9	18·3	15·7
German Democratic Republic	17·3	13·0	11·5
Poland	44·2	34·7	31·4
Romania	65·6	49·3	40·0
Hungary	38·9	26·4	23·3
Soviet Union	38·7	25·2	23·4

From: as for Table 1, pages 393, 395–396.

The comparison of employment in agriculture and forestry with the share in the creation of national income clearly shows a significant difference between individual CMEA countries. Only in the case of the German Democratic Republic and the Soviet Union is there a corellation between the share of agriculture in the creation of national income and the share of this sector in employment; in all other cases – and especially in Poland – the differences are significant. This comparison shows the level of technical equipment in agriculture compared with that in other sectors. This indicates a very general task; that is, intensification of agriculture mainly by increasing the supply of industrial investment goods for this sector.

The share of agriculture products in the total spending of people is large, although during the last few years significant changes have taken place. If we take the share of foods in total sales – services excluded – as a measure, we will see that for the year 1974 this was: in Bulgaria – 43·6 percent; Czechoslovakia – 47·4; German Democratic Republic – 47·4; Poland – 43·1; Romania – 49·6; Hungary – 45·5 and Soviet Union – 54·2.

Though these figures do not show precisely the share of food in the total spending of consumers, they are characteristic enough to show the weight of this group of products in peoples' spending in CMEA countries. So one of the main tasks of CMEA countries, both for individual countries and for the whole integrated group, is to provide a sufficient supply of foods and food products both as to quantity and quality.

Though in CMEA countries both production and consumption of foods per capita is much higher than the world average, it still does not meet the desirable needs. This to some extent determines the production policy and even economic policy in the field of agriculture in individual CMEA countries and within the whole group.

6. DEGREE OF SELF SUFFICIENCY IN CMEA COUNTRIES

Agriculture in CMEA countries is relatively homogenous and significant differences appear only in the case of some districts in the Soviet Union. Agriculture in CMEA countries does not produce tropical products. There are, of course, some differences between agriculture in countries like Bulgaria, Romania and Hungary and countries like Poland, Czechoslovakia and the German Democratic Republic but they are not very significant since in all those countries agriculture is of the same climatic type.

Although the agriculture in CMEA countries reaches world average levels, the supply of agricultural products does not meet the existing demand for non-tropical products.

CMEA as an integrating group is a net importer of agricultural products – both in the form of raw materials and of food products. The figures in Table 3 relate to only two years, 1960 and 1974, but they show the trend characteristics for the 60's and 70's, since there is no reason to expect any changes until the end of 70's. CMEA as an integrating group will not become a net exporter of foods.

TABLE 3. *Import and Export of Agricultural Products – Raw Materials of Agriculture, Forestry and Breeding, and Products of Agriculture and Food.*

	Import		Export		Balance	
	1960	1974	1960	1974	1960	1974
Bulgaria	105·5	666·1	331.8	1127·2	+ 226.3	+ 461·1
Czechoslovakia	673·3	1675·0	200·7	644·6	– 472.6	– 1030·4
G.D.R.	843·8	2367·4	130·3	803·1	– 713·5	– 1564·3
Poland	506·8	2129·3	305·0	1115·0	– 201·8	– 1014·3
Romania	119·2	965·1	256·4	1317·3	+ 137·2	+ 352·2
Hungary	277·7	1199·5	239·5	1315·0	– 38·2	+ 115·5
Soviet Union	1333·8	6380·4	1518·7	5026·2	+ 184·9	– 1354·2
	3860.1	15382·8	2982·4	11348·4	– 877·7	– 4034·4

\+ surplus of export over import
– surplus of import over export
From: Author's calculations based on: *Maly Rocznik Statystyki Miedzynarodowej 1975* Warsaw 1975, pages 182–3, 188, 191.

Although the integrating group as a whole is a net importer of agricultural products the situation in individual countries differs significantly. In 1960 three countries (Bulgaria, Romania and the Soviet Union) were net exporters. The situation in 1974 was similar except that Hungary became a net exporter and the Soviet Union a net importer. The value of the deficiency increased on the scale of the whole group five times but on the other hand the share of agricultural products and food products in the total trade turnovers decreased from 26 percent of total foreign trade turnover in 1960, to 20·2 percent in 1974. Simultaneously the deficit caused by import of agricultural products, which in 1960 reached 6·7 percent of total imports, decreased in 1974 to 6·0 percent.

7. TRENDS IN THE INTEGRATING PROCESSES

The above considerations concerning agriculture, concentrate mainly on the role of agriculture in the national economy, less on the integration processes. Of course, the trends of the integrating processes will be described here very briefly since the description of total activities of CMEA is impossible within one paper. To begin with, I would like to present a general thesis, namely, that the processes within CMEA concern branches complementary to agriculture, much more than agriculture itself.

Michael Kalecki, many years ago, said that foreign trade always constitutes a base for integrating processes. That is because, firstly the structure of exchange shows the possibilities of future specialisation, and, secondly, foreign exchange strengthens certain potential development trends. Because of that, the description of the present exchange of agricultural products and prognosis prepared for this field should be useful. Table 4 shows the share of agricultural products in the total exchange between CMEA countries in the years 1960 and 1970 and the prognosis for 1980.

TABLE 4. *The Share of Agriculture and Food Products in Total Exchange between CMEA countries in 1960–1970 and Forecast for 1980*

	1960	1970	1980	1960	1970	1980
			percent			
CMEA	12·5	9·7	5·4	12·5	9·7	5·4
Bulgaria	5·3	5·4	1·7	48·5	39·7	22·3
Czechoslovakia	22·1	11·1	7·8	3·3	1·6	1·0
G.D.R.	23·9	11·9	8·5	0·7	0·8	0·4
Poland	11·6	9·4	4·6	5·2	4·3	4·2
Romania	4·4	4·5	1·8	18·0	12·7	9·4
Hungary	5·7	4·3	1·9	15·6	18·7	12·6
Soviet Union	8·3	10·8	5·6	17·2	8·7	4·3

From: M. Bogacka. I. Cieniuch, T. Leszek, K. Soczewica, *Foreign Trade in CMEA countries in the years 1950–1980.*

8. FOREIGN TRADE IN CMEA COUNTRIES IN THE YEARS 1950–1980

The common feature of foreign exchange in agricultural and food products between CMEA countries is the fact that the share of this group of articles in total exchange is decreasing. The forecast for the year 1980 also shows that exchange of this group is going to be relatively small. However, Table 4 shows that for some countries this kind of export is going to play a significant role – for Bulgaria, Romania and Hungary. Similarly, import of this group of articles will be of significant value for the German Democratic Republic and Czechoslovakia. The data showing the shares for some countries show a certain specialisation in this field. Though, for the whole integrating group this specialisation cannot be well developed, in the case of individual countries it could be of a great significance. Also the possibilities of agricultural import from the integrating group may be important for some countries.

As I mentioned above, the planned activities are the main mechanism for the integration processes. Among them, co-ordination plays an essential role. In the present development of the integration of CMEA countries, co-ordination of plans took place in the periods: 1961–65, 1966–1970, 1971–1975, and 1976–1980.

For the latter period the main tasks to be achieved by co-ordination of plans are: protection of demand for selected seeds, cattle for breeding, veterinary preparations, biological herbicides, special vehicles for agriculture, fertilizers and some mechanical equipment.

For agricultural food products the most important need is the creation of an international cold storage system and an international system for other forms of protection of food products. In the machinery industry, the base for co-ordination of plans is the demand forecast for tractors, agricultural mechanical equipment, amelioration machines, forestry machines, etc.

Forecasts for the agricultural demand for chemically and biochemically improved breeding stock have been prepared. For the food industry the forecast concerns development of, amongst others, meat, dairying, tinned food production, sugar manufacturing, and the tobacco industry.

In the present practice of co-operation within the CMEA group the most steps have been related to supplies of selected wheat varieties, cattle for breeding, and so on. In the years 1971–75 agreements concerning supplies of selected seeds (78 kinds) were signed. Supplies of cattle for breeding increased significantly – by more than 6 times – compared with 1966–1970 (pigs, almost 5 times; sheep over 5 times) etc. Over 200 different kinds of plant protection preparations were introduced. Another extremely significant form of activity is standardisation of agricultural and forestry mechanical equipment. The first international standards for machines was accepted in 1961. It consisted of 1103 typical sizes of machines. A second one, accepted in the years 1964–1965, consisted of 1133 typical sizes of machines for agriculture and forestry and the third system, accepted in 1971, presented over 700 types of machines and 1750 typical sizes.

Multilateral specialisation agreements relating to veterinary preparations and 12 most important preparations for agriculture were signed too.

An important, although subsidiary, role is played in agricultural development by common scientific research. For example, recently 6 multilateral agreements concerning food research were signed. Apart from common scientific research there is also a highly developed system for the exchange of the results of research carried on in individual countries. The most important results achieved in one country are demonstrated to the others.

9. CONCLUDING REMARKS

CMEA countries are a great producer of agricultural products; especially wheat, root crops, industrial plants and meat products. Although CMEA is never going to be self-sufficient in agricultural products, there is a possibility of ensuring, in these countries, a sufficient supply of those basic products that can be produced in the climatic zone. This fact has certain consequences, both for state policies and policy of the whole CMEA group.

Economic policy in agriculture for CMEA countries may be described by a few main features. Firstly, because of the developed demand for agricultural products in the home market, agricultural and agricultural policy are going to play, in the long run, not only a social role – for example protecting against undue disproportion between sectors – but also an economic one. Therefore, in the integrational processes agriculture must be concerned from the point of view of agricultural intensification and growth of agricultural product. That means attention must be paid to activities for increasing the productivity per ha and per employee. This is an internal task of each country and also a task of the whole integrating group.

Secondly, since CMEA countries have developed productive possiblities in the field, then of course more attention will be paid to agriculture in state programs than in the programs of the whole group. This is because integration at present concerns mainly Section II, and intrabranch specialisation. It is also because the possibilities for specialisation in CMEA agriculture, apart from the Soviet Union, are limited. Thus agriculture to a large extent will remain in the sphere of state activity (economic organisations) which deals with minor economic complexes, producing mainly for the home market, and, what is more important, having the proportions and connections determined by an internal economy.

Such conclusions can be reached to the present economic policy of CMEA countries. Most probably the present character of the policy will remain for the next two years, that is over time horizon that we may now rather precisely define.

Thirdly, integrating activities will take place mainly in branches complementary to agriculture (science, development research, agricultural machinery industry, chemical industry and so on). In these spheres integrating processes will be much stronger than in agriculture itself. Integrating processes appearing in agriculture will play only a subsidiary role, while the main responsibility for protecting the supply of agricultural and food products as thoroughly and for as long as possible will be that of the food and agricultural sector of each country.

V. NAZARENKO *

Integration of CMEA Countries in the Field of Agriculture

The rising and development of the world socialist system has brought into existence a new type of relationship between the countries where socialist society has been built or is shaping. These relationships are based on the principles of internationalism and dialectical unity of interests of the whole socialist system and every particular country of the social community.

For over a quarter of a century the Council for Mutual Economic Assistance has been playing successfully an organizing and coordinating role in cooperation with socialist countries in the field of economics, science and technology.

The CMEA was established in 1949 by the countries concerned. This organization, including its machinery, numerous economic establishments, and financial bodies, is not a superstructure. Its decisions are made at all levels by the interested countries and are not imposed from outside.

The general process of economic internationalization, typical of the present stage of social and economic development, has evoked a real necessity for economic integration of the countries included in the world socialist system. Socialist economic integration is a systematically regulated process of bringing together and mutual adaptation and optimization of national structures within the international economic complex, as well as establishing deep and stable relations in the leading branches of industry, science and technology. Internationalization of industry and economic integration obtains its full and complete development only under socialist conditions. The reason lies in the uniform economic basis of socialist countries, i.e., social property for means of production, the same type of State system (the power in the hands of people headed by the working class), and common ideology (the theory of scientific communism).

The accomplishment of socialist agra-reorganization in the majority of socialist countries, concentration and specialization of production and industrialization of its main branches have served as the base for the development of integration processes in the agriculture of CMEA countries. The large-scale socialist agricultural production clears the way for rationalization of

* All Union Institute of Agricultural Economics, Moscow, U.S.S.R.

reproductional ties between a number of economic branches and the establishment of national-economic agroindustrial complexes (AIC), linked by a common task of producing different kinds of final products from agricultural raw materials. The establishment of the national-economic AIC in CMEA countries has expanded the boundaries of economic integration which gradually involves the whole sphere of agra-industry, as far as internationalization of agriculture itself has very narrow limits.

Forms and types of CMEA countries' co-operation in the field of agriculture as well as integrational processes in the sphere of agra-industry have a number of peculiarities as compared with industrial production and other national economic complexes. These are as follows.

First, the development of integration in agriculture is more difficult than in industry, mainly because of its dependence on soil and climatic conditions and other factors which cannot be reproduced, as well as on seasonal possibilities of production which are different for different crops.

Second, the development of integration between the CMEA countries in this field suffered some difficulties for a long time, because of retarded agricultural development which was connected, to a great degree, with inability at a definite stage to allot sufficient investments and material resourses, to relatively low progress in agricultural machinery construction, chemical industry, and so on.

Nevertheless, co-operation between CMEA countries in the field of agriculture has started its development as far back as the end of the Forties. It was characterized, chiefly, by mutual help in agricultural products, seeds and farm machinery. It was necessitated by the need of rapid restoration of agricultural production which was badly damaged in a number of countries during the World War II. At that period the Soviet Union delivered to these countries grain, cotton and farm machinery in great amounts, despite the fact that our country itself had considerable difficulties connected with the post-war restoration of its economy. In the course of 1945–1955 alone the Soviet Union delivered to the CMEA countries over 100 thousand tractors of different types.

Establishing a Permanent Agricultural Commission within the CMEA framework has played an important role in expanding and deepening cooperation between the community countries. The Permanent Agricultural Commission's activity promotes, to a great extent, better organization and efficiency of bilateral and multilateral relations between the countries.

The main task of the CMEA Permanent Agricultural Commission lies, first of all, in furthering economic relations between the countries and rendering assistance in planning the development of agriculture and other AIC branches, deepening the international socialist share of labour, speeding up technical progress and increasing labour efficiency in the AIC branches, so as to meet most fully the needs of their population in high-quality food products and those of the industry in agricultural raw materials. The countries are represented in the CMEA Permanent Agricultural Commission by their leading agra-specialists. National Departments of the Commission are headed by the ministers of agriculture (in the majority of countries the ministers of agriculture and food industry). Within the Commission there have been organized Permanent Working Groups of Experts whose task is to coordinate cooperation in some subdivisions of agriculture. The Permanent Agricultural Commission members come to their meetings twice a year and Working Groups, as a rule, once a year.

Cooperation in the field of science is of high importance for the development of agricultural production and allied industries in CMEA countries.

Joining the efforts of scientists from several countries has made it possible to decrease the time necessary for solving scientific or technical problems, and to obtain desirable results with smaller material and financial investments.

Scientific and research work in the field of agriculture is conducted according to coordinated plans on scientific and technological problems, which are adopted for every five-year period. From 1970 to 1975 alone, 26 large-scale problems and 86 topics have been worked out at 187 participating research institutions.

Cooperation on selected problems has been carried out according to comprehensive research programmes and is coordinated by agreements, treaties and other documents.

Up to the present time the CMEA countries have signed 6 special agreements on cooperation in conducting important researches in the field of agriculture, and two in the food industry. Specially organized coordinating centers have stepped in to work on the problems in the context of the agreements.

It is impossible to isolate a scientific area of cooperation from a productional sphere and this makes science a real productional force and necessitates its close relation with production.

Active assistance in farming intensification, increasing plant production, by all possible ways, is one of the major lines of cooperation in science and technology between the CMEA countries.

Joint efforts in research and practical work are of great importance. They are aimed at improving soil fertility, developing progressive techniques of growing the grain and improving its quality, organization of breeding work, seed production and new variety testing, exchange of high-yielding seeds.

Active introducing into production of high-yielding crop varieties and hybrids is being promoted by regulated systematic exchange of certified seeds in agreed volumes.

For the years 1971–1975 mutual deliveries of certified seeds and planting material of 78 crops were coordinated.

In 1973 an agreement was concluded on multilateral international specialization in producing certified seeds and planting material of the main farm crops.

International seed testing of promising varieties and hybrids of the main farm crops has been conducted since 1961 in the CMEA countries, and the best varieties have been recommended for the countries where they yield better than local ones. It is evident, from practical experience, that the use of varieties and hybrids, approved in international testing, results in harvesting on average 10–15% higher yields compared with the native standard ones.

Over 160 varieties of farm crops have been recommended for definite regions after international variety testing in the CMEA countries.

A number of varieties has obtained the most wide distribution, these are: Soviet varieties of winter wheat Bezostaya-1, Mironovskaya 808, Avrora and Kavkaz; Czechoslovak and GDR varieties of spring barley Waltizki and Alsa; Soviet varieties of sunflower Peredovik, Mayak, Smena; and GDR and Poland varieties of sugar beet and potatoes. The total area sown to crops of varieties and hybrids developed by the breeders from other CMEA countries exceeds 10 million ha.

Progress achieved in intensification of farming and other branches of agriculture is accounted for, to a great extent, by the level and rate of their 'chemicalisation', and in particular application of mineral fertilizers and chemical means of plant protection.

The CMEA countries have worked out and approved appropriate recommendations concerning rational methods and organization of applying fertilizers and chemical means of plant protection, as well as storage, transportation and application to the soil of liquid fertilizers.

Developing and employing in 1973 a Prospect Programme of cooperation in the field of manufacturing mineral fertilizers, chemicals and mineral feed additives has led to great practical results in agriculture. The Prospect Programme includes the total need of the member countries in the means of agriculture chemization for the period till 1985, with consideration for their assortment, methods of improving physical and chemical properties of fertilizers, procedure and organization of international testing of chemicals and some other problems. The Programme involves also the measures for specialized manufacture of selected chemicals in certain countries, the volume of output, formulation and standardisation questions. An ever growing role in carrying out the Prospect Programme is played by an international economic organization "Interchim", functioning within the CMEA framework. Its activity is concentrated on coordination of specialized and cooperative production within the social community.

The CMEA countries consider of great importance their cooperation for mechanization and electrification of agriculture, which is eventually aimed at the development and further improvement of material and technical basis of agriculture. One may identify considerable success in this field during the recent years. Technical implementation of labour in agriculture is growing rapidly, tractors and other farm machinery are manufactured in greater quantities and of better quality. Mechanization and, in some parts, automation of labour-consuming processes finds its wider application in plant growing and livestock husbandry.

Establishing large-scale animal farms and other enterprises, based on industrial methods of production, has started. Success of cooperative efforts, bent towards the improvement of agricultural mechanization, has enabled the CMEA countries in 1961 to start developing the International Machinery System (IMS) for complex mechanization, in the near future, of agricultural production and forest husbandry.

The development of IMS and its improvement is carried out with consideration of world achievements and successful practice of the CMEA countries in this field, and forecasting investigations as well.

A new improved International Machinery System for complex mechanization of agriculture and forest husbandry was approved in 1971. It includes over 700 models and 1750 pattern dimensions of engineering, the manufacturing and supply of which are of interest to the community countries.

The IMS has a task of promoting the implementation of cooperative measures for the planned manufacturing definite machines, coordination of research work, cooperation in building tractors and basic agricultural and forest machinery, as well as unification and standardization of agricultural and transport machinery. The Agreement involves 9 models of tractors, 61 types of basic agricultural machinery and a complex technological line.

Within the framework of CMEA there is functioning an international economic organization "Agromash", which is coordinating the community countries' efforts in the field of agricultural engineering, i.e., research, developing, designing and specialized manufacturing definite types of agricultural machinery. Analyzing the process of the community countries' cooperation, one will find that its development proceeds from simple to more complicated forms, from mutual assistance and exchange of scientific and technical documentation to interstate specialization and establishing international economic organizations. The present-day stage of cooperation between the CMEA member-countries may be definitely called an international agro-industrial integration.

Working out the guidelines for the development of social integration has found its final expression in a Comprehensive Programme of further deepening and perfecting the cooperation and development of socialist economic integration of the CMEA countries, which was approved at the XXV-th CMEA session in 1971. The Comprehensive Programme for economic integration between CMEA countries is a manifestation of a deep process of the fraternal countries' coordinated efforts at all stages of material production, beginning from forecasting to cooperation in producing and marketing the commodities.

Socialist integration is accepted by all the CMEA countries as a main long-term course of their economic policy.

Further successful development of agriculture and other AIC branches depends, to a great extent, on correct solutions of agrar and economic problems. It is no chance, therefore, that they occupy an important place in the Comprehensive Programme.

Problems of cooperative planning are of particular importance to further improvement of the economic mechanism of international agro-industrial integration.

Planning instruments of the international agro-industrial integration include the coordination of Five-Year Plans for National Economic Development, as well as Development Plans for separate economic branches and industries for the periods of 10–15 or more years, the forecasting of AIC branches development, the cooperative planning of separate branches and industries, and sharing the experience in planning.

Coordination of Five-Year Plans of National Economic Development involves direct or indirect preconcerting the development of all branches included in the AIC. Such scope of planning is sufficient for accomplishing the majority of economic tasks faced by the AIC branches. However, in a number of cases it may be insufficient for separate industries or enterprises. These branches become an object of coordinating the long-term development plans, designed for 10–15 years. The first coordination for the five-year plan period from 1960 to 1965 took place in 1958–1959. Later on such coordinations were conducted in the period of drawing up every subsequent five-year plan.

Great attention is given to the prognosis of agriculture and other AIC branches development. Presently all the CMEA countries have elaborated, according to a coordinated programme, the prognosis of growth of the demand, production and marketing different kinds of products, as well as agricultural and food industry branches, including the development of their material and technical basis, till 1985. Elaboration of a summary projection till the years 1990 and 2000 is in progress.

Quantitative indices of development in agriculture and other AIC branches, derived from projections, are used in cooperatively working out another important problem, fixed in the Comprehensive Programme, i.e., study of conditions and potentials for deeper specialization and implementation of other forms of international social sharing the labour in coordinated branches of agriculture and processing industry.

CMEA countries have been studying the possibilities for joint planning by the interested countries, the growth of production of definite kinds of agricultural produce and means of production used in agriculture. The object of joint planning is cooperation in production, technology, trade and economics, which is accomplished as a set of selected measures and is aimed at solving a definite economic problem.

The idea of joint planning lies in the change from trade relations which are by their nature unstable and incomprehensible to integrational ties between producers and consumers. This form of cooperation is of primary importance to setting up and deepening the international socialist agro-intustrial integration of CMEA countries.

DISCUSSION OPENING – W. J. Anderson, *Canada*

The papers by Dr. Kleer and Dr. Nazarenko both make some points about differences in capitalist and socialist systems. Kleer identifies two ways in which integration under the two systems differ. One factor is the social ownership of resources which means that resource combinations are determined at the macro-level. This, he says, gives a different result than the aggregate of individual decisions when the means of production are privately owned. The second factor is the time horizon, which is longer for society than it is for an individual. Kleer argues that this difference leads to maximization in the longer run in the case of socialist societies and to improved performance because (i) economic surplus is more readily transferred to optimum use without social tensions and (ii) relationships between social and economic activities are more direct.

I would say that Kleer greatly under-rates the effectiveness of incentives and taxes which capitalist societies use to transfer economic surpluses to increase the social benefit. While the relationship between social and economic decisions may be more direct in a socialist system, the capitalist countries are fully conscious of the need to guide private economic decisions with the view to maximizing net social benefits.

The papers emphasize that integration between countries is a systematically regulated process in socialist countries. In market economies, the integration process is much less specific and is designed to provide guidelines in the form of incentives or disincentives to decision-makers. However, both socialist and market systems agree that the exchange of goods and the transfer of technology are mutually beneficial for the participants. However, countries in the market system would disagree with Nazarenko's statement that "internationalization of industry and economic internationalization obtains its full and complete development only under socialist conditions, i.e. social property, power in the hands of the working class and scientific communism".

With respect to the mechanism, Kleer distinguishes between the concept of integration as state and as a process. He argues for the latter concept because integration is a dynamic and iterative process. He lists six main benefits. These are a well thought out list of social and economic benefits from integration.

In carrying out the integration process, Kleer says that there are two main mechanisms. One is the planning mechanism for the integration of national plans which provides for agreements on trade, location of major plants and exchange of scientific and technical information. The other is the money market mechanism which is the process of pricing the products which are exchanged and valuing the local currencies in relation to each other.

In practise, it seems that integration in the context of CMEA emphasizes two dimensions of the benefits which are sought:

(i) transfer of technical knowledge and cooperation in research and development;
(ii) trading commodities under agreements in which prices and quantities are specified.

My impression from the papers is that integration in the area of technological development is well advanced and that the benefits are of considerable magnitude. Exchange of scientific personnel and cost sharing of large scale expensive research projects also takes place. This I can see as particularly beneficial to the low income countries in the group, which must benefit greatly from access to new technology and scientific advancements. Nazarenko points out that technological transfer in the case of agriculture is more difficult than for many other industries because the basic resource, land and climate, varies greatly from one region to another.

I am also impressed by the steps taken to achieve economies of scale in the manufacture and distribution of farm inputs. Nazarenko mentions the manufacture of fertilizers, chemicals and farm machinery as international ventures which attain the benefits of large volumes of output and standardized products. This is also consistent with the principle of obtaining real economic benefits from raising the production function and lowering the costs of inputs.

Under the money market mechanism, Kleer points out that the formation of a proper price system remains one of the most important tasks in economic integration. I can well understand that to be the case because pricing as now carried out appears to be a complex and arbitrary process. For example, CMEA issues a special transfer rouble which is used only for payments in the case of commodities exchanged under bilateral agreements. For exchanges outside these agreements, another currency is used.

Internal prices for the same commodity are not the same as those for which the commodity is traded internationally. I would be interested to know if internal prices are lower or higher than export prices. I suspect that much of the benefit of international economic integration is not being realized due to the very complex pricing system. Such a system must impose the effect of a complex system of taxes and subsidies which probably cancel out much of the benefits of trade arising from exploiting comparative advantage between regions of CMEA.

Kleer concludes with four points on the economic position of agriculture backed up by some statistics. The first is that agriculture's share in the national income in each country is declining. This decline would be expected under conditions of growth in real income per capita. Second, agricultural employment as a share of total employment is declining but remains larger than agriculture's share of the national income. This indicates that labour returns in agriculture are persistently below those of non-agriculture employment. Third, the area as a whole is a net importer of agricultural products. Fourth, the share of agricultural products in trade between the CMEA countries is declining. The evidence therefore seems to be that the terms of trade which emerge from the pricing system CMEA are unfavourable to agriculture and discourage its development. I suspect therefore, that the benefits of economic integration as it operates accrue mostly to those countries which exchange the products of secondary industries for agricultural commodities.

In conclusion, I note the absence to any reference to achieving integration by facilitating the transfer of factors of production, particularly labor. Labor transfer is one of the most effective means of increasing national income and providing more equitable distribution of income among people. I would be interested to know why resource transfers do not play a larger part in the integration policies of Comecon.

DISCUSSION OPENING – A. Kiss, *Hungary*

It is quite sure that the national economies and the agriculture of the CMEA member countries without the co-operation described by Dr Nazarenko and Dr Kleer could not have achieved the pace of development and level of the present day. The importance of this co-operation is well exemplified by the situation of a country like Hungary.

The economy of Hungary is characterized by being poor in raw materials, thus the basic materials and energy necessary to production, as well as machinery investments, come in the main, from imports. Consequently, a great part of the national income is realized through foreign trade; the national economy of Hungary is open. The conditions of agricultural production can be characterized as favourable. 90% of the demands of the country's population are covered by its own agricultural production and 10% is the volume of food which cannot be produced in the country (mainly citrus fruits and tropical produce). More than 10% of the food produced is exported. The share of agriculture in the exports is higher than in production, approx. a quarter of the country's exports consists of food and manufactured food products.

It is quite natural that, under such circumstances, the foreign market is important to Hungary not only as a source of purchases but also as a receiving market. The majority of our exports is to CMEA countries and great advantages of this economic co-operation are stability, security, and being protected against crises over the sale of agricultural products. There is a high turnover with the developed capitalist countries in Europe, especially with the member countries of the Common Market which are our traditional trade partners, but very often great uncertainties in the sale of agricultural products. A good example for this has lately been the restriction of imports of cattle. Since recently the prices in the exchange among the CMEA countries have also changed, the effect of this is, however, lessened.

The other question I would like to discuss, is the future of agricultural co-operation among the CMEA countries. As a result of developments so far, this co-operation has become a basic element in the development of the agriculture of the CMEA countries, representing an important increase of own resources. As it has been pointed out in the papers, the level of technology in the agriculture of the socialist countries has considerably increased, the utilization of materials and assets of industrial origin are becoming more and more important and the use of industry-like production systems (animal breeding and plant cultivation) becomes more wide-spread. All this means that, the increase of efficiency is of high priority. I think Dr. Kleer is right in stating that these circumstances result in increasing development of the co-operation in other branches in contact with agriculture (engineering industry, chemical industry, scientific research, etc.) than in the co-operation in the field of agricultural production itself. since these branches facilitate the growth of efficiency in the field crop cultivation and animal breeding. A basic precondition for the systematic solution of these questions is the co-ordination of the 5 year and long-term co-operation plans.

In conclusion, may I also emphasize that the co-operation in the field of scientific research should be increased; it has to maintain close relation with practice, as it was stated by Dr. Nazarenko. There remains much to be done in this field so that by rendering the relation between research and practice closer and fruitful, the possibilities of mutual co-operation could be more effectively utilized.

RAPPORTEUR'S REPORT – S. Krašovec, *Yugoslavia*

A great part of the general discussion was concentrated on the theory and mechanism of the division of labour within the COMECON. It was emphasised that the main emphasis of COMECIB is not only in the agricultural sector but also in the secondary sector, part of which produces inputs for agricultural production. Questions about the decision process within COMECON were concerned with the underlying goals and criteria for labour division, and also whether a solution is reached by a compromise on the basis of political and economic power, or by the comparison of production costs in different areas. The answer was that there is no supranational body (such as in the Common Market) but the decisions are achieved by consensus of all member countries. What makes this consensus easier is the fact that they have the same ideology, the same political system, philosophy, vocabulary, and last, but not least, often very good personal relationships. All decisions are made on the consensus level – nothing is ever decided against the voice of any country; similarly a decision does not apply to any country which has not joined in voting for it. Each country has its sovereignty and the same weight in voting. It was finally indicated that a lot of research and discussion is underway to develop a theory of division of labour for centrally planned economies which might be based on the concept of production cost – including labour – in different regions.

To the question why the world market is taken as a basis for pricing the transfer of goods within the COMECON, it was answered that COMECON is not on the moon; it is a part of the world market, (and so also is each individual member country) and is to some extent affected and influenced by the world market. So a country may be free to sell produce either to another member country or to a Common Market country if a better price can be got there. The price stability within the COMECON countries was their major advantage in trade. There was discussion on whether or not social ownership of means of production is the real reason for emphasising the long-run instead of the short-run aspects of integration, and that it makes the transfer of surplusses from one subsector to the other easier. It was agreed that the transfer of labour could be a major problem and some conflicts and problems arising from the different level of development of different countries were recognised. Integration is still young, and research and efforts on different levels are being made towards perfecting it.

C. J. VAN DER VAEREN

*Regional Cooperation in Agriculture Amoung Developing Countries – With Particular Reference to Africa**

1. INTRODUCTION

Economic cooperation among developing countries to ease or accelerate their growth is generally recommended in those spheres where the restricted scale of the national market acts as a brake on, or even an obstacle to, the establishment and viability of certain new activities, in particular processing industries intended to replace imports by local products. It is true that industrialization is a difficult process, especially in countries whose economies are small. Cooperation between neighbouring countries rather than competition or a policy of self-sufficiency can help a great deal in achieving this end. Regional cooperation, however, poses serious problems which slow down its practical applications even if the principle is proclaimed loud and clear at numerous international conferences. The main problem is that of distributing fairly the value added, the costs and the risks among all the countries in partnership: the country on whose territory the new industry is implanted and those who offer the preferential openings for its produce.

Regional cooperation in the sphere of agricultural development receives far less attention, except where the development of agriculture in certain river valleys common to several countries calls for coordinated development schemes or even joint investment by these various countries. In such cases, however, the cooperation between neighbouring States is not generally limited to the agricultural sector since the orderly development of the valleys also enables transport facilities to be improved and energy production to be increased.

There are many other fields of regional cooperation touching upon the agricultural sector about which there is less said and which are nonetheless significant for the promotion of agriculture. We shall discuss the spheres in which such regional cooperation is already being practised, how it could be developed in the near future and into which fields it will doubtless be necessary to extend it in the longer term.

Dealing with the various aspects of regional cooperation in agriculture we shall be looking more especially at its present or potential applications in Africa. There is an acute need for regional cooperation on that continent. It is

* Comm. of the European Community, Belgium.

a well-known situation: there is no need to go once more into the state of political and economic balkanization that has resulted from the successive agreements among the great colonial powers and which was accentuated, over the period of internal self-government, by the splitting up of the large territorial units such as French West Africa and French Equatorial Africa and by the dissolution of certain regional economic bodies such as the West Africa Currency Board and Marketing Boards. The barrier that exists because of the different administrative and scientific languages constitutes yet another obstacle to cooperation among African countries.

2. THE AGRICULTURAL DEVELOPMENT OF LARGE RIVER VALLEYS

It so happened because of the routes explorers took or because of conquests and the application of administrative or military criteria that international frontiers were set up that took little or no account of necessities of economic development. Many African frontiers, for instance, have divided large river valleys, either along the length of the river or across it, and cut across the countryside as created by nature. This is the case, to quote but a few examples, along the Falémé between Senegal and Mali, along the Congo–Zaïre and the Ouélé between Zaïre, Congo and the Central African Republic. In the same way the Juba and the Scebeli flow from Ethiopia to Somalia, the Chari from the CAR to Chad. The Zambeze divides Zimbabwe from Zambia and then cuts across the Mozambique frontier. If consideration had been paid to the necessities of economic development, many a frontier would have been drawn along mountain ridges.

In every large valley natural interdependence plays a large part in the development of resources. What happens downstream depends upon what is done upstream, the water resources are affected by what happens in the catchment area, anything affecting the course of the water affects both banks at the same time, and so on. Also, the waterway is very often a trade route, a means of communication, unifying the surrounding countryside rather than dividing it up as a frontier is supposed to do. Very often the two bands belonging to different States are inhabited by people who are ethnically the same.

In areas where rainfall is scarce or limited to one season, agricultural development means control of the water and, consequently, development of the rivers and streams. This intensive development, made possible by water control, is essential, especially since colluvial and alluvial deposits constitute more fertile soils than any others. It is normal , therefore, that regional cooperation should have imposed itself to all intents and purposes on the areas for the development of river valleys such as those of the Senegal, the Niger and the Nile or of lakelands such as the Chad Basin. There is still a lot of effort needed in this direction for other river valleys divided up by frontiers and situated in regions where control of the water is a pre-condition for agricultural development. We are thinking here of the valleys of the Scebeli, the Juba and the Mono.

The regulation of the flow of water for irrigation purposes can, however, pose problems for cooperation among the States along the banks. The effects of undertakings of this nature can be very far-reaching, geographically speaking, and be an economic benefit or a disadvantage for areas situated in States other than those in which the structures are built. The classic example is that of the arable land lost beneath the artificial lake above a dam while most of the irrigable land is downstream, on another State's territory.

A solution must be found so that the distribution of the costs and the benefits among the surrounding countries is not so unbalanced as to render political agreement on the common development of natural resources impossible. Development aid bodies, bilateral and multilateral, have an important part to play in facilitating regional cooperation agreements. If they allocate all their aid to individual States, it is almost certain, and perfectly natural, that each recipient State will give priority to the projects that are sure to procure it the maximum benefit. Projects of regional interest will fall back into second or third place if they are likely to be of more benefit to one or more neighbouring countries. This is why the Lomé Convention* stipulates that a significant, though small, share of financial aid is to be reserved for regional cooperation schemes. The UNDP has been doing the same for many years already. These two cases are, however, relatively isolated among all the sources of aid, including the World Bank.

In order to play their full part in promoting regional cooperation, particularly to develop regions that form a natural whole, aid bodies should not only give a certain priority to this type of operation to offset the national constraints working in the opposite direction, but also be prepared to finance, if necessary, the net costs or loss of profit incurred by some of the partners when such operations are carried out, particularly to compensate a country for the loss of cultivable land entailed when a dam is built that will mainly benefit a neighbouring State.

For many years now the development of river valleys and lakelands has been the subject of sustained effort in the way of regional cooperation in Africa. Two schemes that have been the most fruitful in this field are the Organization for the Development of the Senegal River (OMVS) and the Chad Basin Commission. In the last few years the political will of the States in partnership has finally managed to achieve practical results in the form of decisions to carry out large-scale common schemes, with the support of external financing bodies which have been convinced, it seems, by the recent years of drought that it is important in the long term to go in for large-scale water engineering development projects. The Niger River Commission, by way of contrast, which has also been in existence for over ten years, has done little up to now. Regional cooperation schemes on this great river have been concerned almost exclusively with transport.

3. PASTURE AREAS

The development of large natural regions also necessitates close cooperation among States in one particular sector: stock-farming. In Africa there are many

* Concluded between the European Community and 46 African, Caribbean and Pacific States

examples of complementary areas that form a whole that is most suitable for cattle farming. The areas making up such a whole range from the Sahelian type, where there are wide expanses of pasture available for part of the year, to the coastal regions, mostly very humid, where animal health and a lack of pastureland pose serious problems, but where there are large consumer centres such as Abidjan, Accra, Ibadan, Lagos, Douala and so on. In between are the intermediate areas of the Sudan and Sudano–Guinean type, where agriculture is quite intensive and in the season leaves little land available for livestock, but where the animals can be fattened.

The fact that these different areas scattered over many States complement one another means that agreements are necessary for the development of cattle routes (tracks, watering points, veterinary stations) and the harmonization of regulations and investment, particularly as regards setting up ranches and slaughterhouses. It could be tempting for the meat-consuming countries to set up a costly policy of self-sufficiency; whereas a healthy conception of the collective interest in the region calls for optimum development of the resources available in each State, according to its own possibilities, so that none of these resources are underutilized and the cost of the final product is reduced to a minimum. In West Africa, for example, this cooperation is particularly important for the Sahel which can supply young animals more cheaply to the more fertile areas that are suitable for fattening livestock, but which has hardly any alternative, whereas the areas where nature is more generous can also be used for growing crops. Regional cooperation should not only serve the laudable end of being in the mutual interests of all the partners; it should also contain an element of solidarity to benefit the countries in the region that are less favoured by nature.

Concerted development programmes for such complementary areas are now starting up in West Africa. The States of the Entente have set up amongst themselves a meat and livestock economic community; within this, they are harmonizing their respective measures and regulations on animal health, customs and commercial matters and on the transport of livestock and meat. One striking point is that this agreement does not cover the whole of the natural region concerned since Ghana, a large consumer country, is not a member and Mali, a large producer, is only an associate.

4. CONTROL OF PESTS AND DISEASES AFFECTING FARMERS, CROPS AND LIVESTOCK

Regional cooperation bodies have also been set up and been operating for a long time in another sphere where nature scorns all frontiers, that of the pests and diseases which attack man, plant and beast. In West Africa in particular there is a long-standing joint organization to control locusts and birds: OCLALAV, which is still in operation, despite the vicissitudes of politics, thanks to the common awareness of danger and constant support from external aid.

In a similar field, the Sahel countries, which go in for extensive stock-farming, launched a campaign in the Sixties to eradicate rinderpest. Such joint action is also required to combat other endemic diseases such as contagious pleuropneumonia. In a region inhabited by nomads, control of the major epizootic diseases can be effective only on a regional scale, otherwise the infection will only come back again across the frontiers.

To fight against rinderpest the European Community and US/AID financed the purchase of equipment to supplement the means at the disposal of the local veterinary services and supplied technical assistance whose particular task it was to coordinate the action taken by these services at national level. The result of this campaign was that the disease was practically eradicated and it was possible to increase substantially the yield of the herds.

Unfortunately, much of the beneficial effect of this work was destroyed by successive years of drought; but it must be admitted that, if this disaster had not taken place, the state of health of the livestock would probably have started to deteriorate a few years after the end of the campaign undertaken with the help of external aid. It is very doubtful whether the veterinary services of all the countries concerned would have had sufficient means to maintain the improvement achieved by a coordinated mass attack. The share allocated to the stock-farming services from the national investment and operating budgets of the countries taking part in the eradication campaign was reduced in most of the countries when it was launched because of the inflow of aid connected with the campaign. In addition, the regional coordinating office set up for the campaign was dismantled when the technical assistants had accomplished their mission.

Besides predators and epizootic diseases, agricultural development has to cope with other natural scourges which have to be fought on a regional level when they extend over whole natural areas divided up by political frontiers. Examples are certain endemic diseases that attack man, such as onchocerciasis or bilharzia, and are linked with farming activities or prevent the development of fertile land. It would have been almost useless, for instance, to try controlling onchocerciasis, which blinds so many able-bodied men and women in the villages, in a single country in the infected region, through which flow so many watercourses where the carriers breed. This is why the control and development programmes – both the pilot programme already financed by the EDF and the large-scale programme that is now being carried out with the help of a group of national and international bodies – extend over all the countries affected in West Africa: Mali, Ivory Coast, Ghana and Upper Volta. The fact that this serious disease is now coming under control should enable the men and women freed from its threat to develop hundreds of thousands of hectares in valleys suitable for farming.

The three areas to which we referred earlier show that serious efforts are being made to achieve regional cooperation and that there is a need to increase and intensify these efforts in order to resolve rural development problems facing the developing countries because of the transnational nature of many rural areas, river valleys and lakelands and the fact that "animals" of all sizes – cattle, locusts, queleas, tsetse flies, microbes, viruses and so on – ignore man-made frontiers. Some other areas in which regional cooperation would be necessary, if not vital, receive far less attention – or at least too little considering the wide gap between the State of Africa's agricultural development and its known potential. Generally, these areas are not specific to agriculture, but regional cooperation can be applied and bear much fruit in that sector.

5. RESEARCH IN AGRICULTURE AND ANIMAL SCIENCES

Compartmentalization within Africa is clearly absurd in research, since it inevitably leads to a waste of resources and the results reach far too restricted an audience. The results of research should be rapidly accessible and promptly applicable in all areas complying with the research conditions. Here the difficulties resulting from the existence of State Frontiers are seriously aggravated in Africa by linguistic barriers.

The lack of horizontal transparency – i.e., from one country to another – in research in agricultural and animal sciences has led certain international and bilateral aid bodies to set up the Consultative Group on International Agricultural Research, which supports and finances a complex of research institutes specializing in major types of crop or natural regions. This network of international centres meets a need for and reflects a healthy approach to research. However, it is not the result of cooperation initiated by the developing countries concerned but has been conceived and sponsored from outside. In our opinion there is therefore a serious risk that these international institutes will be too inward-looking instead of acting primarily as a link between the existing national centres and encouraging the latter to expand and intensify their activities.

While regional cooperation is indeed necessary in research, the national institutes still have a vital part to play in ensuring that the research remains effective through to production level. It seems to us that only the national and local research centres are capable of establishing the necessary links with the rural development departments – the authorities responsible for agriculture, stock-farming, rural engineering, etc. – and with the development companies, producers groups and agricultural undertakings. The development departments and the producers are frequently reproaching research in Africa for its failure to respond to the most vital needs of the agricultural sector or its failure to meet all these needs. Efforts should not be devoted solely to improving varieties or breeds under optimum operating conditions; research should also be geared closely to the current capacity of the rural environment and its specific constraints. Consequently, research aimed at perfecting a single crop, generally a cash crop, must also take into account how the new variety to be developed can be integrated in a specific crop-rotation scheme and agricultural calender and how it will effect a host of other factors. The research centres should therefore work in very close liaison with the area concerned; they must be spread over a wide geographical area and then working guidelines must be integrated both with the local environment and the national development strategy.

Regional cooperation in agricultural and animal sciences must do more than simply economize on the available resources and put them to more effective use by widening the sphere of influence of each research project or centre to cover all similar land in various States. It must also see that the results of research are brought before a wider audience and applied in all areas to which they might be of interest, in any country – or even any continent. Lastly, this cooperation should provide the research centres with information on the development problems and needs of the rural environment as experienced *in situ*, in all areas where these problems occur on relatively comparable terms.

6. TECHNICAL TRAINING FOR AGRICULTURE

The geographical, technical and economic arguments in favour of closer regional cooperation on research also apply to training in agricultural science and animal husbandry. Some recent measures taken in this area look encouraging. Since the wave of new national institutes that were created, and which sometimes even temporarily excluded students from neighbouring African countries, specialized establishments common to several States have been set up and are evolving. Good examples of this kind in West Africa are the zootechnical faculty of Dakar and the school for training rural water-engineering technicians in Kamboinsé (Upper Volta).*

7. EXPERIMENTS WITH RURAL DEVELOPMENT METHODS

It is not only strictly in the areas of agriculture and animal sciences that greater regional cooperation is desirable so that the results of research and experiments in rural development can be widely disseminated in all areas which are interested. The same applies to the policy, strategies and methods of development. In this sector, that includes the new institutions or structures which the authorities are endeavouring to set up, mainly to improve the functional links between the rural producers and the economic operators of the modern sector: in trade, industry, national budgets, and so on. It can also involve land-use organizations, agricultural credit arrangements, means of communicating and publicizing new techniques, etc.

In all these areas, which relate more particularly to man, who is the central factor in rural development, experimentation and research are always delicate matters because human reactions are sometimes unpredictable, especially by technicians, and the producers' disappointment can hamper the development process for the duration of the peasant's long memory.

This vast subject cannot be dealt with solely within a laboratory framework. Those who "make" rural development, from the top level (policies and strategies) to the base (action and methods) must therefore be able to communicate the meaning and results of their work to one another. In Africa at least, it is striking to what extent there is ignorance, from one country to another of the same natural region or sharing the same problem, of the new experiments in and approaches to rural development. This also applies to non-agricultural sectors such as education, which have a profound influence on this development. It is true that from time to time there are exchanges of views, symposiums and seminars on a regional basis. They are not very effective, however, for two reasons. Firstly, the delegates are mainly politicians

* These two regional training centres were granted EEC aid for constructing their infrastructure.

and administrators, rather than men of practical experience who know the land and are in daily contact with the realities of rural life. Secondly, the participants at these meetings are too often given the role of representing their State; their task is therefore above all to present fine national successes – or even, as is more often the case, to present superb national plans and intentions – instead of learning from the experience of others and objectively discussing their own experiments and errors, their failures and successes and the causes and consequences thereof.

At this level there are great gaps in regional cooperation.

8. TECHNOLOGICAL RESEARCH

Regional cooperation aimed at establishing closer communication between those concerned with rural development and wider dissemination of relevant information could usefully be extended to cover also the solution of technical problems which are not strictly agricultural or concerning animal husbandry that arise during the rural development process. Sometimes this involves no more than tricks of the trade which do not require great technical knowhow but can solve widely encountered problems easily and at little cost. This applies, for instance, to the various solutions found by the village communities themselves or by technicians working in close conjunction with them to the problems of pumping water, watering market gardens, packing eggs for marketing in the towns, etc.

Important new techniques relating to rural development and worthy of wider dissemination include techniques to make better use of local resources, such as the use of by-products of agricultural or animal origin to resolve certain problems arising in the rural world. Since these are simple and inexpensive techniques no-one is ready to spend money on making them widely known, although vast areas of rural land could profit from this.

Aid bodies could probably stimulate regional cooperation on research in agricultural science and technology and on rural development methods by financing meetings between persons with local experience coming from different countries and above all by financing practical training courses on a regional basis. This would enable those engaged in rural development to observe one another's experiments and methods on the spot and to discuss them not only among themselves but also with the producers concerned. External aid could also contribute to improving the horizontal transparency of research and experimentation by financing an inventory of "finds" and techniques at "grass roots" level suitable for a certain environment and by financing wider dissemination, in the main languages currently used, of information on these techniques and methods and on the necessary data relating to the context and conditions of their use, so that they can be extrapolated more easily. External aid should also help in perfecting these techniques, especially by encouraging the creation of research institutes on peasant farming techniques in the developing countries and regional cooperation between the institutes set up on a national basis. It seems to us that technical and financial support would be better used in this way than on scholarships for advanced courses (preferably for an indefinite period) in educational or research centres in the developed countries, to train specialists in agricultural disciplines who have not yet set foot in the fields.

We have now given a survey of various sectors where we consider that regional cooperation should be expanded: research and experiments in agricultural science, methodology and technology linked to rural development. We consider that at present these areas are not sufficiently developed despite the great need to improve the quality of factors and techniques of agricultural production and to learn more about ways of creating a rural environment which is open to new ideas and imbued with its own dynamism.

There are three other less urgent areas where regional cooperation should be developed. The first relates to trade in agricultural products between developing countries and to promoting those countries' exports to the wealthy countries.

9. TRADE IN AGRICULTURAL AND STOCK-FARMING PRODUCTS

The rapid growth of major urban, mining and industrial centres in Africa has resulted in an increasingly strong dependence on other continents for food supplies: notably on North America for wheat and millet, South-East Asia for rice and even South America for meat. The same applies to certain raw materials of agricultural origin.*

In order to realize their agricultural potential, the developing countries in which this urbanization and industrialization process is taking place should give preference to obtaining supplies from producers in their region so that African farmers will themselves have outlets for the surplus production which they will have as a result of the efforts to modernize agriculture. This would enable the savannah countries to increase their exports of citrus fruit, rice, millet, onions, leguminous crops, cotton, etc. to the forest countries.

However, there is a risk that regional cooperation in this field will come up against the problem of prices since the means and techniques of production used by African peasant farmers are a long way from enabling them to achieve a level of productivity comparable to that of their competitors in the more highly developed economies. Agreement should therefore also be reached at regional level to make the prices paid to agricultural producers – and therefore their income in relation to that of white-collar and industrial workers – sufficiently attractive. This may prove difficult on a short term nationalist basis of calculation but becomes feasible and even vital if the possible benefits are considered on a more long-term regional basis. Some countries have sharply increased the consumer price of basic foodstuffs in recent years in order to stimulate national production; this happened in Senegal for instance with rice. Such a policy should also encourage producers from other countries in the same region.

In some cases the fact that there is no common policy for fixing producer

* Guinea, Sierra Leone and Liberia import cotton from outside Africa while Senegal exports it. Chad and the Central African Republic export cotton to Europe while Zaire imports it from the USA, and so on.

and consumer prices has made it impossible to achieve the objectives of specialization and trade in agricultural products. The Treaty establishing the East African Community, for example, provided for the formulation of a common agricultural policy. In fact trade in agricultural products has developed very little between the three States which make up this Community, mainly because their national marketing offices apply different price-fixing policies.

In some cases what is required at regional level is not only a concerted prices policy but above all the harmonization of regulations governing the production of and trade in food products so as to promote both the outlets for producers and the security of supplies for consumers. A practical example of such an effort is the West African meat community, which is already introducing regional cooperation as regards hygiene, commerce, professional regulations, and so on.

Where it is not regional trade that is involved but rather the exporting of agricultural products to markets outside the region, purchasing in the developing countries is generally in the hands of the big business firms from the industrialized countries. They are in a strong position, given the large number of supplier countries, even where the production of each of the latter is sold *en bloc* through a national agency – which is not the case, of course, for all countries or for all products. Regional cooperation between countries supplying the same product – or a range of interchangeable products, such as oil seeds – may therefore improve appreciably the terms on which those countries sell their produce in the absence of a world market organization for such products.

In particular, there is a potentially substantial out-of-season market in the industrialized countries of Europe for high-value agricultural products: vegetables, fruit, meat, fish, flowers, honey, and so on. This market is at present under-exploited by African countries since supply capacity is still low and transport costs are high. These two limiting factors will probably diminish in the future; but that does not mean that competition should then become undisciplined and destructive. Here, too, regional cooperation should therefore be established, to promote exports of the products in question and to organize in a reasonable manner trade with the markets concerned.

As we pointed out in the case of stock-farming, it is important that regional cooperation with a view to increasing exports to non-African markets should develop in the direction appropriate to long-term solidarity among the producer countries. It is, of course, not only in Europe that countries which are competitors in the short term are interdependent as regards long-term prosperity! The enlightened self-interest of all requires, therefore, that the limited potential of the least favoured should be developed in preference to their competitors which have at their disposal a wider range of export opportunities. In Central Africa, for example, countries such as Rwanda and Burundi, which are entirely land-locked and thus suffer from an undoubted structural handicap, should be able to develop their few high-value products, such as coffee and tea, and should not find themselves faced in their outlets with growing competition from African countries whose geographical position makes exporting much easier.

10. THE MANUFACTURE OF INDUSTRIAL FACTORS OF PRODUCTION

The progressive sophistication of crop and livestock production will entail increased consumption of industrial inputs: machines, fertilizers, etc. Regional cooperation in various forms will be increasingly necessary in this sphere. It is undoubtedly to be regretted that a retrograde step was taken in the specific case of the production of vaccines for livestock. These used to be produced by a central laboratory for each of the vast territories which formerly made up French West Africa and French Equatorial Africa. This regional solution has been weakened by some purely national attitudes. Since this is a matter rather of industrial cooperation, we shall not dwell on the subject. We would only point out briefly that regional cooperation is necessary in the first place to ensure, through economies of scale, that industries producing inputs in the developing countries are better able to compete with their rivals in the industrialized countries. This question of competitiveness arises above all in the initial stages of development of the industries in question, as was clearly seen in connection with the production of fertilizers and agricultural machinery. It should be emphasized, however, that the drive for regional cooperation should not foster only the industrialization of the developing countries; industrialization should in its turn facilitate agricultural development by manufacturing factors of production which, as regards both use and maintenance, are better adapted to the resources and capabilities of the rural environment and to the specific needs of agricultural production in the developing countries concerned. Products from the industrialized countries are often over-sophisticated; they do not make sufficient use of local resources; they often suffer from slow or inadequate after-sales service. The smallness of the national markets, for the next few years at least, therefore requires that these industries manufacturing the factors of production be set up rapidly and that they be programmed on a regional basis.

11. MIGRATION OF AGRICULTURAL WORKERS

Lastly, there is one further sphere in which regional cooperation will become increasingly necessary, particularly in Africa: it concerns the migration of agricultural workers from one country to another. Here we are thinking in particular of the flow of migration from Upper Volta to the cocoa, coffee and other plantations of Ivory Coast and Ghana; the flow from Mali to Ivory Coast and Senegal, particularly for the groundnut and cotton harvests; the flow of Rwandese to the farms and plantations of Kenya; or, to take another example, the migration of workers from northern Nigeria, Chad and the Central African Republic to Sudan.

This is at present a purely spontaneous phenomenon and is subject to hardly anything more than regulations aimed at protecting the national labour market against foreign migrants. But such migrations are economically necessary to both the countries concerned, that is the workers' country of origin and the country where they are employed. Enlightened cooperation between the two sets of countries should permit such migration to continue to be beneficial to both parties, while ensuring fair remuneration and sufficient security of employment for the migrant workers.

Unless the economies of countries of origin, which are less well endowed with natural resources, are adequately compensated for the migrations through remittances of money saved from wages, investments and the acquisition of technical training, they will be inclined to check the migration; the labour-hungry countries could then suffer from under-exploitation of their resources. Even if the material interests concerned are set aside, regional cooperation is necessary in this sphere to protect the interests of the migrant workers themselves, since they come from the poorest countries or areas and therefore run the greatest risk of being exploited for lack of legislation. Regional cooperation would in this instance therefore take on a social dimension, complementing the various economic and technical aspects referred to above.

12. CONCLUSION

We have attempted to indicate the spheres in which rural development is partly dependent on greater regional cooperation, particularly in Africa. There are many and they impinge upon many other sectors: transport,* commerce, education, and so on.

The distinctive feature of regional cooperation in agriculture, however, is its close dependence on natural factors, whether in a complementary sense, thus requiring specialization and inter-trade, or in the sense of convergence, calling for joint research or trade promotion. In this respect, regional cooperation in agriculture differs sharply from cooperation in industrialization or transport, where it is rather geographical proximity or the size of the market that is important.

It is no use trying to superimpose agricultural cooperation upon a regional organization founded upon mainly political affinities. This is the case of OCAM, the Organization Commune Africaine et Mauricienne (African and Mauritian Common Organization), which was much less successful in the agricultural sphere (for example, sugar and meat) than in other sectors such as finance, training or culture.

For crop-growing and stock-farming, therefore, each country should fit into a network of varied forms of regional cooperation, depending on the type of problem to be resolved and the existing natural complementary factors.

In agriculture as in other spheres, however, the concept of regional cooperation is easy to grasp and the potential benefits may easily be analysed and described, but its implementation through practical measures and coordinated policies is slow and difficult. The progress of the European Economic Community demonstrates almost daily that it is not sufficient to have a convergence of political wills "at the summit" on broad – and vague – objectives. It

* Regional cooperation is of special importance for the agricultural development of land-locked countries. Without such cooperation, not only between States but also at the level of transport operators and workers, products for export cannot be sold nor can factors of production be imported.

is when the time comes to apply specific measures that difficulties materialize and the full weight is felt of entrenched situations, ingrained habits, mutual distrust and the obduracy of men with vested interests in established powers.

Economic, social and political structures in the developing countries are, fortunately, not yet as rigid as in the European countries. It should therefore be easier to make the adjustments necessitated by regional cooperation. Nevertheless, poverty itself constitutes a specific obstacle to such cooperation. It is very difficult to divert precious national resources, whether human, technical or financial, in order to risk them in structural changes and long-term adjustments in a multinational context, when those resources are already plainly insufficient to keep the national public services functioning in the immediate future and to meet the essential needs of the people for whom the State has immediate responsibilities.

The support of richer countries can therefore be a useful instrument for providing the developing countries concerned with the means of putting into action their political will to undertake regional cooperation. Where such political will is strong and embraces overall principles and long-term objectives, it would be wise for the two parties – the developing countries and the financing organizations – to come to a joint decision to set aside in advance a portion of the available external resources for regional cooperation measures. Thus, when divergences of immediate interests or technical difficulties raise obstacles between the general will and its specific application, the external support will help to offset these negative factors. The very fact of such external resources being available may even serve as a catalyst in the preparation of schemes and practical measures which will permit more development potential to be tapped through regional cooperation.

ADOLF WEBER AND THOMAS T. HARTMANN*

A Comparative Study of Economic Integration with Special Reference to Agricultural Policy in the East African Community

1. INTRODUCTION

The two predominant economic integration systems in Europe – the European Community (E.C.) and the Council for Mutual Economic Assistance (COMECON) are unique in economic history for two reasons: They represent, as Balassa ([1], p. 7) has defined it, the "liberal" and the "dirigist" ideal of economic integration. Moreover, despite the difference in integration policy, both developed a higher degree of economic integration although the sectors of integration vary widely. This development encouraged further approaches towards similar economic integration systems in Latin America and Africa in the Sixties. The search for a multinational approach to economic development in these two areas was supported by the intellectual leadership of the United Nations Economic Commissions for Latin America and Africa ([28], p. 7). The first United Nations Conference on Trade and Development (UNCTAD), held in 1964 in Geneva, passed a resolution in the Final Act to promote economic integration specifically among developing countries. In 1966, the newly created UNCTAD Secretariat produced a report advocating the assistance and regional association of developing countries in new economic integration systems, to draw the attention of industrialized countries and international institutions to the development needs of such new groupings. Because of the large populations of many Asian countries (China, India, Japan, Indonesia) a similar trend towards economic integration did not occur in Asia.

Within the African continent, the East African Community (E.A.C.) of Kenya, Tanzania and Uganda represents a relatively advanced economic integration system in terms of created legislative, executive and judicial functions ([21], p. 55). Article 12 of the E.A.C. Treaty [4] stipulates in principle a multisectoral integration to provide for the free movement of goods, capital and labour.

However, the trade of "certain agricultural products" is, according to article 13, subject to quantitative restrictions. Article 14 states that, notwithstanding the restrictions outlined in article 13, the "Common Market should

* A. Weber, University of Nairobi. T. Hartmann, Nairobi.

extend to agriculture and trade in agricultural products" and that cooperation and consultation in the field of agricultural policy among the Partner States should be so directed that trade arrangements "between the national agencies or marketing boards of the Partner States may be entered into directly within a single system of prices and a network within the Partner States as a whole of marketing services and facilities".

The prospects of integrating the agricultural sector have been the subject of a recently held seminar [7] where most papers and recommendations outlined the need for further study. Other studies [16, 21] have expressed similar proposals, they also stress institutional and planning aspects. There is, however, a general lack of theoretical and empirical research into the basic prerequisites and expected advantages of further agricultural integration in East Africa.

A general framework for analysis has been developed by Balassa's ([1], pp. 4, 65) concept of the advantages of economic integration: (a) accelerating the rate of growth, (b) economies of scale, (c) mitigating cyclical economic ills and fluctuations, (d) increase of economic welfare by new combinations of factors of production and (e) increase of bargaining power in international trade negotiations on tariffs, commodity agreements, price guarantee schemes etc. These advantages have been formulated for the whole economy of previously independent countries involved in a formal integration process. They relate directly and indirectly to the agricultural sector.

The critical comments made by Helleiner ([17], p. 146) regarding economic integration arrangements and the "extravagant lip-service" paid to them in developing countries reminds us of the realities of the E.A.C. region. By concentrating on the economic integration of the agricultural sector, we have to neglect other forms of integration – political, cultural, social, or technical – which may be more pressing or have higher priorities in the overall goal structure of the countries under examination.

2. EAST AFRICAN COMMUNITY

2.1. *Forms of integration before independence*

Before East Africa fell under European colonial rule, forces of integration had been at work in the coastal region since the 13th century within the Islamic community and with the outside world [23]. East African integration as an administrative endeavour started as early as 1917, when the British administered territories of today's Uganda and Kenya formed a customs union ([28], p. 160). In 1923, Tanganyika joined the customs union and the whole area developed into a kind of common market. During the thirties the region had reached the stage of an economic union with free movement of goods and production factors between the Kenyan, Ugandan and Tanganyikan territories. This "liberal" approach was supported by a commonly shared currency under the East African Currency Board ([22], pp. 43, 75, 79) and by common external tariffs, income taxes and excise duties. The aim was to create the prerequisites for realizing future economies of scale in industrial production by enlarging the domestic market.

TABLE 1. *Synopsis of the World's Main Economic Integration Systems*[a] *with comparison to the German Customs Union*[b] *and the United States of America*

Name of Integration System	German Customs Union	East African Community	Central African Customs and Economic Union	West African Economic Community	Central American Common Market	Common Afro-Malagassy and Mauritian Organization	Council of Arab Economic Unity	European Free Trade Association	Latin American Free Trade Association	Council For Mutual Economic Assistance	European Economic Community	United States of America
Abbreviation	G.C.U.	E.A.C.	C.A.C.E.U.	W.A.E.C.	C.A.C.M.	C.A.M.O.	C.A.E.U.	E.F.T.A.	L.A.F.T.A.	C.M.E.A. or COMECON	E.E.C. now E.C.	U.S.A.
Date of Signing Treaty	1833	6.6.1967	8.12.1964	–	13.12.60	10.3.1960	1957	4.1.1960	18.2.1960	Jan. 1949	25.3.1957	1776
Treaty Became Effective	1834	1.12.67	1.1.1966	1966	1.6.1961	April, 1964	5.6.1962	3.5.1960	12.6.1961	Apr. 1949	1.1.1958	1776
Population (Millions)	31·0	36·2	9·4	28·6	16·0	62·5	77·6	40·5	181·6	361·8	254·9	208·8
Gross National Product per Capita (U.S. $)	240	144	248	161	442	166	269	2805	926	1504	2908	5590
Gross National Product (Billion U.S. $)	7·4	5·2	2·3	4·6	7·1	10·4	20·9	113·5	168·2	544	741·2	1167
Integrated Markets as % of U.S. Market	0·6	0·4	0·2	0·4	0·6	0·9	1·8	9·7	14·4	46·6	63·5	100
Arable Land (Mill. ha)	25·0	22·8	22·1	34·8	4·80	58·1	23·6	12·5	116·0	291·4	51·4	191·1
Persons per ha of Arable Land	1·24	1·59	0·43	0·82	3·32	1·08	3·29	3·24	1·57	1·24	4·92	1·09
Gross National Product per ha of Arable Land (U.S. $)	296	227·5	105·6	132·4	1469	178·3	885·8	9089	1450	1867	14295	6108

[a] Population and Gross National Product data refer to 1972 (German Customs Union for 1834).

[b] Deutscher Zollverein.

Source: Union of International Associations: Yearbook of International Associations. 15th Edition (Publication No. 226). Brussels 1974, International Bank for Reconstruction and Development: World Bank Atlas (Population, per capita product and growth rates). Washington, D.C. 1974; *A. Weber:* Faktorproduktivitat und Technologie in der amerikanischen, europaischen und japanischen Landwirtschaft von 1880 bis 1965. Zeitschrift fur Wirtschafts und Sozialwissenschaften, Vol. 1973. pp. 197–226; FAO: Production Year Book 1973.

After World War II, the East African High Commission was established in 1947 with the express purpose of further integrating the three partners into a viable, independent, economic (and possibly political) federation. But with the proximity of independence other hitherto unconsidered objectives came into the fore: the respective territories discovered their own political identity. The frequently quoted "economic unity and identity" of the East African region was mainly for the administrative convenience of the British Colonial Government, to whose interests the variation of monetary conditions within the region could be relatively easily subordinated. As soon as the metropolitan umbrella of a protected and unified region was gradually withdrawn, the decisive era of regional integration between independent countries began.

The much discussed imbalance of the gains and losses between the three partners during the period of a commonly shared market was not an inducement to move faster towards economic integration. By 1961, when Tanganyika achieved independence, the East African Common Service Organization (EACSO) replaced the East African High Commission.

The whole integration issue now became even more debatable and concentration centred on a fair distribution of the earnings and expenditures of the existing service organizations (railway, harbours, airlines, postal services) when Uganda moved to independence in 1962 and Kenya in 1963, but the political will for integration within these three states was strong enough at the time to overcome these controversies. However, in 1964, when Tanganyika and the small island of Zanzibar formed a political union under the name of Tanzania, the island of Zanzibar still remained outside the customs union and continued to impose its own tariffs and levies.

2.2. *Comparative assessment of the size of economic integration systems*

Various assessments of the sizes of population and domestic markets in various integration systems, measured in terms of Gross National Product in total and per capita and the endowment with arable land per capita, are presented in Table 1. Just how effective the forces of integration have become may be judged from the fact that in 1972 28·5 per cent of the world's population, living in the ten integrations systems listed,[1] produced 40 per cent of the global GNP.[2] However, with the exception of the two European areas of COMECON and the E.C., and the Latin American Free Trade Association, the other integration areas have an extremely small market.

The size of the domestic market in African integration systems barely reaches one or two per cent of the U.S. market, or three per cent of the E.C. market, which means that it has little effect on world economy by force of combined purchasing power, as the U.S. or the two European integration systems have. To earn foreign currency for imports of capital equipment and machinery, the smaller integration systems in Africa have to adjust themselves to the opportunities provided by larger markets. Faced with the attractions of closer integration with these larger markets, it is more difficult to push ahead with regional integration. The regional African integration groupings must constantly weigh the type of integration which is most beneficial to them: regional integration or integration into larger markets.

The historical example of the nineteenth Century German "Zollverein" offers some interesting parallels and contrasts to present day integrated economic groupings in Africa. The Zollverein between German states might be considered as a successful historical forerunner of modern European integration systems, but unlike them, it started on a very small scale. Positive integration effects were probably easier to achieve then, because in those days the Zollverein already had a large domestic market, and high transportation costs favoured regional "trade creation".[3]

However, even if 19th century national income statistics are not, perhaps, 100 per cent reliable, there is ample evidence that even some 140 years ago, the domestic market of the Zollverein was at least no smaller than the present East African Community or other African and Central American Integration systems. Some facts might be helpful in evaluating the statistics.

In 1834, in the Zollverein, per hectare of arable land only 1·24 persons had to be provided with food compared with the 1·59 persons in the E.A.C. in the year 1972. The states of the German Zollverein had one primary goal in their agricultural policy: to increase food production. Capital, labour, industrial goods and agricultural products moved freely in the area without any quantitative restrictions. This was possible, because the various German currencies then in use were convertible. It was the perfect "liberal" approach of integration and not the "planned" or "dirigistic" one so prominent in contemporary meetings, discussions and writings. The fact that the Zollverein lacked a central decision-making bureaucracy to guide overall policy was probably not as crucial then as it would be today, when the demands and expectations placed on policy-makers to increase economic welfare are so high.

2.3. *The role of agricultural policy in various economic integration systems*

The two European economic integration systems, COMECON and the E.C., are both members of powerful military and political alliances. The complementary relationship between political and economic ties might be worth keeping in mind in any subsequent comparisons. The policies of the COMECON and the E.C. towards economic integration represents, respectively, the modern "planned" or "liberal" approach (see Table 2).

The free movement of capital, labour and goods in the E.C. is facilitated by convertible currencies. One of the integration goals stipulated in the Treaty was an approach which included all economic sectors and through which even widely differing agricultural prices were gradually equalized. In contrast, the COMECON Treaty has no provision for a supranational planning authority. The utilization of resources is still determined on a national basis with the result that prices for agricultural and industrial goods differ between member countries. Because of this, a time-consuming bilateral examination of mutual trade prospects is still the starting point of trade negotiations and planned trade flow within the COMECON ([19], p. 151). Furthermore, in spite of its well developed national planning machineries and shared political philosophy, COMECON has as yet been unable to formulate a common agricultural policy ([19], p. 217). Each country still strives to have the highest possible degree of self-sufficiency in food production. However, the benefits

TABLE 2. *Criteria of integration in the World's Main Economic Integration Systems*

Topic	East African Community	Central African Customs and Economic Union	West African Economic Community	Central American Common Market	Common Afro-Malagassy Mauritian Organization	Council of Arab Economic Unity	European Free Trade Association	Latin American Free Association	Council for Mutual Economic Assistance	European Economic Community
Military Alliance	No	No	No	No	No	–	No	(No) (Warsaw Pact)	Yes (N.A.T.O.)	Yes
Monetary System within Community	Commonly at par fixed Exchange Rate	Nationally fixed Exchange Rate	Nationally fixed Exchange Rate	Nationally fixed Exchange Rate	Nationally fixed Exchange Rate	Nationally fixed Exchange Rate	Convertible Currency	Nationally fixed Exchange Rate	Nationally fixed Exchange Rate	Convertible Currency
Movement of Capital	(Regulated)	Restricted	Restricted	Restricted	Restricted	Restricted	Free	Restricted	Restricted	Free
Movement of Labour	(Regulated)	Restricted	Restricted	Restricted	Restricted	Restricted	Regulated	Restricted	(Regulated)	Free
Movement of Goods	(Regulated)	Restricted	Restricted	Restricted	Restricted	Restricted	Free[a]	Restricted	(Regulated)	Free
Integrated Communication and Transport Sector	Yes	–	–	–	–	–	–	–	Standing Commission	Common Transport Policy
Common Industrial Policy	Planned	–	–	In selected sectors	–	–	No	No	Allocated Sector Specialis-ation	Common Energy and Regional Policy
Common Agricultural Policy	No	No	No	No	No	No	No	No	No	Yes
Common Structural Policy/Philosophy	Private or Socialized[b] Agriculture	–	–	–	–	–	Private Agriculture	Private Agriculture	Socialized Agriculture[c]	Private Agriculture
Common Agricultural Price Policy	No	No	No	No	No	No	No	No	No	(Yes)
Common Agricultural Research	Yes	–	–	No	No	No	No	No	Some	No

( ) = Statement needs Specification. – = Not exactly known.
[a] Trade of agricultural products is nationally regulated.
[b] Partly in Tanzania as Ujamaa Villages.
[c] Private agriculture still prevailing in Poland.

of the free movement of agricultural products for agricultural specialization are denied not only in the planned economies of COMECON countries, but also in the European Free Trade Association which has always excluded agricultural products. The same holds true in the Latin American Free Trade Association ([28], p. 82).

Some caution is needed against an extensive interpretation of common agricultural prices in the E.C. It is true that since the second half of the sixties, all agricultural prices are commonly expressed at the E.C. level in so-called units of account. However, the French devaluation and the Dutch and German revaluations in 1969, followed from 1971 on by periods of currency floating, destroyed the painstakingly erected, complex price-determining machinery. The member states hesitated to adjust their agricultural prices immediately to changes in the rate of foreign exchange. To avoid trade distortion and quantitative restriction in intra-community trade created by different agricultural price levels, the governments of the member states are at present again applying taxes or levies on lower priced agricultural imports originating from member countries: subsidies are paid for agricultural exports destined for member countries of the Community which have lower agricultural prices ([25], p. 100).

2.4. *Development of trade in the East African Community*

Economic integration through trade may occur between countries either with formal integration or without integration arrangements. In a formal integration, the process is considered as progressing as soon as the trade between member states increases in real terms per capita. A further measurement of the degree of integration is the percentage of intra-community trade compared to international trade. In the European Community 51·7 per cent of all imports in 1973 originated in the E.C. and 52·7 per cent of all exports were destined for it ([24], pp. 77). In 1965 the COMECON already had an even higher percentage of intra-trade with 63 per cent ([19], p. 114). In the East African Community, a totally different situation prevails.

Figure 1 summarizes the development of total and agricultural trade for the three countries of the Community from 1962–1974. The upper rectangle on the left in Figure 1 shows, in U.S. $ per capita, the net imports in total trade for the three countries. Negative net imports mean that a country exports more than it imports from abroad, in other words, it has a positive balance in total trade, whereas Kenya and Tanzania have been faced with rising net imports since the dramatic rise in oil prices in 1973. This affected their balance of payments and a decrease of foreign exchange reserves occurred, necessitating rising international credits and capital aid to guarantee the continuation of imports.

From the right rectangle, upper part of Figure 1 it might be assumed that all three countries slightly increased their net agricultural exports per capita due to the general price increase in tropical products and raw materials during 1973 and 1974. However, if we could measure the development in constant prices, which are not calculated separately for agricultural products, then

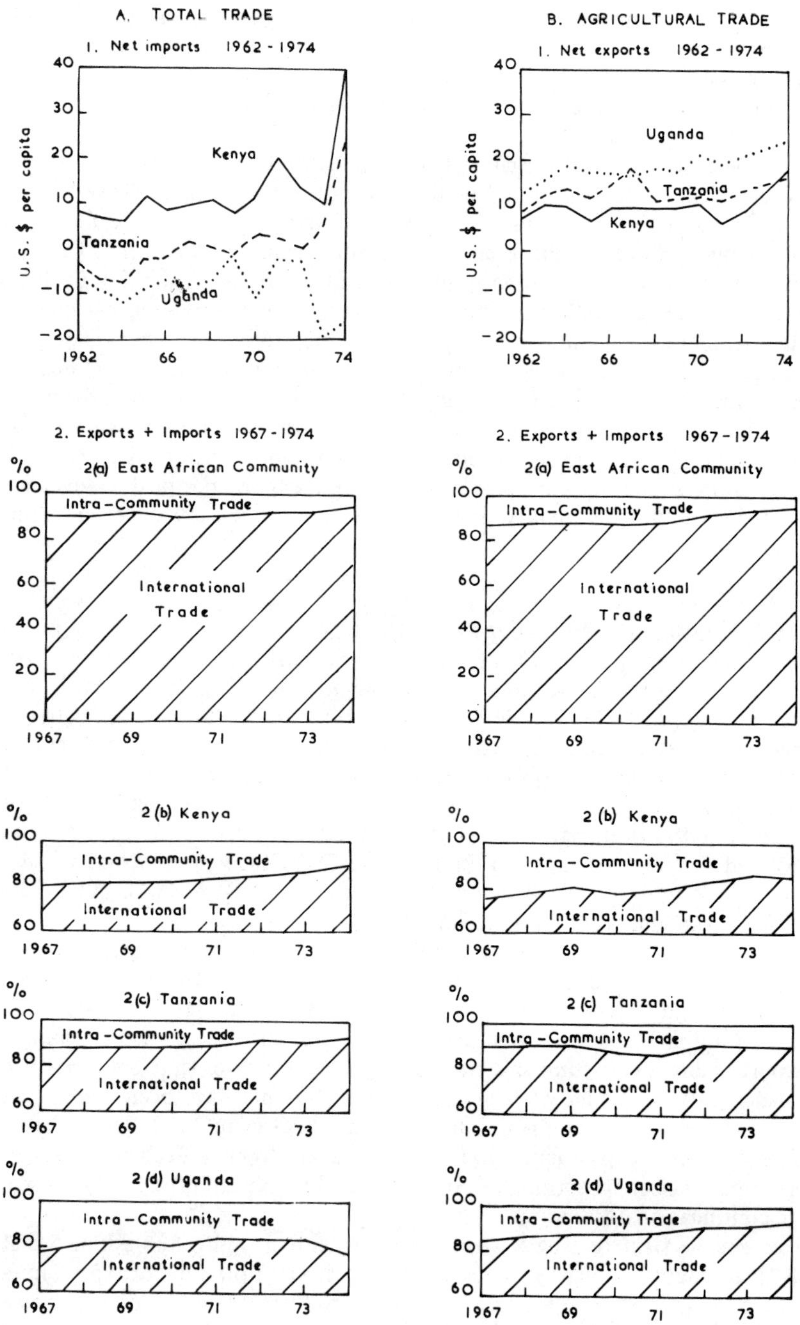
A. TOTAL TRADE
1. Net imports 1962 - 1974
U.S. $ per capita
Kenya
Tanzania
Uganda
B. AGRICULTURAL TRADE
1. Net exports 1962 - 1974
U.S. $ per capita
Uganda
Tanzania
Kenya
2. Exports + Imports 1967 - 1974
2(a) East African Community
Intra-Community Trade
International Trade
2. Exports + Imports 1967 - 1974
2(a) East African Community
Intra-Community Trade
International Trade
2 (b) Kenya
Intra-Community Trade
International Trade
2 (b) Kenya
Intra-Community Trade
International Trade
2 (c) Tanzania
Intra-Community Trade
International Trade
2 (c) Tanzania
Intra-Community Trade
International Trade
2 (d) Uganda
Intra-Community Trade
International Trade
2 (d) Uganda
Intra-Community Trade
International Trade

FIG. 1.

TABLE 3. *Development of the relative importance of the East African Common Market for each partner state in total trade and agricultural trade*[a] *1964–1974 in per cent*

Year	Kenya		Uganda		Tanzania	
	Imports	Exports	Imports	Exports	Imports	Exports
			Total Trade			
1964	13	35	29	13	26	7
1965	12	39	29	14	25	9
1966	9	27	29	16	20	6
1967	11	33	27	14	18	5
1968	10	31	22	12	16	5
1969	9	31	27	12	17	6
1970	10	31	30	10	15	8
1971	8	32	24	8	11	10
1972	7	27	30	4	11	6
1973	6	24	37	4	10	7
1974	3	29	64	3	7	7
			Agricultural Trade			
1967	50	16	61	10	47	4
1968	45	13	62	8	44	4
1969	32	15	66	5	43	4
1970	45	14	72	6	46	6
1971	29	17	61	4	35	7
1972	22	14	60	3	29	4
1973	15	12	[b]	1	36	4
1974	26	10	98	2	35	5

[a] Imports originated in countries of and exports destinated to the East African Community.

Sources: FAO: Trade Year Book 1967, 1972 and 1973; East African Community, Customs and Excise Department: Annual Trade Report for Kenya, Tanzania and Uganda for the Year Ending on December 31st, 1974; East African Community, E.A. Statistical Department: Economic and Statistical Review No. 29, pp. 39, 41 and 42, No. 37, pp. 44, 46 and 48, No. 43, pp. 51, 54 and 57 and No. 52, pp. 69, 72 and 74.

[b] Omitted because trade statistics are not conclusive.

foreign trade in agricultural products in per capita terms could well have been stagnant or even declining.

The comparatively low degree of integration in total intra-community trade in the initial stages did not improve during the period 1967–1974 when the international market increased its percentage from 92·1 to 95·3 per cent. In 1974, Uganda was the only country which had higher imports from the Community than in 1967.

In agricultural trade, due to the increase in international demand, the percentage rose from 91·4 to 94·3 per cent. The share of intra-community imports of agricultural products declined for Kenya and Tanzania and seems to have increased for Uganda (see Table 3). The share of intra-community exports of agricultural products fell for Kenya and Uganda and was stagnant for Tanzania. It may, therefore, be concluded that during the past seven years in the E.A.C. there has been no "trade creation" as assumed by the integration theory.

The members of the East African Community have on average an agricultural trade (net exports) of 20 U.S. $ per capita. A break-down of this figure reveals that not more than 5 per cent, or one U.S. $ per capita, goes on agricultural products exchanged in the E.A.C. During the period of 1950–1970, primary exports (excluding petroleum) expanded on world markets by 3 per cent per year only, in quantity and value terms ([2], p. 89). The possible effects of future export development and agricultural specialization may be assessed from Chenery's and Syrquin's pioneering study [2]. Using data from 67 countries, they observed an "average expansion path of primary and industrial exports". As the GNP per capita rose from 100 to 500 U.S. $, primary exports expanded from 16 to 60 U.S. $. These data refer to 1965, when the three countries had an average per capita income of 85 U.S. $. In 1972 the average per capita income rose in current U.S. $ to 150 U.S. $. Even if a threefold increase in GNP per capita to 500 U.S. $ (or more) and an increase in the share of intra-community agricultural trade of 5 to 10 per cent is assumed, not more than 6 U.S. $ per capita could be exported to the E.A.C. This small amount of expected intra-community trade does not, according to the authors' view, justify the creation of a complex decision-making machinery to administer common agricultural prices for the next decades.

The present low level of trade in agricultural products in the E.A.C. therefore follows the general development pattern of other low income countries analysed by Chenery *et al.* The similarity of agricultural production conditions in the E.A.C. which are competitive and not really complementary, does not favour any substantial specialization in agriculture between the countries.

At their present stage of economic development the E.A.C. countries are forced to exploit their comparative and complementary advantages, and this will be so for decades to come. The export of tropical products such as coffee, tea, sisal, cloves, etc. and of seasonal products (vegetables, fruits, flowers during winter) to the high-income countries of the northern hemisphere is more crucial than regional integration. Table 4 illustrates, as a direct result of the present export market situation, the absolute necessity of concentrating on export expansion to those overseas complementary markets which alone can provide the urgently needed foreign exchange to pay for imports of machinery, chemicals, minerals and fuels.

Another element which will probably influence long-term, intra-community trade prospects is the Lome-convention, by which the European Community, the first major world trading block to do so, has promised 46 African, Caribbean and Pacific countries to remove unilaterally tariffs on all industrial products ([13], pp. 35). Hence the East African Community represents only a small part of a subsystem of a huge integration system. The quantitative restrictions of the E.C.'s agricultural policy on the northern hemisphere's agricultural products remain, but the trade prospects, especially for industrial products, are brighter than at the time the East African Community was

TABLE 4. *The structure of the East African partner states' principal export commodities: 1973 in per cent*

Products	Kenya	Tanzania	Uganda	East Africa
1. Coffee	29·2	22·1	67·7	38·7
2. Cotton	1·1	15·8	16·0	10·2
3. Tea	13·8	2·4	5·2	7·4
4. Petroleum products	10·6	3·8	–	5·1
5. Sisal fibre	3·9	9·6	–	4·6
6. Cloves	–	10·4	–	3·4
7. Hides & skins	4·2	2·1	1·6	2·7
8. Diamonds	–	7·5	–	2·5
9. Cashew nuts	0·4	6·3	–	2·3
10. Animal food	0·7	2·4	2·1	1·7
11. Copper	–	–	5·2	1·6
12. Meat	3·0	0·8	–	1·4
13. Sisal manufacturers	0·4	3·2	–	1·2
14. Pyrethrum extract	2·3	0·7	–	1·1
15. Tobacco	–	2·5	0·5	1·0
16. Soda ash	2·2	–	–	0·8
17. Cement	2·0	–	0·1	0·8
18. Beans, peas etc., dried	0·9	1·0	–	0·7
19. Wattle extract	0·9	0·7	–	0·6
All above products	75·6 (= 1,874 Mill. sh)	91·3 (= 2,031 Mill. sh)	98·4 (= 2,074 Mill. sh)	87.8 (= 5,978 Mill. sh)
All other products	24·4 (= 579 Mill. sh)	8·7 (= 207 Mill. sh)	1·6 (= 32 Mill. sh)	12·2 (= 818 Mill. sh)
Total Domestic Export	100·0 (= 2,453 Mill. sh)	100·0 (= 2,238 Mill. sh)	100·0 (= 2,106 Mill. sh)	100·0 (= 6,796 Mill. sh)

Source: [5], p. 47.

originally established. If the expected increase in trade between the East African Community and the European Community is supported by capital and technical aid, then the forces of integration might be stronger than the present institutionalized ties within the East African Community. Any attempt to review the E.A.C. Treaty will be faced with the "trade creation" of the E.C. and other markets in industrial countries. The recent experience of the E.A.C. reveals a further deficiency in present integration theory, because it does not treat the relative size of integration systems as a decisive structural factor for success.

TABLE 5. *Comparison of population and Gross National Product between the European Community (E.C.) and the 46 African, Caribbean and Pacific countries (A.C.P.) 1972*

	Population millions	Gross National Product	
		Mill. U.S. $	U.S. $ per capita
European Community (E.C.)	254·930	741,210	2907·50
46 African, Caribbean and Pacific countries (A.C.P.)	265·784	40,820	153·58
A.C.P. as percentage of E.C.	104·3	5·5	5·2

Source: F.A.O: Commodity Review and Outlook 1974–1975. Rome 1975, p. 35; International Bank for Reconstruction and Development: World Bank Atlas, Population, Product and Growth Rates. Washington, D.C. 1974.

3. OVERALL ASSESSMENT

3.1. *Concepts of integration*

As already noted there are two basic concepts for the multinational integration of the agricultural sector: the "liberal" and the "dirigistic or planned" approach. The "liberal" approach uses the market and the "dirigistic" approach uses the plan to achieve a higher degree of integration. Both models have certain prerequisites, which must be complied with before any integration policy becomes a meaningful instrument. In the "liberal" approach to regional integration of sovereign states, the intracommunity market forces, after the removal of tariffs and other barriers, allocates, via a convertible currency, the resources between countries and economic sectors. In the "dirigistic" approach, a supranational planning body has to allocate national resources to reach production goals.

Theoretically, both models are capable of reaching a high degree of economic integration between countries and economic sectors. In practice, however, no overall supranational planning body has evolved in the planned economies of Eastern Europe, because all countries consider the risk of a lessening in national sovereignty as too great.

Despite the generally accepted axiom that market forces determine prices, those for most agricultural products are still regulated by the state, even in classic market economies. To some degree with exception of the E.C., no integration system has been able to fix agricultural prices at a community level. A complex planning, proposing and decision-making machinery is needed to design and to administer a common agricultural price system for several countries. It has to account for spatial and seasonal differences to promote the flow of commodities. The development of such a sophisticated system is beyond the present research and administrative capacity of most integration systems in developing countries. It is, therefore, no surprise that the East African Community has not succeeded in establishing common agricultural prices as proposed in article 14. It might even be doubted, considering the limitations of its administrative capacity, whether such a goal is economically feasible,[4] even if very desirable in the long-run.

3.2 *Constraints to integration*

a. *Political and institutional constraints*

The countries of the E.A.C. – like other developing countries – need an intensification of the political process. A complex network of interaction of goal-setting and goal-controlling between and at the local, regional and national levels has to be established by positive and negative feed-backs between the various elements of society. The mobilization of internal political forces for nation-building is of higher priority than regional integration, which at present has no clearly defined, feasible goals, and no well co-ordinated political machine to achieve them.

Another political constraint to further integration in East Africa which should be mentioned is that the basic political and economic philosophies in all three East African countries are drifting further and further apart. The agricultural sector *per se* has structural approach to farming in the region which makes any trend toward free agricultural trade highly unlikely. This departure from traditional patterns is such a basic policy feature, with so much depending on its success, that it will probably be protected from any competitive distortion from intra-community trade.

Following the Arusha Declaration of 1967, Tanzania decided to nationalize wholesale trade and most major economic activities. In Uganda, the expulsion of British passport-holders of the Indian Community affected the entire national economy. In Kenya and Uganda, both Governments tend to acquire majority or dominant participation in several enterprises. In a recent article Ghai [15] has analysed the reasons which make both intra-community and international trade in the E.A.C. more and more difficult:

(a) continuously changing regulations in licencing imports in each member state;

(b) the lack of a common trade policy and a continuous neglect of the trade and monetary agreements stipulated by the Treaty. Imports into Tanzania from Partner States, for example, require import licences for all 26 agricultural products listed in article 13, annex III. Similar measures now exist in Uganda (for 10 agricultural products) and Kenya (4 agricultural products) [13]. The control of import licences may have been used to reintroduce quantitative restrictions;

(c) the purchasing policy of parastatal marketing organizations which engage in "one channel marketing", forbidden under article 16 as discriminatory practice;

(d) the decisive factor which is causing disintegration is that due to the hidden changes in the value of the East African shilling, intra-community trade is now being paid for in foreign convertible currency.

b. *Monetary constraints*

As mentioned above, the most severe and persistent constraint to the trade of agricultural products with the E.A.C. is the increasing scarcity of convertible foreign exchange. On balance of payment-grounds, more and more barriers have been erected. The process of disintegration has now (1976) reached dimensions which are no longer of a short- or medium term nature. As long as the exchange rates for the three countries in the E.A.C. are not adjusted individually and according to their real value, the agricultural sector will face further setbacks in the process of economic integration.

The common management of the commonly owned Service Corporations in the East African Community forces the countries to maintain the monetary fiction that all currencies are correctly fixed at par. Otherwise the assets and liabilities held by each country would immediately have to be adjusted to the newly agreed exchange rate, which could well lead to the eventual dissolution of the Corporations. Because clear deviations from the official exchange rate are obvious, serious structural imbalances within the E.A.C. must exist. The Kenyan Shilling seems undervalued with in the Community and the Tanzanian and Ugandan Shilling are overvalued.[5] As long as political considerations impede a re-adjustment of currencies no economic assessments of comparative advantages, and therefore of future intra-community trade prospects for agricultural products, can be made.[6]

4. CONCLUDING REMARKS AND POLICY RECOMMENDATION

The obstacles to further integration of the agricultural sector in the E.A.C. have increased over the years. In any review of existing Treaty arrangements and of the experience of other integration systems, one conclusion is inescapable: the agricultural sector can neither be the leading force toward integration, nor can any economic integration succeed, where basic economic and political philosophies differ, where the search for national identity and sovereignty has the highest priority and where the necessity to earn convertible foreign exchange dictates and overshadows all foreign trade policy.

Confronted by these numerous problems the realities of the situation must be grasped: the goals of the Community as regards the economic integration of the agricultural sector, must be re-considered and re-written if they are to be at all relevant. Modest goals [13] such as improvement of agricultural statistical services, export marketing research, commodity research within the Community, standardization of agricultural products, and in the exchange of economic information (prices, harvest forecasts and yield monitoring, storage coverages, etc.) must be achieved before deliberate steps towards economic integration are feasible and meaningful to the governments and peoples of East Africa.

5. SUMMARY

The study presents a brief synopsis of the size and conditions of the world's ten major integration systems with special emphasis on the East African Community (E.A.C.) which also attempts at integrating the agricultural sector.

By using the external trade statistics, the study points out that total trade, as well as agricultural trade, has actually decreased within the E.A.C. since the Treaty of 1967 came into operation, amounting in 1974 to less than 5 per cent. This decline of integration is no surprise, since (a) the capacity of the internal market for agricultural products is too small compared with the markets of the northern hemisphere; (b) trade expansion into those markets is the major source of foreign convertible currency; (c) the administrative capacity and decision-making structures within the E.A.C. are at present too inadequately organised to implement a common agricultural price policy as in the case of the European Common Market. Special emphasis placed on the divisive results of the fiction of exchange rates fixed at par within the E.A.C., which hamper an increase in intra-community trade.

Finally, in the light of these current institutional and economic constraints, it is proposed that more modest goals should be attempted – standardization of farm products, expansion of harvest and price monitoring, as well as common marketing research, if, as planned, a revision of the present Treaty is to be agreed on.

NOTES

[1] Due to a lack of consistent data we have not included: the Caribbean Free Trade Association, the Andean Common Market and some smaller regional groupings such as Economic Cooperation among Maghreb Countries in North Africa or between Pakistan, Iran and Turkey [28].

[2] 47·7 per cent of the world's population (China, India, U.S.A., Indonesia, Japan), producing 44·4 per cent of the World's Gross National Product, live in countries with more than 100 Mill. inhabitants (USSR as part of COMECON has been excluded). They are in general large enough to guarantee *per se* economies of scale in industrial production without requiring the benefits of regional integration. Without the U.S.A. and Japan, the population percentage of the remaining three Asian countries drops to 40·3 per cent and the GNP more drastically to 10·1 per cent [18].

[3] Dumke ([3], p. 25) has recently resurrected an interesting argument of 1836, which explains how the founders of the Zollverein achieved agreement between the very small German princedoms on a customs union. The revenues from customs were distributed per head in the Zollverein, thus increasing the revenues of the smaller federal states at the expense of their larger associates.

This singular effect of revenue sharing on a per capita basis is due to the fact that the smaller the area, the larger the ratio between its borders and its area; consequently the administrative costs of tariff collection and smuggling controls are higher. However, this does not apply in the case of Zanzibar, where controls and tax collection can be much more effectively organized in comparison with the former very small, land-locked German princedoms. Even more to the point, if the same principle of central distribution of revenue tariffs per capita were applied in the E.A.C., Zanzibar would be deprived of its main source of revenue.

Undoubtedly Germany then had more material resources per capita than the E.A.C. today, if production from pasture, forest and huge coal reserves is added to yields obtained from arable land. The lower German population growth rate must also be considered. Between 1850–1870, the rate was 0·7 rising to 1·38 per cent in the period 1900–1913, compared with the present 3·3 percent population growth rate in the East African region; the required annual GNP growth rate (in the Zollverein) to achieve per capita income increases was correspondingly much lower. Today scientific developments have made available many natural and social remedies which could be utilized to redress the imbalances of the comparisons in E.A., but it may be worth mentioning that at that time in Germany, human resources were comparatively well developed. Compulsory primary education had been in force for several decades and extended even to the rural areas.

[4] The E.A.C., with 1·7 Mill. km^2 extends over a larger area ([16], p. 4) than the E.C. with 1·5 Mill. km^2. The E.C. has a considerably higher traffic and communication density ([24], p. 96) which also reduces the administrative control costs per unit of marketed agricultural production.

[5] The official exchange rate for all three Community currencies sets one U.S. $ at 8·16 shillings. But on the stock exchange in Zürich, one U.S. $ buys 10 Kenyan, 30 Tanzanian and 50 Ugandan shillings (Newsweek, January 5th, 1976, p. 61). Although we do not argue that these rates represent the "equilibrium" exchange rates, they nevertheless indicate the monetary fiction of exchange rates fixed at par for the East African Shilling within the Community.

[6] The last available agricultural price data for E.A.C. cereals cover the year 1973. According to these studies Kenya had higher farm gate prices for maize and wheat but considerably lower retail prices for maize meal and white bread than Tanzania and Uganda ([8], IA, p. 10, IB, p. 9). In 1973 in Kenya, farm-gate and retail prices of rice were only 54 and 83 per cent respectively of Tanzanian prices ([8], IC, p. 11).

REFERENCES

[1] Balassa, B. (1961), *The Theory of Economic Integration.* Homewood, Illinois.

[2] Chenery, H. and Syrquin, M. with the assistance of Elkington, H. (1975), *Patterns of Development 1950–1970.* A World Bank Research Publication. Oxford.

[3] Dumke, R. H. (1974), *The Political Economy of Economic Integration. The Case of the German Zollverein of 1834.* Discussion Paper No. 153. Institute of Economic Research, Queen's University, Kingston, Ontario.

[4] East African Community (1972), *Treaty for East African Co-operation.* Arusha, February.

[5] East African Community (1974), *Review of Economic Integration Activities within the East African Community.* Common Market and Economic Affairs Secretariat. Arusha.

[6] East African Community, Custom and Excise Department (1974), *Annual Trade Report for Kenya, Tanzania and Uganda for the Year Ending on December 31st.*

[7] East African Community, Common Market and Economic Affairs Secretariat (1975), *Report of Seminar on East African Co-operation in Agricultural Development.* Held at the East African Community Regional Headquarter, Nairobi, January 13–22, 1975. Arusha, June.

[8] East African Community, Common Market and Economic Affairs Secretariat (1975), The East African Statistical Department, *Commodity Situation Reports, No. 1A (Maize), No. 1B (Wheat), No. 1C (Rice).* Nairobi.

[9] East African Community, E.A. Statistical Department, *Economic and Statistical Review No. 29, 37, 43, 52.*

[10] FAO: *Trade Yearbook.* Rome 1967, 1968, 1972, 1973.

[11] FAO (1974), *Production Yearbook 1973.* Rome.

[12] FAO (1975), *Commodity Review and Outlook 1974–1975.* Rome.

[13] FAO (1974), *Report to the Governments of Kenya, Tanzania, Uganda on Agricultural Co-operation in the East African Community*, based on the work of T. J. Aldington, Rome.

[14] FAO (1975), *Monthly Bulletin of Agricultural Economics and Statistics*, Vol. 24.

[15] Ghai, D. P. (1974), State Trading and Regional Economic Integration. The East African Experience. *Journal of Common Market Studies*, Vol. 3.

[16] Hazelwood, A. (1975), *Economic Integration. The East African Experience. Studies in the Economics of Africa.* London.

[17] Helleiner, G. K. (1972), *International Trade and Economic Development.* Penguin Harmondsworth.

[18] International Bank for Reconstruction and Development (1974), *World Bank Atlas, Population, Per Capita Product and Growth Rates.* Washington, D.C.

[19] Kaser, M. (1967), *Comecon. Integration Problems of the Planned Economies.* 2nd ed. London, New York, Toronto.

[20] Kyeyune, L. L. (1975), *Economic Integration and National Development in East Africa. The Agricultural Sector.* Social Science Conference, 17/18 December, University of Nairobi.

[21] Mutharika, B. W. T. (1972), *Toward Multinational Economic Co-operation in Africa.* Praeger Special Studies in International Economics and Development. New York, Washington, London.

[22] Ndegwa, P. (1968), *The Common Market and Development in East Africa.* East African Studies 22, Nairobi.

[23] Oliver, R. and Fage F. D. (1973), *A Short History of Africa.* Penguin. Harmondsworth.

[24] Statistisches Amt der Europäischen Gemeinschaften: *Grundzahlen der Gemeinschaft. Vergleich mit verschiedenen europäischen Ländern, Kanada, den Vereinigten Staaten von Amerika, Japan und der Union der Sozialistischen Sowjetrepubliken 1973–1974.* Luxemburg, Without year.

[25] Swann, D. (1972), *The Economics of the Common Market.* 2nd ed. Penguin, Harmondsworth, England.

[26] Union of International Associations (1974), *Yearbook of International Association.* 15th ed. (Publication No. 226). Brussels.

[27] Weber, A., Faktorpreise, Faktorproduktivität und Technologie in der amerikanischen, europäischen und japanischen Landwirtschaft von 1880 bis 1965. *Z. für Wirtschafts- und Sozialwissenschaften*, Vol. 1973.

[28] Wionczek, J. S. (ed.) (1969), *Economic Cooperation in Latin America, Africa and Asia.* A Handbook of Documents. Cambridge, Mass. and London.

J. A. AKINWUMI AND A. J. ADEGEYE*

*Prospects and Problems of a Common Agricultural Policy Among West African States**

1. INTRODUCTION

In Europe, America, Africa and Asia, the Twentieth Century has witnessed movements towards economic unions and/or common trading arrangements. The fact that early movements towards common trading policies were among advanced countries, indicates that even economies at such advanced stages of development see ample benefits to be derived from such associations. The early theories of economic co-operation centred on trade creation and trade diversion are said to be static and are therefore inapplicable to the situation in developing countries. This is due in most cases to their low volume of trade in the world and even between the developing states themselves. Hence economic co-operation here is seen in the light of its contribution to economic growth and/or structural transformation of these developing countries. Thus alliances like those for cocoa and coffee are seen as a means of obtaining high export prices to provide the foreign exchange necessary for economic development.

The movements in West Africa which culminated in the setting up of an Economic Community of West African States (ECOWAS) had their origin in various groupings, both during the colonial era and post independence period. Such movements include the Federation of French West Africa (AOE) which enjoyed close links in trade, finance and common services. In 1959 there was the West African Customs Union (UDAO) which included Dahomey (now Republic of Benin), Ivory Coast, Mauritania, Niger, Senegal, Mali and Upper Volta. This soon ran into trouble over allocation of import duties. There were others like the Ghana–Volta customs union 1961, Ghana–Guinea agreement of 1961, and Senegambia in 1962.

In 1966 at Niamey, and in 1967 in Ghana articles of association for a West African Economic Community were considered. However, it is the ECOWAS which has included all the countries of West Africa. These countries are listed in Table 1. The treaty setting up the community includes articles on most facets of socio-economic policies. But this paper considers only two

* J. A. Akinwumi and A. J. Adegeye are Lecturers in the Department of Agricultural Economics, University of Ibadan, Nigeria.

TABLE 1. *ECOWAS: Arable land and labour resources*

	Arable land	Total population	Rural population	Arable land land per rural person
	'000 km2[1]	million[2]	million[3]	km^2
Dahomey	86·72	2·69	1·88	0·046
Gambia	8·47	0·36	0·25	0·034
Ghana	178·90	8·64	6·05	0·096
Guinea	183·77	3·92	2·74	0·067
Guinea Biscau	36·13	0·56	0·39	0·093
Ivory Coast	241·85	4·31	3·02	0·081
Liberia	83·53	1·52	1·06	0·079
Mali	930·00	5·05	3·54	0·263
Mauritania	773·03	1·16	0·81	0·954
Niger	950·25	4·02	2·81	0·338
Nigeria	692·83	55·07	38·56	0·013
Senegal	147·14	3·93	2·75	0·053
Sierra Leone	53·81	2·55	1·79	0·030
Togo	42·00	1·97	1·38	0·031
Upper Volta	205·65	5·38	3·77	0·054
Total	4,614·08	101·13	70·80	

Scources: (1) Olayide S.O. (A).
(2) U.N. Demographic Yearbook. 1973.
(3) Assumes 70% of total population is rural.

of the articles dealing with Agriculture. These articles are 34 and 35. Article 34 states that

> Member States undertake to work towards the harmonisation of their internal and external agricultural policies in their relation with one another. While article 35 states that "Member states undertake to take all measures of a common policy especially in the fields of research, training production, processing and marketing of the products of agriculture, forestry, animal husbandry and fisheries.

Though at present no measure has been announced as regards such common policies, this paper attempts to evaluate the problems and prospects of common policies in the areas identified by the treaty. This exercise is embarked upon mainly because the authors feel that policies on these lines are likely to have greater effects on the population of West Africa when compared to policies relating to other sectors of the economies such as industries. First, the bulk of the population (between 70 and 90 per cent) depend on agriculture for a livelihood. The proportion depending on agriculture is less than 20 per cent for most EEC countries. Secondly, the GDP contributed by agriculture ranges between 20% in Liberia to 51% in Niger. While in the EEC, it ranges between 3% in U.K. to 15% in Ireland. Even in a country like U.K. where the contribution of agriculture to GDP is small and the percentage of people in agriculture is only about 10 per cent, the working of a common EEC agricultural policy has featured most prominently in British

TABLE 2. *G.D.P. contributed by agriculture in some countries of ECOWAS and EEC*

West Africa (1969)	Percentage	Europe (1972)	Percentage
Niger	51	Ireland	15
Ghana	47	Italy	8
Nigeria	44	Denmark	7
Togo	43	France	6
Upper Volta	37	Belgium	4
Ivory Coast	30	Germany	3
Liberia	20	U.K.	3

Scource: U.N. Yearbook of National Accounts Statistics, 1973.

TABLE 3. *Estimates of population, total and per capita domestic product, among members of the ECOWAS 1970*

Countries	Population(1)	Gross Domestic(2) Product GDP	Per capita GDP
	million	*million US*	*$*
Dahomey	2·69	221	82
Gambia	0·36	46	128
Ghana	8·64	2214	256
Guinea	3·92	320	82
Guinea Bissau	0·56	127	227
Ivory Coast	4·31	1495	347
Liberia	1·52	417	274
Mali	5·05	267	53
Mauritania	1·16	180	155
Niger	4·02	298	74
Nigeria	55·07	7711	140
Senegal	3·93	852	217
Sierra Leone	2·55	451	177
Togo	1·97	264	134
Upper Volta	5·38	347	65

Sources: (1) Population: *U.S. Demographic Year Book,* 1975.
(2) G.D.P., *U.N. Yearbooks of National Accounts Statistics,* 1973.

negotiation. This is because that country is concerned about the welfare of her few farmers.

It seems certain, therefore, that the articles relating to agriculture are the most important as far as the majority of citizens in West Africa are concerned and a common agricultural policy on the lines stated in articles 34 and 35, needs careful examination in order to identify prospects and problems of getting a workable agricultural policy. This exercise, we hope, would not only be of use to those who would see to the realisation of these objectives but could be of value to other developing areas in Africa, Asia, and Latin America where agriculture is the mainstay of the economy and there are drives towards economic unity.

Table 3 shows the countries in ECOWAS together with their population, total and per capita GDP in U.S. dollars. GDP per head is low and ranges between 65 U.S. dollars in Upper Volta to 347 in the Ivory Coast. A large proportion of the GDP in West Africa is derived from agriculture. This is shown in Table 2 which also gives the comparable situation for Europe.

This further underlines the importance of agriculture in the West Africa sub-region. In other words, every agricultural policy is bound to have very significant impacts on the majority of the people living in the region.

In the rest of this paper, Section 2 discusses the agricultural economy of West Africa, Section 3, the resource endowments, Section 4, trade, Section 5, gains and problems of integrated agricultural policies, while Section 6 presents the summary and conclusions.

2. THE AGRICULTURAL ECONOMY OF WEST AFRICA

West Africa is a primary producer of agricultural commodities both for domestic consumption and for exports. The majority of the population in each country (70–80 per cent) lives in rural areas and is engaged in agriculture. The crops grown are similar, running across the area from east to west along the same latitudes. The rainfall pattern, with heavy rains along the coast, gradually thinning out as one moves northwards, dictates the vegetation and cropping pattern. There are, however, small pockets whose extra rainfall or special soils support the growth of some specific crops, such as rubber in Liberia and parts of Nigeria.

2.1. *Crops grown by member countries*

Table 4 shows the main crops grown by each of the 15 countries listed in the ECOWAS charter. The proportions of these crops produced by each country are presented with the total actual tonnage so that the relative importance of each country as a producer can be seen while the absolute magnitude of its production in 1971 can be estimated. The table indicates that for major food grains, the leading producers are Nigeria for sorghum and millet, followed by Niger, Mali and Senegal. Nigeria tops the list in maize followed by Ghana and Ivory Coast. The largest volume of rice produced came from Nigeria, Guinea, Sierra Leone, Ivory Coast and Upper Volta. The smallest total production for millet and sorghum came from Gambia, Ivory Coast and Mauritania while three countries – Guinea Bissau, Liberia and Sierra Leone had no production figures for 1971. Dahomey produced negligible quantities of paddy rice, while Mauritania and Togo came close to nothing, too. No estimate of maize in Gambia and Guinea Bissau were available while Mauritania and Niger had very little output.

For the roots and tubers, Nigeria produced the largest crops of yams and potatoes as well as cassava followed by Ghana, the Ivory Coast and Togo. Mauritania, Senegal and Sierra Leone had no production figures while Guinea Bissau and Upper Volta produced the least cassava.

Senegal led the production of groundnuts followed closely by Nigeria, Niger, Mali and Gambia. Ghana is number one producer of cocoa, Nigeria was second and Ivory Coast third in 1971.

For several other crops, total production was negligible and individual countries' shares were small. Several countries had no production of some crops as shown in Table 4. It is however possible that some quantities were produced but FAO had problems with the estimation. Such crops include

TABLE 4. *Share of total output of crops and livestock produced by each ECOWAS member in 1971, percentage*

Crop/Animal	Dahomey	Gambia	Ghana	Guinea	Guinea Bissau	Ivory Coast	Liberia	Mali	Mauritania	Niger
Maize	8·0	–	17·0	3·0	–	11·0	1·0	3·0	0·1	0·08
Millet and Sorghum	0·6	0·4	2·0	1·4	–	0·5	–	8·0	0·8	10·0
Rice Paddy	–	2·3	2·7	15	1·3	13	3·5	5·7	0·5	1·5
Yams and Sweet Potatoes	3·8	–	5·2	0·5	–	9·0	–	0·4	–	–
Cassava (Manioc)	9·1	0·05	12·0	3·6	0·3	4·3	2·8	1·2	–	1·4
Dry Beans	2·5	–	–	–	–	–	–	0·9	1·0	15·0
Other Pulses	–	–	–	–	–	–	–	–	–	–
Groundnuts	2·2	4·4	2·7	0·9	2·5	1·6	–	5·4	–	9·7
Cocoa	0·9	–	43·0	–	–	19·4	0·2	–	–	–
Coffee	0·9	–	1·7	0·4	–	87·4	1·5	–	–	–
Palm Oil	–	–	9·0	–	3·0	9·0	–	–	–	–
Palm Kernel	12	0·4	7·0	–	2·3	3·6	3·2	–	–	–
Cotton Seed	18·5	–	–	–	–	10·8	–	16·2	–	3·0
Cotton Lint	–	–	–	–	–	18	–	23	–	3·7
Rubber	–	–	–	–	0:07	9·4	54·0	–	–	–
Citrus Fruits	–	–	52·0	48·0	–	–	–	–	–	–
Bananas	–	24·0	–	24·0	–	52·0	–	–	–	–
Pineapples	–	–	14·0	6·0	–	80·0	–	–	–	–
Onions	–	–	26·0	23·0	–	–	–	–	–	50·0
Copra	–	–	21·0	–	–	58·0	–	–	–	–
Tobacco	–	–	–	–	–	–	–	–	–	–
Dates	–	–	–	–	–	–	–	–	75·0	25·0
Cattle	1·7	0·8	2·6	5·2	0·8	1·1	0·08	15·5	7·5	12·4
Sheep	2·0	0·5	5·0	1·7	0·2	3·0	0·6	21·0	9·9	9·9
Goats	1·5	0·2	3·8	1·1	0·4	1·7	0·3	12·7	5·4	14·0
Poultry	–	0·1	4·8	1·8	–	3·4	0·7	5·9	1·06	2·96
Pigs	6·4	56·0	3·7	0·5	2·8	3·1	1·5	0·62	–	0·5
Asses	–	–	1·05	0·1	0·1	–	–	20·0	9·8	16·2
Horses	–	–	0·4	–	–	–	–	18·2	2·4	18·9
Camels	–	–	–	–	–	–	–	15·9	51·1	30·4

TABLE 4. (*continued*)

Crop/Animal	Nigeria	Senegal	Sierra Leone	Togo	Upper Volta	Total Production '000 metric tons
Maize	48·0	2·0	0·7	4·0	2·0	2,529·5
Millet and Sorghum	59·0	6·0	1·2	9·0	–	10,613
Rice Paddy	21·0	3·9	14·2	0·8	14·1	2,626
Yams and Sweet Potatoes	78·0	–	–	6·6	0·3	17,421
Cassava (Manioc)	56·0	–	–	9·0	0·3	13,136
Dry Beans	70·0	2·2	–	2·0	9·0	9·96
Other Pulses	27	–	–	7	67	120
Groundnuts	32·3	35·0	0·05	0·7	2·6	2,629·3
Cocoa	33·0	–	0·7	3·0	–	924·7
Coffee	1·3	–	3·0	3·9	–	306·7
Palm Oil	73·0	–	7·0	–	–	689·4
Palm Kernel	59·0	–	9·0	3·4	–	524·6
Cotton Seed	30·0	8·0	–	5·0	9·0	259·0
Cotton Lint	36·0	–	–	5·5	14	109·0
Rubber	36·5	–	–	–	–	137·5
Citrus Fruits	–	–	–	–	–	166·0
Bananas	–	–	–	–	–	358·0
Pineapples	–	–	–	–	–	218·0
Onions	–	–	–	–	–	64·0
Copra	14·0	–	–	7·0	–	14·6
Tobacco	96·0	–	–	–	4·0	18·3
Dates	–	–	–	–	–	20·0
Cattle	32·7	7·6	0·7	3·4	8·2	35,527·0
Sheep	29·0	9·9	0·2	–	7·0	28,142·0
Goats	53·0	–	0·4	–	5·9	44,587
Poultry	35·1	–	1·27	0·8	4·2	236,134·0
Pigs	15·7	3·3	0·6	2·6	2·6	5,355
Asses	35·8	8·3	–	0·09	8·3	2,288·2
Horses	31·4	21·2	–	0·09	7·3	953·9
Camels	1·5	0·5	–	–	0·5	1,349·0

Source: Africa South of the Sahrar, 1974: Europa Publications Ltd., London.
Note: Figures are FAO Estimates. Where no figures are given, either the production is negligible or an estimate was impossible.

citrus, of which Nigeria produces a small quantity, bananas which also grow in the country but have not been shown in the production estimates.

Although Nigeria featured prominently as a major producer, her production of several crops is very small compared to many other ECOWAS members when calculated on a per capita basis.

The similarities in production patterns limited the exchange of food crops among the countries and the poor infrastructural development and lack of processing/packaging further reduced the chances of trade.

2.2. *Livestock development*

The eco-systems favour the production of livestock in countries farther away from the rain forest belt. These areas are free of trypanosomiasis which cause death to livestock particularly beef cattle. These livestock, including sheep, goats and beef cattle are sent down to the coastal areas for sale. From all indications the supply of livestock is falling short of demand. The reasons are many and include the increased demand for livestock by the coastal countries whose incomes have risen in recent years. Secondly, the nomadic Fulani system of management restricts the offtake of slaughtered cattle to as low as 15 per cent per annum. Furthermore, droughts have destroyed almost half of the cattle in the Sudanian and Sahelian savannahs of West Africa. It is generally believed that the rate of livestock production can be increased in West Africa if there can be stratification whereby cattle reared in the savannah are finished in the medium to heavy rainfall area (given that the susceptibility of the zebu cattle to trypanosomiasis can be overcome). A common agricultural policy which aims at achieving this objective would have the effect of increasing livestock production in the sub-region.

As shown in Table 4, Nigeria is the largest single producer of cattle, sheep goats and poultry. Mali and Niger are next and Mauritania comes fourth. Mali occupies the second place in poultry production and Ghana is third. Gambia leads the way in pig production with Nigeria coming second. For horses, asses and camels, which are work and transport animals, production is largest in countries which have expansive savannah terrains. Nigeria, Mali, Niger, Mauritania and Senegal feature prominently in these. The production figures tend to mislead unless placed on per capita basis. If this is done, Nigeria appears to produce the least, given its large population. This has implications on intra ECOWAS trade in livestock.

Inter-state trade in livestock, especially cattle goes on among ECOWAS members in the north–south direction. Thus Niger, Mali, Upper Volta with higher per capita production sell livestock to Nigeria, Ghana and Ivory Coast, where per capita incomes are reasonably higher.

A harmonisation of processing and distribution of livestock products would help to reduce the high disparities between prices prevailing in countries of West Africa. For instance, some of Nigeria's imports of cattle from Chad, Niger and Dahomey are later sent enroute to Ghana – the price of beef in Ghana is sometimes twice what it is in Lagos–Nigeria.

The nomads would pose problems in any effort at harmonising livestock policies in West Africa. At present most nomads move freely across national boundaries in search for food and water for their cattle. Even when individual countries are aware of the need for settling nomads, the problem is not easy. A possible solution will have to provide for the establishment of common grazing reserves across national frontiers. These would provide necessary amenities for these nomads irrespective of where they come from. The cost would have to be borne by the community.

Another aspect of the problem of livestock development is that the Fulanis rarely sell their young animals. Hence the possibility of increasing meat output through stratification in production between the north and south in West Africa is severely limited. Therefore, ranches would have to be set up in some of the southern countries, where trypanosomiasis resistant breeds of cattle can be reared. But then the problem of the nomads would still need to be tackled in the context of the whole West Africa.

All these suggestions imply a complete submission of individual countries to an agricultural policy-making body which would have the power to dictate where, how, and what quantity of different products must be produced. The chances of such total submission are very slight.

3. RESOURCE ENDOWMENTS

3.1. *Land*

As shown in Table 2 there is abundant supply of arable land. A total of 4614·08 thousand km^2 is distributed as shown in the table. Some countries are more thickly populated than others and some areas within the same country have a higher concentration of population. This leads to wide variations in per capita arable land available. But, given that agricultural progress will further release people from the land, there is an adequate supply of land for agricultural production.

3.2. *Labour*

Column 2 of Table 1 shows the total population of the sub-region while column 3 shows the proportion of the population that is assumed to be rural. By 1970, the sub-region had a population of about 100 million on land area of 4·6 million square kilometers. At least 70 per cent of the population live in rural areas. The rate of growth of population in this sub-region is put at between 2·5 and 3 per cent per annum. Average population density ranges between 70 persons per square mile along the Coastal countries to less than 5 persons per square mile in the inland countries like Niger, Mali and Upper Volta. These disparities have led to the existence of pockets of over-cultivated land with low marginal productivity. There is also the question of age distribution of farming populations. The general trend is for the youth to obtain some education and drift to the cities or industrial centres in search of clerical jobs. This has gradually raised the mean age of active farmers and consequently lowered productivity as their capacity to work declined.

A common phenomenon of labour in West Africa is its migratory nature, both within a country and between countries. The causes of migration are diverse but the most important seems to be the quest for economic opportunities particularly in agriculture. In fact, it is estimated that over 90 per cent of migrants from Niger, 84 per cent from Upper Volta and 74 per cent from Togo were involved in agriculture. In the 1960's Upper Volta, Niger and Mali recorded between 17 to 42 per cent of their total population migrating to other areas as farm workers. For instance, most of the migrants from Upper Volta were found in Ivory Coast where they are employed on plantations. Nigerians also went to Equitorial Guinea to work on plantations.

A factor which has eased the flow of labour from one country to another is the existence of ethnic affinity between citizens of two or three countries. Fulanis, for instance, are found in practically all countries of West Africa. The Yorubas who are the major ethnic group in Western Nigeria are also found in the Republic of Benin. Again, Ghana and Togo have similar situations just like Gambia and Senegal. The existence of job opportunities is easily identified and communicated to relatives wherever they may be. This situation would probably continue on a larger scale if within the community no restrictions are placed on labour movement so that labour could move to areas where its marginal productivity is higher. This could have been an ideal situation because of shortages of labour are often complained of in rubber, coffee and cocoa plantations. The shortages of labour in these plantations are due to a number of factors. Among them can be named the recent migration of workers to urban areas in search of real or imagined employment, the emancipation of women and free primary education, all of which have deprived the farmer of his traditional sources of labour supply. This has been worsened in recent times by the existence of the ageing people in agriculture.

Can the present treaty make for easy migration of labour particularly when it is known that some countries impose restrictions on migrants from neighbouring countries? For instance, the Republic of Benin in 1975 closed its borders with Togo, the main reason being that there was a fear that Togolese nationals were coming to the Republic of Benin to cause political trouble. In the same way Nigerians were sent packing from Ghana and recently from Equitorial Guinea, where they had been working on plantations for years. This is a major challenge to the policy-makers in the ECOWAS union.

If agreements could be reached towards a reasonable migration policy, it would serve a lot of beneficial purposes. First, agriculture would have sufficient labour, which it is hoped would help to increase its productivity. Again, since most of the people who move are unable to get employment in their respective countries, it would go a long way to solving unemployment in these countries, thereby making the maximum use of labour resources.

3.3. *Capital*

There is very little capital investment so far in the region's agricultural enterprises. The implements are still the simple hoe and cutlass; farm structures, machinery and equipment are almost non-existent. In a few cases, plantation agriculture has resulted in some investment in processing equipment, such as oil mills and rubber processing equipment. Cotton ginnery and textile mills are also important sources of capital investment in agriculture.

If the existing permanent crops (cocoa, oil palm, rubber, cashew, etc.)

were capitalised, substantial amounts of capital could be estimated. Although water dams are primarily for hydro-electric power, they could be turned to agricultural investment through irrigation.

3.4. *Moisture and water*

This is a critical factor determining the pattern of agriculture in West Africa. Rainfall is high near the coast and gradually thins out towards the north. Much of the rain water runs off into the rivers and eventually enters the Atlantic, carrying with it a lot of rich surface soil. Lack of water critically limits production and forces large numbers of farmers to seek alternative jobs or remain idle for parts of the year. The production and annual yields of such crops as rubber, swamp rice, oil palm and citrus, depend on having considerable amounts of moisture.

At present various member countries are setting up dams with the intention of retaining some of the water and using it for irrigation so as to develop, among other things, market gardening, and to have year round production of some crops. These projects generally require heavy capital expenditure to harness the water.*

3.5. *Technology*

The production technology in the region has remained simple and disturbingly unchanging for centuries. The maximum productivity possible with this technology can be assumed to have been attained. New technology must be applied in order to raise productivity and reduce the drudgery of farm work. Since agriculture provides employment for up to 70 per cent or more of the population, the policy-makers generally fight shy of introducing capital-intensive technologies which would rapidly displace millions of farm workers. However, they forget that the next stage to move into is the processing, packaging and distribution of these products such that the new agro-industries can complement farming and absorb much of the released man power. With the stock of horses, donkeys, asses, bullocks and camels, the animal draft power can continue to be used at an intermediate technology level in parts of West Africa. The displacement of labour from the land is less rapid under such conditions. But the most important aspect of technology is to employ such new methods that will minimize total production costs so as to make the West African region competitive with other parts of Africa as well as the world, given the transportation costs for agricultural commodities.

4. TRADE

4.1. *Trade between ECOWAS, the rest of Africa and the World*

Tables 5 and 6 indicate the values of imports and exports of some ECOWAS members in 1973. Imports from the rest of Africa ranged from \$5·4 million in Togo to \$37·3 million U.S. dollars in Ivory Coast. Imports from the entire world ranged from \$49·7 million in Upper Volta to \$1,505·03 million in

* The Kanji dam project cost Nigeria close to 200 million Naira.

TABLE 5. *West African States: imports from Africa as percentage of total imports 1970*

Countries	Imports from Africa (1)	Imports from the world (2)	1 as percent of 2
	'000 U.S. $	*'000 U.S. $*	*percent*
Dahomey	9,901	76,323	12·97
Ghana	24,384	433,639	5·61
Ivory Coast	37,294	398,148	9·35
Liberia	9,016	162,228	5·56
Mali	11,571	59,265	19·52
Mauritania	9,476	62,993	15·04
Niger	7,576	52,991	14·30
Nigeria	14,623	1,505,028	0·97
Senegal	28,565	218,001	13·10
Sierra Leone	15,470	113,121	13·6
Togo	5,372	70,024	7·67
Upper Volta	12,835	49,660	25·85

Source: U.N. International Trade Statistics, 1973.

TABLE 6. *West African States: Exports to Africa as percent of total Exports 1970*

Countries	Exports to Africa	Exports to the world	1 as percent of 2
	'000 U.S. $	*'000 U.S. $*	*percent*
Dahomey	4,808	41,938	11·46
Ghana	5,703	341,415	1·67
Ivory Coast	49,699	455,571	8·93
Liberia	642	98,482	0·65
Mali	18,570	36,012	51·57
Mauritania	3,501	93,895	3·73
Niger	14,059	38,373	36·64
Nigeria	27,244	1,793,171	1·52
Senegal	12,926	34,887	37·05
Sierra Leone	2,896	222,374	1·30
Togo	3,220	56,354	5·71
Upper Volta	8,580	15,863	54·07

Source: U.N. International Trade Statistics, 1973.

Nigeria. These imports included processed agricultural commodities such as sugar, canned beef, frozen meats, tobacco, textiles, tomato puree and some vegetable oils.

The 1973 values of ECOWAS exports to other African countries ranged between $0·64 million in Liberia to $3,412·9 million in Senegal while exports to the entire world ranged from $15·9 million to $1,793·2 million in Upper Volta and Nigeria respectively. It is feasible that as processing and distribution of agricultural commodities among ECOWAS members increases, the gap between their imports from and exports to the outside world will narrow.

Looking at exports only to the traditional metropolitan countries of

TABLE 7. *Intra-State trade among the ECOWAS members: exports F.O.B. 1971 '000 U.S.*

Countries	Mauritania	Senegal	Mali	Ivory Coast	Dahomey	Gambia	Liberia	Niger
Mauritania	–	874	–	–	–	–	–	–
Senegal	–	–	–	–	–	–	–	
Mali	487	2900	–	9797	50	–	63	654
Ivory Coast	–	8447	4848	–	1404	–	499	1832
Dahomey	–	667	–	150	–	–	–	447
Gambia	–	–	–	–	–	–	–	–
Liberia	–	23	175	284	–	–	–	–
Niger	–	53	50	948	1706	–	–	–
Nigeria	–	2158	–	6623	90	–	145	247
Upper Volta	–	13	590	6094	49	–	–	153
Guinea	–	–	–	–	–	–	–	–
Ghana	–	–	77	396	152	–	126	13
Togo	–	30	–	273	776	–	–	155
Sierra Leone	–	6	3	–	263	91	–	–
Guinea-Bissau	–	–	–	–	–	–	–	–

TABLE 7. *(continued)*

Countries	Nigeria	Upper Volta	Guinea	Ghana	Togo	Sierra Leone	Guinea-Bissau
Mauritania	–	–	–	65	–	–	–
Senegal	–	–	–	–	–	–	–
Mali	–	–	314	2835	–	–	–
Ivory Coast	1267	5207	–	727	529	–	–
Dahomey	2055	–	–	333	547	–	–
Gambia	–	–	–	–	–	–	–
Liberia	209	–	233	418	–	97	–
Niger	9817	184	–	43	381	–	–
Nigeria	–	–	–	11670	–	4380	–
Upper Volta	–	–	–	1610	69	–	–
Guinea	–	–	–	–	–	–	–
Ghana	2836	32	–	–	475	–	–
Togo	1	15	–	351	–	–	–
Sierra Leone	54	1	–	–	–	–	–
Guinea-Bissau	–	–	–	–	–	–	–

Europe, Ghana, Niger, Dahomey, Senegal and Nigeria had the highest percentages of exports while Sierra Leone, Upper Volta and Mali had the lowest. Those with high export percentages can expect the formation of ECOWAS to affect their export earnings significantly.

4.2. *Intra-West African trade*

What, if any, has been the trade pattern among the countries constituting ECOWAS? There are no adequate data on traded agricultural commodities and the general similarities in food crop production reduces the trade potentials in these. However, ethnic affinities (e.g. Yorubas in Dahomey (now Republic of Benin) and Western Nigeria) as well as North–South differentials in the pattern of production (crops versus livestock) have led to substantial trade in agricultural products.

Table 4 illustrated the proportions of various crops and livestock produced by the member countries. Food crops have been essentially for home consumption but sorghum and millet have moved south as major ingredients for livestock feeds, especially in Nigeria. Cattle, goats and sheep have played prominent roles in the southward movement of products. The lack of processing and packaging has denied the centres of high population the consumption of many animal products such as milk which is a major by-product of cattle, goat and sheep industry in the northern parts of the region. Expansion of distribution of animal products along the East–West coast of West Africa must await the development of convenient packaging, refrigeration and efficient handling and haulage.

Tables 7 and 8 indicate the intra-West African trading activities in 1971. Trade balances can be deduced by combining these two tables. The value of imports and exports include re-exports of many processed agricultural commodities. If ECOWAS materialises, the trade among members can be expected to increase with the removal of some artificial barriers and the development of infrastructural facilities.

Ilori [2] showed that 98 per cent of Mali's intra-West African exports were made of agricultural commodities. Niger came second with 91 per cent while Upper Volta had 87 per cent. Ghana had 34 per cent and Nigeria had 20 per cent. Liberia's share was the least, only 2 per cent. For imports Ghana topped the list with 87 per cent, Togo had 40 per cent, Upper Volta 31 per cent and Nigeria and Niger had 6 and 2 per cent respectively.

In general, agricultural products have accounted for increasing proportions of Intra-West African trade. But the exchanges are not even over the countries and this could have far reaching effects on the ECOWAS. For example, with the exception of Ghana, Mali, Niger and Upper Volta, agricultural share of intra-State trade is less than 50 per cent in the ECOWAS. This is significant in view of predominance of agricultural commodities in the region's production.

4.3. *Direction of trade*

This can be examined at the intraECOWAS level, inter-African States level and the level of trade with traditional trading partners in metropolitan Europe. Olayemi and Olayide [3] have constructed the matrices for these three levels and they indicate movement of export trade from Ivory Coast to Mali, Senegal

TABLE 8. *Intra-State trade among the ECOWAS members – Imports C.I.F. 1971 '000 U.S. $*

1 Countries	Mauritania	Senegal	Mali	Ivory-Coast	Dahomey	Gambia	Liberia
1 Mauritania		4,900	3	155	1	1	936
2 Senegal	33	–	1030	10709	908	92	1
3 Mali	267	3,724	–	6231	1	–	17
4 Ivory Coast	1	7,185	1434	–	121	–	9
5 Dahomey	1	2,207	43	1904	–	–	21
6 Gambia							
7 Liberia		478	148	743	–	–	–
8 Niger	3	1,053	8	2848	380	–	2
9 Nigeria	–	879	–	996	294	–	47
10 Upper Volta	–	782	2241	1761	11	–	–
11 Guinea							
12 Ghana	41	61	3268	461	509	14	133
13 Togo	–	300	–	719	690	–	–
14 Sierra Leone		533	–	199	2	16	105
15 Guinea Bissau							

1 Countries	Niger	Nigeria	Upper Volta	Guinea	Ghana	Togo	Sierra Leone	Guinea Bissau
1 Mauritania		–	–	–	–	–	–	–
2 Senegal	1	2367	3	5	1	2	123	–
3 Mali	32	30	189	111	513	–	7	–
4 Ivory Coast	977	6816	204	–	210	1	87	–
5 Dahomey	4	1373	–	–	39	3220	15	–
6 Gambia								
7 Liberia	–	203	–	2 3	418	–	259	
8 Niger	–	1253	115	–	98	80	–	–
9 Nigeria	109	–	–	–	1403	104	140	–
10 Upper Volta	142	86	–	–	420	72	–	–
11 Guinea								
12 Ghana	11	7823	1946	14	–	962	18	–
13 Togo	144	520	119	–	1806	–	–	
14 Sierra Leone	–	6530	–	24	151	–	–	
15 Guinea Bissau								

Sources: [1] *U.N. Year Book of International Statistics*, 1974.
[2] *U.N. Stat. & Econ. Info. Bulletin for Africa.* E/CN/14SEC, 1973.

and Upper Volta, then from Mali to Ghana, Upper Volta to Ghana and from Niger to Nigeria. The major import directions are from Ivory Coast, Mali, Nigeria and Upper Volta to Ghana, from Ivory Coast to Mali, from Ghana Togo and from Ivory Coast to Upper Volta. In the subregion, therefore, Ivory Coast, Mali, Senegal, Upper Volta, Ghana, Niger, Nigeria are the countries with the strongest trade relations and they will be expected to form the strongest trading poles. As shown in Table 6 of Olayemi and Olayido [3], five countries – Gambia, Ghana, Mauritania, Sierra Leone and Togo generally had negative trade balances in the period 1965–1970.

In summary, trade among West African countries has grown and a few of them have sought new trading partners among the other countries of Africa while the trade with established traditional partners in Europe has weakened considerably, due in part to the problems created by the EEC.

5. GAINS AND PROBLEMS OF INTEGRATED AGRICULTURAL POLICIES

The overall aim of ECOWAS should be to develop a large internal market for manufactured or processed agricultural commodities. When the market is large and all artificial barriers to exchange of products and or resources are removed, economies of scale can be enjoyed in processing, transportation and distribution of these commodities. Efficiency in location of processing centres can be obtained through the application of transportation models and pooling resources and identification of major centres of production of specific commodities even when these cut across state boundaries.

The undisputable gains with little or no political hindrance involve common research effort. This was effectively used prior to independence so that we had WAIFOR, WACRI, WAISER etc. serving the English-speaking West African Countries. Today more than ever before, the duplication of effort and the consequent poor use of scarce research personnel, ought to be avoided. Part of the agricultural research effort is at present being carried out at the International Institute for Tropical Agriculture (IITA). More physical facilities can be provided for research based on ecological zones and crop-specific research to be strategically located in various parts of the region.

5.1. *Agro-Allied industries*

The location and efficient operation of fertilizer manufacturing plants, agricultural implement manufacture, jute bags production, textiles, livestock and various crop processing centres can be based on the expected demands of the regional market, the sources of supply of raw materials and the export potentials for these processed commodities. Each member of the ECOWAS has carried on substantial trade in processed food, textiles, agricultural machinery, etc. Import substitution within the region will now take the form of export expansion for the particular area in which each processing plant is located, since it will be expected to serve the neighbouring members of ECOWAS.

In this way a nitrogenous fertilizer plant can be based on the natural gas industry of Nigeria, while a phosphate plant can be established in Senegal or Togo where phosphates are known to exist. Large scale production of small implements and farm transportation equipment can be developed in step with an iron industry in Nigeria and possibly in Sierra Leone, Liberia or Mauritania. Certain countries have specialised in the production of identifiable qualities of textiles, such as Akwete and Ankra from Ghana, the Monrovia print from Liberia and Aso Oke from Nigeria. Large scale production of these for distribution within the region will help create employment and import substitution.

Sugar is one of the most highly demanded commodities in the region. There are areas with great sugar growing potentials where plants of optimum sizes can be erected to serve the region. In all these developments, the road, rail, water and air transportation complement present enormous opportunities for expanding employment for the population. Supportive businesses such as banking, credit houses and insurance, are bound to spring up. Tourism within the area would expand considerably.

5.2. *Development assistance by ECOWAS*

There is bound to be differential development among member states. The agricultural policy secretariat will be called upon to formulate policies that minimize the differences. This will be partly fulfilled through making available free the research results from centrally organized research pools. Then more developed countries will be called upon to subsidize the less advanced members in their programmes. The overall aim would be to ensure increased welfare for all citizens of West Africa. Whether such development assistance is seen as an advantage or disadvantage will depend on who gives and who receives it. The point may be stretched farther by arguing that internal assistance among ECOWAS members would relieve the rest of the world of part of its commitments to assist the region.

5.3. *Pricing agricultural products*

This is a problem area. The EEC sets and maintains an artificial price for several of its agricultural products with the aim of protecting the internal farmers and preventing dumping by outside producers of the same commodities. The same would be expected of the ECOWAS, but this is difficult to operate where there is massive lack of education and the infrastructure is still so poorly developed that pockets of surplus exist simultaneously with pockets of acute scarcity within the region. There is the aggravating problem of lack of accurate statistics. For the export crops, a common policy-making body should present a stronger bargaining position when facing the buyers, who in the past tended to dictate the prices and to make producers' association ineffective through driving a wedge between them.

For food crops, the guaranteeing of a steady supply may follow the ability to process and store with research determined shelf lives that allow for reasonable transportation and storage between producer-consumer locations. The most serious criticism against ECOWAS common agricultural policy can be based on the fact that all the countries produce similar food and cash crops. There is little exchange among them while they tend to direct their exports towards the same countries outside the region. This has been true till today but changes are likely to occur if the transformation of products into processed and packaged, refrigerated commodities take place. A substantial proportion of intra-West African trade is made up of re-exports which include processed agricultural items like sugar, dairy products, confectionary items, and textiles. It is the contention of the authors of this paper that such re-exports can be replaced by primary exports of products whose processing value-added will accrue to indigens of some members of ECOWAS.

5.4. *Problems of over-production*

If one goes by the experiences of some highly developed countries, like the U.S., the suggestions presented above can lead to over-production, the control of which will be one of the most thorny political problems facing ECOWAS agricultural policy-makers. No matter how good is the research on storage and preservation, there is a limit to the shelf life of agricultural commodities. And even when it is possible to store products, the storage costs increase with time and quantity stored. The carrying costs may completely destroy the expected gains from storage. Even if one argues for insurance against disasterous events such as the drought of 1972 to 1974 in the region, it is difficult to predict such recurrence. Much politics goes on between the producers, who want the government to buy and store when there is surplus but are unhappy if government attempts to unload reserve to reduce the price effects of scarcity. There is bound to be serious conflicts of interest among the members.

5.5. *Retaliatory measures from importing and exporting nations*

Even when cotton, cocoa, rubber and groundnuts were exported directly to feed the processing plants of Europe and America, these countries worked diligently to find substitutes for these commodities. The recent ganging up and mounting of massive effort to find alternative sources of energy as a result of OPEC action on petroleum is a lesson in regard to what ECOWAS can expect if there is unification and a strong single voice among the members. Many EEC and other countries will intensify their search for substitutes and eventually try to leave our agricultural products to us. The short-run action they will take may be to encourage production in other parts of the world – oil palm and rubber in Malaya, for example. This will be in addition to political manoeuvres to break up or prevent the formation of ECOWAS. The Third World people must by now have wondered how genuine is the cry for their development, and aid towards it, coming from the so-called developed world. If our own people continue to be unemployed or underemployed while we create the demand for the industrial products of the developed world, something must be wrong with our thinking and politico-economic orientation.

5.6. *Smuggling between states*

Smuggling between West African States sometimes reaches such alarming levels that law enforcing agencies are sometimes stationed along borders to prevent it. Smuggling takes place without the condemnation of the culprits even when the offended country makes protest about this. This attitude is seen as a sign of official approval if not of support for those concerned. Of course no one can know for certain the value of commodities smuggled but in some cases like Nigeria and Dahomey (Republic of Benin), sometimes as much as one quarter of the cocoa produced near the common border finds its way to the Republic of Benin, particularly in years when producer prices are low in Nigeria as compared to those in the Republic of Benin.

Now with a common agricultural policy, will smuggling be stopped? It may or it may not, depending upon how much a country gains by smuggling in pre-common policy period. For small countries, like the Republic of Benin and Gambia, smuggling means much to their economy. In fact, it is sometimes said that smuggling is the greatest industry in Gambia. Such a union can prevent smuggling if it is prepared to compensate for revenues lost. But smuggling is not recognised as a legitimate means of trade and the amount of goods smuggled are not recorded and as a result compensation cannot be asked for. Now the ability of a country to determine what it expects from a union is a reflection of the domestic political setting in that country. If people who direct the affairs of the state are also the businessmen in smuggling or supporters of such businessmen (there is a positive correlation between wealth and political power irrespective of whether it is a military or civilian regime) then it might be difficult to come to a solution.

But a solution to smuggling can come if countries adopt a common price policy. For instance, if export prices for cocoa are the same in, say, Nigeria and the Republic of Benin, then there will be no urge to smuggle. But soon it will be realised that smuggling of cocoa from Nigeria to the Republic of Benin is not due to its price differential alone. Often goods imported to Dahomey are cheaper than those imported to Nigeria because of the higher import duties in the latter country. Hence smugglers use the proceeds from smuggling to import goods like textiles, tomato puree and tobacco to Nigeria. So, in addition to common prices for export crops, such countries must also have the same rates for imports. Now the Republic of Benin will be called upon to raise import duties on these goods but cannot afford to do so because the masses in that country will not be able to afford them. Nigeria cannot afford to reduce import duties of these items because the higher import duties (say on tobacco, cigarettes, and tomato puree which feature most prominently in goods smuggled to Nigeria), are meant to protect the home industries producing these commodities. Will the Republic of Benin be prepared to take Nigerian goods and rely less on France for trade? Can the prices be brought down through increased production and reduced transportation costs? Moreover, smuggling may be difficult to curtail because of the existence of the same ethnic groups in different countries in West Africa. Sometimes such trades are termed to be normal since they are carried on between relations in both countries. From the point of view of countries concerned, of course it is smuggling. There must be a way in which smuggling can be prevented but how it can be done is highly debatable and this will be an important question for common agricultural policy-makers.

6. SUMMARY AND CONCLUSIONS

West African states have responded to the general world trend towards formation of economic blocs by the formation of ECOWAS. The great similarities in production patterns and levels of economic development in the region make one wonder what the chances are for ECOWAS success. Any successful policy must contain a well planned agricultural component since agriculture is the mainstay of the sub-region. This paper therefore examined some issues bearing on the chances of success of a common agricultural policy among ECOWAS members.

The agricultural component of intra-West African trade is not easy to estimate but can be considered to be small except in regard to livestock. The lack of processing, packaging and distribution was suggested as a factor responsible for the limited exchange of agricultural inputs and products. The creation of a large internal market will hopefully improve the planning, siting and operation of agro-industries.

A joint agricultural policy will have great advantages with respect to co-ordination of research efforts and dissemination of findings. But production and marketing policies will require total submission of member states to a central policy-making body with the authority to dictate what, where and how much to produce.

Other problems to be tackled include pricing problems, development assistance to slower developing members, possible retaliatory measures from the rest of the world and the question of smuggling between states. In the face of political differences, diametrically opposed interests and differences in affiliations to traditional trading partners, one may be tempted to say that a common agricultural policy will not be workable. But if looked at from the point of view of development effort for the region, rather than the usual trade creation–trade diversion theory of customs union, one can see the hope for success. If European countries which are individually well developed formed the EEC to enhance their mutual economic positions, there is no reason why ECOWAS should not be given a trial.

REFERENCES

[1] Olayide, S. O. (1972), 'Regional Cooperations and Food Production Under Rapid Economic Development In West Africa', *West African Journal of Agricultural Economics* Vol. 1, No. 1.

[2] Norman, D. W. (1967), *An Economic Study of Three Villages in Zaria Province, I.* Samaru Miscellaneous Paper No. 19, Zaria. Institute for Agricultural Research, Ahmadu Bello University.

[3] Ilori, C. O. (1972), 'Trade in Food Products and Economic Cooperation in West Africa', *West African Journal Of Agricultural Economics* Vol 1, No. 1.

[4] Olayemi, J. K. and Olayide, S. O. (1973), 'The Structure and Pattern of Intra-West African Trade', *West African Journal of Agricultural Economics* Vol. 2, No. 2.

DISCUSSION OPENING – J. B. Yankey, *Barbados*

The presentation of the East African Community experiences showed quite clearly what is likely to happen in an economic grouping where the component countries reached unequal levels of development. The evidence showed that the movement and benefits of trade flowed specifically to the most developed of the group. At the same time, if the agricultural production patterns and systems are the same, because of similarity of natural resources, the more developed will produce more of its own needs and trade less with the others. It will have both the ability and capability to mobilise its resources at a much faster rate, thereby enjoying a rate of progress above that of its partners.

Kenya undoubtedly has illustrated this pattern if one examines very closely the tables which are shown in paper by Weber and Hartmann. However, the economic community developments which have occurred and are being advanced are very relevant to developing countries which share frontiers or seas. The case is justified on the basis of the benefits which can accrue from an enlarged market as well as the exploitation of natural resources and sharing of common services, thereby taking advantage of economies of scale and the relative scarcity of trained manpower in the developing world.

However, because of the negative effects from the unequal distribution of gains, developing countries willing to participate in any economic integration movement are usually only willing to experiment and not to be too involved in a deepening and widening process. This is the experience of the relatively young Caribbean Community – made up of Guyana, Trinidad and Tobago, Barbados, Jamaica, Belize, St.Kitts-Nevis-Anguilla, Antigua, Montserrat, Dominica, St. Lucia, St. Vincent and Grenada. Of these, five are independent countries, (two Republics), and the others are Associated States and Colonies of the United Kingdom.

As a result of these difficulties, the Caribbean Community has developed a special regime of measures for the advancement of agricultural and industrial growth of the relatively less developed group. These measures, however, have to be administered, institutions have to be set up to push them forward and, as in all such cases, the problems of location of offices, economic activities, rising administrative costs and bureaucratic procedures, all help to delay whatever results may be expected. Developing world people have acute rising expectations, and Governments can ill afford delays in achieving the benefits envisaged from such integration movements. As a result, because of the gaps between expectations and actual performance, if integration movements are to work and find popular acceptance, there must be much more than a trade arrangement, there must be policies and programmes designed to foster the growth and delivery of benefits from these economic integration institutions in the developing world where inequality among partners is still significant.

Regional co-operation among developing countries is also an area of great significance. The paper dealing with this subject has adequately covered the areas which I consider crucial if we are to maximise the use of scarce resources in the developing world by meaningful and effective cooperation in agriculture.

Like Africa, rural development in the Caribbean is dependent on some measure of regional cooperation. We have areas with high land pressures for farming, whereas in Guyana and Belize the situation is the reverse. There is an opportunity for re-settling people and specialising in specific agricultural enterprises on a scale that will provide increased supplies of food to cut drastically the Caribbean food import bill.

Because we produce the same crops and types of livestock under similar tropical conditions, thereby encountering the same agronomic problems, and in addition, since we market in the same metropolitan countries, there is always the need for joint research and trade promotion. The Caribbean has taken the crucial steps to set up a joint agricultural research and development institute and has always marketed or bargained jointly, as in the case of such crops as bananas and sugar.

Undoubtedly, regional co-operation in agriculture is of special significance for the agricultural development of not only land-locked countries as in Africa, but of small states sharing the same sea basin, with the same natural characteristics and with people originating and having similar cultural identities.

RAPPORTEUR'S REPORT – A. J. Adegeye, *Nigeria*

The discussion was concerned in the main with various specific points arising from the papers. On what might be termed the nature and content of co-operation it was argued that co-operation should be problem-solving in its orientation. Academics tended to concentrate too much on trade matters and underplayed such activities as co-ordinated research and pest and disease control. Governments seem reluctant to found research institutions for themselves: collaboration decisions may be more promising. More explicit identification of the desired types of co-operative activity was needed. At first sight language problems might seem of some gravity, but the EEC seemed to be managing with French and English and in West Africa Yoruba and Hausa are very widely understood in trading affairs.

Among more specific points it was thought that the paper somewhat over-emphasised the significance of smuggling of food. Farmers should be the gainers under a common market system in a way which they were not under a Marketing Board system. The constraints imposed by trypanosomiasis on potential collaborative development in beef production were noted.

Views appeared to differ about the influence of the differing political complexion of governments on the prospects for successful economic integration. An optimistic view was that the flexibility of political structures so far in developing countries gave scope for adjustment to further co-operation. However, the essential reasons for political differences in structure between neighbouring countries were not analysed.

P. C. VAN DEN NOORT

Integrated Rural Development Projects: Decision-making on the Regional Level. Developed Countries/EC Integration Concepts and Experiences

The assignment "Integrated (Multi-Package) Rural Development Projects: Decision-Making on the Regional Level for Developed Countries/EC-Integration: Concepts and Experiences" is too pretentious. It cannot be substantiated in a short discussion so we first have to delimitate the problem area.

We have to realize that we are dealing in the first place with economic policies and the way in which decisions in this field are made. A popular definition of economic policy may be "What do we want of the economy and how do we get it?". It is a familiar description of a "means and goal"-model in which the word *We* is more or less disregarded, much to our disadvantage because the question: "Who is meant by we?" is very essential in the field of sector/regional/national/international economic policy. What is the we-group? Is it the feudal caste? The mercantile group of dominating merchants? Is it another special group or is it the generality of all citizens? And if so, who are the citizens? Who is in and who is out? Who do we have in mind? This is an important question because the ultimate goal of our economic policy depends completely on the definition of our "we-group". We could think of many alternatives, e.g., the interest of a special group of farmers, of all farmers in a special region or nation, or all citizens of the Common Market. In the real political decision-making process we will meet all kinds of interests, and hence of definitions of "we-groups", though almost always very carefully concealed behind a facade of the very broad terms of "general interest".

The analysis of the decision-making process can be done in many ways. The method I will use here is the action scheme. In Fig. 1 we see a governmental agency as an Actor that pursues an End by selecting appropriated Means compatible with Conditions.

Now there is a kind of hierarchy of Ends and Means, that is to say, (1) one goal might be of more importance than another one, and (2) one goal might be regarded also as a Means of policy depending on the level of the responsible Actor. A simple example can illustrate this: the agricultural price level might be regarded as a *Condition* for some conservation policies, as a *Goal* for agricultural price policies and as a *Means* for agricultural income policy. So there is no clear-cut difference between Conditions, Means and Ends. In fact, there might be a system or hierarchy of Ends, in which a so-called

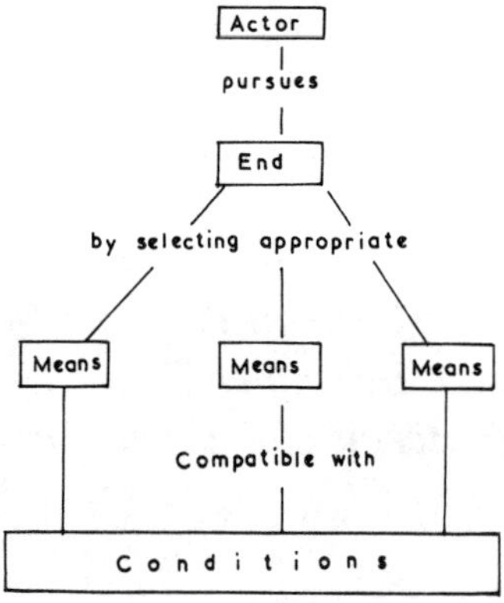

FIG. 1. Diagram of an action system

Lower End can be considered as a Means to reach the *Higher* End.

The *ultimate* End can be described as the "general welfare" and eventually measured with the *general welfare function* **Ω**. The higher Ω the better the situation is for the We-group or community. We can see different aspects of the general welfare, e.g., (a) Economic welfare and (b) Cultural, social, political welfare. Let us concentrate on the first.

Economic welfare can be improved via a higher social product and/or goals of economic policy, but also means to reach the ultimate goal, according to the following scheme:

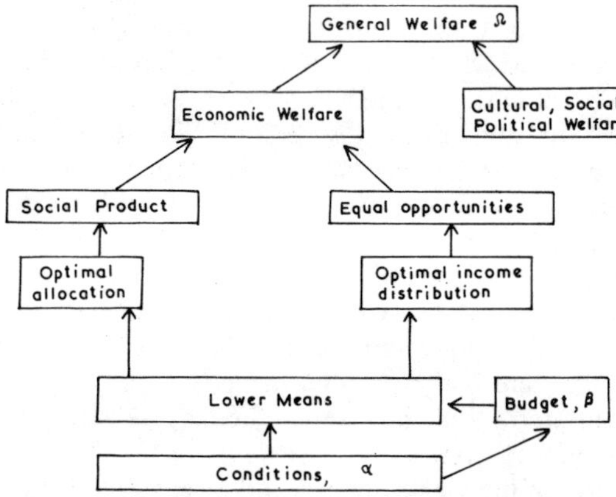

FIG. 2. The hierarchy of Means and Ends.

This whole scheme would collapse if we could not agree about Ω and this depends also on our definition of the "We-group".

If we are not completely honest about this we could end up with very conflicting material, only useful in the framework of regionalism, nationalism, various ideologies or combinations of them.

The second point we have to clarify is the place of agricultural policies in this general scheme. I will start with the scheme that is most appropriate in our conditions to-day, and that is the action-scheme of economic policy in the *"national state"*.

Given the ultimate goal of general welfare (Ω), we have the *major* objectives of economic policy:

(1) social product (GNP);
(2) employment;
(3) income distribution;
(4) food availability;
(5) balance of payment;
(6) foreign relations;
(7) population.

These major objectives may well be served by many measures under agricultural means. We *assume* that these last means are also the objectives of the national agricultural policy. But it is clear that some see the agricultural policy as their ultimate goal of activities that can even be pursued at the expense of the welfare of others in the State. This is also a consequence of a difference in definition of "we-group".

The agricultural objectives can be divided into *higher* and *lower* ranking goals. Lower indicates that such an objective may be used as a means to reach the higher objective.

The lowest ranking items are called the programs These are policy measures that serve the agricultural objectives

(a) *directly*. They act upon the high ranking agricultural objectives e.g. direct income payments to farmers, regulation of food consumption

(b) *indirectly*. They act upon the lower ranking agricultural objectives e.g. land improvement, land consolidation, subsidies on investments, or premiums for grubbing up fruit trees.

It is clear, however, that improvement for agriculture or agricultural regions can also be reached by other means, e.g., industrialisation, a policy of opening up the countryside by new and improved roads via new alternative employment. These non-agricultural programs may have an influence on (1) the direct and/or the indirect agricultural programs, (2) the conditions under which the programs must operate and/or the decision-making process must take place.

The third point of clarification regarding the title of the assignment is to find the so called "packages". We have found already three important types of program packages:

(a) direct agricultural programs;
(b) indirect agricultural programs;
(c) non-agricultural programs, with effects on agriculture or agricultural regions.

We see in Table 1 that almost none of the well-known agricultural programs belong to only one column. Nevertheless we can consider all programs

TABLE 1. *Types of programs and their immediate effect on some ends.*

Program	Effect of Program on							
	1 Export import	2 Price output	3 Price input	4 Effici-ency	5 Quality of factor	6 Quantity of factors	7 Direct programs	8 Non-agricul-tural programs
Vocational training			X	X	X			
Agricultural research				X	X			
Agricultural Advisory Service				X	X			
Land improvement			X	X	X			
Land reclamation			X	X		X		
Land consolidation			X	X	X			
Soil conservation			X	X	X	X		
Soil bank			X	X		X		
Enlargement of farms			X	X	X			
Resettlement			X	X	X			
Leasing regulation				X	X			
Rent regulation			X					
Credit arrangements			X			X		
Interest subsidies			X			X		
Grubbing up schemes for fruit trees				(X)		X		
Slaughtering schemes		X	X			X		
Giving up farming schemes				X		X		
Pensioning of farmers				X		X		
Sanitary regulations	X	X	X	X	X	X		
Food aid	X	X	X					
Intervention		X	X					
Quantitative restrictions	X	X	X			(X)		
Variable import/export levies	X	X	X			(X)		
Fixed import/export levies	X	X	X			(X)		

TABLE 1. *Continued.*

Program	Effect of Program on							
	1 Export import	2 Price output	3 Price input	4 Effici-ency	5 Quality of factor	6 Quantity of factors	7 Direct programs	8 Non-agricul-tural programs
Direct income payments							X	X
Social allowances							X	X
Fiscal policies for agriculture			X			X	X	X
Industrialisation								X
Opening up the country-side		X	X			X		X

that are represented in a specific column as a "package". *In* these packages we can find that the various programs can act upon each other, either "positively" or "negatively", or that they are neutral to each other. *Between* those packages we can find the same relationships.

Each of the programs (packages of programs) can only be executed if there is a legal basis and a budget for it. So we get a somewhat more detailed action scheme in which the ultimate goal is pursued by selecting appropriate programs, compatible with economic, social and political conditions. These conditions will be of enormous importance for the program decided upon and for the budget available for it. Our scheme for national agricultural policies is now:

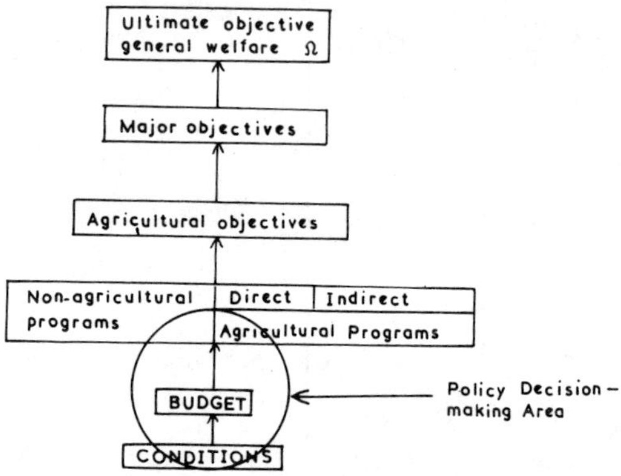

FIG. 3. The place of agricultural policy in the action scheme

If we have to look interregionally there is here also an international policy analysis – the picture becomes more complicated. The general welfare function is then not the national welfare function but the *Community welfare function*, and they are by no means identical. In the idealistic situation of economic integration the national interest has to bow to the interest of the Community as a whole. In the process of integration we will have to solve various problems connected with:

(1) setting up the Community's General welfare function
(2) setting up a political Actor for the programs
(3) setting up the budget for the Community
(4) antimony between the national programs and the programs of the Community.

In order to analyse the decision-making process we *assume* that the political decision-maker acts in a rational way, that is by trying to increase the welfare function. In case this function is not known we could add "the general welfare function as the politicians see it".

Whatever decision they want to make they should have to evaluate the effects of a specific program on the major objectives or on the general welfare

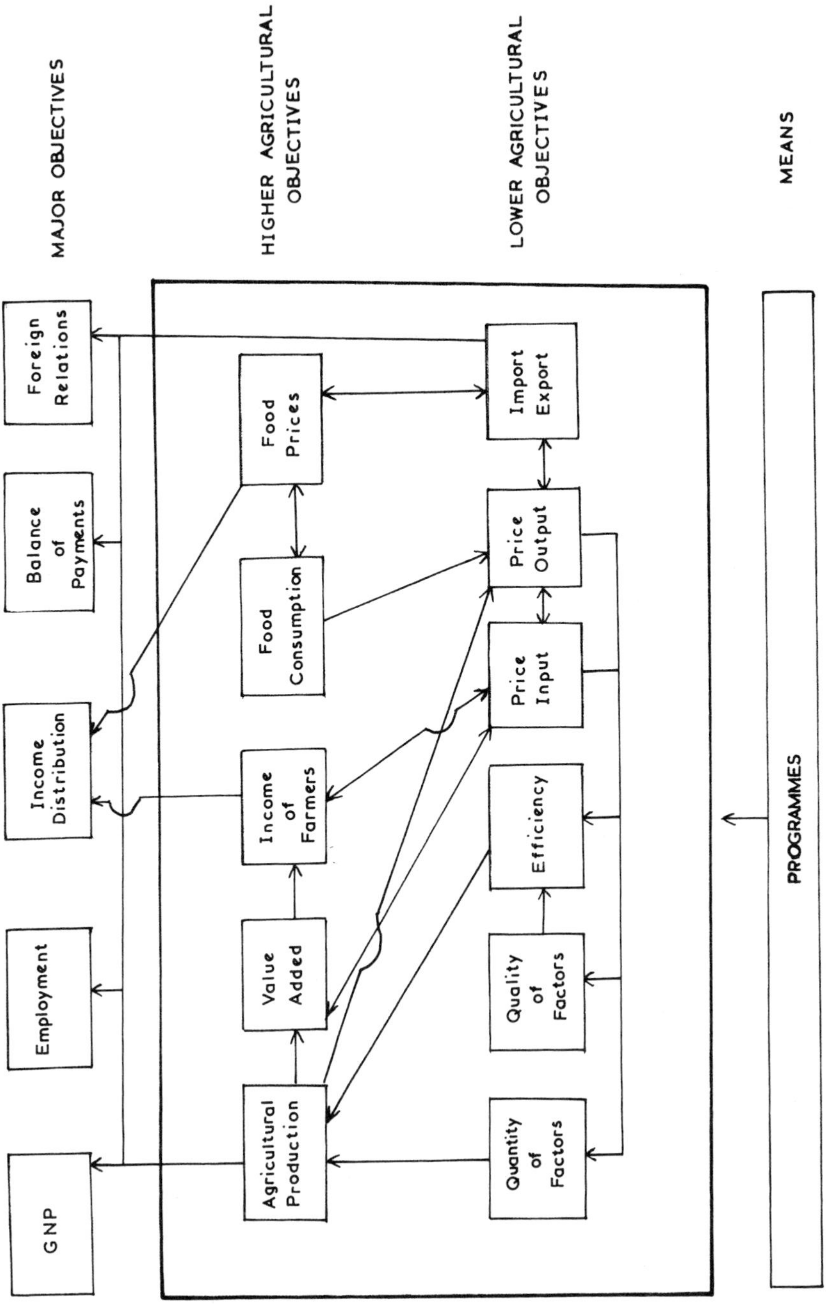

FIG. 4. A scheme of the economic "machinery" (relationships) in the hierarchy of ends and means.

function. To be able to do this we need to know the effects of a program on the agricultural and major objectives. This effect is not easy to detect because of the many interactions. To find these, we need some knowledge of the economic "machinery". See Figure 4.

This scheme gives a rough idea of the effects to be expected. The vectors in this diagram give the direction of the effects. We see that there are one side effects (→), mutual effects (↔) and "internal circuits". As a consequence of this a lot can happen in the machinery, without any effect on the major objectives or general welfare. Let us take some examples.

(a) In order to improve the income position of the farmers we decrease imports of feed grains (in a net-importing country)→Prices of output increase→higher production→higher value added (= agricultural income) →higher prices of inputs, e.g., feeding grains or land. If the farmer is not the owner of the land or there is no rent regulation (as there is in the Netherlands) we can expect no clear-cut improvement of income of all farmers, not even of all the farmers who were producing grains themselves.

(b) Again, to improve farmers' income: increase in efficiency by research and extension → higher production → lower farmer prices → lower agricultural income → lower rent and/or farmers' income and perhaps a decrease of consumers' prices, so the result may be somewhat different than expected when the program was started.

Not only the direction of the effects is interesting but also the "intensity". Notwithstanding the extensive agricultural economics literature there are some gaps in our knowledge. There is still surprisingly little known about the welfare effects of programs. This is even evident in the rather loose way we use the term "cost of a program". This can have four different meanings: effect on the cost of living, effect of the budget cost, the resource cost or the welfare costs. The nasty thing here is that too many do not realise the very different way the word "cost" is used in this phrase, quite apart from the meaning these "costs" may or should have for the decisions of the politicians, who are sometimes not trained economists. There are some conflicting means in the scheme for we should realize that we have policies:

(a) to stabilize output prices;
(b) to protect output prices;
(c) to regulate input prices;
(d) to improve the "state of farming" and the structure of agriculture at low or no cost or even with subsidies for the farmers;
(e) to bring about an economic union.

We might get the impression that the director of the agricultural machinery at the same time tries to accelerate, to use the brakes, to change the machinery and install it in another "mill". This opens fields of conflict in the agricultural policy in respect of:

(a) The general welfare function of the community against various regional, sectoral and national interests;
(b) This conflicts about "who is the actor and provides for the budget".

Decision-making about programs of agricultural policy is at the same time making decisions about governmental budgets for such programs. These budgetary problems can be approached in two different ways:

(a) According to a normative theory about policy decision-making – the best known example is the so called "p.p.b.-system".*

(b) According to a positive theory about policy decision-making in which more "empirical" knowledge about the decision-making process is involved.

In the normative approach the goals of economic policy are carefully formulated in quantitative as well as qualitative sense. Moreover, clear priorities are given to the various goals. Alternative means (programs) to reach those goals are proposed, and by a careful benefit-cost analysis the best program is sought. This program gets the appropriate part of the budget and is accorded a special plan over time.

This straightforward method does not always work within European countries in which *coalitions* are predominant. The essence of coalition is the presence of a set of groups with different goals and values. In order to survive as a coalition it must try to minimize conflict. This has as a consequence that clear cut goals, priorities, benefit-cost analysis and programming cannot be set up, because this would highlight the sometimes very large differences within the coalition, only leading to skirmishes and collisions which would bring the coalition to an end. That straightforward method is not available to the European Commission either because this is in essence also a coalition that can only rule with unanimity – it is indeed a very strict form of a coalition government. Setting up well defined goals, priorities, benefit-cost analysis, etc., would expose the real controversies, which a politician working in such framework could only wish to prevent.

In this coalition structure of many parties we find special interest groups within each party, e.g., the "green lobby or farm bloc". It is well-known that a well organised farmers' organisation can have influence within various parties. In each party we will find specialists in various fields, including agricultural policy. A remarkable feature of the political structure in many European countries is that the farm specialists of the various parties do not disagree very much on matters of agricultural policy. So they can pursue their end in a common effort. The same happens to other special interests as, for example, public housing, education, health or defence. These special groups do not usually interfere in each others "business". So the decision-making process is a special type of negotiation between (1) specialists in parliament, (2) the minister for that special field, (3) the bureaucracy and (4) minister of finance. It has been proven that in many cases the specialists, the special minister and the bureaucracy can agree rather easily and the only one who can oppose the decisions is the minister of finance. We can also see that the influence of the bureaucracy in the European Community is even stronger than on a national level, whereas the influence of Parliament is much weaker. The process is not always that simple, e.g., elections and or a break-down of a

* Program–Planning–Budgetting System.

coalition-government can change the priorities. The bureaucracy is also not always a homogeneous body. There are controversies about priorities between the directors of the various programs and a lot of infighting may go on, e.g., to get an alternative program approved.

It has been proven that the welfare functions of the bureaucrats and the politicians do not necessarily coincide. The problem is that the equilibrium they reach need not correspond to the highest Ω or the highest general welfare. The bureaucracy is not a neutral agent in the decision-making process. Here is a very important problem area for every democratic policy-making procedure, not only for agricultural or regional policies. But in relation to these policies we have to realize that the planning and execution is done not only on the national level but also on the regional and international level. This will complicate the process, for there are more actors and different procedures to observe.

One of the differences between the positive theory of decision-making and the normative one is that the latter assumes that the optima of the various groups coincide with the highest Ω. This still needs a proof as well theoretically as empirically.

An additional difficulty in the decision-making process is the changing aspects of the whole scheme: the conditions, the political process, the economic machinery, the objectives and even the general welfare function are not constants but are changing over time.

In the course of the last 50 years we have got a lot of agricultural programs in Western Europe. Some of them have been mentioned already in the schemes. There are some factors that make it very difficult to remove a program. Around such a program other programs may be built up to strengthen the policy. In this way new packages of programs may develop.

As illustration, the Netherlands started the land consolidation programs in 1923. In the depression of the Thirties, land consolidation was thought of as a public work providing employment. The program got a broader scope, all kinds of things were combined: land improvement, drainage, reclamation, reallotment, resettlement, construction of roads, bridges, electrification. They become an integrated measure for modernizing the countryside. This set of integrated programs is now one of the most important structural policies for agriculture in the Netherlands. The problem was, and still is, the small-scale of many farms and in such a situation modernisation has not the effect it might have on larger units. So programs for pensioning off farmers were set up, as well as providing other farmers (who wanted to give up farming in an area where such a "land consolidation program" was enacted) with money or land in the new polders. Of course, this package of programs has had a large impact on agriculture, the countryside and the landscape. Moreover, this interlocks with the fact that in the Seventies the city-man wants to visit quiet rural places, see the romantic countryside and wants to preserve the landscape. The Netherlands, moreover, is a densely populated area and therefore needs careful town- and country planning, that is, physical planning. No wonder that it is now trying to develop a set of integrated modern multi-package (rural) development projects in which farm and non-farm interests in land use will be served. It stands to reason that this is a very difficult task.

The land consolidation issue was debated heavily because of the low rate of return for the national economy as a whole, but as a matter of fact these calculations had no significant influence on the budget provided for this integrated program in the Fifties or Sixties. Moreover such programs must be executed in certain areas and become therefore a part of all kinds of policies for small regions and even of regionalism. In the decision-making process the influence of the region and regionalism may play an important role, hence complicating the policy decisions. We are not sure that the aims of the different actors on different levels are the same. Some may pursue farm interest, others non-farm interest, one group may see the land consolidation policy as a national one, others want to use it as part of their regionalistic policy. Disentangling these influences has not been tried, nor the systematic use of welfare functions-analysis. It is clear, therefore, that non-economic influences also play an important role. Land consolidation programs are available in many other EC-countries, but nowhere to the same extent in the technical and economic sense, as in the Netherlands.

The second example of a multi-package project in the Netherlands is the famous New Polders in the formerly Zuydersee. These polders were built on the bottom of the sea from the Nineteen-Twenties on. Originally three factors were of importance (1) to get a higher degree of security by shortening the coastline, (2) to get new agricultural land and (3) to get a reservoir of fresh water especially for agriculture and horticulture. In the polders whole new communities had to be created using the insight of modern sociological theory. It is important to notice how far from the original plans we in fact now are, as a consequence of the use of the car, the declining agricultural population, the new agricultural technology and also the new demands of city-man for more space and green prospect to enjoy. The whole project is now closely linked or integrated with physical planning in the Netherlands so that it is hardly possible any longer to speak of *agricultural* programs and goals in these very special kind of regional policies. They are of national importance now. It is remarkable that we find complicated problems here about who is the Actor, who provides for the budget, who decides about changing goals and priorities. It is also remarkable that economic considerations (e.g. benefit-cost calculations) played only a subordinate role in the decision-making process.

The third example of multi-package projects in EC-countries is the well-known combination of programs (1) to improve the technical state of agriculture and (2) the programs to protect prices at relatively high levels. This multi-package development scheme creates large difficulties because it results in surpluses. These have a bad political reputation, although they are more a sign of affluence than of economic failure and their costs are exaggerated.

Tecnnical assistance and price protection normally works best for the large farmer, leaving the peasant farmer in the cold. Sometimes it is also very nice for landowners. In the Netherlands this aspect of the income distribution is handled by official land rent policies. It is surprising, however, how little is known about the real cost of these (combined) policies. There has been interest in the budgetary aspects, there is an awareness of the influence on the cost of living but of resource costs or welfare cost not much is known, so we can conclude that these costs cannot play a significant role in the decision-making process.

There is no use going into the details of this multi-package deal for the EC-countries and regions, it may be sufficient that this package of programs has had many critics. The alternatives that are available stem from opinions about the functions that prices should serve. As we know, farm prices have at least two functions: directing agricultural production and generating income for the farmers.

(1) One extreme position is to change drastically the structure of European agriculture, and modernise farming completely in a short period. It is expected that the problems of the present multipackage program would then have gone for ever. This simple thought is the core of the well-known Mansholt Plan. Typical of this plan was the technocratic approach, economic analysis of the plan not being carried out by the European Commission. So it was only hoped that the Plan would be worthwhile and accepted by the politicians, the farmers and the public. This hope was idle indeed. The Plan was not sound economically and was socially unpleasant. It did not fit into the ideologies of farmers' interst groups. Typically for such a situation, almost no benefit-cost calculation or other types of economic analysis were used in the discussions. This is an interesting observation for an economist. The discussions were very emotional, and urged drastic changes in the Plan. In the end we got some directives based on the results of the Council of 25th March 1971 concerning:

(a) modernization of farms;
(b) encouraging farmers to give up farming and allocating farmland with view to improving farm structures;
(c) vocational training.

Some Community criteria were formulated, but the action would be implemented through measures taken by Member States, partly financed by the EAGGF. The fact is that only certain countries actually have such measures. The interest of farmers nowadays is confined to the first point in order to get an interest subsidy. The other measures are not widely used because of the high rate of unemployment. So this alternative to the traditional multi-package ended up, not as an alternative to it, but as an amplification and strengthening of it.

(2) The other extreme position was to change the price policy from a protective one into a policy for stabilization of farm prices on a low level only. The prices should only direct agricultural production. The second function of the farm prices is that the generation of income should be eliminated and replaced by direct income payments to farmers for a *limited period of time* (e.g., 20 years) and *only* for farmers who were already productive when that plan was constructed. This Plan was heavily debated but benefit–cost calculations, again, were scarcely used in the analysis. In fact the Plan was not accepted as a workable alternative.

In the Netherlands a Ministerial Committee concluded, however, that direct income payments to farmers under certain conditions could be a fruitful addition to the existing farm program. This could be the case when we have older peasant farmers (with small uneconomic farms) in a situation of a high overall rate of unemployment. The traditional farm programs would not work out satisfactory then either from a social or an economic point of view. Such a supplementary income policy has only been used once in an emergency-situation. The rules of the Economic Community do not provide for it, so it has tried to bring that group of farmers under the general social regulations. This means shifting the problem from the Minister of Agriculture to the Minister of Social Affairs.

The idea of direct income allowances is also part of the policies for farmers in the Alpine regions. In order to conserve the Alps we need farmers. Farmers can only do this and live if they have a reasonable income and an acceptable form of community life. In order to arrive at this goal a special policy based on the Resolution of May 1973 concerning agriculture in disadvantaged areas was set up. This policy can also be regarded as a part of environmental policies or regional policies. There is not much experience with this type of program in EC-countries.

(3) All this leads to non-agricultural programs for improving the conditions of the agricultural population, that is via opening up the country side, providing farmers' children with general education, industrialisation in rural areas, physical planning and administering general social policies. This non-farm-package has been important in the Netherlands and elsewhere in the EC and had an enormous impact on our social and economic conditions. The agricultural programs can now even be considered as special measures to assist farmers in adapting their farms and their lives and communities to the newly created social, economic and physical conditions. This type of policy mix fits into the process of capitalistic development, that is to increase productivity of agriculture, and of the country as a whole but – and that is essential – without permitting the social tragedies which accompanied such development in the last century.

This was also the leading optimistic philosophy in the Fifties and Sixties. But nowadays we are more pessimistic, not only with respect to the future, but also because we realise the high social cost of progress, and the problems in special areas. Even in a small country like the Netherlands there are regions that became backward, or were damaged in the process: the agricultural northern part is lagging behind the metropolitan west, the textile area in the east is in deep trouble, not to mention the troubles in the former mining area in the south. How much larger are the differences and the frustrations on a European scale. This leads to various kinds of political philosophies: to increase or decrease the power of the state in the social and economic field, to extend the Community or to break it up, to concentrate on national welfare or on regional interest. There is much confusion. As far as my assignment is concerned; the fragmentation of Europe, the nationalism and, even more, the regionalism could become a menace for Europe. It could become a fatal disease, seen in terms of the general welfare in Western Europe as a whole. From this fear stems the hope that multi-package rural development projects and/or programs will *only* be used to increase the general welfare of the Community and will not be used in narrow-minded regionalistic policies.

REFERENCES

An assessment of Investments in land reclamation, Report of the Dutch "Agricultural Investment Commission", Wageningen (Holland), 1960.

Bergmann, D. (1975) *Matériaux et réflections pour une réorientation de la politique agricole*, Paris.

Boulding, K. E. (1958) *Principles of Economic Policy*, Englewood Cliffs.

Boulding, K. E. (1970) *Economics as a Science.*

Buchanen, J. M. (1967) *Public Finance in Democratic Process.*

Cairncross, *et al.* (1974) *Wirtschaftspolitik für Europa, Wege nach vorn*, München.

Fennel, R. (1973) *The CAP; A Synthesis of Opinion*, Wye College, University of London.

Hathaway, D. E. (1963) *Government and Agriculture, Public Policy in a Democratic Society*, New York.

Hofstee, E. W. (1957) *Rural Life and Rural Welfare in the Netherlands*, The Hague.

Horring, J. (1970) *Twee recente kosten-baten analyses van door de overheid gesubsidieerde cultuurtechnische investeringen en wat daarme verband houdt*, Openbare Uitgaven, 2, 3.

Jacoby, E. H. (1959) *Remembrement en Europe*, Wageningen, Holland.

Jöhr, W. A. and Singer, H. W. (1955) *The Role of the Economist as Official Adviser*, London.

Johnson, D. Gale. (1973) *World Agriculture in disarray*, London.

Josling, T. E., *et al.* (1972) *Burdens and benefits of farm-support policies*, London.

Koopmans, L. (1973) *Beheersing van overheidsuitgaven*, Deventer.

Mehler, F. (1970) *Ziel-Mittel-Konflikte als Problem der Wirtschaftspolitik*, Berlin.

Niskanen, W. A. (1971) *Bureaucracy and Representative Government*, Chicago.

Nooy, A. T. J. (1971) *"Het landbouwbeleid"*, in, *Beleid Belicht*, Alphen, Holland.

O.E.C.D. (1974) *Agricultural policy of the European Economic Community*, Paris.

Louwes, S. L. (1972) *Landbouwpolitiek; failliet van een monodisciplinaire landbouw-economische benadering*, Haarlem.

Priebe, J., Bergmann, D. and Horring J. (1972) *Fields of Conflict in European farm policy*, London.

Schmidt, P. and Schmitt, G, (1975) *Auf dem Wege zu einer positiven ökonomischen Theorie der Agrarpolitik*, Berichte über Landwirtschaft, p. 521.

Schickele, R. (1954) *Agricultural policy*, New York.

Schreiner, G. (1954) *Fazit einer dreisig jährigen Agrarstruktur politik in der BRD*, Berichte über Landwirtschaft, p. 455.

Self, P. (1975) *Econocrats and the Policy Process, the Politics and the Philosophy of Cost–Benefit Analysis*, London.

Tindemans, Leo (1976) *De Europese Unie*, Commission of the E.C., Brussels.

Tracy, M. A. *Fifty Years of Agricultural Policy.*

Van den Doel, J. (1975) *Demokratie en Welvaartstheorie*, Alphen, Holland.

Van den Noort, P. C. (1969) *Inleiding tot de landhuishoudkunde*, Haarlem.

Van den Noort, P. C. (1971) "Un modèle de developpement et des incertitudes à la base (Le Plan Mansholt)", in, *Cahier du Ceneca*, Paris, p. 148.

Van den Noort, P. C. (1974) *Benefits and Costs of Various Agricultural Policies*, Agricultural Economics Institute, Oxford.

Van den Noort, P. C. (1975) *A comparison of two systems of agricultural policy for the E.C. – The case of Dutch agriculture*, Institute Interantional de Finances Publiques, Saarbrücken, p. 108.

Van den Noort, P. C. (1976) *Some ideas about agricultural Sector Economics*, Wageningen.

Van Riemsdijk, J. F. (1973) *Directe Inkomenstoeslagen*, Wageningen.

DISCUSSION OPENING – C. Vazquez-Hombrados, *Spain*

Two main threads in Dr. van den Noort's paper can be distinguished: (1) a description of the relationship between programmes and the political economy; the way in which the role of agricultural programmes fits in, with the aim of increasing welfare, is particularly well explained. (2) a review of some of the EEC and Netherlands policies; the future of agricultural economists' programmes is felt to be unclear, as these are not, or only partially, implemented.

EEC structural policy is discussed in technical and social terms rather than being subjected to economic analysis. Price policy is discussed at length; but with hardly any cost-benefit analysis it is decided to give subsidies to elderly farmers. In the Netherlands a land re-allocation programme has been set up to counter unemployment rather than to solve an agricultural problem.

In spite of van den Noort's pessimistic view, in all the cases cited the programme devised has played an important part in decision-making.

In the Gualdalquivir area of Southern Spain a multi-disciplinary team has devised and implemented an irrigation programme; no-one is satisfied with its results, even though it was applied with close adherence to its plan.

If this is what happens with a small-scale programme, with no political complications (land reform having already been carried out), what happens (road blocks, pigs in Brussells, etc), when a more inflammatory programme is being considered?

When only the results from the technical experts involved in a programme are reported, this only represents a fraction of those affecting decision-making, and thus such programmes are only partially implemented. This problem will perhaps be solved when all the interested parties are fully involved.

Finally, as a Spaniard, from a country on the threshold of the EEC, is the impression gained from this paper a valid one that the Community is moving towards a break-up? Some comment on this would be welcomed.

DISCUSSION OPENING – J.F. Heavey, *Ireland*

My first reaction to Dr. van den Noort's paper is that it is extremely realistic in that, having examined the theretical framework in which rural development decision-making might be carried out if an economist were in charge, he then contrasts this with the realistic world of decision-making by politicians.

In effect, the latter say, when challenged, "Never mind the theory – this is how we actually do it". Strange as it may seem, my sympathies are with the politician. The paper describes how a programme of land consolidation was carried out in Holland without any regard for benefit/cost ratios; how the New Polders were reclaimed from the sea; and that, "benefit-cost calculations played only a subordinated role in the decision-making process", which I take to mean that benefit-cost calculations were completely ignored. Perhaps we could admire the politicians who decided to get on with the job whilst economists came up with a wide variety of benefit-cost ratios and almost certainly came up with widely-differing solutions.

But we are talking about concepts and experiences in rural development in EEC integration. This brings me to the question of planning and the differentiation between ends and means which the author outlines in the paper. Ends and means can very easily become confused, especially in the area of indicative planning, where targets are set as a means of achieving a desirable end. As often as not the target is not achieved exactly, and very soon the debate begins, not on whether the actual achievement was an improvement on the original position, but on the divergance from the set target. The problem then becomes the relationship of the target and in no time at all the target has become an end in itself. A good example of this is where full employment is an end and a phased reduction of inflation by target levels is a means. Where these targets have not been met the achievement of the target, probably arbitrarily set in the first instance, becomes the end, and people accept rising unemployment with submission.

I would like to have the author's views in more detail on the concept in the EEC package of directives, concerned with "a comparable income", and the related concept of "a farm worth developing". The whole question of the relationship between farm and non-farm incomes have come up several times during this conference. Professor Ohkawa says in his paper that "in developing countries in the pursuit of an equity objective, the income level of the non-agricultural sector is taken as a standard". This wording leaves several important questions unanswered. Does "the income level" mean "the average income level"? Most countries use that definition and certainly that is the meaning set out in EEC directives. But I do not, at this stage, want to make an issue of whether all or part of the non-agricultural sector be included. The next question is what does the average income level mean as a standard? In the case of EEC directives this is what is referred to as "the comparable income" which all "farms worth developing" must be able to achieve at the end of a 4–6 year development plan. Firstly, income distributions are notoriously skewed to the right. Yet the implication of the EEC directives is that each and every "farm worth developing" must be capable of achieving an average level of income which only a small proportion – much less than half – of the non-farm population is capable of achieving. We must ask then if these concepts are consensus concepts of equity and if they are acceptable in achieving the kind of Society which "We" referred to by Dr. van den Noort, wish to see brought about. Or are they arising because farm organisations constantly harp on another concept of developed countries – "the income gap"? Incidentally, it is my experience that those farm organisations who express most concern about the wide disparity between farm and non-farm incomes take little or no interest in the huge disparity which exists between farm incomes themselves.

There appears, however, to be an underlying assumption behind these concepts of parity, that all "farms worth developing" are farmed by people who really do want an income level equal to non-agricultural incomes or, put another way, that only those farmers who want this equality are to be regarded as "worth developing". There are many other aspects of farming beside income aspects and these cannot be ignored by agricultural economists in their calculations or in the conclusions they draw from farm income data.

Finally, I would like to say a word for that other concept, the consumer/taxpayer, who is constantly being asked to provide funds for rural and farm development programmes, through higher food prices and higher taxes. It seems to me that the aspects of welfare economics here are quite simple. The consumer/taxpayer, in general, regards the farmer, as someone whose car needs a push to get it started. He is quite willing to give the farmer's car a push up the hill to take advantage of the downslope – on one condition, namely, that the farmer helps with the pushing and doesn't insist on sitting in the car while it is being pushed. If he does so insist, then the consumer/taxpayer feels that, in equity, he is entitled to say "to hell with him". And I feel we could be sure of a consensus on that point without appeal to theory.

Report of the general discussion

The discussion on Professor van den Noort's paper seemed to strike a somewhat pressimistic note. The divergence between programme objectives and achievements which he referred to was confirmed by speakers in the discussion. One underlined a consistent tendency for benefits aimed at a low income group in practice to improve conditions for a better-off group who had more competence, better organisation, greater political leverage, etc. – attributes which had resulted in their being already in more comfortable circumstances.

The Mansholt plan was seen differently by different speakers. It was held that Professor van den Noort's judgement was unduly harsh, and gave too little credit for the depth of wisdom behind it and that its provisions were, in fact, deliberately chosen to be provocative, as a matter of political tactics, because he did not believe the necessary political decision could be reached without a jolt to public opinion. Some of the options from which Dr. Mansholt made his choice came close to the final decisions of the Council of Ministers a few years later.

The position of agricultural economists in this area was touched on by several speakers. They were reported to have taken part in the preparation of the Mansholt plan options – presumably during involvement with the EEC organisation. The question was raised, however, – following Professor van den Noort's reference to the constraints on goal setting arising from fear that the exposure of conflicting national positions would disrupt the coalition – whether "outside" agricultural economists should also feel similar constraints in criticising EEC policy or whether their professional ethics required them to speak out.

In connection with agricultural economists' contributions of this kind it was argued that, without explicit study of policy decision-making under real world conditions – and the implication was that such studies were scarce – statements to the effect that politicans' welfare functions diverged from the "real" ones were invalid. Moreover, the economist had no justification for criticism unless he was prepared to state clearly his own value judgements.

About the only contribution to theoretical aspects of the main paper spoke of the need for plans to include proposals for control and correction. Control of goal fulfilment and efficiency in means were seen as the basis for feed back for the correction of means – and of goals, too.

Changes in conditions, and the greater difficulty of implementing programmes than of formulating them, make these control measures important. Moreover, they are open to logical analysis and match the common practical solution where a policy programme is introduced and successively improved by experience.

This line of thought perhaps offered a hope of guarding against the risk of benefits being diverted from target groups to others less in need which was the plea of an earlier speaker who left the means unspecified.

Participants in the discussion included: C. P. Baillet, *France*; D. K. Britton, *U.K.*; L. Folkesson, *Sweden*; M. E. Mlambiti, *Tanzania*; D. Paarlberg, *U.S.A.*; U. Renborg, *Sweden*.

P. C. van den Noort (in reply)

The bringing of cows into Brussels offices and the blocking of roads by farmers illustrates my point that politics is an enormous factor in the economic decision-making process. I agree that, odd though these actions may seem, they are important for the politican who risks having his office invaded by cows but not for the econometrician working in the University where farmers do not know their way about.

Regarding fragmentation of the EEC, after the war there were very high hopes for a United Europe. In the early years they only succeeded in getting three small nations to join up. After a lot of negotiations six larger nations came together in 1958 but a lot of difficulties subsequently emerged. Now the group is enlarged with Britain, Ireland and Denmark. However, each time a specific national interest is harmed it is clear that EEC is not a Community but a coalition. Each acts for his own benefit. That is not to say that a nation should not act in that way, but it is an indication that it is not yet a community – it is a sign of fragmentation. Another sign of fragmentation is our inability to implement economic and social policies for Europe as a whole – for example in relation to inflation and monetary problems which are highly frustrating. The third indication is that Communists can take over in one country to the great annoyance of ministers in others, not to mention the separists movements.

In view of this state of fragmentation it is remarkable that negotiations for Greece to enter the Community were opened a few days ago. Turkey will soon follow and, perhaps, Spain also, with Portugal and Turkey in the rear. That can only mean a lot of strife because these countries differ so much from the rest of the countries in the North Western part of the EEC that I cannot imagine how to reconcile those interests. I therefore fear that the coalition will break down sometimes – but that is only a personal opinion.

I agree with others speakers that the social cost of some measures is not properly allowed for. It is no use putting farmers out of agriculture as a means of getting larger farms without taking into account what becomes of those who are displaced. If they have no employment it is not a good policy, it is social and economic waste. I agree, also, that part-time farming is only a step in the process of agricultural adjustment, both in Europe and in other countries. It is one stage between being a peasant and being an industrial worker.

There are two aspects to the question of income parity, one a value judgement and the other factual. I will start with the latter. Income parity was referred to in discussion as each farmer having the same income as the average citizen in the rest of the economy but that is not the way it is operated in my country or in Germany or in the EEC. There it is not every farmer, it is not the marginal farmer, it is not the average farmer, that is to have parity income but a specially well-defined type of farmer. The second point is that the comparison is not between the income of the average farmer and the average citizen, or worker, but the comparison is with a worker with the same type of skills and, moreover, living in the same conditions – in the same region or village, so that he has the same cost of living. The third point is to make clear that a goal of income parity does not mean that this situation will hold for all areas and all farmers. The fourth point, which is sometimes forgotten, relates to the value of being your own boss, to having freedom to live in the countryside and to have the way of life that you are used to. The fifth point, which is also sometimes forgotten, is that there is a large variation between incomes inside the farming sector.

Now turning to the value judgement namely whether such a farmer should have a target income. I do not know the answer, for it is a political decision. It is more or less based on the idea that, in a smoothly operating economy, things would work out in that direction. Everyone would have the same real income for the same kind of work; that is the true value basis for the concept of income parity for farmers.

Turning now to the question of high food prices and taxes, high food prices are bad for the consumer but high food prices do not necessarily mean high farm prices. In Europe people can better afford to pay high food prices than people in developing countries. However, permanently high food prices would stimulate agricultural production all over the world and that might be a positive good. So I cannot give a final statement whether high food prices are good or bad. The Tax aspect has already been discussed in earlier sessions. What concerns taxpayers, to my mind, is not simply that they have to pay to supplement prices but that they must pay also to support projects, electrification, housing and various other items. I would say that taxpayers would be helped very much if careful cost-benefit calculations were carried out and projects only put into practice if the benefit exceeded the cost.

I agree with the contention that the target group is not always accurately reached, and that groups can bring pressure to bear so as to get a larger part of the cake than perhaps was originally intended. But I would also point out that the outcome also depends on the kind of programme you choose. In the United States, as in Europe, Price Programmes are the choice. These give the larger farmer a larger income protection because he sends more produce to the market. Under another programme, e.g. negative income tax, or direct income supplement, aggressiveness would not offer the same gain – except, perhaps, in trying to circumvent the tax authorities; but that is another story.

I fully agree that Implementation, Control and Correction are very important. I omitted them partly for brevity and partly because there are substantial complexities arising from the fact that, once you have put a programme into action, then conditions in which you live and make decisions afterwards are changed.

I agree also that the Mansholt Plan was not an econometric study but was a tactical proceedure and that proves the point that economic philosophy we find in the text books is limited in terms of application to real policy-making. Unless text books improve politicians will have to choose another tactical and strategical base for action. As a political plan the Mansholt Plan was a very good one because it started a lot of discussion about the CAP.

In other connections I would stress that it does not harm us as economists to have a long debate but a politician must get something through. You have to see the different positions of the decision-making philosopher studying the process and the man acting in the process. I think the politicians must take the view that it is more valuable for the Community to get a law passed than to have quite a lot of debate in the Parliament and nothing accomplished.

Regarding the criticism that my contention that the function of politicians and the general welfare do not necessarily coincide is unproven and therefore it was unscientific to put it into my paper, I would say that I can prove it, not personally. In my list of references I give the contribution of Mr. Niskanen of the United States who wrote a book about beaurocracy and representative government, published in Chicago in 1971.* In my opinion at least he proves the contention quite clearly.

* Niskanen, William A. Jr. *Bureaucracy and representative government* (Chicago: Aldine) 1971.

ING. JAROSLAV KUNC *

Regional Social Planning in Agriculture

The planning of economic development in socialist countries has achieved substantial results. On the regional level the planning now extends to widespread elements of well-being of the agricultural population. The state managing authorities in CSSR approached the so-called social plans on the enterprise and the regional level. Social planning increases the possibilities of integrating numerous activities previously organised somehow for the people's benefit, yet with less consideration for conception and priorities. The social policy of the socialist countries has, even before this, provided the main features of life and social security to every citizen. Nevertheless, social planning becomes a special instrument in attaining state social policy goals. Special importance within the whole system of social planning in CSSR is given to regional planning.

In connection with the development of regional social planning there arises – with respect to the horizontal area character of agricultural production – the necessity of solving the relation of social plans on supra-enterprise level to plans of social development on the regional level. At the basic supra-enterprise level the agricultural management link is provided by district agricultural boards; the attention is primarily focused on the social plans of this district management level. The district social agricultural plan is developed in coordination with the District National Committee and the District Committee of the Union of Cooperative Farmers, as well as with the district bodies of the Trade Union of Agricultural Workers. The social agricultural plan of the district becomes part of the social plan of the whole region which is compiled by the District National Committee as the representative authority of state administration.

Perspective and complex plans of the socio-economic development of agriculture originate at the district level based on social plans of agricultural cooperatives, state farms, partner-agricultural cooperative and state enterprises, agricultural services and other production units. Of course this plan is not merely a summary of enterprise plans. Social plans on the supra-enterprise level manifest the influence of state social policy. From the social plan of the

* Institute Director, Economic Research Institute for Agriculture and Food, Prague.

district agricultural management body are projected goals and the orientation of social development of the whole agriculture and agricultural industries into the regional plan of socio-economic development.

The importance of social planning in agriculture increases with intensified concentration, specialization and cooperation in agriculture through levelling up of cooperative and state membership. The central authorities focus, in the first place, on ensuring the system of social welfare, incomes, adult professional education and other complex tasks. On all agricultural management levels principles are applied covering collaboration of territorial, departmental and professional bodies and organizations. In the hierarchy of the social management system the social plans of higher level bodies are obligatory on the bodies on a lower level.

The concept of the regional plan of social development, especially the long term programme goals, involves a base derived from detailed complex analysis of social problems of the region. The analysis contains detailed objective information on the main social problems of the region, the solving of which is a decisive prerequisite for general social development (population development, its stabilization, work opportunities, housing, etc.), including information on individual spheres of regional social life (e.g., leisure time, cultural and sport activities, participation in public and other activities of inhabitants, etc.).

The social analysis is based on extensive data collection covering the social situation of the region. It is based on three main data and information sources. The first source is composed of official statistics (e.g., census of population, housing, etc.) and of other current data investigations by authorities and organizations. The second source constitutes the social plans of individual enterprises and organizations in the region. The third source represents the sociological research complementing missing and necessary data. The sociological research is usually performed on representative samples. Besides these main sources utilized other additional information is used (e.g., projections of the productive-technical base, orientative regional plans, etc.).

Methodically the social analysis is based on the confrontation of the present state of development in the region and evaluations by the inhabitants of the region. Considerable attention is devoted to the opinions, attitudes, aspirations and preferences of the region's inhabitants, as they substantially influence the people's satisfaction, trends to stabilization in the region and active participation to create the most suitable conditions of work and life.

In the planning of trends in the CSSR three principal methods of social planning are mainly used – the methods of normatives, balances and optimalization. A degree of rational knowledge in social planning and their quantification requires adequacy of social terms, i.e., to link social terms to the mathematical ones. Finally, this adequacy means the construction of social quantities and variables and their qualitative interpretation.

At present the method of normatives is mostly used which is the most simple and enables plan development even in the present conditions of unsatisfactory aspects of social terms. It is possible to ascertain some social elements through simple summarization, others by means of standards and equivalents where broad application is applied to so-called social norms (standards). Social norms differ substantially from the norms used in economic planning. Detailed definition is necessary for the determination of the utility function of the plan. The utility (objective) function of the plan is understood as a certain state of social elements considered as optimal.

Also balancing is used for planning social processes. Through this method various problems are solved and tasks coordinated.

Finally, there is the use of mathematical programming extended to social planning, especially linear programming with optimalization from possible alternatives, including utility functions and limits. Leisure time is very often optimalized. Practically it means searching for an optimum between working and leisure time, minimizing the working time and maximizing the leisure time. Of course, alone a utility function constructed in this way is not entirely satisfactory because in this way leisure time becomes the indicator of all well-being. In practice there are further problems and even the definition of leisure time is not always clearly determined. Optimalization of the social plan assumes reaching maximal results for any indicator under given limits.

The present attention given in CSSR to methodological and theoretical problems of social planning contributes further to solving some questions in the application of mathematical methods in social planning and to the further effectiveness of social plans.

Functions and goals of social plans, based on social and economic characteristics of the region, are the following:

(i) planning of changes in social structure with special reference to professional and geographical mobility of the labour force,

(ii) planning of further development of the conditions of work and living – including working environment, technical change in work characteristics, improvement of hygiene and safety of workers in enterprises, incomes, transport, recreation and welfare, living environment, infrastructure, housing etc.,

(iii) planning of special care of selected groups of workers and their families – youths, women, pensioners and persons with lower work ability,

(iv) planning of social activities, participation and education – various forms of adult-education, socialist competition, cultural and sport activities, etc.

Finally, the goal of social planning in the conditions of Czechoslovak agriculture is the levelling up of the working and living standards of country and city populations. Now, when income parity between agriculture and industry has already been achieved, social planning is aiming at other aspects of work and life. Regional social planning is above all orientated to living conditions and the sphere of working conditions is, first of all, integrated at enterprise level. With respect to living conditions in the country the main problems relate to the living environment, housing, public health, education and other facilities and possibilities of cultural life. Within the village the trend is to separate agricultural production units and dwellings of inhabitants. Large scale agricultural production units are being built as closed complete and compact units in places where living environment will not be disturbed by agricultural emissions. The dwelling part of the village consists of dwelling houses and public facilities serving both the agricultural and non-agricultural village population. In the country, individual family houses predominate even though in many villages there are also multistorey dwelling houses. Housing construction is conducted according to territorial planning ensuring that all aspects of urban practice and architecture are respected. The extent of public facilities (infrastructure) of a village depends on the classification in which the village has been included. In villages forming historically and economically central points of areas and having about 2000–5000 inhabitants all kinds of public facilities are provided, e.g., nursery schools; kindergartens; health, education and cultural facilities; shopping centres and other services. In smaller villages the main kinds of infrastructure only are provided for. The classification includes also the very small villages with decreasing local population which gradually are intended to come to be used for recreation purposes for city inhabitants.

Priority in the development of public facilities in social planning is concentrated on design and construction of nurseries and kindergartens. In the construction of these facilities special features of the village are respected – lower capacity, use of the landscape, etc. Further shopping centres, various service-points, cultural and sports facilities, etc. are designed and constructed. Public facilities are mainly built by means of associated financial funds of national committees and enterprises in the region. Many good results can be attributed to the initiative of local inhabitants who participate in their leisure time in constructing the various facilities. The overall appearance of the village is complemented by green and park areas in the neighbourhood of the productive and non-productive units.

Generally speaking, regional social plans as they have been built up for the period of 1976–1980 also distinctly contribute to further development of dwelling and productive capacity of the village and represent further steps towards equalizing living conditions for country and city populations, while at the same time, the special values of country life are preserved, gaining more and more importance in the context of industrialization of the whole society.

DISCUSSION OPENING – A. Maris, *Netherlands*

In relation to Dr. Kunc's paper my first question is – which are the state social goals and how does the state know what is good for the well-being of the many different social groups with different culture patterns and behavioural patterns?

Is the goal of the Government to get one uniform culture based on several normatives or a variety of sub-cultures with individual values?

My second question has to do with the remarks about three principal methods – normatives, balances and optimalization. It is not quite clear to me what is to be understood by the balancing method. It would be helpful also to have some further details about the words "Socialist competition."

"Social planning" is given at the start of the paper as being "a special instrument in attaining the state social policy goals". It is stated in the paper that income-parity is reached between agriculture and industry. Does the author mean that income-parity is reached on national and regional level? What is known about the personal income-distribution within the different sectors of the economy?

My final question relates to large-scale farming in, for instance, the East European countries and the modernised small scale farming in Western countries. The productivity in agriculture in the countries of the Common Market which started with mainly small-scale farming shows an annual increase of about 5%. Apart from these economic results, small-scale farming is an example of decentralisation and distribution of power, initiative and capability. The personal initiative and capability are fully used. Is it not a danger of large-scale collective farming that these forces are suppressed at the expense of economic and social goals?

MONTAGUE YUDELMAN*

The Role of Agriculture in Integrated Rural Development Projects. The Experience of the World Bank

During the last decade, the World Bank has made a substantial adjustment in the scale and direction of its assistance for agriculture and rural activities. The Bank is at present the largest single source of external funds for investment in the rural sector, contributing some five percent of the annual investment by the public sector in the agriculture of its client countries. Since 1946, when its lending operations began, the Bank has lent nearly $8 billion for agriculture and rural development. In as much as each dollar invested by the Bank is generally matched by an equivalent investment from domestic resources, the Bank has in the past 30 years contributed to and participated in the financing of some $16 billion in project-oriented programs in agriculture and rural development. Although this is a substantial sum, it represents only a marginal contribution to the total investment in agriculture that was made during those years. The contribution of the Bank has increased substantially in recent years, however, rising from an average of $120 million a year in the mid-1960s to more than $1·6 billion a year in the mid-1970s.

Through the years the Bank's approach to the rural sector has changed, as have the nature and design of Bank-financed projects, reflecting the significant changes which have taken place in thinking about and attitudes toward agricultural and rural development in particular and economic development in general. In the early 1950s the Bank made few loans for agricultural development, because the development process was assumed to revolve around the transfer of real resources from rich nations to poor nations, with emphasis on closing the foreign-exchange gap; it was presumed that investment in the modern, capital-intensive sector was what would bring about rapid growth. Few investment opportunities were recognized in agriculture except in the export sector, and the capital requirements for agriculture were believed to be small in any event. Further, agriculture was given a low priority because a lack of demand was assumed to be the major constraint upon increased agricultural output; it was reasoned that increased investment in the nonagricultural sector would be required to generate higher nonagricultural incomes,

* The World Bank. The views expressed in this paper are those of the author and do not necessarily represent the views of the World Bank.

which in turn would increase demand in the agricultural sector, thereby stimulating underemployed capacity to greater output.

Subsequently, agricultural growth was recognized as a necessary, though not sufficient, condition for economic development. At the same time, there was a growing awareness of shortage of capital in the rural sector. New technological advances made it clear that there was scope for high-yielding investments in agriculture. The Bank responded to these new views by increasing both the volume of its investments in agriculture and the share of its total investments that went into agriculture. Lending for agriculture rose from a total of $451 million during the period 1948–63 to $621 million during the period 1964–68, or from 8·5 percent to 12·3 percent of all lending. The emphasis, though, continued to be on closing the resource gap, with consequent stress on investments in capital-intensive projects; thus 75 percent of all investment was allocated to large-scale irrigation projects.

It soon became apparent that capital works, however well engineered, did not in themselves lead to increased output. Agricultural development necessitated complementary changes at the farm level, and the promotion of agriculture required dealing with complex systems both within and outside the rural sector. The Bank therefore broadened its lending, shifting its direction toward the encouragement of technological change at the farm level. Its program now included the financing of research, extension services, training facilities, marketing, credit, rural transportation, and small-scale irrigation.

The emphasis was thus shifted from lending for "off-farm" infrastructure (such as irrigation projects) to rural credit as a means by which farmers could obtain capital to finance "on-farm" investment. In most instances loans were channelled through specialized agricultural development banks or established commerical banks. Ideally such projects fulfilled three related aims: they transferred resources, they led to increased production, and they contributed to the overall growth of the economy.

These loans also had their limitations. A sample of five loans made during this period for livestock development indicates that the average loan per farmer was in excess of $25,000. The loans were accompanied, however, by insignificant employment effects, the average number of beneficiaries being only one or two families from each sub-loan. Concern with issues related to employment contributed to the next stage in the evolution of lending by the Bank in the rural sector, which was a shift in emphasis toward rural development. This shift was influenced greatly by the work of social scientists – many of whom are here today – who pointed out that while economic growth had taken place in most countries, large groups of persons, particularly in rural areas, were not sharing in the benefits of that growth. Rural poverty appeared not to be diminishing but rather to be increasing. Demographers and labor economists made it clear that the developing countries of today could not in the short or medium term repeat the pattern of development that most of the developed world had followed – namely, that of increasing the intensity of capital investment in agriculture, thereby contributing to the displacement of labor. Capital was too short and alternative employment opportunities were too limited for the developing countries to follow such a pattern.

Analyses of the distribution of capital expenditures for social services and facilities indicated that – in all countries – the rural areas were far less well endowed than urban areas, according to the usual indicators of social welfare. Rural areas had fewer schools, hospitals, teachers, and doctors per capita than did urban areas. Despite migration, furthermore, rapid increases in population, combined with low levels of both on-farm and off-farm investment in rural areas, were bringing about greater underemployment, and substantial numbers of unemployed, landless laborers were emerging, especially in Asia.

Fortunately, at the time these studies were undertaken, there was a growing literature by agricultural economists on the theory of small-farm production and its relation to national development. The analyses of these economists showed that small farms were both labor-intensive and productive. All this led to a call for a new approach to agricultural and rural development. The new approach would have to be focused on the traditional sector and on ways of providing the credit and the technical and material inputs which would make rural labor more productive. New high-yielding varieties of such staples as rice and wheat seemed to provide a technological basis for this change, since they could be used by the small farmer both for his subsistence and for the generation of cash income. The traditional, small-farm sector would have to become the producer of an agricultural surplus rather than the provider of surplus labor, as it had been in the past.

The Bank thus adopted a new approach to one aspect of the problem of rural poverty to guide its lending in the rural sector. The emphasis was shifted from efforts aimed at promoting simple growth to an attempt to improve the output of the small farmer, principally by increasing those investments that would benefit lower-income groups in the rural areas. This effort has become the main focus of the present approach of the Bank to rural development, although loans are still made for larger, more capital-intensive agricultural undertakings when they can be justified on the basis of other criteria development.

The Bank's present approach to rural development has a sound economic basis: recognition of the fact that the mechanism implied by the traditional labor-surplus, two-sector model is incapable of solving the problems of the rural sector in the short-to-medium term. The problems of poverty in rural areas cannot, of course, be solved in the rural areas alone; a dynamic, expanding non-agricultural sector is essential for providing employment opportunities for an increasing population. So long as the small farmers and low-income groups in the rural areas are unproductive, however, they can contribute little toward solving problems of malnutrition or inadequate food production, in either rural areas or the towns – and according to our estimates there are more than 100 million holders of less than five hectares of land and close to a billion rural inhabitants with average per capita incomes of less than \$100. This is why increasing the on-farm productivity of small-holders remains a central – though not the exclusive – element in the Bank's lending program in the rural sector.

1. THE CONCEPT OF INTEGRATED RURAL DEVELOPMENT

Most of those who live in rural areas are still to varying degrees dependent on agriculture for their livelihood. Consequently, any approach to development in the rural areas has to be concerned largely with agriculture. Rural development on the other hand, means more than just agricultural development; exactly what it does mean, though, has been the subject of much debate and little agreement – a fact that can be illustrated by recent experience within the United Nations.

A recent study made for the UN indicated that five of the ten principal agencies of the United Nations concerned with various aspects of economic and social development had no practical definition of rural development. Only two of these ten agencies had any means of monitoring the percentage of their activities that might be classified as "rural development" or of assessing the results of those activities. It was clear that the term "rural development" has different meanings in different institutions; the study indicated that the term tends to evoke involuntary responses that vary widely from institution to institution and even among individuals within the same institution. This finding applies with even greater force when the adjective "integrated" is appended to rural development. The word "integration" seems to be subject to interpretations that range from broad philosophical concepts to micro-economic issues of linkages within projects.

The World Bank has devoted a great deal of thought to the concept and meaning of rural development, as is reflected in the policy paper on rural development prepared in 1974. In the course of the preparation of that paper, and in subsequent discussions, it was agreed – in the light of the analysis mentioned earlier and of subsequent research – that any approach to rural development must be addressed to the most urgent problem in the rural areas, the problem of rural poverty. It was therefore agreed that concentrated attention to projects which directly and in very large measure aid the rural poor must be a common element of all approaches to rural development. In its policy paper the Bank defined the rural poor – the target group to be aided – as those placed by their annual incomes in the lowest income groups defined on the basis of a mixture of absolute and relative poverty criteria, according to the special circumstances of each country. How one defines and approaches the rural poor varies greatly from institution to institution, but in the present-day vocabulary of international institutions the term "rural development" now implies, among other things, a focus on poverty. Thus we have an interesting phenomenon in that the definition of a functional subject centers around an income criterion.

The Bank's rural development policy paper included another concept which has gained broad acceptance, namely that the rural poor represent a reservoir of untapped talent which has been left out of the mainstream of development. In order for this group to contribute to the development process, they have to be more productive than they have been in the past; increased productivity and higher incomes for the target group are implicit in the rural development program. By the same token, rural development also implies giving the target group the opportunity to enjoy the benefits of development: improved education, health, and nutrition, among other things.

The notion of "integrated" rural development adds a further dimension to the focus on poverty. In a broad sense the process of rural development must integrate the rural poor into the social, political, and economic life of a country. Our concern here, however, is rather with the project-oriented concept of integration. A mix of components in rural development projects is usually justified on the grounds that a mix of investments in any given project produces a synergistic effect – the old idea that the whole is greater than the sum of its parts. Perhaps this can best be illustrated in technical terms within agriculture. The use of improved seed, fertilizer, and controlled water supply have an established relationship to plant production. It is easily demonstrated that the simultaneous use of these three inputs can produce an output greater than, let us say, the sum of the outputs that would result from using each input separately, or from using two of the inputs but not the third. In general, agricultural projects financed by the Bank have a holistic cast; increasingly they include research, extension, and provision of credit and other inputs to farmers, each component reinforcing another.

Rural development projects extend beyond agriculture, for they include such nonagricultural components as potable water supplies, shelter, rural electricity, health and educational services, and roads. A rural development project can involve only a single sector – transport, education, or agriculture, for example – or they can be multisectoral, but in either case the key features are explicit and direct orientation toward the target group. Any holistic effects of a multisectoral approach to investment would presumably arise from the interaction between investments in human capital and those having the purpose of raising rural incomes, primarily by means of increased agricultural production. The case to be made is not that healthier, better educated people use productive opportunities more effectively than would less healthy, less well educated people; it is rather that a particular pattern of investments represents a wiser use of resouces than would some different mix of investments and that part of the test has to do with the benefits realized from the interaction of the components.

Most multi-sectoral rural development projects involve a package of investments to be made within a defined area; there is a methodology derived from our theoretical considerations for providing the underpinning of the economic allocation of resources to maximize the benefits of integration. It is difficult though, to calculate any benefits of integration because of the problems of measuring any contribution of integration and attributing benefits to the various project components. This applies with special force when the mix of investments includes those that can increase output in the short run as well as those intended to improve human capital in the long run.

A close analysis of the components of integrated projects reveals a further problem in developing criteria for a optimal mix. Most integrated projects usually consist of two sets of components chosen on the basis of two different sets of criteria. The first component, which might be called the directly productive element, is intended to raise incomes in the short and medium term.

This component usually includes such items as land clearance, minor irrigation, credit to finance technological improvements in farm operation, and so forth. In practice, the other project-related components are usually included on an entirely different basis largely because of the difficulties of measuring the contribution of each component within a project. Thus, multisectoral projects which include supplying potable water or health and educational facilities tend to include these investments in a program, on the basis of providing services to reach some standard established in relation to the resources available either nationally or regionally. This standard is usually based on a "minimum need" criterion. These projects, then, are not fully integrated in the narrow sense of the term, especially since the components are included on the basis of more than one set of criteria.

Once the criteria of need are seen to be part of a multisectoral project, there is little merit in using arbitrary international guidelines in developing the most appropriate mix of components within such a project. The mix between production and social overhead in a project can well vary according to national levels of income. Some countries have developed much lower cost methods of delivering services, such as health, than others so that some poorer countries have been able to provide a high level of services at a lower cost than some of the wealthier countries. Nonetheless, richer countries will, in general, be able to aim for a much more favorable achievement as registered by social indicators than will poor countries – a larger number of doctors or schools per region, for example. This disparity is further compounded by the extent and availability of skilled and semi-skilled persons and by the existing rural infrastructure. The experience of the Bank has indicated clearly that the mix of components in a country such as Malaysia should differ considerably from that in a country such as Mali; in Malaysia social infrastructure and a relatively high standard of rural housing services are emphasized; in Mali the emphasis is on functional literacy and increased farm production. The simple fact is that the poorer countries must be more concerned than other countries with providing basic necessities for their peoples and allocating resources to those activities in the rural areas which are immediately productive.

A further project-related difficulty is that comprehensive, or multisectoral, rural development projects are more difficult to design and to administer than single-sector projects. Our experience also indicates that they are more difficult to implement. Most government departments are organized along functional lines, such as agriculture, forestry, fisheries, health, public works, education, and small-scale industry. Few governments have operational responsibilities for multisectoral rural development. Consequently, projects that deal with rural development tend to require some horizontal integration of administrative effort, thus necessitating a break with traditional patterns, in many instances, and generating conflict between line departments that provide sectoral services and those that are interested in promoting intersectoral coordination at the project level.

Clearly there are conceptual qualitative and administrative problems which tend to conflict with the idea that an integrated multisectoral project leads to a more efficient use of resources by virtue of a holistic effect. These problems should not be allowed to obscure the value of a comprehensive approach to development of the rural areas. Such an approach, though, would presumably be in the form of a national or regional plan or program with a mix of soundly oriented investments not based exclusively on criteria related to any presumed short-run effect on production. There should be no need to rely on the presumed gains from integration of components for justification of an investment the purpose of which is to raise the level of living in a backward or poor area. These investments should be justified on other grounds, such as the meeting of acceptable minimum needs or the removal of a particular obstacle to increased production, and should be based on an integrated approach. Undoubtedly, a multifaceted or multisectoral approach has the merit of quickening the pace of change, which is the essence of rural development, but the gains from integration do not by themselves seem to justify such an approach.

To summarize the Bank's approach to rural development, it is a process intended to raise the output and incomes of the rural poor. It extends beyond agriculture, and it has the purpose of bringing the benefits of development to the rural areas, with emphasis, once more, on aiding the low-income groups or those who have been passed by in the growth process. While the Bank supports multisectoral projects, some uncertainty remains about justifying such an approach solely on the grounds that a mix of components will lead to better results because of the effects of integration on the use of resources. In our view, a well-designed rural development project should reach large numbers of low-income producers; should be able to raise the incomes of this group; and should be replicable, especially with regard to the costs involved. It should be comprehensive in scope where it is clear that the nonagricultural components are consistent with national, regional, and sectoral guidelines.

2. BANK EXPERIENCE

2.1. *The General Environment*

The World Bank, like many other international agencies, is a relative newcomer to the field of rural development. Other external donors, notably some of the bilateral agencies, have had much wider experience in this general area than the Bank has had. Many governments, such as those of India, Tanzania, Mexico, and Korea, were active in providing poverty-oriented comprehensive rural development for many years before the Bank took any special interest in doing so. As latecomers we have been able to learn from the experience of others. Perhaps the most important lesson that we have learned is the importance of including productivity and opportunities for raising incomes in projects as a means of ensuring their viability and gaining popular support for them. We have also learned from our own experience in some projects. While it is too soon to assess their full impact, we believe that we have gained from them some insights into problems of rural development, especially problems of raising the incomes of small-scale producers, which are the focus of the following discussion.

Our experience so far indicates that projects seldom bring the desired results if the environment is not conducive to success. The environment is conditioned by a host of social, economic, physical and cultural factors. I would like to emphasize two of these factors; the first is related to the low-income farm sector, and the second influences the entire agricultural process.

The first and most significant factor influencing the general environment for rural development is political (and hence national) commitment to a policy of making the rural sectors more productive and, especially, of involving small farmers in development. One indication of the degree of such commitment is the amount of the budgetary allocations for rural development – though it is difficult to isolate these allocations when looking at budget documents. It appears, though, that most governments fall into one of three categories: those that are committed to assisting low-income groups in the rural areas; those that are receptive to the idea of doing something but are not in a position to mount a program; and those that deny that there is any need of special assistance for the low-income groups. An analysis of budget allocations in the principal developing countries suggests that perhaps half the governments are increasing their investments in rural development.

Another element of commitment that affects the environment of a project is the attitude of the local bureaucracy. This is a vital factor in most rural development projects, since they usually involve local action through community participation. Most programs and projects involve decentralized decision-making and administrative functions. Consequently, a great deal of agricultural and operational responsibility tends to devolve upon bureaucrats at the field level. There is ample experience to indicate how significant the attitudes of local officials are in influencing the course of rural development projects. Such officials usually have responsibility for initiating activities, for organizing and mobilizing local resources, and for helping in the management of operations. One mark of the commitment of governments to rural development is, or course, the extent to which they assign capable officials to work on rural development and give them adequate support in the form of attractive terms of service – something that only a few governments, notably those of Mexico, Korea, and India, do. Thailand, to give another example, gives special allowances to field-level personnel of its Institute for Rubber Replanting.

Our experience indicates clearly that economic incentives determined by the terms of trade for agriculture and related price and fiscal policies constitute another factor in the outcome of all agricultural projects including rural development projects. The best-organised projects have encountered difficulties when the structure of costs and prices has not provided adequate incentive to producers, both large and small, to increase their output. This is true in both centrally planned and free-market economies. There is some evidence that an increasing number of governments, especially in Asia, are adopting price policies that are more favorable to all agriculture (and rural development). Nonetheless, there are still many governments, both in Latin America and Africa, where the prevailing price policies are creating a poor environment for agriculture and rural development of any kind.

2.2. *The Project Approach*

Much of the effort in rural development projects of the Bank has been directed towards increasing the output and incomes of low-income producers, often by means of the introduction and expansion of technological change at the farm level. The assumption underlying this effort is that three basic conditions must be met if changes are to be brought about: producers must know how to increase their output, they must have access to the means of increasing their output, and they must have the incentive to make the effort and accept the risk associated with increasing their output. Agriculture is atomistic in the sense that there are many producers, each with little influence over the prices they receive (though improved marketing techniques can often raise prices at the farm gate). Consequently, most projects tend to be focused on cost-effective ways and means of delivering to farmers the goods and services that they need. These include the inputs that investment in infrastructure, such as water and transport, will provide.

Most programs and projects designed to help farmers depend to a large extent on support services provided by institutions away from the farm – institutions that provide information, credit, technical inputs, and marketing and transport services, for example. In many countries – indeed in most countries – these institutions are organized in such a way that they serve large-scale producers primarily. Frequently this is a politically determined matter of policy; also the larger farmers and the suppliers of agricultural inputs often have common interests. Sometimes an institutional bias towards larger producers exists simply because management finds it to be easier, safer, less costly, and less troublesome to have a clientele of larger producers. Be that as it may, the poorer producers with small holdings are seldom the primary beneficiaries of programs designed to aid farmers in becoming more productive. Thus the heart of any program aimed at helping small farmers to be more productive must be the creation of institutions that will undertake those tasks which are of assistance to small producers.

The Bank expects to increase substantially the volume of funds committed to projects designed to strengthen critical rural support services, thereby breaking the contraints on production and, in many circumstances, permitting more effective use of physical infrastructure which is already in place. The attention given by the Bank to command area development in already irrigated areas is part of this effort. Gains in production from such programs could be substantial. In India possibly as many as 15 to 20 million hectares of irrigated lands are producing at only half their potential of perhaps two tons per hectare. In large parts of the rainfed grain-producing areas of West Africa, an increase in output of less than one percent a year, attainable with known technology but requiring stronger support services to farmers than they now receive, would permit increased self-sufficiency in a wider range of crops which can be produced locally.

Programs and projects intended to help low-income producers must be designed and implemented with that end in view. A first requirement for any such program is that there be a clear understanding and identification of the target group. It is relatively easy to establish an objective measure of a poverty datum line; those measures that have been used range from refined estimates based on the critical level of consumption and minimum nutritional requirements of a family to crude estimates, such as half the average family income. Once a poverty datum has been determined, it can be translated into some appropriate yardstick – the size of a holding under various conditions, for instance which can generate a size-of-holding criterion for the minimum income level of the target group. Other criteria which might be used are: a measure of a farmer's assets or the number of laborers employed. Whatever criterion is used will provide the upper limit of income of the target group and permit a complete enumeration of the group in the project area.

In some situations structural change may be a necessary precondition for rural development. The most important kind of structural change in the rural sector is the redistribution of land. It may be necessary to change the land-based power structure in a society before much can be done about political commitment to rural development and the scope for local participation in it. From a more limited perspective, there are also situations in which the redistribution of land can bring idle land into production by making it available to underemployed rural laborers. While our experience indicates that substantial rural development projects can be undertaken to the benefit of large numbers in the target group even in situations in which the distribution of land is skewed, the Bank encourages and supports the efforts of borrowers to improve tenurial arrangements.

Experience confirms the fact that tenurial arrangements in the project area have an important bearing on the success of projects. The reason is that all projects alter the distribution of gains from the land; unless there are adequate safeguards built into the project the weakest landholders are invariably at a disadvantage. A project which is in an area of smallholders who own and operate their own land or who have secure rights to land tends to benefit the producers. In areas of absentee ownership, however, in which the land is operated by sharecroppers or by tenants who do not have security, the benefits from increased investments and higher output are not necessarily realized by the producer. Experience has shown that when projects have provided profitable opportunities, tenants without secure leases have been evicted and replaced by day labor or machines. Some projects in neighboring Ethiopia provide examples of this phenomenon. Also, the introduction of new technologies under standard crop-sharing arrangements which are based on profit-sharing or fixed proportional crop shares has often – though not always – led to a disproportionate increase in returns to the landowners.

If there is support for the objectives of rural development, however, it is possible to design projects that will benefit smallholders (substantially if not exclusively) without the necessity of substantial structural change. This is especially true in situations in which land is being newly developed and resettled. Examples of such projects include:

Irrigation projects. In recent years, increasing attention has been given to designing irrigation projects in such a way that large numbers of smallholders can benefit. Many governments now specify limits on the amount of land that can be held by a producer who stands to benefit from the irrigation system.

The public costs of these projects tend to be higher because of the necessity of providing a more elaborate and extensive distribution system for carrying the water to a large number of small farms.

Smallholders schemes. The Bank has participated in the design and implementation of projects for the production by smallholders of crops which were formerly grown almost exclusively under the plantation systems; these include rubber, tea, sugar, and oil palm. The focal point of many of these projects is the processing plant, which requires a guaranteed supply of raw materials for full utilization of its capacity and realization of economies of scale. Many had come to believe that only large plantations could ensure a steady flow of raw materials to these plants, but it has been demonstrated through experience in many projects – including the Kenya Tea Development Authority and the Malaysian Federal Land Development Authority – that smallholders can produce quality products that were formerly considered the province of centrally managed estates.

Land settlement projects. The Bank has financed a fairly large number of settlement projects, principally in Malaysia and South America. Settlers from low-income groups have benefited from these projects, many of which provide all the infrastructure, such as housing, roads, and social facilities that the settlers need. The cost per beneficiary family varies from US$850 to US$28,000 (in 1975 dollars); the cost per hectare is 56 times greater in the most expensive project than in the least expensive.

As a general rule, though, these projects do not reach large numbers of rural people. Functional projects and area developments are the most widely used among projects of the general type with the cost ranging from $1,000 to $33,000 per family, averaging around $1,300. A *functional project* is one which provides an input or technical package to the target group. A program to provide credit either in kind or in cash to producers is an example of a functional project. In this instance, the target group is usually defined by size of holdings or fixed assets and income (often expressed in multiples of the minimum wage). In the case of *area development projects* the target group as such is not specified. Instead, these projects are for the purpose of developing large areas, which are chosen because per capita incomes are below average and the rating by social indicators is low. The underlying assumption of the area approach, which is typified by the introduction of infrastructure and rural services, is that most of the benefits (though not all) will go to the target group. In the main, most people who live in areas in which average incomes are low and which are generally deprived of social infrastructure tend to be poor themselves.

One of the most ambitious national programs that is directed towards rural poverty areas and supported by the Bank is the $1·2 billion effort now underway in Mexico. The program of the Government is to channel additional investments into microregions by improving the administrative operations of its existing planning and execution system through the establishment of coordinating machinery in the Ministry of the Presidency. The approach is three-fold:

(1) Increase investments and services of existing agencies and direct

them towards selected poor rural microregions with productive potential;

(2) Decentralize planning and, especially, execution so that both are performed at state and local levels; and

(3) encourage participation in the planning and execution process at the village and *ejido* level.

To this end, a new system of planning and delivery has been developed that embraces existing government agencies. Through this system, extra resources are allocated to agencies in poverty areas – areas which have traditionally been bypassed by the "nominal" programs of these agencies. The basic unit of these selected poor areas is defined, for purposes of investment programming, as a microregion comprising from two to seven contiguous municipalities within a single state. The average microregion contains about 50,000 persons living in three municipalities.

The experience of the Bank with the range of projects in which we have participated has led us to several conclusions. The first is that it is possible to design projects that will assist large numbers of rural poor to expand production and increase their incomes; there are significant qualifications however:

(1) Farmers and producers of all kinds will use new technology only when it is profitable for them to do so. The new high-yielding varieties of rice, maize, and wheat give high returns when they are used with fertilizer and water; many small producers have therefore adopted this technology with gratifying results. We have become very much aware, however, that no such technology is available in many ecological zones of the world, especially in the rainfed areas which contain most of the world's farmers and where less common cereals and root crops are important. Furthermore, technologies developed for a specific location require adaption before they can be employed elsewhere, even when environmental differences may not appear significant. Many Bank-supported projects involve "stabilizing" agriculture by persuading producers to change from shifting cultivation to sedentary agriculture. Much of this is being done under conditions in which considerable risk is involved because of the limitation of the available technology. It is thus important that there be no diminution of efforts to develop high-yielding technologies for commodities grown in the rainfed areas. Appropriate technologies for dealing with problems such as soil depletion also need to be developed.

(2) Many of the poor farmers of the world live in areas in which the resource base has been eroded by deforestation and overexploitation. There is little in the existing array of technologies that can deal with this problem effectively.

(3) It is extremely difficult to develop and sustain low-cost delivery systems for agricultural support services. Some progress is being made, however, in the organization of extension services, for example. One approach which we consider to be promising and which is now embodied in some 15 Bank-assisted projects, involves close links between research and extension, careful selection of high-yielding agricultural practices, concentration of efforts in favorably situated agricultural areas, precise work programming of field-level staff, and regular meetings with selected contact farmers.

(4) Much more thought and effort is needed in order to find cost-effective ways of delivering goods and services to large numbers of low-income producers. Low-cost, effective credit programs are urgently needed. Lending to large-scale producers is in general less costly than lending to small-scale producers; the unit cost of the loans may be the same but the return is much higher on a large loan than on a smaller one. It is estimated that the administrative costs of loans to large producers are often in the neighborhood of 3 or 4 percent of outstanding loans; administrative costs of loans that pass from the government through such groups as cooperatives to small producers tend to range from 10 to 20 percent. These costs do not include any charge for the use of capital, so an interest rate for small farmers that would cover both capital and administrative costs could be as high as 30 percent.

Despite all efforts, including the shifting of the administrative burden at the local level to cooperatives, there seems little prospect that costs can be reduced to levels comparable to those of handling loans to larger farmers. Thus, governments (and bankers) interested in helping small farmers through credit schemes will have to reconcile themselves to subsidizing the administrative costs of these schemes if they continue to lend through systems designed to help large-scale producers.

Many innovations have been made in the effort to reduce administrative costs by taking account of the fact that the major clients will be small-scale producers. Some banks have eased their requirements for small borrowers, basing creditworthiness on reputation rather than requiring collateral with the associated costs of recording titles. Other lending institutions are providing credit to villages as units; the villages then take responsibility collectively for the repayment of loans. Our experience is that costs can be reduced when there are especially tailored procedures to facilitate lending to small producers.

Another issue is that of local participation. In practice villagers are rarely consulted. Commitments to rural development are initiated at central levels. Budgets are centrally determined. Decisions are made by central planners. Obtaining full participation, not only by villagers but also by local and state bodies, often means reversing trends toward centralization. There are some signs that these trends are being reversed in a growing number of countries. In our view local participation is essential if a project is to have the support of the community and if the benefits of the project are to be realized by those for whom they are intended. Local participation in decision-making is often the principal safeguard against leakage of funds; in the final analysis, though, once funds have been committed, who has control over the use of them frequently depends on the nature of the local power structure. This in turn brings me back to the importance of involving the local bureaucracy in all matters relating to rural development.

A further point is that rural development projects have the purpose of providing opportunities for the target group to become productive. Most projects are designed with the assumption that the numbers of those who will avail themselves of opportunities to raise their incomes will increase steadily. This has not always been the case. The difference in response to the opportunities that have been provided is one of the more interesting and puzzling aspects of the implementation of rural development projects. In some situations the opportunities are seized by 20 percent of the producers in the first year and by 10 percent more in each of the next four or five years. In other situations, the pattern to be observed is the familiar one of a slow beginning followed by a rapid acceleration of adoption of new technology. In still other situations there is a slow but steady increase, with roughly thc same number of additional producers taking up the new technology each year.

The rate of take-up and the associated issue of risk and uncertainty in decision-making are interesting in themselves. Clearly the element of risk in varying situations in rural development projects needs to be analyzed and understood better than it is now. The rate of take-up, moreover, has a bearing on the rate of increase in output and incomes. This in turn influences the rate of return on a project. Thus, in many respects this aspect of rural development is the crucial one. It is also the aspect about which the least seems to be known It is therefore appropriate that the question of the way in which decisions are made in agriculture is the topic of this conference.

Finally we must consider the fiscal impact of projects. Few governments embarking on large-scale rural development programs are aware of the long-term budgetary effects of their programs. In many cases, major portions of the initial investment are from external sources. But detailed forecasts for financing and administering the operation and maintenance of such investments are rarely made, and no resources are allocated for maintenance. Thus canal systems silt up, rural roads wash out, water pumps break down, health clinics are not staffed, extension vehicles break down, and deliveries of fertilizer are not sustained. Our experience is that if fiscal and administrative problems are to be overcome, rural development programs must from the outset be designed to maximize the local financing and maintenance of rural development investments. By way of illustration, in several countries local committees for the collection of fees for the operation and maintenance of the water supply, hire a local villager and pay him to maintain the pumps and reticulation system. When he needs to he can call upon the staff of the central water-supply agency for technical help, but the principle of local operation and maintenance is nonetheless established.

In conclusion, I would like to review three major themes of this paper and to mention some areas in which we think that additional research and attention to policy is needed.

The Bank's lending for agriculture has shifted in emphasis and is now focused on lending for small farmers. This is part of the Bank's policy for promoting rural development where rural development is defined as having an antipoverty focus.

It is necessary that there be a suitably encouraging environment before there can be successful rural development. Such an environment requires that both political and economic conditions be favorable. Given a suitable environment it is possible to design and implement projects which will fulfil the objectives of raising the output and incomes of low-income producers. These projects include irrigation schemes, programs for smallholders, settlement schemes, and functional and area projects with the more general provision of rural credit and infrastructure.

The Bank's experience indicates we need to examine very critically the notion that a multi-sectoral project generates gains to the rural community above those which would arise from an integrated approach of single sector projects. We need to find practicable methods to measure what happens to rural welfare when we alter the relative sizes of individual project components; we also need to examine the proposition that local participation during the project identification stage both reduces project preparation costs and results in a better project mix.

The experience of the Bank so far indicates that there is still a need for the development of suitable yield-increasing technology for agriculture in many parts of the world and that there are few opportunities for lowering the costs of delivery systems when these costs are built upon systems intended to help large producers. We need to study alternative systems and to learn which are the most cost effective among the array of systems now being tested in many different projects. We also need to know much more about the principles and patterns of behavior of small-scale producers and what influences the rate at which they accept innovations.

Rural development is a slow process and we need to learn from our experience. To do this we need to develop mechanisms for monitoring and evaluation which can provide guidance for effective implementation of projects and can help all of us learn more about what constitutes a viable rural development project. We are finding that the whole subject of designing suitable project monitoring systems is a challenge to us and to the academic community at large.

DISCUSSION OPENING – A. A. Abdullah, *Bangladesh*

The title of Dr. Yudelman's paper is somewhat misleading. It is too unassuming for it comes close, in fact, to being an exposition of a theory of rural development. It is also misleading in that it gives rather little space to the problems of defining, formulating and implementing an integrated programme.

A selective reconstruction of Dr. Yudelman's central themes might be as follows.

He starts by identifying a double shift in the Bank's development philosophy – first, from industry to agriculture, and second, within agriculture, from projects concerned solely with output growth to projects specifically directed to the alleviation of the poverty of some defined target group. This target group is defined in terms of income, as the "rural poor". In his words "Any approach to rural development must be addressed to the most urgent problem in the rural sector – the problem of rural poverty".

By "integrated" programmes the author seems to understand programmes which have components that cannot be directly justified in terms of their contribution to output growth. Thus to justify these components one has to resort to arguments based on indirect productivity effects – good drinking water prevents manhour losses and lowered efficiency through disease – or one has to base oneself on minimum-need welfare criteria. In the author's opinion the second course is preferable as being more honest and more logical.

Dr. Yudelman recognizes that under certain circumstances the power structure at the local as well as national level may obstruct the implementation of such a development policy, and that substantial restructuring of society – involving, in particular, land reforms – may be a prerequisite for meaningful development effort.

One must thus identify the following major theme in Dr. Yudelman's exposition

(a) rejection of the ideas of industry as the "engine of growth";
(b) rejection of a commitment to output expansion that does not explicitly address itself to the problems of the rural poor;
(c) identifying a "target group" in terms of income levels and designing all programmes so as to have a direct impact on this group's well-being;
(d) integrated programmes as tools of social welfare; and
(e) the role of local and, perhaps, national power structures in blocking and distorting developments efforts.

In the following I shall discuss each of the above in turn.

The primacy of agriculture

The industry versus agriculture controversy is an old one and has appeared already in this conference. It was the pivot, for example, of the Soviet industrialization debate. Dr. Yudelman treats this debate as essentially resolved in favour of agriculture, at least for the purposes of short and medium-term policy formulation. I do not think it is a meaningless compromise formula to speak of an integrated approach to development, in this respect, also one that would stress equally both industry and agriculture, and stress in particular the linkages between them. While industrialization can be brought up short by an agricultural sector that refuses to get moving, there is also the possibility that agriculture will stagnate if faced with a stagnant industry, unless agriculture becomes essentially world-market orientated, thus giving rise to what Samir Amin has called "disarticulated development".

From output expansion to poverty eradication

An affirmation of the human content of development, is certainly welcome – a better life for people is what development is all about. However, on this point there is a certain ambiguity in Dr. Yudelman's exposition. He seems to be talking about two different things, but believing that he is talking about the same thing. The two things are:

(a) a shift from an "elite farmer strategy" to one based on the small farmer, and

(b) a shift from a pure growth strategy to an anti-poverty strategy.

It appears indeed that for the author, and one presumes for the Bank also, the action consequences of (b) are exhausted by (a). Now, as has been pointed out, the small farmers are poor, but they are not the poorest. A true concern with poverty should lead one to concentrate, not on the small farmer, but on the landless wage-labourer – a growing multitude in most poor countries. True, there is not much that the Bank can do to initiate projects that will help this class – but this is equally true of small farmer programmes.

My own belief is that the shift to the small farmer strategy comes, not so much from an increasing concern with the rural poor, as from a growing realization that the elite farmer strategy would succeed in raising output and producitivity only in very limited areas where conditions were already propitious for the emergence of a robust capitalist sector in agriculture. Given the small average size of holdings and the fragmented nature of holdings, and also, perhaps, a tendency among well-off farmers to satisfice rather than maximize, the base for a take-off via the elite farmer strategy simply did not exist in most poor countries. In any case, what is relevant is not the question of motives, but of workability. Is the small farmer strategy going to produce growth at an acceptable social cost?

The "target group"

I have reservations about this notion of "target groups", particularly when it is produced to define the group by a minimum income criterion. I think Dr. Yudelman is wrong when he says, "It is relatively easy to establish an objective measure of a poverty datum line." (p. 452) I would suggest that all poverty lines are arbitrary. One can, of course, easily name an income level so low that anyone earning less than that must be poor, and another level such that anyone having a higher income cannot be poor. We would have these two coincided, but in most cases there will be a very large gap between them.

More fundamentally, defining a class of "rural poor" represents an act of illegitimate aggregation – illegitimate from the viewpoint of social theory and analysis. The small farmer may be as poor as – or even poorer than an agricultural labourer, or the artisan, petty trader, school teacher, and so on. To lump them together in one category will not help to illuminate social reality but to obscure it. It may be quite legitimate for short-term relief or welfare measures, but not for the planning of rural development. To plan for rural change we must first understand the laws of motion of the particular rural society we want to plan for. For this we need well elaborated social theory.

Integrated programmes

The idea of an integrated programme has several aspects as spelt out or implied by Dr. Yudelman. The following may be identified:-

(a) integrating – that is putting under one co-ordinated programme the supply of inputs and services needed by the small farmer to achieve higher productivity;

(b) integrating production-orientated projects with welfare projects;
(c) integrating agricultural with non-agricultural activities; and
(d) "In a broad sense, the process of rural development must integrate the rural poor into the social, political, and economic life of a country." (p. 448).

Clearly a somewhat protean concept, to be used with care. In practice "integrated" has mostly referred to sense (a).

Item (d) represents a noble wish – one with which we can all agree – but it remains, however, a wish. Whether and under what conditions it can be fulfilled remains problematic. Perhaps we should all be more sparing of our use of the verbs "must" and "should".

Social Structures as barriers

The recognition of the role of power, vested by certain social structures in certain classes of people, in inhibiting or furthering development processes is a relatively recent phenomenon among economists. Dr. Yudelman suggests that land reform may be needed in many areas precisely to deprive certain classes of their power.

To close the circuit we should add that in situations where landed interests are an important inhibiting factor, they are also quite likely to be well represented in the government – or at least they may have the power to hold the government hostage. This is a vicious circle whose breach may involve some unpleasantness. Well-meaning pleas for reform are likely to fall on deliberately deafened ears. This happened, for example, in Bangladesh at the time when the first Five Year Plan was being drafted. Clearly there is nothing here for the World Bank to do – except perhaps to refuse to finance projects whose benefits are likely to be preempted by such entrenched landed classes. This would probably involve closer monitoring, than recipient governments are likely to welcome.

Clearly, development economics has come a long way since the days when it was believed that if you took care of industry, agriculture will take care of itself. This progress is reflected in bank terminology and to some extent, as Dr. Yudelman shows, in Bank policy. We can only hope that the process will be a continuous one.

DISCUSSION OPENING – A. Brun, *France*

Yudelman's paper is particularly welcomed for its retrospective look at the way the World Bank has modified its objectives to keep pace with thinking on economic development.

Three points are made with reference to research in agricultural economics:

(a) what is meant by an "integrated" development model? (in the context of the integrated development project) – if "integrated" means "global", what modifications are made to fit the model to a micro-region (with its problems of social change), and what direct and indirect immediate and long-term results can be expected from the alternative institutions (i.e. forms of control) set up? The data on which macro-economic models applied to a whole economy are based (social accounts, input-output model, global equilibrium relationships, etc.) are not suitable for the micro-region, particularly in terms of income distribution. Systems analysis could be used to advantage to examine the decision-making process, and local institutions, and would be worth further discussion.

(b) Two preconditions need to be fulfilled before implementing such macro-economic and micro-regional models.

(i) Consider the human element: those for whom the projects are designed. Not only are the rural poor in the developing world a reservoir of unexploited talents. There is a pool of underutilized knowledge, particularly of relevant sociological factors, which it is difficult to make the best use of. However, some work has been done in France on individual projects which is relevant to the integrated project question.[1]

(ii) interdisciplinary co-operation, since other disciplines have worked at the micro-regional level (biologists, ecologists, etc.) e.g. the research on the Vosges region in France, the main motivation for which, and results, came from an agronomist. This approach may require the economist to develop some degree of intellectual humility. His ideas and recommendations should be shared both with those for whom they are intended, and with colleagues in other disciplines. The risks involved are still less than those taken by poor farmers.[2]

(c) On evaluation and control of development projects, adequate criteria for evaluation and feedback between evaluation and the fundamental aims of the project are needed. It is too tempting to evaluate the obvious and immediate; the less tangible may be more difficult to evaluate, but also more important.

Those implementing projects may be tempted to go for results in quantity rather than quality, partly in order to justify the costs involved. This may be self-defeating, since lack of popular consultation and participation may create conditions in which these goals cannot be attained.

Finally, there is a common basis to the problems of developing countries and those of rural development in developed countries. The very existence of rural poor in the latter shows how inadequately we have achieved at home what we are promoting elsewhere. Progress to be made in analyzing and implementing rural development policy is similar in both cases. Perhaps this common element may encourage a better understanding of the relations between production and the distribution of wealth.

1. Equipe INRA-ENSSAA. *Conditions du choix de techniques de production et evolution des exploitations agricoles.* INRA-SEI 1976.
2. BRUN, A. La participation des agriculteurs au développment. *Bullétin Technique d'Information* June 1976.

Report of the general discussion

Discussion and questions centered on two main subject areas. First there were various contributions stemming from the factual content of the paper rather than its conceptual implications, mostly reinforcing the picture of the changed aims described. However, it was also stated that some countries saw a different path, with agriculture regarded as a way of life in rural areas and the question was asked how the Bank would evaluate integrated rural development projects that fell within that alternative approach.

The Bank – and other national and international agencies – were still seen as often lending too much towards capital intensive projects; probably more bottom-heavy organisations were needed to handle such projects. Emphasis on land as security, stemming from Bank operations, barred many farmers from access to credit who appeared to be in target groups. The Bank's continued insistence on conventional criteria for judging rate of return was stated to make for difficulties in application to small, subsistence farmers. Learning by doing is acceptable for "infant industries" – why not for the small farm group? Considering the contribution of livestock to income improvement, both for subsistence owners and the landless, they got too little attention from the Bank.

However, though the emphasis was given in the paper to the problems of the rural poor and the elimination of rural poverty, it was held that the main emphasis was towards smaller problems, leaving the landless labourer unaided unless employment opportunities happened to result from farmer-focussed programmes. Would there be a fourth stage of the Bank's evolution which would give emphasis to solution of the problems of the landless labourer?

Some unease about operating effectiveness seemed innate in several comments and questions. Currently the Bank's monitoring of projects did not match its control in project preparation. Thus a project designed to assist small farmers might come to be operated for the benefit of larger ones because they could meet credit-worthiness criteria. Looking forward to emphasis on the landless labourer Dr. Yudelman was asked whether he felt that the Bank could adjust its style, staffing, planning and monitoring to handle a programme focussed on his problems. Considerable satisfaction was expressed that for once the "learning process" – usually so ill-documented – had been so well described for the World Bank case.

The theme which contributors came back to again and again was the central place of the household in decision-making. The almost uniquely integrated character of the farm household was stressed and the grave lack of information abouts its operations.

The second area of emphasis was on investigational work which was judged to be needed. The main subjects were – the farm household, the role of women, inter – and intra socio-economic strata decision-making – and indeed social areas generally were particularly mentioned, again with special attention on the landless labourer. More information on the socio-economic differentiation of the rural households was asked for in order to support classification systems which would allow target groups to be effectively defined. Target groups might differ greatly from area to area. Since any rural development process implied social change – and we know all too little about social change – this was clearly seen as an important facet of research in this area.

Though the points were posed in various ways it would probably be reasonable to say that research in aspects of organisation was still seen to be needed. For example, in the way by which the diversity of organisations, some very powerful within their field, can coordinate effectively. A plea was also made that the large volume of critically valuable data in World Bank files might be made available for research purposes.*

Participants in the discussion included: H. Albrecht, *Fred. Rep. of Germany*; S. Biggs, *U.K.*; E. D. Brandao, *Brazil*; P. L. de las Casas, *Peru*; I. Hasain, *Pakistan*; K. Ohkawa, *Japan*; M. Petit, *France*; T. W. Schultz, *U.S.A.*; C. J. van der Vaeren, *Belgium*.

* This discussion was continued in Special Group No. I but no report of the further discussion was made.

ADOLFO FIGUEROA*†

Agrarian Reforms: a Framework and an Instrument of Rural Development

1. INTRODUCTION

The economic and social conditions in which peasants live are the principal feature of less developed countries. Rural underdevelopment is therefore one of the main problems in these countries. And before anything can be said about economic policies to overcome rural underdevelopment, economic analysis must be applied in order to explain this reality.

Land reform has been applied in several countries and there seems to be an agrarian reform project in almost every country. The question is, however, what can land reform really do for rural development.

The objective of this paper is to evaluate the effects of land reform in the changing rural economic situation. For this purpose, a discussion of the rural economy in various countries is presented. Then, in the light of this, a section is presented summarizing some hypotheses for explaining rural underdevelopment. In the context of these hypotheses, land reform is evaluated, not only as an instrument but also as a framework for rural development. The paper is based on the Latin American situation and experiences in land reform.

2. LAND TENURE AND RURAL POVERTY

All the existing empirical evidence asserts that poverty in less developed countries is more dramatic in their rural areas. In Latin America, a sample of seven countries during the decade of the sixties indicates that the mean money income of the urban population ranges from nearly two to three times the rural mean (see Table 1). The usual argument that differences in money income exaggerate differences in real income since cost of living is cheaper in rural areas seems to neglect two aspects of the issue. Firstly, the urban vector price is not greater (in the mathematical sense) than the rural vector price.

* Catholic University Lima, Peru.

† The author wishes to thank his colleagues Ivan Rivera for his valuable comments and Ruben Suarez and Liliana Rojas for research assistance. This paper was written while the author was doing research on income distribution in rural Peru with a grant from the Social Science Research Council.

TABLE 1. *Urban-rural mean income and inequality in selected Latin American countries*

	Year	Mean income urban/rural	Gini urban/rural	Gini rural	Gini Nation
Colombia	1964	1·7	0·98	0·56	0·58
Colombia	1970	2·3	1·28	n.a.	n.a.
Costa Rica	1971	2·2	1·22	0·30	0·37
Chile	1967	1·9	1·01	0·47	0·50
Mexico	1963	2·3	1·09	0·47	0·54
Peru	1961	2·7	1·03	0·48	0·62
Puerto Rico	1963	1·9	1·06	0·41	0·46
Venezuela	1962	3·0	n.a.	n.a.	0·44

Sources: Figueroa and Weisskoff (1976), Table 2.3; except Venezuela taken from CORDIPLAN (1964), p. 68.

The fact that there are many manufactures and urban products which are imported into the rural economy and which are more expensive in rural areas shows that adjustment for differences in cost of living is complex. Secondly, social overhead is concentrated in the cities so differences in money income underestimate the differences in standards of living due to the unequal access to services such as health, education, water system, electricity and so on.

But rural populations are not a uniform mass of poverty; on the contrary, the degree of rural inequality is significant. Gini coefficients shown in Table 1 validate this assertion. Also these data suggest that income concentration is lower in rural areas, and that countrywide inequality is always greater than rural inequality reflecting thus the importance of differences in urban–rural mean incomes. The logical consequence of these results is that rural population is under-represented in the upper tail of the countrywide income distribution, that is among the rich group who live in a poor country. On the other hand, rural population is over represented in the lower quartiles of national income distribution, that is among the extremely poor group who also live in a so-called poor country. Hence, poverty in Latin America is concentrated in its rural population.

Two implications can be derived from the existence of a significant degree of inequality within the rural sector. The first refers to economic theory, one needs to explain the mechanisms of rural underdevelopment, its low mean income and inequality in income distribution. The second refers to economic policy, it is not sufficient to be concerned with rural development projects, these projects must be selective if one wishes to reach the poor groups.

Agriculture is the most important component of the rural economy. To a large extent rural problems can be understood by examining the agricultural sector. Two sub-sectors have often been distinguished in the agriculture of most of the less developed countries, namely modern and traditional. The existence of a highly skewed distribution of labor productivity levels (value added per worker) among farms provides the usual criteria for that distinction. Data for Colombia in Table 2 illustrates the case. The source of the greater productivity in the modern sector is the concentration of most capital stock

TABLE 2. *Colombia: value added per worker and farm size, 1960*

Farm size (Hectares)	Value added per worker (1000 pesos)
– 3	1·67
3–5	2·08
5–10	2·71
10–50	3·47
50–200	5·35
200–500	8·61
> 500	15·07
Total	3·71

Source: Berry (1972), p. 406.

TABLE 3. *Farms with more than twelve permanent workers; labor and value of output in selected Latin American countries, 1962–1963 (proportions)*

	Farms	Labour	Value of output
Argentina	0·8	6	15
Brazil	4·7	21	36
Colombia	1·3	4	15
Chile	6·9	38	57
Guatamala	0·1	7	21

Source: CIDA data, taken from Furtado (1969) pp. 73 and 75.

in a small segment of the rural population, as can be seen in Table 3. Modern sector workers are better equipped than traditional sector workers.

Therefore, highly capitalized farms belong to the rural modern sector, and even to the modern sector of the country. Certainly, this is the case for most plantations, where capitalist social relations of production prevail. The other type of large estate is the latifundia, which must be included in the traditional sector on the criteria of value added, and where pre-capitalist social relations are dominant. A large proportion of the rural economy is constituted by subsistence farming – the minifundia – which is obviously a traditional sector.

There are two typical production arrangements in the economics of latifundio, which will be considered throughout this paper. First, part of latifundio land is given to peasants as subsistence plots and they must work on the latifundio land a number of days per week with no remuneration (share-laboring). Second, part of latifundia land is given to peasants and they must deliver half of the output produced on those plots and also work a number of days in the latifundia land (share-cropping).

The highly skewed distribution of productivity levels has clear influences on rural inequality. A significant proportion of agricultural output is produced in the modern sector. Table 3 indicates that this is the case in several Latin

American countries. In Mexico two thirds of the total value of agricultural output is produced in only 15% of the farms (Solis, 1970, p. 179). A high proportion of farm property income is also generated in the modern sector, but most of it is not rural income since landlords live in urban areas. On the other hand, labor income is greater in the modern sector relative to the income of small farmers.[1] All this picture implies that the peasants of the traditional sector are the largest portion and the poorest group in the rural population and, consequently, the most important social group at the base of the nation's income pyramid.

Highly capitalized farms are also unevenly distributed by regions within a country. This fact originates unequal regional development; areas where the latifundio–minifundio complex is dominant are generally more depressed regions.

3. HYPOTHESES RELATING TO RURAL UNDERDEVELOPMENT

Underdevelopment in the traditional sector needs an explanation before any economic policy can be designed and, so far, such explanation has not been provided. Some hypotheses have been advanced but they need further empirical test to become theories of rural underdevelopment.

Four hypotheses will be reviewed here. Although they do not include all the existing hypotheses, they seem a helpful subset of hypotheses to start the economic analysis of rural underdevelopment.

3.1 *Traditional agriculture is inefficient*

This hypothesis says that by reallocating resources and changing the organization of production, with the same resources and technology, agricultural output – or market value of output – can be increased substantially. Traditional agriculture is therefore full of technical and economic inefficiency.

3.2 *Traditional agriculture is over-populated*

This hypothesis, due to Lewis and others, says that marginal productivity of existing labor in agriculture is either zero or below subsistence income. Since all the population get a share of output, the average income is lowered. The implication of this hypothesis is that an important fraction of the labor force should be transferred to other sectors of the economy; as a consequence, agricultural output would not decrease and average income would rise in the rural population.

The hypothesis of over population in agriculture has been challenged in Latin America. Professor Schultz, for instance, provided several cases in which agricultural output fell after some labor had been transferred to other sectors. This fact does not imply that overpopulation is not applicable in other regions of the world. As Professor Georgescu-Roegen argues: "The situation of most Latin American countries is not identical with that of the East European or Asiatic countries, although they all have this in common: they are underdeveloped. Although overpopulation is always accompanied by underdevelopment, it is neither a necessary nor the only cause of it. The underdevelopment of Latin American countries may have other bases than overpopulation. Overpopulation, therefore, cannot provide the basis for a *general* theory of underdeveloped economies, but only those economic realities beset by it (Georgescu-Roegen, 1967, p. 372).

3.3 *Traditional agriculture is poor but efficient*

This is the well-known hypothesis due to Professor Schultz, according to which there is no economic inefficiency, nor overpopulation in the traditional units of production. Farmers base their decisions on marginal costs and returns considerations. Thus, "...the community is poor because the factors in which the economy is dependent are not capable of producing more under existing circumstances. Conversely, under these simplified conditions, the observed poverty is not a consequence of any significant inefficiencies in factor allocation" (Schultz, 1964, p. 48). Hence, economic policy should concentrate on modernization, for it is not possible to get more output from the existing resources and there is no incentives to accumulate more of these traditional factors given their very low rate of return return. In other words, "...in traditional agriculture the factors of production on which a community depends are expensive sources of economic growth" (p. 97). Thus traditional agriculture is poor because it depends on traditional factors, not because of allocative inefficiency. Hence, "Economic growth from the agricultural sector of a poor country depends predominantly upon the availability and price of modern agricultural factors" (p. 145).

3.4 *Traditional agriculture is exploited by the rest of the economic system through market mechanisms.*

Since traditional agriculture is connected to modern agriculture and the rest of the economy through the market mechanism, underdevelopment of the traditional agriculture cannot be treated in isolation from development in other parts of the economy and consequently nor even disconnected from the world economic system.

The latifundia and medium size farms, which provide the commercial part of traditional agriculture, are certainly connected to the rest of the economy through product markets. In this relation terms of trade keep real incomes depressed. Although landlords may find economic compensations through institutional control, workers are the most injured.

On the other hand, small farmers are also connected to the market. The common idea that they are self-sufficient, and outside the market – as some hypotheses on economic duality have tried to show – seems to overlook their resource endowments. In fact, with one or two hectares and some cattle it is almost impossible to have a self-sufficient family economy. Many commodities must be imported from the urban economy for which cash income is needed. To finance their imports, they must export goods and labor. If crops are too few to sell, livestock may be a source of cash income. The fact that seasonality in the agricultural production process creates a significant seasonal unemployment for such a small unit as a minifundia, means that seasonal migration is another source of income. *Minifundistas* are so poorly endowed that generally their land has no irrigation; this fact increases seasonal unemployment since they have only one crop yearly.

Subsistence farmers seek jobs in medium size farms, in latifundia, in modern farms, as well as in urban activities (construction, mining). In all of them they usually face a segmented labor market. Their salary is lower than the salary obtained by permanent workers, by a larger margin that possibly productivity differentials may justify. The action of worker unions and minimum wage policies are responsible for this segmentation. Thus the surplus generated in these firms is created by permanent and seasonal labor; and yet surplus value is extiacted in a greater proportion from subsistence farmers through lower wages. This mechanism can operate in this manner because seasonal salaries need not cover all the subsistence income of a small farmer because part of the commodities required for his support are obtained from his minifundian. Thus, contrary to the hypothesis of economic duality, minifundia are not disconnected from the economic system but an integral part of it.

Market mechanism works, therefore, against agriculture in general, with the possible exception of exporting farms. But it works particularly unfavorably for subsistence farms; they operate in product and labor markets and yet they get exploited in both. Price policies must then be designed so as to reverse this pattern. For instance, doubling the price that subsistence farmers get for their crops and doubling the salary they get on seasonal migration, and no change in other prices, would have a substantial effect on their income.

In reality, there are four components in the problem of rural underdevelopment: (a) the level of output and efficiency in production so as to reach the production frontier in agriculture; (b) accumulation and technical change, so as to increase the production frontier; (c) terms of trade and (d) income distribution. Each of these components may favor or impede rural development. To these components have been directed the hypotheses reviewed here. Hypothesis (1) and (2) are contradictory to (3). If traditional farmers make decision based on marginal costs and returns there cannot be a significant degree of inefficiency.

The exploitation through the market hypothesis is, however, complementary to all the others. That this is the case with the first two hypotheses is clear enough to deserve more comment. Compared with Schultz's hypothesis two aspects must be pointed out. First, to his contention that traditional farming is endowed with poor factors of production one must add that the market mechanism aggravates this initial inequality in the ownership of resources. Second, Schultz takes into account the market mechanism in his analysis of modernization, but he refers mainly to the space of factors of production. The trading of final commodities is the other component in the social relations through product markets; and, one must add, the relations through labor markets. For this reason, gains in output due to the incorporation of modern inputs, which in turn was stimulated by an appropriate policy of these factors, can be transferred to the urban economy through unfavorable terms of trade in consumption goods.

The three first hypotheses derive economic policies that are in some sense technocratic. The exploitation through the market hypothesis calls for policies that are more economical to apply (require less investment) but involve social conflict, such as redistribution of power, because price policies are designed to conform with the "national" development strategy of the ruling classes.

In the following sections, all these hypotheses are related to land tenure systems. Land reform is then evaluated as a means to attack rural underdevelopment.

4. AGRARIAN REFORM AS AN INSTRUMENT OF RURAL DEVELOPMENT

4.1 *Income redistribution*

Inequality of incomes in agriculture is substantial. Gini coefficients shown in Table 4 of several countries in Latin America confirm this view. Since this inequality is explained in part by the unequal distribution of land, agrarian reform is expected to have a tremendous impact on raising the income of the poorest groups in the rural population.

TABLE 4. *Mean income and inequality in agriculture in selected Latin American countries*

	Year	Mean income Non-agr./agr.	Gini coefficient agriculture
Argentina	1961	1·31	0·48
Brazil	1970	2·73	0·43
Brazil	1970	3·51	0·53
Chile	1967	1·89	0·40
Colombia	1960	n.a.	0·58
Colombia	1970	2·10	0·43
Mexico	1963	1·98	0·50
Puerto Rico	1963	1·70	0·41

Source: Figueroa and Weisskoff (1976), Table 5 and Berry (1972), p. 403.

But, how much income do agrarian reforms transfer? In the first place, in all Latin American countries the value added per worker in agriculture is very low compared to other sectors, yet this is the most important single sector from the employment point of view (see Table 5). This aspect of the economic structure sets limits to an important redistribution effect. Even if workers were to receive *all* the agricultural income, their income levels per worker would still be very low.

TABLE 5. *Agricultural labor force and value added per worker as a proportion of national totals in selected Latin American countries, 1970*

	Labor force	Value added/worker
Bolivia	65·7	0·26
Brazil	44·2	0·34
Colombia	44·4	0·66
Chile	21·6	0·34
Ecuador	55·9	0·45
Mexico	39·2	0·26
Peru	45·1	0·42
Venezuela	25·1	0·31
Andean Group	41·5	0·40

Source: Andean Group, JUNAC (1974), Tables D-6, AE-1, AE-1f Brazil, Mexico, ILO. *Anuario de Estadísticas del Trabajo* Capítulo I-A and *América en Cifras*. Tomo 4.

The fact – shown in Section 2 – that mean income in urban areas is nearly two to three times the rural mean is already an indication that more equality around a lower mean does not have a tremendous effect on poverty. Data for the agricultural sector yields the same conclusion. As Table 4 shows, non-agricultural mean income is also nearly two to three times agricultural mean income. Therefore, land reform implies more equality around a low mean income. From the outset, land reform shows a limited capacity to solve rural problems.

The income redistributed through land reform refers not to all value added but mostly to property income. For Peru some estimates indicate that between 20–24% of agricultural income is property income. This structure is unlikely to be far different in other countries. Most of property income originates in the few modern agricultural farms. Property income per worker in the latifundia may not be very high, – as shown for the Peruvian case[2] in Table 6 – although the appropriation of this income by one person (or family) gives the impression of a large property income per worker.

TABLE 6. *PERU: income by sectors, 1970 (1970, US dollars per worker)*

	Modern	Traditional	
		Urban	Rural
Percentage of labor force	21·6	33·1	45·3
Values added	3590	810	600
Property income	1070	80	90
Labor income	2520	730	510

Source: Webb (1973), Table 6.

On the other hand, not all land is subject to transfer; usually agrarian reform programs do not affect the so-called medium and small size farms conducted directly by owners, even if they use permanent workers. In Mexico by 1967 a total of 54 million hectares of land had been transferred since 1916; this is almost a third of the total 1960 agricultural land. In the case of Bolivia only 13% of the total land existing in 1950 had been transferred in ten years (1953–1963) of agrarian reform. Another figure given by the 1964 MNR congress is 5·4 million hectares which implies 17%. (In this figure 16% is arable land, 31% pastures and 50% non-arable). In Peru this proportion for the period 1963–1976 is also 30%, although it is programmed to reach to 10 million hectares by the end of the program in 1977, which will raise this percentage to 43%. In any case, these three countries where land reform has been massive involve a transfer of only a part of total agricultural land (see Table 7).

With respect to beneficiaries the conclusion is very similar: only a fraction of the agricultural labor force receives land. In Mexico 44% of the peasants are beneficiaries of the land transfer; in Bolivia this proportion is only 20% of the 1950 agricultural labor force, and in Peru close to 20% (see Table 7). The target in Peru is to benefit 400,000 families which would raise this percentage to 27%.[3]

The limited reach of land and labor force involved in land reform programs clearly implies that only part of agriculture property income is redistributed. If land is going to be expropriated, then the price of land has to be deducted from the income transfer. Actually if compensation to landlords for the expropriated land is correctly calculated there will be no redistribution of income at all; in this case land reform would be a process of land ownership redistribution only, a compulsory process of massive buying and selling of land with state intermediation.

TABLE 7. *Land redistribution in Mexico, Bolivia and Peru*

	Mexico		Bolivia		Peru	
Land (Millions of hectares)						
Total	(1960)	169·1	(1950)	32·7	(1972)	23·5
Transferred	(1916–1967)	54·1	(1953–1963)	4·4	(1963–1976)	7·0
Agriculture labor force (Thousands of families)						
Total	(1960)	6,000	(1950)	672	(1972)	1,500
Beneficiaries	(1916–1967)	2,640	(1953–1963)	133	(1963–1976)	280

Sources: Mexico: L. Solis (1970), p. 155; Franco (1974), p. 155.
Bolivia: A. García (1965), p. 417 and 1950 Census.
Peru: Official reports and 1972 Census.

For a better understanding of what has been said it would be useful to examine a specific country. The author did a study in which it was shown that in Peru income redistribution through land transfer amounts to 1–2% of national income. Property income is 20–24% of total agricultural income, but because the annual payment for the compensation of expropriated land amounts to 6% of agricultural income, transferable property income is between 2–3% of national income, since the agricultural share was 18% in the 1966 national income. The proportion of land transferred is 43% which implies that income redistribution must be between 1–2%, as stated before (Figueroa, 1975, pp. 167–170).

The Peruvian case provides a clear example of what land reform can do for raising standards of living in rural populations through its redistributive effect. All this effect depends on the economic structure of a country, and this structure in Latin American countries is not the most favorable for substantial redistributive effects. The Mexican experience leads to a similar conclusion: "The redistribution impact of land reform modified income distribution and was quantitatively very small since the agricultural share, towards 1930, was 15% of national income" (Solis, 1970, p. 154).

On the other hand, income redistribution through land reform does not occur over-night. In Mexico, it took half of a century, in Bolivia ten years and thirteen years in Peru, although the rate of expropriations was not uniform during these years.

An income transfer to the poorest groups of the society, however small, would represent an important impact of land reform on poverty. However, land reform is usually segmented. The principle: "land for those who work on it" leads to giving the more productive land to workers of the modern sector who are not at the base of the rural income pyramid, leaving the more traditional farms (of low productivity) to the poorest groups. Hence, the more depressed regions where agriculture is traditional get the smallest fraction of income redistribution. This has been particularly the case in Peru: modern sugar farms and coastal haciendas have been given to the workers (wage-earners) of these units, whereas for the sierra region beneficiaries, specially for the southern sierra which is the most depressed region, traditional latifundia are transferred; and most subsistence farming areas receive no land at all.

Clearly the most depressed rural areas will not experience substantial increase in standards of living as a consequence of land reform. In these areas traditional agriculture is dominant and therefore there is no important property income to be transferred. Yet those workers benefited by the reform receive traditional factors of production.

The segmented redistribution scheme associated with land reform is evidently of little impact on traditional agriculture. Redistribution through land reform is redistribution of poverty. However, what is true for a part need not hold true for the whole. For example, this would not be the case if income from the top percentiles of the nation were transferred to the poorest 30–40% of the rural population, this would have a tremendous impact on their income over-night. Land reform – and any sectoral reform for that matter – will not do this job.

The common argument that market size for industrial goods may be increased substantially via land reform and contribute to industrialization does not follow from the economic structure of most underdeveloped countries.

4.2 *Level of output*

Are the distinct land tenure systems efficient ways of organizing production? If the answer is in the negative, land reform will have a positive effect on the level of agricultural output with existing resources and technology. Land tenure would then explain in part rural underdevelopment.

For plantations there is no question about their allocative efficiency; with land reform this efficiency may be affected significantly if they are fragmented.

Schultz's hypothesis of poor but efficient applies to subsistence farmers. Although he does not make it explicit, it seems that these farmers will remain efficient under different land tenure systems. However, Schultz does not consider the case of latifundia. So, it seems necessary to analyze some features bearing on the efficiency of the latifundian system.[4]

(1) There may exist little land within latifundia which would be put to work once property is transferred to peasants.

(2) The monopsony that latifundia imposed in labor, through "captive communities", implies less employment and output compared to the more competitive labor market that would emerge with land reform as labor mobility would be increased by this reform. Nevertheless, if large estates prevail some degree of monopsony will continue even if there is a transfer of ownership; the new owners (peasants) will be employers of the non-beneficiary peasants, as in the case of forming cooperatives in the expropriated farm.

(3) In the feudal-type of system where the peasant has to give half of the crop produced in the plot which the landlord allows him (sharecropping), the peasant considers as its revenue only half of the actual product price which leads him to produce less than in the situation of land ownership.[5] With land reform he gets that ownership and also gets total revenue (which is equivalent to an increase in product price), so production in the plot will be increased.

(4) There is no incentive for peasants to work hard in the latifundia, since they are somehow forced to work, whereas they may be very efficient in working the plots given to them. Thus, by changing the latifundian system output may also be increased.

One may argue that the size of minifundia, is anti-economical and they are even more inefficient because they are fragmented in many parcels. This may be true in many cases. But in others one must look at the geographical context. For example, fragmentation in the areas with high altitudes (like the sierra of Ecuador, Peru, Bolivia) is a way to have vertical control over different ecological levels. This vertical control over ecology becomes very important because it allows the peasant to have access to complementary resources (maize-land, barley-land, potato-land, and natural pastures as one moves from the valley towards the highland) in order to gain more self-sufficiency. Additionally, this vertical control over ecology plays an important role in reducing the risk in agriculture by means of a diversified choice of portfolio which includes different crops and micro-climates of various types. A land reform that rearranges these holdings structure and pays no attention to this problem may disrupt an economic organization of subsistence farming without replacing it by a superior system.

For the countries which have experienced a significant land reform there is no clear evidence of its impact on production. The *pure* land ownership redistribution effect is not easy to evaluate since many things change simultaneously.

To sum up, from the previous analysis there are indications that the latifundian system impedes attainment of a higher level of output, even with given resources and technology. So, land tenure does explain, in part, rural underdevelopment. The inefficiency hypothesis is applicable in this case.

4.3 *Capital formation*

The very first problem of capital formation that arises in the application of a land reform program is the possibility of an immediate decrease in the capital stock. Landlords may react to expropriation by selling and dismantling the existing equipment, installations, cattle, forest, etc., given the gradualness in the procedure of expropriation. This has been a clear case in Peru where many farms that were transferred to peasants had no capital goods except a house (with no doors and windows) and the land. No reference to this problem appears in the literature on land reform in Mexico and Bolivia, but one may guess that the pattern must have been very similar.

The effect of land reform on the rate of savings is another aspect to consider. Schultz's hypothesis is that savings are very low in traditional agriculture. In modern agriculture savings may be higher, but they are invested outside agriculture also. Landowners usually have a diversified portfolio. There seems to be no evaluation of this problem in countries where land reform was applied. In the case of Peru, the cooperatives are obliged to distribute any surplus, after provision for amortization, agrarian debt and social benefits as follows: not less than 10% to the Reserve Fund, not less than 5% to the Education Fund, not less than 10% to the Social Security Fund and not less than 5% to the Cooperative Development Fund. The total of these funds must not exceed 70% of the surplus. The balance is distributed among members. However, not less than 25% of this distribution will be capitalized in the form of savings and the rest paid in cash or in kind.

These provisions show that self-financing of investment and expenditure on social services are promoted by the Peruvian agrarian reform law. However, a cooperative's surplus depends on the wages the members decide to assign to themselves.

The agrarian debt has the effect of reducing the saving rates of the new units. Again, in the case of Peru, 3,600 millions of soles has been paid in cash for equipment, installations and cattle and 12,000 millions of soles has been paid in bonds for the land; that is, the total value of expropriation amounts to 15,600 millions of soles (almost 300 millions of American dollars). As this amount should be repaid in 20 instalments by beneficiaries, annual payment is around 800 millions of soles of 1970–1971; this amounts to 1971 public investment in agriculture. Since the State pays to landlords for the expropriated land, which in turn the State will collect from peasants, the landlord is allowed to use these funds to invest in industry, not in agriculture, not in rural industrialization. The agrarian debt is then a mechanism to transfer investment funds from the rural to the urban economy.

In Mexico, as it is known, land reform was not paid for. In Bolivia's case compensation was not applied, either. But some peasants actually paid landlords for the land they acquired. On the other hand, peasants usually paid the direct costs of expropriation. In the majority of cases this expenditure was quite high, particularly when landlords' appeal to courts created a complicated legal battle (Dorner, 1974, p. 179). Hence, even in this case, land transfer was not free for peasants, thus reducing the capacity to accumulate in the new units of production.

In any case, given the relatively small income redistribution impact of land reform, the effect on increasing rates of savings may be of little significance. In other words, the effect of land reform on savings and accumulation depends on the amount of income redistributed to peasants and their propensities to save. In China, for example, land reform redistributed between 15–30% of national income, on the first round, among the peasant population. This proportion is much more significant than the experience in Latin America, and the transfer occurred in a relatively short period (1949–1952). Government then intervened, through tax and price instruments, to capture much of this income and redirect it toward capital accumulation. This policy played a major factor in raising the ratio of gross domestic capital formation to GNP from 7·4 percent in 1931–1936 to 24% in 1952–1957 (Reynolds, 1975, p. 420). Reynolds reminds us that this was also the case with Japanese government Strategy in the early Meiji period, when diversion of income from the previous landed nobility into government channels helped to raise the national savings rate. Compare the case of Peru, where these ratios were 23–24% between 1950–1967 but decreased to 18% in 1968–1973 which was a period of massive land reform.

4.4 *Modernization*

The fact that traditional agriculture is dependent on traditional factors of production which are very expensive sources of growth, as Schultz's hypothesis goes, needs to be related to land tenure systems. Since a significant increase in agricultural output will come from the introduction of new factors of production – the lack of which explains poverty in traditional farming – one needs to consider the effect of land tenure systems on supply and demand for those modern inputs. Moreover, since the availability and price of modern agricultural factors will not be significantly affected by land reform it is the demand side of the problem which needs a closer examination.

In general, the latifundian system does not favor the introduction of new factors of production. This is clearly the case when half of the output must go to the landlord, all the costs being met by the farmer. This system does not favor the accumulation in new factors of production since the farmer gets only half of the additional output whereas he incurs in *all* the additional cost; hence, the incentive to the farmer to accept or not to accept a new factor is only half of the true profitability. That is "Tenure arrangements . . . that determine how landlords and farmers share costs and returns can block the acceptance of factors that would be highly profitable under more appropriate arrangements" (Schultz, 1964, p. 168).

The latifundista himself has no incentives for using modern inputs because most of them are land-savings innovations and land is the more abundant factor in the latifundia. Moreover, land is the basis of the landord's power and they try to make it a scarce factor. As de Janvry (1974) puts it: "since monopoly of the land is the basis of the social power of these elites, in the domain of technology only those changes that are not substitutes for land will be fostered" (P. 17).

The latifundia system therefore does not favor the diffusion of modern inputs and thus contributes to rural undervelopment. Land reform will have an important positive effect on the introduction of new factors of production. Since the factors of production that land reform transfer in the traditional sector are traditional factors, the increase in the propensity to accept modern inputs is a very remarkable effect of land reform. This is not to say that modernization depends entirely on land reform, nor that this is the most important factor for it. The relative importance of land reform for developing new inputs may be very low compared to, for instance, factor endowments if one considers that technical change is induced by them. What land reform does is to help the diffusion of modern factors of production, while the effective modernization of agriculture depends on other variables.

In summary, the arguments presented in this section lead to the conclusion that land tenure systems, in particular the latifundian system, explain to a certain extent rural underdevelopment because they lead to inefficiency in the organization of production and they do not favor the acceptance of modern inputs. By changing the latifundian system the incentives are greater for increasing output and for the modernization of traditional agriculture, but the actual use of modern inputs will depend on their availability and prices, as well. Finally, the effect of land reform on income redistribution is very modest, given the economic structure of Latin American countries. This effect, in turn, sets limits to a significant increase in the savings rate. Therefore, land reform as an instrument for overcoming rural undervelopment has very limited reach.

5. LAND REFORM AS A FRAMEWORK

If land reform alone cannot help to develop the rural economy, complementary policies are called for. Moreover, land reform provides a favorable framework to carry out those complementary policies that are needed for a more significant rural development.

Before land reform, large landowners were the main beneficiants of rural development projects. With land reform, landlords are eliminated from the rural social map and the reach of the benefits of projects will be much larger. Besides, some projects that were inconsistent with the old social system may now appear as a possibility. The extent to which government uses the new framework in the direction of favoring rural population will also show the real nature (and purpose) of a government applying land reform.

5.1. *Expansion of agricultural frontier*

The fact that land is a very scarce resource is brought out most clearly when land reform cannot provide sufficient land for all peasants. For landless peasants it is possible to augment land supply through irrigation programs and through colonization; for peasants with insufficient plots improvements in the water management system will be an advisable policy.

In Mexico, in fact, after the massive land reform that took place in the period 1934–1940, a process of road construction and irrigation began. "At the end of the thirties and during the first half of the forties the agricultural situation was critical; the latifundian system of production being destroyed it was necessary to assure an efficient functioning of the new system of production. This pushed expenditures in irrigation high. The irrigated land was given in property to farmers in plots of 5–100 hectares, which were the orign of the present commercial agriculture" (Solis, 1970, p. 160).

In Bolivia the choice was colonization; "Colonization was an integral part of the Bolivian land reform. Actually, colonization completed the reform which was carried out in great scale in the highlands and traditional valleys. The government's target was to place 100,000 families during the decade 1962–1971 in the colonization area. By 1970 only 30,000 families had been settled. This limited number of families indicates the failure of colonizations programs in absorbing the increase in rural population" (Dorner, 1974, p. 177–178).

In Peru, irrigation projects are of large size (which in most cases include energy). By 1975 the cost of irrigation schemes in progress amounted to close to one billion dollars, 90% relating to only three projects; these large projects are on the coast. For small irrigation projects in the sierra financial efforts are very small. All these projects will improve the water management in 223,000 hectares and in 100,000 hectares of new land; all this within 12 years. In terms of colonization the efforts have been less intense, as only 18,000 families have been settled on 400,000 hectares in the jungle area during the land reform period.

Given the limited reach of irrigation and the concentration in large projects, technological dualism will be reinforced in the rural economy. Relatively few families will benefit from these projects. Solis has illustrated this tendency for the Mexican case very clearly: the irrigation projects applied after land reform (cited above) helped to reinforce technological dualism. He adds: "It can be thought that land reform objective is to increase equality in income distribution; but this seems not to be the case. On the contrary, it is possible to say that land reform originated a set of dynamic forces, creating a commercial agricultural sector with high productivity where income increased considerably with respect to the technologically stagnant substistence agriculture" (Solis, 1970, p. 149–150).

The effect of large irrigation programs is not only to increase income disparities within the rural economy, but the concentration of development in certain regions. More developed regions tend to improve while more depressed regions are confined to the direct impact of income redistribution associated with land reform. This effect, as was argued before, is not substantial. Thus, irrigation programs concentrated in large projects tend to reinforce uneven regional development.

In sum, large irrigation projects are inserted in the growth pattern of the entire economy, which has a very limited absorptive capacity. That is, as resources are concentrated in a small proportion of the population and – given the capitalist rules of income distrubution – income is bottled up in this small segment of population; policies for income redistribution or for more spread in investment are called for in order to remedy the consequences of technological dualism.

5.2. *Policies for modernization*

It was suggested before that land reform may increase the demand for modern factors of production in traditional agriculture. But actual modernization depends upon the availability and price of these modern factors. Professor Schultz (1964) has made this point very clear: "The suppliers of these factors in a very real sense hold the key to such growth. When they succeed in producing and distributing these factors cheaply, investment in agriculture becomes profitable, and this then sets the stage for farmers to accept modern factors and learn how best to use them". It is also an inducement to increase savings and to develop institutions to provide credit for financing investment in such factors" (p. 145). The contention derived in the previous section is that with land reform there will be a shift in the demand for modern factors which will increase acceptance in a larger proportion for a given supply conditions. Schultz's argument is directed to shifting the supply conditions as well.

Therefore, the processes by which new agricultural factors are to be supplied becomes of tremendous importance. However, much of what can be done in order to affect supply conditions depends on decisions outside of agriculture. If anything, it requires a considerable government expenditure to be invested in supplying new agricultural factors (dams, irrigation canals, roads) and also in improving the capabilities of the farmers to use modern inputs (credit, extension research). Also import policies must be taken into account, since an important proportion of modern inputs (new seeds, new breeds, fertilizers, agrochemicals) are imported.

In spite of its great need, as a complementary policy, public investment allocated to agriculture is usually low compared to the needs for modernization. In Peru, less than 10% of public investment has gone to agriculture during 1960–1975 of which almost half has gone to irrigation projects; in the 1975–1978 Plan, 75% of public investment in agriculture will go to irrigation projects, which are very large projects in the coast. To improve water management in the sierra region (the more depressed region of Peru) where most agriculture is not irrigated has a very low priority. Only 0.6% of the public investment in agriculture is allocated to research, which has even decreased from the 1971–1974 Plan where this proportion was 2.3%. All this occurs in a country where land reform is massive compared to historical standards.

The same problem applies to credit. The banks created to provide credit in a land reform framework have had limited coverage. The Mexican *Banco de Crédito Ejidal* covers only a minimum fraction of *ejidatarios*, with even a tendency to diminish from 30% in 1936 to 16% in 1940 and 14% in 1959. Those *ejidatarios* that get no credit from the Bank must obtain it from intermediaries to whom they must sell their crops. (Franco 1974, p. 157). In Peru, the *Banco Agraio* gives priority to farmers benefited by the land reform, which means that credit is still directed mainly to coastal modern cooperatives. According to 1972 Census only 7% of Peruvian farms obtained any credit from this bank.

A recent study on the peasants of the Bolivian *altiplano* shows that 44% of the peasants get technical assistance, 10% by potato seed and only 41% use the variety that government favors as the most efficient ("Sani Imilla").[6] Also 35% use insecticides and/or disinfectant and 48% use fertilizers, but all these elements are used in insufficient quantities to have optimum results (Urioste, 1975, p. 128, 178). All this picture of 1974, corresponds to a period after land reform had taken place. In Peru, the 1972 census showed that 4% of the farms received technical assistance, 19% used fertilizers, and 24% bought seeds.

The effort needed to supply new factors of production to a substantial segment of the farmers seems considerably above the existing resources and priorities in most Latin American countries. And because of this, subsistence farmers will be the last to incorporate modern factors of production, which means that technological dualism, with all its consequences, will still prevail. Land reform *per se* can do little to change this pattern.

5.3. *Price policies*

Another complementary policy to be considered in the context of land reform is price policy. Agricultural relative prices can influence in two ways the land reform inpact on traditional farming. First, gains obtained through income redistribution may disappear in thin air or may be increased substantially, depending on price policies. Second, incentives to adopt modern technology are made higher with land reform, but the acceptability of a package of modern inputs, water and agricultural services can be increased further if more favorable prices for agricultural products accompany that package. Thus, an appropriate set of prices may increase substantially the initial limited impact of land reform. This set of prices may never come because it is inconsistent with the priorities and patterns of development desired at the national level.

In Peru, price policy has become even more discriminatory in the years of massive land reform. Import subsidies on food that compete with domestic agricultural products, subsidies to gasoline but not to fertilizers, price control on food, are among the package of measures used during 1972–1975. At the same time land was being redistributed. From this and previous sub-sections one can conclude that complementary policies have not accompanied land redistribution in Peru. Official pronouncements indicate that these policies will come after the process of redistribution is completed.

Complementary policies were not provided in the case of Mexican land reform either, as one reads in Solis' (1970) study: "Although the majority of rural population was not favored with public investment or credit, nor could make use of modern technology, they received land freely instead" (p. 179). In Bolivia the pattern has been similar: "The principal areas in which sufficient attention was not paid refers to the creation of new services and institutional arrangements so that peasants could exploit their plots more efficiently, for instance: extension services, credit, supply of fertilizers, pesticides, new seed varieties, and irrigation systems and transportation. Moreover, price policies and institutional arrangements to sell production surplus were also needed" (Dorner, 1974, p. 180).

5.4. *Rural industrialization*

Given the seasonality in the agricultural production process, there exists in the rural economy an important feature of seasonal unemployment. This unemployment is more important in the case of subsistence farmers. The use of this off-season time is diverse and depends on the land tenure system. "Captive communities" have to work for the landlord's amenities and social position; in the latifundian system seasonal unemployment is consumed as services. For "non-captive communities" an alternative is off-farm migration into segmented labor markets. But all subsistence farmers utilize part of the seasonal unemployment in rural activities: craftsmanship, construction, repairs, and so on, which are of relatively low productivity. These potential services of labor could be utilized in a more productive manner by means of establishing complementary activities in the same region. Land reform also provides a favorable framework for that.

Complementary activities to agriculture must be taken to mean processes which have a seasonality just opposite to that of agriculture or processes which have no seasonality, when one may start and end at the process whenever one pleases. That is, the rural economy must be a mix of processes in *parallel* with processes in *line,* in the terms of Professor Georgescu-Roegen, if idleness is to be avoided in a given rural economy. Processes which exactly match in seasonality are rare and this is why it is excluded as an alternative.

These complementary activities may include (i) exploitation of new resources, like mining, forestry; (ii) improvement of resources, like terracing, reforestation; (iii) more processing of products, like dairy products, wool, wood; (iv) some services, like repair of machinery and (v) industries which may operate efficiently on small scale. Most of these activities imply rural industrialization.

Landlords have had little interest in developing these new activities, so with land reform more resources may be put into production. But, again these rural complementary activities may be limited by the financial capacity of the peasants benefited by land reform and of the State. In both cases, rural investment will be restricted unless national priorities are modified, and in this respect government may not be interested in rural industrialization either.

Yet, rural industrialization is an important component in any rural development program. Industries which can operate efficiently on a relatively small scale and producing not only for farm use, would have an important impact on rural development, particularly because it is a better economic alternative to off-season activities.

6. SOME CONCLUSIONS

Rural poverty is the most important component of the poorest social groups in most less developed countries, where susbsistence farmers are at the base of every national income pyramid.

Land reform has been a major rural policy in several countries and attempts as well as projects to change land ownership exist in almost every country. However, when the average labor productivity is very low in agriculture – as is the case in most countries – the impact of land reform on income redistribution and capital formation is very modest. Therefore, given the present economic structure no significant solution to rural undervelopment can come by attacking the problem within the rural sector alone. Yet, this is what agrarian reforms pretend to achieve and this is why they show no important effects as an *instrument* for rural development. Although it is necessary, land reform does not seem to be a sufficient condition for rural development.

Changes in land tenure originate a new institutional *framework* in which complementary policies may be inserted. This new framework is clearly more favorable for a greater impact of rural projects. In particular, the elimination of rural elites opens the way to new projects which were inconsistent with those elites, and the reach of a given project is much more widespread since the role of those elites as principal beneficiaries has also been reduced with land reform. In addition the incentives to increase production and to incorporate the use of modern technology are much higher in the farms once the latifundian system has been modified. In sum, land reform as a framework shows great possibilities for more significant rural development.

However, much of the complementary policy lies outside agriculture. The allocation of public investment, price policies, credit policy, import policy are all important complementary policies and yet they are subject to exogenous decision making with respect to agriculture. That is, policies favorable to agriculture cannot be chosen arbitrarily; they must be drawn from a feasible set given by national objectives and priorities of the ruling class. For instance, price policies favorable to agriculture are inconsistent with incentives to industrialization. Land reform cannot change the "national" priorities and patterns of development, unless national political power is also redistributed.

If land reform has a limited impact as an instrument of rural development and if exogenous decision making does not favor the needed complementary policy to get a more substantial improvement in the rural population situation in a land reform framework, what are the real objectives of agrarian reforms conducted by ruling classes? and for that matter what can sectoral reforms achieve in a market economy where technological dualism prevails?

These questions are difficult to answer, but they are legitimate questions in the light of the results presented in this paper.

NOTES

[1]If capital–labour ratio is higher in the modern sector one could expect that relative return to labour is also higher in this sector; moreover relative return to capital should be higher in the traditional. However, this is not the case because traditional agriculture is largely endowed with traditional capital which consequently has low returns, as will be discussed in the next section.

[2]Webb's definition of rural traditional sector includes many modern plantations, for only sugar farms are included in his modern sector. Hence, the 90 dollars of property per worker must be much less under the concept of traditional agriculture used here.

[3]For Chile until June 1972, 600,000 hectares of irrigated land – 33% of total irrigated land – were transferred to 60,000 peasants of 760,000 agricultural labor force (Arroyo, 1973, pp. 5, 16).

[4]On peasant's economic behaviour Schultz's hypothesis will be assumed in the rest of the paper.

[5]The argument can also be developed without reference to market prices. The only condition is that marginal cost and returns be evaluated by peasants.

[6]Another problem of diffusion of new varieties of seed is that usually they are more suitable for urban consumption than for peasants' consumption. New varieties are sought as elements of solving urban consumption problems and not peasants'. This is also a reflection of urban strategy of growth.

REFERENCES

Arroyo, Gonzalo (1973) 'Después del latifundio qué?', *Revista Interamericana de Planificación* (Bogotá), Vol. VII, No. 26, Junio.

Berry, Albert (1972) 'Farm Size Distribution, Income Distribution, and the Efficiency of Agricultural Production: Colombia', *American Economic Review,* Vol. LXII, No. 2, May.

de Janvry, Alain (1974) 'The Political Economy of Rural Development in Latin America: An interpretation', Giannini Foundation of Agricultural Economics, University of California, December.

Delgado, Oscar (editor) (1965) *Reformas Agrarias en la Amèrica Latina.* Fondo de Cultura Econòmica, Mexico.

Dorner, Peter (editor) (1974) *La Reforma Agraria en Amèrica Latina.* Editorial Diana, Mexico.

Figueroa, Adolfo (1975) 'The Impact of Current Reforms on Income Distribution in Peru', in A. Foxley (ed.) *Income Distribution in Latin America,* Cambridge University Press.

Figueroa, Adolfo and Weisskoff, Richard (1976) 'Traversing the social Pyramid: A Comparative Review of Income Distribution in Latin America', *Latin American Research Review* (forthcoming, Summer 1976).

France, Alberto (1974) *Reforma Agraria en Amèrica Latina,* Centro Regional de Ayuda Técnia, Mexico.

Furtado, Celso (1969) *La Economía Latinoamericana desde la Conquista Ibèrica hasta la Revoluciòn Cubana,* Siglo XXI, Mexico.

Georgescu-Roegen, Nicholas (1967) 'Economic Theory and Agrarian Economics', in *Analytical Economics,* Harvard University Press, Cambridge, Mass.

Junta del Acuerdo de Cartagena (1974) *Grupo Andino. Algunos Indicadores Socioeconòmicos.* Lima, November.

Reynolds, Lloyd G (1975) 'China as a Less Developed Economy', *American Economic Review.* Vol. 65, No. 3, June.

Schultz, Theodore, *Transforming Traditional Agriculture,* Yale University Press, New Haven.

Solis, Leopoldo (1970) *La realidad econòmica mexicana: retrovisiòn y perspectivas,* SigloXXI, Mexico.

Urioste, Miguel (1975) *Conducta Econòmica del Campesinado e Incorporaciòn de*

Tecnologia Moderna en el Proceso Productivo, La Paz: Instituto de Investigaciones Socio-econ6micas, Universidad Católica Boliviana, Documento de Trabajo 06/75, Septiembre.

Venezuela, CORDIPLAN (1974) *Primera Encuesta Nacional de Ingresos y Gastos Familiares en Venezuela.* Caracas.

Webb, Richard (1973) *Government Policy and the Distribution of Income in Peru, 1963–1973,* Woodrow Wilson School, Princeton University, Princeton, New Jersey, September.

DISCUSSION OPENING – B. Dasgupta, *India*

An underlying assumption in Professor Figueroa's paper is that there is incompatability between redistributive aspects of land reform and production. In India, at least, empirical evidence does not support this. There is evidence that inputs used in large-scale farms are not always indivisibles and therefore arguments of large-scale economics on large farms are not always valid. The comparative advantage of large-scale farmers lies in their social and economic advantages which give them access to best quality inputs.

Land reform offers much scope for improvements of income of the poorest group – the landless labourer – and those with very small holdings – by giving them access to land. This further tends to reduce unemployment.

There is already serious unemployment in the urban sector and, therefore, rural unemployment must be solved in the rural sector. As the author argues, land reform destroys the traditional village power structure and therefore gives the deprived access to inputs, thus reducing the comparative advantage of the richer landlords.

It follows from these facts and arguments that the reasons underlying the lack of success of land reform programmes in improving the conditions of the poor rural people were political. Politicians viewed reforms as necessary political manouvres. Politicians did not have the will to confer benefits to the poor. He therefore felt that land reforms could be an effective instrument for improvements of the rural poor. The paper concentrated too much on the analysis of obstacles and not enough on the effectiveness of the instrument and means of increasing it.

DISCUSSION OPENING – S. Sawada, *Japan*

The Land Reform in Japan, a conspicuous feature of the immediate aftermath of World War II, may have some lessons for this discussion.

The Land Reform, in the sense of giving farmers ownership of their farmland was almost completely carried out for arable land throughout the country during the period 1945–1949. The rented arable land was compulsorily bought by the Government, and sold to tenant farmers who had cultivated the land. Even the arable land owned and cultivated by owner-farmers was bought and sold by the Government within a certain limit if the cultivating area was over about 3 hectares per farm for the most part of the country. Each landlord was allowed to keep only about 1 hectare of rented land; except land of absentee landlords whose land was all bought and sold. Thus, the area of rented arable land decreased from about 46 per cent in 1945 to about 10 per cent in 1950 The number of tenant farmers, including those having a tiny area of their own land, decreased from about 47 per cent of all farmers to only about 9 per cent during the same period. Besides such a drastic change of ownership, the terms of tenancy were placed under strict legal restrictions in favour of tenant farmers. Thus, the rent of paddy decreased from 40–50 per cent of the yield, paid in kind, in the prewar period to 4–7 per cent, paid in money, around 1950. Such a drastic reform was carried out under a strong leadership of the occupation army.

About a quarter of a century has passed since the Reform took place. What is the present situation? The number of owner-farmers, including those renting small area of land beside their own land, stood at about 90 per cent of all farmers in 1950. They have remained as they were up to now, though slightly decreased in number in terms of households, due to the absorption of labour from other industries. Their farms have remained as small in scale as ever. The average scale of all farms was a little smaller than 1 hectare at the time of the Reform. Now, it is a little larger than 1 hectare. The number of farms of over 2 hectares is about 3 per cent of all farms.

However, the growth rates of Japanese agriculture and its productivity in the postwar period, excluding the early convalescent period, were really large, comparing with those of the prewar period. According to the estimates of Hayami and others, the growth rate of real output of agriculture was 3.6 per cent per annum in the period, 1955–1965, compared with 1.6 per cent in the period, 1880–1935. Such a high rate as 3.6 per cent was never seen in any ten-year period since the abolition of the feudal system in 1868. How was productivity? The compound growth rate of the productivity of total inputs was 2.7 per cent per annum in 1955–65, compared with only 0.69 per cent in 1880–1935. (Hayami and others: *A Century of Agricultural Growth in Japan,* 1975) The yield of paddy rice per hectare increased by about one third, and attained the level two or three times that of other countries recently. A tremendous number of machines began to be used on farms along with the absorption of labour from other industries. The number of power-tillers was less than 10,000 in the prewar period. Now, it increased up to over 3 million. Small tractors and combines began to be used widely, too. These several years transplanting machines diffused throughout the country, and made farmers free from the toilsome labour of transplanting. Biological and chemical technologies also advanced greatly.

Some critics have insisted that the Land Reform lessened the mobility of the ownership of arable land, making small farmers stick to their own farm land, and accordingly hampered the formation of large and efficient farms. It might be true. However, the growth rate of output and productivity increased remarkably in the postwar period as stated above. The theory putting "large" and "efficient" as equal does not seem to apply to the development of Japanese agriculture. In this sense, Japanese agriculture seems to be treading on a "unimodal" (B. Johnston) path of development.

Some other critics argue that for some years after the Reform, the levels of income and livelihood went up side by side, and accordingly investment in their own farms could not occur sizably from their own savings. It might be true during the decade just after the Reform. I think, however, it was not the case for the subsequent period. It has been said that land reform in less developed countries brought first the increase of farmers' food consumption through income effect. (T. W. Schultz) It was really the same for Japanese farmers regarding the increase of consumption, including other items than food. I would say that such an increase of consumption itself must have had a close relationship to the elevation of productivity through giving vitality and willingness to improve their farms further.

It should be noted that the Land Reform was never the sole factor in the development of agriculture in the postwar period. Particularly, the manufacturing industries played a great role in development, providing agriculture with many advanced kinds of inputs, creating demand for agricultural products, and absorbing much redundant labour from agriculture. Agricultural policies, centring on those for rice production and marketing, must have also had a large effect. However, it should be noticed that such external factors could have directly effected agriculture through the Land Reform. In many countries, the largest problem of Land Reform is not in the difficulty of finding merit in the Reform, but in the difficulty of decision-making towards the Reform. Even in Japan, it could be carried out only at the time of struggle for existence.

RAPPORTEUR'S REPORT – W. Nguyo, *Kenya*

Part of the discussion focussed on the distinction between land reform and agrarian reform which included other ameliorative measures, such as improvements to the infrastructure and supply of inputs. Agrarian reform was seen as an essential adjunct to land reform if the benefits of the latter are to be realised. It was felt that the author did not draw the distinction clearly enough – but the author subsequently rebutted this. Moreover, it was arguably contradictory in, on the one hand, giving little credit to land reform in improving rural incomes and on the other, seeing it as facilitating introduction of innovations.

The possibilities of the co-operative operation of land, not owner-operation, after land reform, were also noted.

In his final comments the author emphasised that the first part of the paper sought to give a factual account of Latin American land reform experience – which indicated that politicians' expectations of it are rarely realised.

Overall, discussants' differences between themselves and with the author seemed to underline sensitivity to the balance of emphasis which individuals thought appropriate in this important emotive subject.

UMA LELE*

Considerations Related to Optimum Pricing and Marketing Strategies in Rural Development

1. INTRODUCTION

Pricing policies towards food crops in most developing countries have been aimed at keeping the price of food low to the urban sector. Given that supply of domestic food production has generally not kept pace with the growing demand, this has meant a policy of price and supply controls, leading to considerable inefficiencies. The inefficiencies arise from malallocation of government's fiscal and administrative resources to implementation of price controls, albeit ineffectively; from adverse effect of the price controls on growth of food production by increasing risk and uncertainty and reducing return to investment in food production; and finally, from bypassing the very considerable potential of the traditional marketing systems to carry out the pricing and marketing function, and through a misplaced emphasis on the role of the public sector. The equity effect of the policy is usually also questionable.

In this paper, therefore, the considerations that should enter formulation of a more desirable agricultural pricing and marketing policy are discussed. A case is made for a price and supply stabilization program involving fixation of guaranteed minimum and maximum prices and buffer stocks – as a way of reconciling the objectives of agricultural growth and urban political stability. The factors that should enter fixation of minimum and maximum prices are discussed. It is argued that the direct role of the public sector should be confined to implementation of the price stabilization program, including purchases to support prices at the minimum and sales to counter price increases over the prescribed maximum, and that the remaining marketing function should be left to the private and the cooperative sector, with the latter growing only gradually to ensure its long-term success. The very substantial role that the government must, however, play in improving the efficiency and equity in pricing through investment in physical infrastructure, market intelligence and regulation is outlined.

* The views expressed in this paper are those of the author and do not necessarily represent the views of the World Bank. I am grateful to Mr. Jayanta Roy for his comments on an earlier draft of this paper.

There are three reasons why the discussion is mainly focussed on agricultural crops produced for domestic consumption and refers to situations of excess demand. First, in the past food deficits were most often the feature of densely populated Asian countries. However, they have become increasingly common in a substantial number of African and Latin American countries as the domestic food production has failed to keep pace with the demand generated by rapidly growing populaton and urbanization. Second, much experience has accumulated with regard to conceptualization and implementation of pricing and marketing policies in developing countries from which useful lessons now can be drawn to tackle problems, particularly involving growth of the low income food crop producing sector. Finally, pricing and marketing of food crops pose the most difficult conflicts among economic, sociopolitical and administrative considerations and, therefore, pose the most challenge in their resolution. Despite its focus on food crops, and excess demand, many of the observations made in this paper, however, also apply to traditional export crops and to marketing in situations of excess supply.

2. OBJECTIVES OF PRICING AND MARKETING POLICIES

2.1. *Reducing marketing margins*

Markets are expected to perform the function of signalling prices to consumers and producers thus bringing about an efficient allocation of available supplies among consumers and of resources by producers. Such price signalling is normally accomplished within the existing distribution of income. Through its effect on effective demand income distribution, of course, has a profound impact on allocation of supplies among consumers and of resources by producers. An efficient allocation is, therefore, not necessarily equitable.

Ignoring for the moment these important equity considerations discussed later in the paper, the popular view in low income countries has been that the traditional markets for agricultural commodities do not perform the function of price signalling effectively, and that there are substantial differences between prices paid by consumers and those received by producers, both over time and space, that are caused by monopolistic and profiteering elements in private trade.

Introduction of alternative marketing channels in the public and/or the cooperative sector is, therefore, often considered necessary on the ground of reducing market margins so as to improve the prices to the farmer and/or to reduce the price of food paid by the consumer. Improving competition through different channels of marketing can potentially have a significant impact on pricing efficiency. However, the alternatives of government or cooperative marketing have not been very successful, particularly in food grain marketing, despite considerable government support. As will be documented later this is because the available evidence frequently does not support the popular view about private trade; nor does it provide the basis for the belief that the inefficiencies that do arise in traditional markets are merely the result of monopolistic elements. It indicates that inadequate development of physical infrastructure, poor dissemination of market information, insufficient regulation of marketing practices and uncertainties arising from basic imbalances in demand and supply result in inefficiencies. Through pricing and marketing policies of the type discussed in this paper, governments can play a very important role in each of these areas to reduce marketing margins. However, the route of a takeover or discouragement of private trade that is often followed in LDCs is neither necessary nor desirable to fulfil the objective of reducing market margins.

2.2. *Reducing price and supply instability*

Even if the traditional markets are reasonably efficient, they may be unable to cope with, and may even exacerbate, the year-to-year instability and uncertainty in the marketed supplies and prices that often result from fluctuations in production. These pose a particular problem for the governments especially when there is only a marginal balance between supply and demand. In such a situation, the marketing and pricing policies, of course, have to serve different and often conflicting functions in the urban-industrial and the rural-agricultural sector, as pointed out below.

2.3. *Ensuring low cost supply of food to the urban sector*

The relative income effect of a given price change in the essential food items is greater on lower than on upper income groups.[1] Their precise magnitudes and hence the need for an assured low cost food supply at various income levels, however, remain unknown because of the generally poor empirical knowledge about demand schedules for food in widely diverse conditions of income levels, scope for substitution, and consumer preferences that exist in different countries, and often, within a large country, among regions.[2]

Despite the disputed need on economic grounds, governments often feel obliged to ensure low cost food supply to all sections of the urban population albeit on different grounds. In a situation of income inequality that prevails in most cities of developing countries, left to themselves, low income urban classes are often even unable to acquire the necessary minimum supply of food for domestic consumption in years of supply shortages and high prices. Ensuring supplies to this class, therefore, seems essential to governments on welfare grounds.

Because agricultural commodities and, in particular, food crops constitute an important item in the consumption expenditure of industrial wage earners, supply shortages and high prices almost always lead to demand for increased wages, providing justification for increased substitution of capital for labor in industrial production over time.[3] Because of the likely adverse effect of high wages on the pace and pattern of industrial growth and, therefore, on the growth of industrial employment, governments are often under pressure to ensure low cost supplies of agricultural commodities to the low-middle income industrial laboring class.[4] This pressure is, of course, greater in countries that have a significant industrial labor force, as in India, than in less industrialized African countries, such as Tanzania, Ethiopia or Malawi.

Relative to the wage earners, the middle and high income classes, including the civil servants, spend a lesser share of their budget on agricultural items.

Nevertheless, given their administrative influence, this class often plays a substantial role in ensuring low cost food supplies for the urban sector as a whole.

An even more important reason for the policy of price controls, of course, is that the high concentration of population in urban centers makes political unrest much more imminent there than in the rural areas, reinforcing the inclination of most governments to maintain prices and ensure overall supplies to cities. Also, targeting food distribution towards specific low income groups is administratively highly demanding, as the information of the size of the population falling in various income classes is rarely adequate. There are also numerous practical difficulties, including the political pressure to broaden the program, where the target group cannot be easily delineated, particularly as the socioeconomic classes are not obviously and highly stratified in most cities. In situations of overall shortages, the scope for evasion of the restricted distribution program is, therefore, considerable. Thus, for a combination of social, economic, political and administrative reasons, a substantial commitment of the government's budgetary and administrative resources is frequently made in many countries to controlling the level of prices and for ensuring food supplies to the urban population.

This commitment may take three forms. In many cases, governments fix prices ranging from the farm gate to the consumer. These controls are often combined with restrictions on the sale of surplus to private agencies and movement of grain among administrative boundaries on private account. Purchases and interregional transfers of grain are made on government account at statutorily controlled prices.[5]

Paradoxically, such restrictions are justified largely because governments do not aim at evening out prices and supplies among years by purchases in good crop years and sales in bad crop years. Despite legal fixation of prices, in absence of effective implementation of price controls, farm gate prices frequently drop precipitously even with a slight increase in production. Poor implementation results from lack of adequate administrative machinery to purchase the quantities necessary to support prices. Frequently governments also lack the inclination to support prices. In situations of surpluses, left to themselves, the traditional markets are often unable to dispose of the market arrivals without a major price drop because of the poor transport, storage and market intelligence facilities with which they operate. The situation is usually worsened by (i) movement restrictions and (ii) by the fact that the effective demand rarely extends beyond the large urban centers even if there is food deficit and even if supplies can flow easily to the deficit areas as rural income levels are usually very low in such areas.

The experience is the opposite in bad crop years. The controlled prices at which government purchases are attempted often tend to serve as the floor rather than as the ceiling price. Consequently, governments are unable to acquire control of a sufficiently large share of the marketed surplus early in the season to influence domestic prices and urban supplies. When forced withdrawal of a share of the total surplus is attempted in the public sector through monopoly procurement, there is a great deal of evasion. Market prices of the produce passing through the non-governmental channels increase even more than they would in the absence of such withdrawals. The effective weighted average price of the controlled and uncontrolled marketed surpluses is thus not necessarily lower with government intervention than without.[6]

Imports are the other major instrument used by governments to stabilize domestic prices and supplies. Such a policy may be justified when low cost concessionary imports are available at a short notice and/or the balance of payments situation is favorable. In such a situation, the real cost of imports to the government may frequently be significanly lower than that of carrying stocks from one year to another. In some cases, concessionary imports have even been a major source of revenue for governments.[7] Reliance on imports for supplies is also administratively less demanding than acquiring control of the domestic production.

However, the uncertainties of timing of imports have often led to disruption of the domestic market, exacerbating rather than reducing speculation, increasing market inefficiency and through increasing risk adversely affecting the domestic incentive to increase production for the market.[8] In absence both of good market intelligence on domestic production and a well-conceived pricing and marketing policy, some countries have even followed trade policies contrary to those required by the domestic situation, importing grain when domestic surpluses existed and exporting in times of deficits.[9] The changing world food situation has, of course, changed the economics of imports vis-a-vis that of the domestic buffer stocks, and once again stressed the need for domestic self sufficiency.

Influencing prices through retail distribution has been the third major policy instrument used by governments. In countries such as India, Bangladesh and Pakistan, although governments have carried out "fair price distribution" on a massive scale, it has been confined largely to urban centers. Low income rural populations have benefited only in situations of extreme shortages verging on famines. This is despite the fact that market dependence of low income rural households has been increasing in recent years.[10] Even the urban distribution has been more significant in years of shortages than of surpluses.[11] In China and to some extent Sri Lanka, on the other hand, the governments have undertaken to assure supply of basic food items at a fixed price both to the urban and the rural population at all times. In Africa, governments have played a relatively minor role in retail distribution mainly because of their limited manpower to administer such programs.[12] Even wholesale distribution to deficit regions has been avoided by many marketing boards as being too costly and administratively infeasible, as for instance in Malawi.

Contrary to the usual contention, government intervention in trade has been usually destabilizing rather than stabilizing. In the short run, controlling prices at levels below the market prices often results in withholding of marketed surpluses by farmers and thus makes the flow of marketings unpredictable.[13] The capacity of farmers, particularly of those who contribute most to the marketed surpluses, to withhold has increased considerably in recent years with increases in their incomes.[14] In the long run price controls affect incentives to produce for the market in various ways as will be illustrated later in this paper. Low food prices often also lead to clandestine flight of surpluses across borders to deficit regions and countries. In the latter case this means loss both of marketed surpluses and of government revenues.

Also, contrary to the popular view, the marketing margins incurred by government and parastatal agencies are almost invariably higher than those incurred by traditional traders.[15] This is because of their high overhead costs and poor management. The latter results in greater handling and storage losses and is frequently associated with corrupt practices.[16] Government agencies usually also have a poor record in timeliness of services in purchasing from the producer and selling to the consumer. In addition, the illicit trade caused by government regulations in the private sector frequently increases the chain of intermediaries, reduces the scope for bulk marketing and thus increases costs of marketing.[17]

Controlled prices are also less equitable than they appear on the surface. Where attempts are made to procure market surpluses in situations of supply shortages through a compulsory levy at controlled prices, as in India, the large commercial farmers are greater evaders of the procurement system than the politically less powerful small farmers. The lower effective weighted average price earned by small farmers through more regular sale of their marketed surplus to the government seems particularly paradoxical as the surpluses are usually distributed to the relatively high income urban consumers. Apart from adverse equity effects in the short run, price controls also result in inequitable incentives to increase production among large and small farmers in the long run. Because the size of the marketed surplus controlled by the government usually varies from year to year, the effective weighted average price received by farmers is also unpredictable leading to considerable uncertainty as to the return to production. The ramifications both of uneconomic and unstable agricultural prices to agricultural growth among small and large farmers are discussed later in this paper.

Given the shortcomings of the policy of controlled prices, there is need for a different pricing and marketing policy which encourages growth of the domestic food supply through assuring minimum prices for the major food crops, that allows greater scope for the market forces to work so as to minimize the drain on the government's budgetary and administrative resources, but that nevertheless allows the government to play an important role in stabilizing prices to meet its political and developmental objectives.

2.4. *Assuring a minimum return to resource use*

The size of production of any particular crop depends on its profitability. Even if new technologies assure a much higher physical response of output to inputs than realized before and, therefore, reduce costs per unit of production, if the input and output prices do not assure a certain return, farmers are unlikely to be interested in increasing production. To the extent that technological change results in a substantial shift in physical response, little or no increase in existing input/output prices may be necessary to provide incentive, particularly if existing producer prices fully reflect domestic demand. However, for the reasons discussed earlier, the greater the increase in production from such technical change, usually the greater the possibility of a substantial decline in prices. A pricing policy that assures a minimum return to the producer is, therefore, necessary even if a substantial increase in the ratio of output to input prices from the previous level may not be called for.[18]

(a) *Ensuring a balance among crops.* Whether or not new technologies exist, through reallocation of existing resources among crops, the relative profitiability of different crops influences the composition of agricultural production. The evidence with regard to yield responsiveness to changes in relative prices is poor. This is partly because of inadequate research in this area, but also because in addition to the level of inputs – yields are determined by climatic factors and management practices, the contributions of which pose difficult problems of estimation. Much more research, therefore, is needed in this area. Considerable evidence, however, exists to indicate a significant response of acreage to changing relative prices.[19] Evidence also exists with regard to response of labor allocation to return among crops.[20] Relative price changes can, therefore, be an important means of allocating inputs and thus of ensuring the necessary balance in supply between food and export crops or among food crops between cereals and pulses, and between crops and livestock.

Increase in the acreage and labor allocated to export crops in the last three decades in the smallholder sector in many African countries is, among other things, a result of the relatively higher prices that such crops have enjoyed in comparison with the controlled prices of food crops, leading to both substitution away from food crop production as well as bringing of additional labor and land into production.[21] Farm gate prices are frequently attractive despite the high costs of marketing incurred by marketing boards and cooperatives that usually handle export crops and despite their having been a major source of fiscal revenue for governments. In contrast to food crops, export crops have usually also enjoyed an assured channel of marketing at the farm gate through greater investment in marketing facilities and lesser control over their movements. The relative differences in pricing and marketing have resulted in (i) disparities in incomes between producers of food crops and export crops, (ii) slow rate of growth of marketed surpluses of food crops, (iii) growing dependence on imports to meet the rapidly increasing urban demand and (iv) growing income disparities between the subsistence agricultural sector and the urban industrial sector. In establishing relative pricing of crops, the following specific objectives, therefore, need to be considered if long-term overall development is the objective: (i) realizing domestic self sufficiency in food, (ii) maintaining or expanding agricultural export earnings and (iii) striving for more equitable benefits of growth in the agricultural-rural and the urban-industrial sector.

(b) *Increasing aggregate agricultural production.* Whether the objectives of increasing production of individual crops and of increasing overall agricultural production are mutually consistent or conflicting will depend on the extent to which the overall supply of agricultural commodities responds to relative changes in prices between the farm and the non-farm sector. Empirical evidence on aggregate supply response to price changes among sectors is very limited.[22] However, a number of observations can be made on *a priori* grounds. The aggregate supply response will depend on the extent to which additional resources can be brought into agriculture in response to changes in relative prices among sectors without substantially increasing costs of production. The possibilities for additional resource mobilization, of course, vary considerably among countries depending on the scope for exploitation of additional resources.[23] Where a single crop already occupies a very major share of the cultivable land and much of the cultivated land is already being farmed, as is the case with rice in many Asian countries, the elasticity of input use with respect to change in relative prices can, of course, not be very significant. In such cases, production increase must come about largely through use of increased modern inputs made profitable by technological change. Relative prices can facilitate the use of modern inputs necessary in the new technology. In contrast, where a variety of crops are grown, each constituting a small share of the total agricultural production, the elasticity of input use with respect to price changes among sectors is likely to be greater than in a monoculture situation and even more so where scope exists for additional resource exploitation.[24] These conditions exist in several African countries. In such a situation, agricultural prices both relative to each other and relative to the non-farm sector may bring about a significant increase in overall production. Increase in overall production arising from increase in the quantities produced of major individual crops, therefore, needs to be distinguished from the increase in the *value* of total production through substitution of high value crops for low value crops. As pointed out earlier, much of the increase in production in African countries has come about in this latter way.

How minimum and maximum prices are determined within which government may buy and sell, will depend on how comprehensive the government aims are in its policy along the lines described above, and equally important, how much weight is attached to minimizing use of government resources and to the ease of implementation in the choice of a pricing and marketing strategy.

3. ESTABLISHING MINIMUM PRICES

Minimum prices should be set for only a few crops that are important in terms of consumption and as substitutes in production. Setting prices for a large number of crops is neither necessary nor desirable. It is administratively demanding, and also defeats the purpose of incentive effect of individual guarantees.[25]

For the reasons outlined below, minimum producer prices may be determined on the basis of a combination – rather than being based on only one – of the following criteria: (i) average total cost of production, (ii) paid out costs of production, (iii) parity with the non-farm and/or international prices.

Average total cost of production in a normal year is the most logical basis for formulation of floor pricing if a certain net positive return to agricultural production is to be ensured. Use of average total cost criterion, however, poses a number of problems. These may be classified into three categories:

conceptual, empirical and budgetary. Only some of the more important questions from each of these categroies are raised here to illustrate the nature of the considerations that must enter establishment of minimum prices.[26]

Estimating what the real opportunity cost of inputs is to the farmer is often a problem as a number of inputs used in production frequently do not enter the market economy. Domestic labor provides a good example as it is usually the most important element of costs. Its pricing raises questions as to whether to value labor at the opportunity cost in alternate uses within the agricultural sector or whether to use the opportunity cost in non-farm employment, particularly in cases where prospects exist for migration to obtain urban employment. In arriving at costs, if wages are not valued even at other alternative productive pursuits in agriculture, through reducing return to labor, this may divert labor away from production into leisure or other agricultural and nonagricultural employment.[27] In countries where seasonal labor bottlenecks pose a major problem, as in many African countries, inadequate valuation of domestic labor in determining producer prices, albeit implicitly, has often had a substantial adverse influence on the flow of labor, particularly into food crops, as pointed out earlier. On the other hand, establishing statutory minimum wages of hired labor too high by governments has often increased cost of hired labor particularly for the small farmers to whom the alternative of increasing output through mechanization is not open. Valuation of labor at alternate employment opportunities may, however, also raise the floor price to a level which many governments may be unable to support with their scarce budgetary resources. The shadow wage rate and a minimum agricultural wage rate may, therefore, be viewed as the lower and upper bounds in costing labor to establish minimum prices.[28]

The price of labor is also highly variable among regions and seasons depending on the degree of the labor constraint, posing a problem as to which levels of costs should be adopted as a guide to policy. The regional variability in costs has important implications for promoting interregional specialization or regional self sufficiency through the pricing policy. Choosing a high enough level of costs would promote production even in the least suited regions; a low level of costs on the other hand would discourage production for the market in regions with high costs of production. In arriving at prices, it is important to take into account not only the costs of transportation from low cost to high cost regions – which may be quite high where transport is poorly developed – but also whether the market structure, either traditional or organized, is likely to have the capacity to distribute in rural areas. Low prices may promote regional specialization before infrastructural and market development justifies this course, with substantial adverse implications for equity among regions, particularly when poorer regions do not have incomes high enough to attract surpluses by bidding prices high in competition with the cities.[29]

Valuation of the land input also poses problems particularly where a land market does not exist and/or the rental charge either does not exist or is an imperfect index of the opportunity cost. In this case, opportunity cost of land in its alternate uses is the logical basis for evaluating the land input.

Costs of production also vary considerably depending on the technology used. Setting prices low on the basis of new technology will discriminate against farmers using more high cost traditional technology with adverse equity effects, particularly if new inputs are highly subsidized and/or if the bulk of the farmers do not have access to that technology. Once food production has reached the level of domestic self sufficiency, prices may however have to be lowered to discourage further increase in food production and to diversify the composition of domestic production. A careful analysis of costs is, therefore, necessary by farm sizes, types of technology and regions on a regular basis to determine the level of floor prices.

Parity with industrial prices is an additional index of pricing which may be considered the basis for determining floor prices. To some extent the cost of production approach already takes these prices into account to the extent that industrial goods and services constitute a direct input into production. However, prices of goods and services utilized in consumption may also affect incentives in agriculture through their effect on the cost of living in the rural sector, hence on real rural incomes and on the desire of the rural labor force to be employed in the rural or the urban sector. Once again setting parity with urban incomes or industrial wages may significantly raise the level of prices. Many governments may prefer not to – or may not afford to – go so far in transferring resources to the agricultural sector.

As mentioned earlier, direct taxes on rural incomes or land and indirect taxes on goods consumed or used in production in agriculture reduce net rural income. Subsidies, on the other hand, increase incomes. Taxes and subsidies, therefore, affect the incentives to increase agricultural production, and need to be taken into account in establishing minimum prices.

If input subsidies exist, minimum prices may not have to be established at as high a level as otherwise to assure a certain minimum return. Nevertheless, for the reasons discussed above, if the net profit is lower than in growing other crops there may still be substitution in favor of those crops. Input subsidies are, therefore, not an adequate instrument for increasing input use in a particular crop. The relative demands on budgetary resources of input subsidies and support prices also need to be examined in establishing minimum prices. A general input subsidy may involve greater commitment of government resources than would price support.

Governments frequently do not have the necessary professional manpower to develop so comprehensive a criterion of average cost of production and to carry out collection of reliable data and analysis on a regular basis to formulate and review prices on this basis. Such capacity can, however, be developed over time, as examples of India and Pakistan illustrate.

In the interim period, to simplify planning and implementation, governments may adopt a less complete and fiscally less demanding approach of using paid out costs as the basis for setting prices. To simplify the approach even further, they may only set prices of fertilizer, as it usually constitutes the most important cash cost, along with floor prices of selected crops. Variations in the fertilizer output coefficients among regions, size and type of farms must, nevertheless, be considered in determining fertilizer/output price ratios. As such an approach does not take into account all paid out costs and understates the total costs of production even more, an adjustment in the fertilizer/crop price ratio may be made to allow for these other costs.

Import parity criteria is often favored over the cost of production approach for formulating prices. This is because, if pursued regardless of world prices, the cost of production criterion may distort domestic production by valuing various alternative inputs and outputs either more or less than they would cost (or could be sold for) in the world market. This has further consequence of reducing a country's foreign exchange earnings, as farmers will use more inputs than they should for maximum net production if the inputs are underpriced in relation to the cost of acquiring them internationally and produce less output if the output is similarly underpriced. By the same token, they will use less quantity of inputs if inputs are overpriced and produce more output if output is overpriced.

There are a number of reasons, however, why import parity criterion should not be used as the sole basis for setting prices. First, the considerable fluctuations in world prices of fertilizer and food in the recent years emphasize that import parity is not more sacrosanct than are other criteria for allocating a country's domestic resources. Used by themselves, parity prices may lead to a much greater instability in the domestic prices and supplies than can be afforded by most countries. The fluctuations thus raise a question as to which international prices to use. Even an average of five years if not properly selected may lead to price levels so high as to involve a considerable subsidy, placing burden on the budgetary resources with various far reaching implications of the type discussed below. On the other hand, in times when international surplus exists, prices implicit in concessionary imports may frequently be far below domestic prices, implying considerable taxation of the domestic agricultural sector. In case of crops such as rice, where internationally traded amounts constitute a very small share of the world production, the extent to which international prices are a reasonable index of the supranational production function is also questionable.

The extent to which import parity is used as a criteria also has to be examined in terms of whether imports serve the function of stabilizing production fluctuations arising largely due to weather factors or whether they fill a gap created by chronic excess in demand. Even in the latter case, whether raising prices to parity level will lead to increase in domestic supply will depend on the aggregate supply response that, at least in some cases, may depend more on technology than on prices as pointed out earlier. Further, even if overall production is highly responsive to prices, there may not always be sufficient effective demand to absorb the surplus.[30] Therefore in general, rather than using a particular criterion by itself, the combination of considerations involved in the various criteria discussed above should be the basis of formulating floor prices depending on the relative importance of these factors in individual cases.

4. PRICE STABILIZATION

Apart from determining the level of minimum prices, the range within which governments may allow prices to fluctuate and thus the upper price bound needs to be determined. In this context, three interrelated considerations have to be taken into account: (a) costs of holding a buffer stock vis-a-vis other alternative uses of government resources, (b) effect of price stability on production incentives and hence on growth in the industrial and agricultural sector, and (c) the incidence of taxation and/or subsidy by sectors and classes.

A narrowly defined price range has costs similar to those already incurred by many governments by fixing prices below the market. To implement it effectively, it requires a substantial control of stocks. If imports are not available at a short notice, and at concessionary prices (as years of low production domestically may be associated with high international prices and high cost of imports), control of stocks may mean building buffer stocks, involving between 15 to 20 percent of the value of stocks in annual carrying costs. Depending on the size of the stock, this may mean a substantial commitment of the government's budgetary resources.[31] Because most governments are unable to reduce their recurrent budgetary expenditures, the expenditure on buffer stock will usually be at the cost of resources that would otherwise be invested in the public sector. If the marginal propensity to save in the public and the private sector is the same, investment in buffer stocks and the consequent control of inflation may only transfer savings from the public to the urban private sector with no significant effect on growth (assuming that both would invest in the industrial sector). With buffer stocks, the rate of investment will be relatively steady as prices will not increase in short crop years to a level to reduce private savings and investment. Without buffer stocks, however, timing of the investment will vary considerably from year to year, assuming that instability has no adverse effect on industrial investment. It is, however, likely that price and wage instability may lead to considerable political unrest and/or disincentive to invest, with adverse effect on viability of governments and/or industrial growth. A narrowly defined price range may, therefore, seem desirable to the governments from both these points of view with considerable commitment not only of budgetary but administrative resources to such a program.

The need for a narrowly defined price range is, however, much less clear in the case of growth of the agricultural sector, although it is often argued that price instability reduces farm investment and supply, and that larger levels of instability may be associated by farmers with larger risks.[32] How important price stability is to agricultural growth will depend, among other things, on the degree of stability of yields, the proportion of the production that is marketed and the price elasticity of demand.

In cases where yields are stable, even if there is considerable price instability, if the magnitude of the lowest prices that are likely to be reached are more than compensated by a certain *assured* increase in yields, the innovation may still be profitable and attractive. The yield stability or guaranteed minimum prices may thus minimize the importance of price stability.[33] However, in situations where price instability is combined with yield uncertainty, it may increase the risk and, therefore, discourage adoption of new innovations. Given the dependence of the traditional agriculture on the weather and the consequent tendency for fluctuating yields, some degree of price stability may therefore be important in ensuring return to new innovations. However, where scope exists for reducing yield instability, as for instance through investment in irrigation, this latter may be a better alternative in the long run, particularly if the costs of price stabilization program are comparable with those of investment in yield stabilization.[34]

Given everything else, the more inelastic the demand with respect to price the greater the income instability, if the farmer markets all his produce, as is likely to be the case with larger commercial farmers. A price stabilization program may, therefore, be useful in stabilizing incomes of such farmers and, therefore, in increasing their investment. However, the smaller the portion of the total produce that a farmer markets, the more elastic is his marketed surplus likely to be with respect to production, assuming that the demand for food for domestic consumption is inelastic with respect to production. In such a situation, even the price inelastic demand may not be as income destabilizing. A price stabilization program may, therefore, be less necessary or beneficial in the case of small than large farmers, as a way of inducing investment.[35]

However, in considering the government's role in this context, distinction must also be made between price instability and price risk. Risk is usually defined to mean the difference between expected and observed prices. It is argued that in some cases observed prices could be highly variable, but if expected prices vary with observed prices, risk may still be small. Particularly with a linear investment function, the same level of investment would result over a period of time under stable and unstable income streams provided that average incomes were the same in two cases. Only the timing of investment would be different. The risk factor, on the other hand, is said to have considerable adverse effect on investment.[36]

The conclusion of these complex considerations is that, in the case of the agricultural sector, fixing minimum and maximum prices within a certain range may be necessary both for commercial and commercializing farmers so as to reduce risk; but that unlike from the viewpoint of the urban industrial sector, it may be less necessary or desirable for agricultural growth to stabilize prices within such narrow ranges as to come close to fixing a single controlled price of the type that is currently followed by most governments.

If set at too low a level, a small price range also comes close to being a tax on agriculture of the type currently imposed by many governments, whereas if set at a high level it implies a considerable subsidy to agriculture. A broader price range, on the other hand, can pass on the gains of price increases to the farmer without taxing the government's budgetary and administrative resources excessively.

Where the resources for price stabilization come from will determine the net incidence of the cost of the program among classes and sectors. If they come from revenues earned through concessionary imports of the type discussed earlier, then, of course, there is no domestic burden of price stabilization. However, resources may also come from imposition of additional direct or indirect taxes in either or both sectors, or through foreign assistance with complex and highly variable net effects on the various classes and sectors. These must be analyzed on an individual country basis to arrive at judgements as to the desirability of the net transfers implicit in price stabilization.

Throughout the previous discussion, arguments have been made to indicate why the government's role in pricing and marketing should be confined only to that defined by the size and scope of the price stabilization program, and why the price range itself should not be excessively narrow. In the sections below, a more positive reason has been advanced for this contention, namely, the very considerable potential of the traditional sector to perform a number of important functions with regard to pricing and marketing, provided certain preconditions are ensured by governments.

5. CHOICE OF MARKETING INSTITUTIONS FOR OPTIMAL POLICY

5.1. *Traditional trading systems*

Despite differences among countries both with regard to the stages of market development and to availability of documentation, considerable evidence has accumulated with regard to the working of traditional markets dispelling stereotypes about the degree of oligopolistic tendencies and spacial and seasonal price differences.

It is generally believed that there are only a few intermediaries at each level of marketing and that as a result collusive tendencies and price fixing is rampant in traditional markets. It is also argued that because there usually is a long chain of intermediaries from the producer to the consumer, it results in high costs of marketing. Evidence, however, indicates that entry in traditional trade is generally free and that there is overcrowding and significant competition at each level of marketing.[37] Most rural traders work with small amounts of capital, on low margins and earn a meagre income.

Few traders may, nevertheless, be seen to handle a large share of the marketed surplus in many markets.[38] Evidence indicates overwhelmingly, however, that they are not able to influence prices through collusive action if transport and exchange of market intelligence among producing markets and between producing and consuming markets are effective.[39]

Intermarket price differences are usually small and commensurate with the costs of transportation and handling. For instance, contrary to the general view even small farmers in village markets in India are observed to receive prices commensurate with those prevailing in larger wholesale markets.[40] Excessive price differences among markets arise because of:

(a) Poor dissemination of price information and poor communication facilities which do not allow transmittal of knowledge of price disparities effectively among markets,

(b) inadequate and unreliable transport facilities that often result in accumulation of surpluses in producing areas and shortages in consuming centers, particularly in the immediate post harvest period,

(c) poor transport and handling facilities that result in losses in the quantity and quality of crops during movement of grain,

(d) lack of implementation of standard weights, measures and/or marketing charges that allow scope for cheating, and

(e) lack of enforcement of an open bidding system.

There are also reasons why markets seem more inefficient than they actually are including: (i) price differences among varieties and grades of crops which lead to the impression of monopolistic elements in trade if price data in the two markets being compared do not refer to the same grade or variety, (ii) movement restrictions imposed by the governments that increase price disparities, and (iii) speculative tendencies that are caused by frequent overall shortages, and often exacerbated by sudden imposition and withdrawal of government restrictions on trade.[42]

These factors suggest that the solution to monopolistic practices is not to discourage trade through overt or covert means, as is done by many governments, but rather through governmental action to remove the above conditions which lead to monopolistic practices. The former course usually only exacerbates exploitation, by reducing the number of intermediaries and thus by increasing opportunities for monopolistic practices.

Facilitating efficiency in traditional trade is necessary as rural traders perform a number of important functions that cannot be replaced by government or cooperative agencies, without incurring substantially greater costs in administrative manpower and finances than is implicit in allowing the private sector to operate.[43] Traders function in the remotest and least accessible areas where government or cooperative machinery frequently does not reach, thus performing an important function of providing a market channel for the rural surpluses, and in many cases also fulfilling the consumption needs of the rural communities by selling consumer goods, thus providing further incentive to produce for the market.[44] They meet an important credit need of the small producers, including the need for consumption credit. Government agencies are unable to administer credit in small amounts, particularly for consumption, without incurring substantial administrative costs and defaults.[45] The timely availability of the much needed credit is an important reason why farmers continue to turn to private traders even if alternative marketing channels are available. Timely provision of services, including their presence when marketed surpluses are available, is of course the other reason for reliance on private traders. These various needs of the rural communities are generally greatly underestimated by policy-makers in replacing or discouraging traditional trade.

The other usual allegation about the private trade is that farmers sell their surpluses in the immediate post harvest period when prices are low because of (i) their heavy indebtedness to the village traders and moneylenders, (ii) their need for cash for paying taxes and debts and (iii) inadequate storage facilities at their disposal. Off-seasonal prices, on the other hand, are said to rise significantly higher than costs of storage and normal profit, allowing traders to make excessive profits.

The evidence is once again much less supportive.[46] It indicates that there is considerable variability in the pattern of seasonal price movements and that this pattern is usually difficult to predict leading to considerable risks in storage until the off season. The profits of traders when averaged over a number of years are not excessive, as high profits in some years are compensated by low profits (or even losses) in other years. Given this pattern it is questionable whether transferring the storage function from traders to farmers will necessarily result in farmers earning a significantly higher price unless there is an overall price policy that reduces risks. Thus it is not so much the low price received by farmers, but the lack of choice open to them as to when to sell, which has been the unacceptable feature of traditional trade. With increases in income and awareness of markets, this pattern has already begun to change significantly in many cases. Farmers are beginning to sell less in the post harvest period and to hold more stocks until the off-season to improve the price they receive. A number of steps of the type discussed in the final section of this paper, nevertheless, need to be taken by governments to further improve bargaining power of the farmers, none of which involve so drastic a step as discouragement or abolition of private trade.

5.2. *Marketing cooperatives*

Cooperatives have been promoted in many countries, through substantial commitment of finances and manpower, as an alternative to private marketing channels to augment competition and thus to improve the bargaining power of and the prices received by farmers. Several other steps have also been taken by governments to encourage growth of grain marketing in the cooperative sector. For instance, in many countries marketing cooperatives are appointed agents of the government in procurement of marketed surpluses or certain functions, such as paddy processing, have been completely delegated to them.[47]

Marketing cooperatives have been generally effective in dealing with traditional cash crops such as sugar, cotton, tobacco and coffee. They have, however, failed to make a headway in marketing of food crops.[48] Because the difference in the experience has been so striking, and because marketing cooperatives continue to receive enthusiastic support from a very broad range of interests, it is important to understand the reasons for the differences in the cooperative experience in the case of export and food crops. These differences indicate that developing grain marketing cooperatives is indeed a difficult task and must be handled gradually and carefully. It is counterproductive to push cooperative development too rapidly as it usually backfires, frequently making cooperatives the haven of government subsidies and a barometer of inefficiency, in which the relatively more efficient private trade can survive relatively easily, defeating the purpose of augmenting competition among channels of marketing.

Because many cash crops require further processing, unlike food crops they cannot be used in domestic consumption or sold in rural markets. A centralized marketing facility is, therefore, relatively easier to organize in such cases than for most food crops. For this reason, in the case of cash crops, the centralized marketing is usually also easily integrated with provision of credit. Credit has been much less well organized in the case of food crops, leading to the scope for considerable competition from private traders, as pointed out earlier. Second, the value added in processing of cash crops is usually substantial, also providing scope for economics of scale and thus providing greater opportunity for cooperatives to be viable as processing as distinct from marketing entities. Besides in the case of export crops, the price of the final product is usually sufficiently high to reflect the value added. Therefore, even if marketing and processing is not efficiently organized, cooperatives frequently are able to offer a farm gate price that is high enough to assure an attractive return to the farmer. When these conditions – in particular of scope for easy centralization of marketing and/or high international prices – do not hold, even cash crop cooperatives flounder as exemplified by the cotton cooperatives in Kenya or the cocoa cooperatives in Cameroon.[49] Third, the buoyant international market for some cash crops has permitted cooperatives to perform the relatively simple administrative function of procuring surpluses at a given price rather than requiring them to do the more difficult task of trading, involving decisions as to when and how much to buy and sell at various prices.[50] As pointed out earlier, because of the generally high degree of instability and uncertainty in the domestic food market, dealing with food crops involves considerably greater risk and, therefore, requires greater skill than the inexperienced management of most cooperatives can provide.

For these various reasons, development of food crop cooperatives must, by necessity, be more gradual and viewed as part of an overall policy package involving a well conceived and implementable price policy, reforms related to overall marketing of the type discussed below, and a well thought out training program. In initial years, cooperation among farmers may best be confined to relatively simple activities, such as their participation (i) in insistence on the use of standard weights and measures by private traders, (ii) on disseminating information on prices prevailing in other markets, and (iii) in construction of storage facilities where they may wish to store for the off-season, particularly if they are advanced credit for a portion of value of their crop as outlined below. Purchase and sale of surpluses on account of cooperatives may come as the next stage, when a more effective price policy has been formulated and when cooperative managers are trained and gain experience in management as well as in trading. In absence of such a gradual approach, cooperatives frequently become a marketing alternative in slogan more than in practice.

5.3. *Role of the public sector*

A far more broad-based and positive role for the public sector is implicit in the approach suggested in this paper than currently followed by most governments. This is not only because of the role envisaged for them in stabilizing food supplies and prices from year to year, through guaranteed prices and buffer stock operations, but also because of a range of public sector investments in market development that are considered essential to improve equity and efficiency in pricing. To follow this course, decisions need to be made on a routine basis as to the level of minimum and maximum prices, the size of the marketed surplus that the government should control, and both the administrative and the financial means to acquire and distribute the marketed surplus as necessary. An investment strategy to develop the necessary physical and institutional infrastructure also needs to be developed so as to facilitate formulation and implementation of government decisions with regard to its direct role, and through improving performance of the trading sector to minimize its direct interventions. These steps are outlined below.

6. POLICY IMPLICATIONS

6.1. *Developing professional and institutional capability for formulation and implementation of the policy*

(a) *Improving data base.* Because a large part of the pricing and marketing problem arises from poor knowledge of the size and sources of food production and marketable surpluses, there is no easy substitute for improved data on yields and acreages under various crops by areas and by farm sizes as a way of estimating production and marketings. Developing a reliable data base requires considerable professional input and the training and developing of the staff for data collection and analysis. It is, therefore, necessary to begin this process as early as possible without expecting dramatic short-term results.

Expertise and institutional capability must also be developed within the government to collect and analyze information on (i) factors that should enter formulation of floor and ceiling prices, (ii) size of buffer stocks, and (iii) actual behavior of prices and marketings by crops and by seasons. All these sets of data will, of course, have to be collected by geographical locations. The considerations that should enter data gathering and analysis with regard to minimum and maximum prices and buffer stocks are already discussed in the main body of the paper. Those with regard to collection of actual prices and marketing are discussed in the section on market regulation.

One of the reasons why administrators are often wary of recommendations involving further data gathering is the very little feedback that available information usually has in policy or program formulation and implementation.[51] The solution to this problem is, however, not to avoid information gathering but rather to examine how it may provide an input into decision-making. This should be an important consideration in determining functions of the institutions to be responsible for data collection and analysis. On the one hand, they have to have the necessary pressure to be responsive to the policy and operational needs of the agencies responsible for pricing and marketing, in their gathering and analyzing data; on the other hand, they need the professional independence to carry out reliable data collection. Whether such data gathering units are established within the ministry of agriculture or planning or as autonomous entities will depend on how well they can do these two functions from within and outside the normal governmental structure.

(b) *Policy implementation.* In addition to creating the various data gathering units and a central policy formulating unit, an agency will need to be given the responsibility of carrying out the task of purchases and sales and the maintenance of buffer stocks involving staff, storage and transport facilities both at headquarters and in major producing and consuming areas.

6.2. *Regulation of traditional markets*

Governments have spent few, if any, resources on improving performance of the traditional markets. The Indian experience with regulated markets provides one of the few exceptions from which a number of useful insights can be derived as to the possible lines of approach in this very important area.[52]

(a) *Declaration of market centers where marketing practices are regulated.* Governments may declare many of the important existing assembly centers as "regulated" markets where standardization of weights and measures, fixed marketing charges and traders' commissions, open auctions, standard methods of payment and grading, etc. may be enforced through a regulated market committee consisting of representatives of farmers, traders and government officials. Ministry of Agriculture may establish a Directorate of Marketing and Inspection to carry out this function.

(b) *Collection and dissemination of market information.* The Directorate of Marketing and Inspection may collect and disseminate information on prices and market arrivals by appointing an investigator in each market committee for recording and forwarding the necessary statistics. In addition, as and when necessary, it may have to conduct farm-level surveys on an ad hoc basis through its own staff or through use of universities to assess the size of marketable surpluses. Given the erratic behaviour of farmers, the knowledge of farm-level stocks is extremely essential for formulating government policy with regard to public sector stocks.

To increase the market consciousness of farmers, information on prices prevailing in the various important producing and consuming centers and the floor and ceiling prices at which the government will buy and sell may be disseminated routinely and actively by the Directorate through all possible means including radio, newspapers, bulletin boards in marketing centers and announcements in villages.

6.3. *Construction of storage facilities*

Losses in handling and storage in many developing countries are significant and lead to considerable inefficiency in marketing. However, there is a great deal of myth about excessively high storage losses at times leading to the conclusion that the food problem can be solved if only improved storage facilities were constructed.[53] It is important to ascertain the precise nature and magnitude of losses by stages of marketing, crops and areas as fancy high cost storage construction may not always be justified by savings in losses.

Expansion of storage facilities is necessary not only to reduce losses in storage by farmers and in the trading sector, but also to improve staying power of the farmers and to be able to hold public sector supplies in strategic places in producing and consuming areas to effectively implement the buffer stock policy. Storage may be built at three levels of marketing.

First, farmers who individually or collectively wish to build relatively simple, scientifically constructed storage facilities at the farm or village level should be assisted with credit and knowhow on construction.

Second, government may either provide credit or participate partially or fully in constructing storage facilities in the major assembly market centers in producing areas for the use by farmers and by the public, private and cooperative trading agencies. One of the most effective ways of improving staying power of the farmers is to advance credit to them for a portion of the value of their marketed surplus (70 to 80 percent of the harvest price) on the basis of the voucher for the surplus deposited in such public warehouses. As appropriate, commercial, cooperative or government agencies may be given the responsibility to advance credit, with proper attention to ensure that the credit activities of the various agencies are well coordinated. With cash available for payment of taxes and debts, the farmers will then have the option of holding stocks until the off season to fetch a better price. Similar facilities may also be provided to the private and cooperative trading agencies. The government may then use the percentage of the value advanced in the form of credit as a way of controlling the supply of credit and thus of manipulating stocks in the farming and the trading sector to avoid excessive speculative stocking.

Warehouses for storage of surpluses in the public sector in strategic consuming centers, including in deficit rural areas, is the third type of investment necessary both for timely release of stocks when distribution is deemed necessary as well as to stabilize speculative price increases that arise in periods of critical shortages. Although considerable storage facilities are usually constructed at ports and in large urban consuming centers, in most developing countries little attention has been given to assessing needs and developing facilities for holding stocks in deficit rural areas for distribution and price stabilization. From all indications, Mainland China appears to be one of the few exceptions where each production unit is encouraged to hold sufficient stocks to alleviate shortages arising from normal production fluctuations in every defined area.[54] Such a policy would, of course, also require that not all the surpluses be mopped up for distribution in the cities.

6.4. *Development of a transportation network*

How effective the interventions outlined above are will depend on how well producing areas are connected to assembly markets and the assembly markets in turn to the consuming areas. This means investment not only in trunk roads, but in feeder roads connecting rural areas with each other and with market towns. In general, governments have spent few resources on development of rural infrastructure. The main emphasis has been on connecting large urban centers with each other and with major producing centers.

7. CONCLUDING REMARKS

This paper emphasizes the following important points. First, it stresses the need for guaranteed minimum prices for a few key crops the production of which needs to be increased. Second, it illustrates the complexity of considerations that need to enter formulation of minimum prices and stresses the need for developing the domestic professional and institutional capability, both of a conceptual and statistical nature, to determine appropriate price levels. Third, it outlines the nature of considerations that should enter formulation of maximum prices, and the fiscal and administrative implications, particularly of a stabilization program that is narrowly defined in terms of the price range. Fourth, it stresses the need for the implementing capability to effectively carry out the government role in pricing and marketing and the nature of investments necessary to improve that capability. Finally, it emphasizes the very important role that the traditional trading sector can play in carrying out the pricing and marketing function, particularly until the cooperative sector develops and, therefore, the way to augment that potential.

NOTES

[1]See Mellor [47] for empirical evidence on income effects of changes in prices of food grains in India among various classes and for extensive review of literature on related issues.

[2]See Mellor [45] for further discussion of this issue.

[3]This, of course, holds only if the cost of labor increases at a rate higher than the cost of capital. The policy of capital subsidization followed by most governments usually ensures that this is the case, a tendency that is reinforced by the enforcement of minimum wages, particularly in the case of industrial labor.

[4]See Mellor [48] for an excellent discussion of the role of wages goods in industrial strategy.

[5]See Lele [38] for the discussion of the history of government controls on Indian grain trade. Also see Anschel, K. R. [4], Helleiner [17], Jones [25] and Moyer and Hollander [49] for discussion of the government controls in some of the African countries.

[6]See Lele [38], pp. 104–108, for evidence on intermarket wheat prices between Punjab market and Bombay during periods of controls and free trade.

[7]In India concessionary imports, at times, constituted up to 95 percent of the quantities distributed by the public sector domestically. These not only provided counterpart funds which could be used for domestic expenditure, but also led to revenues for the Government through the difference in the c.i.f. and retail price of PL 480 surpluses, allowing subsidization of domestically procured grain for urban consumption in the sixties. See Lele [39].

[8]See Lele [38, 39] for discussion of this problem in the Indian context.

[9]See Lele [41] pp. 107–108 for discussion of the import and export policies of the Ethiopian Grain Corporation and their effect on adoption of innovations in the Chilalo Agricultural Development Program.

[10]See Parthasarthy [54] for an excellent review of literature on increased market dependence in India. See Lele [41], Chapters II and III for discussion of the effect of export crop development on market dependence in many African countries.

[11]See Lele [39], Table 1, for quantities of food distributed in India during 1952–53 and 1973–74.

[12]Tanzania, for instance, considered a takeover of retail distribution but abandoned the idea due to its infeasibility.

[13]This pattern of market withdrawal was noted as early as 1957–58 in India; (see Lele [38] for discussion of market arrivals.

[14]See Parthasarthy [54] for additional review of evidence.

[15]See Government of India [16] for evidence.

[16]See Lele [37] for evidence.

[17]In India in years of controls the grain that would normally be handled through wholesale trade is brought to cities in headloads. Although retail price of this produce may be between two to three times that sold in fair price shops operated by the Government, a small share of it goes to the farmer, due to the clandestine nature of trade and a large number of intermediaries. This is in contrast to periods of free trade. See Lele [38].

[18]It is important to note this distinction as there is frequently a simplistic tendency in literature on agricultural development to argue that pricing policy is not as important as technological change, a superficial interpretation which leads to undermining the importance of a guaranteed price support program. See for example Mellor [45] and Parthasarthy [54].

[19]See Krishna [30], Mellor [47] and Parthasarthy [54] for excellent review of literature.

[20]See Lele [41], Chapter II, for review of literature and discussion of issues related to labor allocation.

[21]See Lele [41], Chapter II, for discussion of the factors explaining growth of export crops in Africa.

[22]See Krishna [30] and Parthasarthy [54] for review of literature. A noteworthy exception is the excellent study of the role of intersectoral terms of trade in growth of Taiwan's agriculture. See Lee [33].

[23]How profitable such investment is must, however, depend on what value is placed on increased food supply, not only in the short run but in a more dynamic sense of how the increased food production may be used to serve an employment-oriented growth strategy. See Mellor [48] for discussion of this issue.

[24]See Krishna [30] for support of these various arguments.

[25]Once some prices are regulated the temptation to extend regulations and controls to all crops is, however, usually substantial and especially needs to be avoided.

[26]For discussion of several other issues see Krishna [30].

[27]See Lele [41], Chapter II, for discussion of these interactions.

[28]Leaning towards a shadow wage will, of course, be less favorable to the growth of the subsistence agricultural sector than opting in favor of the minimum wage.

[29]See Lele [41], Chapters II and VI, for discussion of these issues.

[30]The food pricing must, therefore, be viewed as part of an overall development strategy. A more employment-oriented strategy that generates effective demand may have to be combined with increased support prices. In absence of such a strategy, raising prices up to the parity level may, however, not always be justified.

[31]For instance, a very conservative buffer stock of 10 million tons of grain for India (i.e., about 8 to 9 percent of its annual domestic production) would involve an annual carrying cost of approximately $250 million to $300 million (assuming annual interest cost of 10 percent, storage losses of 5 percent and the investment costs in storage facilities amortized at 5 percent of the value of the 10 million tons of grain at $120 to $150 per ton). This expenditure is equivalent to the planned annual financial allocations of the Indian Government to irrigation investment of all types during the Fourth Plan Period, about 35 percent of the annual World Bank assistance and 50 percent greater than the total US assistance to India, both at the current rate. If such investment is made in expansion of irrigation facilities, as for instance in the eastern Gangetic Plain, this may have a substantial impact on increasing and stabilizing yields. Most governments may have to have buffer stocks which are not too dissimilar to the above, in terms relative to their domestic food production and current investment in agriculture, as the normal fluctuations in weather lead to annual production fluctuations of between 10 to 20 percent. See Lele and Mellor [34] for discussion of levels and fluctuation in production.

[32]See Tomek [63] for discussion of these issues.

[33]This may, however, mean considerable fluctuations in incomes, the effect of which on investment needs to be considered as discussed in the text below.

[34]This may explain the emphasis in Mainland China on developing areas of high and stable yields as one of the major focal points of the agricultural pricing policy. See Timmer [62].

[35]This, however, does not minimize the importance of a guaranteed minimum price for small farmers, particularly in its importance in enhancing certainty of return to

investment. See Schluter [58, 59] for the importance of risk and uncertainty in adoption of innovations by small farmers.

[36]Where price fluctuations arise from weather factors rather than from unpredictable government interference, it could be argued that the difference in the expected and actual prices and, therefore, the risk, may generally not be high. To the extent that the government policy increases predictability of price movements, it may reduce risk and facilitate production even without a narrowly defined price range.

[37]See Anthonio [5], Bauer [6], Cummings [8], Dewey [11], Farruk [12], Gadgil [13], Galbraith and Holton [14], Hermann [18], Hill [19], Hirsch [20], Jasdanwalla [23], Jones [24, 25], Kriesel [27], Kulkarni [28], Lele [35, 36, 38, 39, 41], Manig [44], Nightingale [52], Rao and Subbarao [55], Satyapriya [57] and Thoday [60, 61].

[38]In 1967, six major sorghum markets in the Marathwada Region in Maharashtra State, India, had between 71 to 281 commission agents and wholesalers of type "A" alone. And yet, only 10 percent of the traders handled over half of the total volume of jowar traded in the markets. See Lele [38]. Investigations in Punjab in 1967–68 also indicated that despite the nearly 40 to 50 traders that were registered in each of the five large wholesale markets, only two traders in Barnala, four in Jugraon and five in Kotapura handled over half the wheat arrivals in those markets. See Lele [38]. Similar concentration has been noted in a number of other studies. See Lele [39].

[39]During free trade, covering the 1955–1965 period, the difference in the wheat prices of comparable varieties between Moga in Punjab and Delhi was less than Rs5·00 over 90 percent of the time, i.e., barely adequate to cover costs of shipment of a quintal of wheat between the two markets (see Lele [38], p. 108). Such high degree of price integration has been found in a number of studies on India, which are listed in footnote 37 above. See in particular Cummings [8], Gadgil [13], Kulkarmi [28], Lele [35, 36, 38, 39], Rao and Subbarao [55] and Stayapriya [57]. Although – for a variety of obvious reasons, including lack of data on standard prices – price movements show less close relation between markets in African countries, prices in primary markets are generally observed to move in line with those in consuming centers. See Jones [24, 25] and Manig [44].

[40]The study of village pricing of paddy in Cattack district in Orissa, West Godavari district in Andhra Pradesh and in Thanjavur district in Tamil Nadu in India by Rao and Subbarao is of particular interest because it relates to the 1967–68/1969–70 period when new paddy varieties had been introduced in many areas, generating additional surpluses and reducing bargaining power of the producers somewhat. Among other things, the study compared the prices received by large farmers with those by small farmers. Their findings indicate relatively minor differences ranging from 0·5 to 6·0 percent between village and primary market prices. The differences on the high side, i.e., of 5·0 to 6·0 percent, arose in the case of the flooded zone in Cuttack and the waterlogged old delta in Thanjavur, where village to market transport was poorer in comparison with other markets. In the Thanjavur villages there was also greater inequality in the land ownership and thus poorer bargaining power of the small farmers vis-a-vis the moneylender/traders. Price differences were negligible in cases where infrastructure was well developed. Prices received by small farmers were somewhat lower, but the difference between those received by large and small farmers was insignificant either by timing, which itself was not significantly different in the case of large and small farmers, or by location of their sale, or by the agency (i.e., moneylender vs primary market trader) to which the sale was made. The study concluded that the "fixed margin (2·5 percent of the price) realized by the moneylender from the small farmers may be attributed essentially to the diseconomies of small-scale purchases, as the time taken in collecting the small quantities of produce for their disposal in the bigger market, transit losses and the discount on the quality of produce owing to its heteroegeneity would be greater (per unit of output) than in the case of bulk sales by a few large farmers". See Rao and Subbarao [55] p. 28.

[41]This is the overwhelming concensus of the studies cited in 37 above and many others that are not cited here for lack of space.

[42]See Lele [38] particularly Appendix 2, p. 238, for detailed discussion of the importance of price specification in examining market integration.

[43]In the Chilalo Agricultural Development Unit in Ethiopia, the costs (excluding overheads to the project management) of marketing wheat between Asella and Addis Ababa were more than twice as much as those incurred by private traders. Many other agricultural projects in Africa have encountered problems due to undertaking similar high cost market interventions. See Lele [41], Chapter VI, for discussion of these issues.

[44]After much experience in procurement, the senior management of the Food Grain Corporation of India recognized that this is a very major limitation of government agencies in purchasing market surpluses and results in substantial administrative costs. (Based on personal discussions in 1969–70.) Where private trade is very active, as in Punjab and Haryana, the FCI has, therefore, relied on traders to act as commission agents of the Corporation in its procurement operations. Many observers argue that although this reduces FCI's administrative costs and problems and in short crop years, to some extent, shifts the political burden of procurement to traders, it does not serve the function of creating competition for traders by providing an alternate channel of marketing to the farmers. In any case, the example highlights the problem of even an experienced and well-staffed government agency being able to carry out the marketing function satisfactorily except in a marginal role. See Lele [39] for further discussion of the procurement problems. Also Lamade [31].

[45]This is one of the reasons why the management of the Wallamo Agricultural Development Unit in Ethiopia began to provide consumption credit to farmers. See Lele [41], Chapters V and VI.

[46]See Lele [35, 36, 38], Cummings [8], Jasdanwalla [23], Jones [25], Kulkarni [28], Rao and Subbarao [55], and Satyapriya [57].

[47]See Lele [37].

[48]Despite much support, cooperatives on the average handled only 3 percent of the total market arrivals of grain in India in the sixties. See Rao and Subbarao [55]. Also see Lele [41], Chapter VI, for discussion of cooperatives; also Anschel [4] and Widstrand [65].

[49]See Lele [41], Chapter VI.

[50]This is one of the major reasons of the success of cotton cooperatives in Tanzania. See Lele [41], Chapter VI; also Collinson [7] and Maguire [43].

[51]See Lele [41] for how this affects quality of planning and implementation.

[52]See Lele [38, 39] for discussion of the role of regulation in Indian grain trade.

[53]Contrary to these assertions of high storage losses, made largely by international agencies such as FAO and USAID, the studies on grain trade indicate relatively minor losses. particularly in the traditional trading sector. See Cummings [8], Kulkarni [28], Faruk [12] and Lele [34, 36, 38, 39].

[54]Timmer [62].

REFERENCES

[1] Abbott, J. C. (1967) 'The Development of Marketing Institutions', pp. 364–398 in Southworth and Johnston (eds.), *Agricultural Development and Economic Growth*, Cornell University Press, Ithaca, N.Y.

[2] Advisory Committee on Overseas Development, *Farmer Cooperatives in Developing Countries*, Washington, D.C., October.

[3] Alvis, V. Q. and Temu, P. E. (1968) 'Marketing Selected Staple Foodstuffs in Kenya', Department of Agricultural Economics and Office of International Programs, West Virginia University, Morgantown, IP-25, March.

[4] Anschel, K. R. (1969) 'Agricultural Marketing in the Former British West Africa', in K. R. Anschel, R. N. Brannon and E. D. Smith (eds.), *Agricultural Cooperatives and Markets in Developing Countries*, Praeger, New York.

[5] Anthonio, Q. B. O. (1968) 'The Marketing of Staple Foodstuffs in Nigeria: A Study of Pricing Efficiency', Ph.D. Dissertation, University of London.

[6] Bauer, P. I. (1963) *West African Trade*, Routledge and Kegan Paul, London.

[7] Collinson, M., 'Tanzania's Cooperative Movement and Farmers Credit in the 1960's, unpublished paper.

[8] Cummings, R., Jr. (1967) *Pricing Efficiency in the Indian Wheat Market*, Impex, New Delhi, India.

[9] Dandekar, V. M. (1968) 'Agricultural Price Policy: A critique of Dantwala', *Economic and Political Weekly*, IV, March 16.

[10] Dantwala, M. L. (1967) 'Incentives and Disincentives in Indian Agriculture', *Indian Journal of Agricultural Economics XXII*, April–June, pp. 1–25.

[11] Dewey, Alice G. (1962) *Peasant Marketing in Java*, Free Press of Glencoe, New York.

[12] Farruk, M. O. (1972) *The Structure and Performance of the Rice Marketing System in East Pakistan*, Cornell International Agricultural Development Bulletin 23, Cornell University.

[13] Gadgil, P. A. (1975) 'Marketing of Tumeric in Sangli District', Ph.D. dissertation, University of Poona; Subbarao K. (1973) *Market Structure in Indian Agriculture: A Study of the Economic Efficiency of Paddy/Rice Marketing System in West Godarvi District, Andhra Pradesh*, Ph.D. dissertation, University of Delhi.

[14] Galbraith, J. K. and Holton, R. H. (1955) *Marketing Efficiency in Puerto Rico*, Harvard University Press, Cambridge, Mass.

[15] Good, G. M. (1970) *Rural Markets and Trade in East Africa: A Study of the Functions and Development of Exchange Institutions in Ankole Uganda,* Research Paper, No. 128, Chicago: University of Chicago, Department of Geography.

[16] Government of India, Ministry of Food and Agriculture, *Report of the Agricultural Prices Commission on Price Policy for Kharif Cereals for 1965–1966 Season*, 1965.

[17] Helleiner, G. K. (1964) 'The Fiscal Role of the Marketing Boards in Nigerian Economic Development, 1947–61', *Economic Journal*, Vol. 74, No. 295, September.

[18] Hermann, L. F. (1964) *Considerations Relating to Agricultural Price Policy in India, With Special Reference to Rice and Other Foodgrains*, Report to the Government of India, Ministry of Food and Agriculture.

[19] Hill, Polly, (1963) 'Markets in Africa', *Journal of Modern African Studies*, Vol. 1, No. 4, December.

[20] Hirsch, L. (1961) *Marketing in an Underdeveloped Economy: The North Indian Sugar Industry*, Prentice Hall, Englewood Cliffs, N.J.

[21] Hodder, B. W. and Ukwn, N. I. (1969) *Markets in West Africa: Studies of Markets and Trade Among the Yoruba and Ibo*, Ibadan: Ibadan University Press.

[22] Hussain, M. A. (1965) 'A Field Investigation Into the Management of Village Cooperatives in Comilla Experimental Area', (mimeo).

[23] Jasdanwalla, A. (1966) *Marketing Efficiency in Indian Agriculture*, Allied Publishers, Bombay.

[24] Jones, W. O. (1970) 'Measuring the Effectiveness of Agricultural Marketing in Contributing to Economic Development: Some African Examples', *Food Research Institute Studies on Agricultural Economics, Trade and Development*, Vol. IX, No. 3, pp. 175–196.

[25] Jones, W. O. (1972) 'Marketing Staple Foodstuffs in Tropical Africa', Cornell University Press, Ithaca, N.Y.

[26] Kriesel, H. C., Laurent, C. K., Halpern, C., and Larzelere, H. E. (1970) 'Agricultural Marketing in Tanzania, Background Research and Policy Proposals', Michigan State University/USAID, June.

[27] Kriesel, H. C. (1969) 'Cocoa Marketing in Nigeria', Consortium for the Study of Nigerian Rural Develpment Paper No. 21, East Lansing, Michigan State University Press, May.

[28] Kulharni, A. P. (1962) 'The Behavior of Prices of Groundnut Pods in Some Regulated Markets in Maharashtra', Ph.D. dissertation, University of Poona.

[29] Knight, Sir Henry (1954) *Food Administration in India*, Stanford University Press, Stanford, Calif.

[30] Krishna, Raj. (1968) 'Agricultural Price Policy and Economic Development', in Herman M. Southworth, Bruce F. Johnston (eds.), *Agricultural Development and Economic Growth*, Cornell University Press, Ithaca, N.Y., pp. 497–540.

[31] Lamade, W. (1968) 'Marketing Boards in Tanzania', *Z. ausi, Landw. Frankfurt a.M.*, Vol. 7, No. 4, pp. 334–348.
[32] Laurent, C. K., 'Cotton Marketing in Tanzania', unpublished paper.
[33] Lee, T. H. (1971) *Intersectoral Capital Flows in the Economic Develpment of Taiwan, 1895–1960*, Cornell University Press, Ithaca, N.Y.
[34] Lele, Uma J. and Mellor, J. W. (1964) *Estimates of Change and Causes of Change in Foodgrains Production in India 1949–50 to 1960–61*, Cornell International Agricultural Development Bulletin 2.
[35] Lele, Uma J., 'Working of Grain Markets in Selected States, India, 1955–56 to 1964–65', Occasional Paper 12, Department of Agricultural Economics, Cornell University USAID Prices Research Contract.
[36] Lele, Uma J. (1968) 'Traders of Sholapur', in John W. Mellor, Thomas F. Weaver, Uma J. Lele and Sheldon R. Simon *Developing Rural India*, Cornell University Press, Ithaca, New York, pp. 237–294; and an excellent survey article by Abbott, J. C.
[37] Lele, Uma J. (1971) 'The Modern Rice Mill in India', Occasional Paper No. 49, Department of Agricultural Economics, Cornell University-USAID Employment and Income Distribution Project.
[38] Lele, Uma J. (1971) *Food Grain Marketing in India: Private Performance and Public Policy*, Cornell University Press, Ithaca, N.Y.
[39] Lele, Uma J., 'Marketing and Pricing of Food Grains in India', in C. N. Vakil and C. H. Shah (eds.), *Agricultural Development in India: Policy and Problems*, forthcoming.
[40] Lele, Uma J. (1974) 'Role of Credit and Marketing in Agricultural Development', in *Agricultural Policy in Developing Countries*, Naral Islam (ed.), London, MacMillen.
[41] Lele, Uma J. (1975) *The Design of Rural Development: Lessons from Africa*, Johns Hopkins University Press, Baltimore, Md.
[42] Livingstone, I. (1973) 'Production, Prices and Marketing Policy for Staple Foodstuffs in Tanzania', in *Agricultural Policy Issues in East Africa Africa*, V. F. Amann (ed.), Kampala: Makerere University, May.
[43] Maguire, Gene Andrew (1969) *Toward 'Uhuru' in Tanzania: The Politics of Participation*, Cambridge University Press, London.
[44] Manig, W. (1973) *Marketing of Selected Agricultural Commodities in Bako Area, Ethiopia*, Occasional Paper No. 66, Ithaca, N.Y., Cornell University, Department of Agricultural Economics, December.
[45] Mellor, John W. (1968) 'Functions of Agricultural Prices in Economic Development', *Indian Journal of Agricultural Economics*, XXIII, Jan.–March, pp. 23–27.
[46] Mellor, John W. (1973) 'Accelerated Growth in Agricultural Production and the Intersectoral Transfer of Resources', *Economic Development and Cultural Change*, Vol. XXII, No. 1, October, pp. 1–16.
[47] Mellor, John W. (1975) *Agricultural Price Policy and Income Distribution in Low Income Nations*, World Bank Staff Working Paper No. 214, September.
[48] Mellor, J. W. (1976) *The New Economics of Growth: Strategy for India and the Developing World*, Cornell University Press, Ithaca, N.Y.
[49] Moyer, R. and Hollander, S. (1968) *Markets and Marketing in Developing Economics*, Homewood, Ill., Richard D. Irwin.
[50] Mutti, R. J. and Atere-Roberts, D. N. (1968) 'Marketing of Staple Food Crops in Sierra Leone', University of Sierra Leone, AID Contract No. csd-801, March.
[51] National Cooperative Development Corporation, *Marketing by the Farmer for the Farmer*, The Gondals Press, New Delhi, January 1967.
[52] Nightingale, Ray W. (1969) *The Modernization Decision in Indian Urban Fluid-Milk Markets*, Cornell International Agricultural Development Bulletin 15, Cornell University.
[53] Ogunsheye, A. (1966) 'Marketing Boards and Domestic Stabilization in Nigeria', in *Review of Economics and Statistics*, Vol. 48, No. 1, February.
[54] Parthasarthy, G. (1976) *Dilemmas of Marketable Surplus: The Indian Case*, a

paper prepared for the Seminar on Agricultural Price Policy held at Cornell University, Ithaca, N.Y. February.

[55] Rao, C. H. Hanumantha and Subbarao, K., *The Impact of the Agricultural Marketing System on the Small Farmers in India: A Case Study of Rice*, unpublished paper.

[56] Reserve Bank of India, *Development of Cooperative Marketing: A Survey Report*, Bombay, 1968. (Shri V. M. Jakhade, Adviser.)

[57] Satyapriya, V. S. (1973) 'Marketing of Arecanut in Mysore State', Ph.D. dissertation, Gokhale Institute of Politics and Economics, University of Poona.

[58] Schluter, Michael (1971) 'Differential Rates of Adoption of the New Seed Varieties in India: The Problem of the Small Farm', Occasional Paper No. 47, Department of Agricultural Economics, Cornell University, USAID-Employment and Income Distribution Project.

[59] Schluter, Michael and Mellor, John W. (1972) 'New Seed Varieties and the Small Farm', *Economic and Political Weekly*, Review of Agriculture, Vol. VII, No. 13, March 25, pp. A-31–A-38.

[60] Thoday, A. R. (1968) 'Marketing of Staple Foods in Western Nigeria', Stanford Research Institute, Menlo Park, Calif., AID Contract, AID/csd 601, 3 Volumes, March.

[61] Thoday, A. R. (1969) 'Marketing of Grains and Pulses in Ethiopia', Stanford Research Institute, *SRI Project IV-6350*, Report No. 16, April.

[62] Timmer. Peter (1975) *Food Policy in China*, December, unpublished paper.

[63] Tomek, William C. (1969) 'Stability for Primary Products: Means to What Ends?', Occasional Paper No. 28, Department of Agricultural Economics, Cornell University, USAID Prices Research Project.

[64] Weiss, Joseph S. (1971) *The Benefits of Broader Markets Due to Feeder Roads and Market News: Northeast Brazil*, Dissertation Series No. 24, Latin American Studies Program, Cornell University.

[65] Widstrand, C. G. (ed.) (1970) *Cooperatives and Rural Development in East Africa*, Uppsala: Scandinavian Institiute of African Studies.

OPENING DISCUSSION – A. J. Badillo, *Venezuela*

This paper is one of the best contributions to this Conference. It goes straight to the major role that agriculture should play in any market economy pursuing a strategy for industrial growth, namely, to provide basic consumption goods – food and fibres – at the lowest possible prices.

Mrs. Lele argues that in countries where agricultural prices tend to rise as a result of permanent supply-demand disequilibria, this major agricultural role is ordinarily operated through governmental pricing and marketing control mechanisms, leading to inefficient use of government and private resources. These conditions seem to fit most of the Third World economies. Because of this strictly economic failure of the usual government policies, alternative pricing and marketing mechanisms are necessary if agriculture is to play its role efficiently.

My comments deal mainly with the thesis she develops in order to explain the need for governmental mechanisms of price control and her interesting analysis of current pricing and marketing policies through the Venezuelan case.

Mrs. Lele describes the main objectives of pricing and marketing policies and evaluates the performance of current governmental mechanisms with respect the efficiency and degree of success in meeting those objectives:

(a) In respect of *reducing marketing margins*, according to Mrs. Lele, the alternatives of government or co-operative marketing have neither been very successful in guaranteeing narrower margins nor in improving competition. She argues that bypassing the considerable potential of the traditional marketing systems – in a clear reference to traditional mechanisms of marketing, with decentralized and multiple decision-making – explains at certain levels the failure of government involvement. I think that this fact cannot be generalized. Once a country is advanced enough in its first stage of industrialization (or import-substitution) strategy – as are some of the Latin-American countries – monopolistic marketing patterns arise with the setting up of monopolistic urban agribusinesses covering the whole domestic market. They may have high unused capacity. These industries exert price control through direct mechanisms (quotas, quality control, future price and other contractual arrangements) or indirect mechanisms such as governmental pricing policies or government marketing of agricultural inputs. Through these monopolistic or administrative mechanisms, prices are kept fixed, or growing at a lower rate than industrial prices. Thus, new systems with a high degree of centralization are involved, not traditional ones.

Commercial profiteering or speculative groups appear also in the marketing of unprocessed foods (food staples) in governmental sponsored central or terminal markets. These are known as speculative "roscas" in Venezuela.

Thus, it is not a traditional marketing system which sets a problem, but one whose objectives are in contradiction to those on which an industrial strategy is supposed to rest. Monopolistic or oligopolistic food processors and traders do not guarantee low margins but high and increasing ones.

In both cases, direct governmental control of agribusinesses (first-stage industries) and marketing margins is necessary, in addition to other mechanisms leading to productivity increases (agricultural and marketing infrastructure adjustments, research, technical assistance, etc).

Improving competition can have a significant impact on pricing efficiency. Direct mechanisms to check excessive margins and improving competition through co-operatives, or agribusinesses owned by groups of farmers, seem to be practicable when diversification of capital toward more profitable financial, commercial and industrial areas is operative.

In the meanwhile a contradiction arises between the low food prices needed and the high margins sought in monopolistic marketing and agribusinesses. This contradiction can be reconciled only under the following conditions (i) when profiteering commercial groups are not emphasized as necessary in an industrial strategy; (ii) when agribusinesses start to lose attractiveness to owners of capital who are eager to invest in more profitable areas.

(b) *Price and supply stability* – I agree that the only effective mechanism is a governmental one.

(c) *Ensuring a low cost supply of food to the urban sector* – Ensuring supplies of basic commodities to marginal or unemployed sections of the population has not been a major government objective in most Latin-American countries. Keeping marginal population from starvation is important in pure governmental or entrepreneurial economic terms and not only on welfare grounds, but what seems relevant to an industrial strategy and to capitalistic industrial groups is the wage rate and the effect of food prices on it.

Agricultural supply shortages and high prices almost always lead to demands for higher wages, but Mrs. Lele's reasoning on price control gets around the point. Price increases lead to potential wage increases; the possibilities of making them effective depend on the specific labor-relation conditions and potential losses in capital gains or profits. That means that profitability of urban investments can be upset by pressures for higher wages, with the consequence that the pace and pattern of economic growth may be affected. Thus, price control is not only justified on pure welfare grounds but on governmental or industrial economic reasons, also.

If possible, industries will tend to substitute capital for labor, with consequent adverse effect on the employment rate. Thus, worsening of labor-capital conflicts arising from increasing food prices and potentially increasing wages are avoided through governmental price control.

RENE BENALCAZAR R*

New Techniques, Agricultural Extension Services and Credit Facilities as Instruments of Economic Development, with Special Reference to Latin America

1. FOCUSING THE PROBLEM

In writing this paper one can be misleading because the best policies in the interests of rural development do not always coincide with the best policies for the development of the economy as a whole. Therefore, I have amplified the scope of the paper, using the concept of economic development instead of rural development. It is convenient first to point out the general agricultural policy objectives in the furtherence of the economic development of a country.

2. AGRICULTURAL POLICY OBJECTIVES

The following objectives could be suited to any country whatever its degree of development; although priorities would differ with the particular social, economic, and political conditions.

(i) To produce food in the quantity, quality and on the timing appropriate to the needs of the population so far as it is economically possible.

(ii) To produce agricultural raw materials, in terms appropriate to the demand of national industry, in so far as it is economically feasible.

(iii) To produce food and agricultural raw materials for export when the level of international prices makes it profitable to produce them.

(iv) To maintain an increasing rate of farm productivity.

(v) To generate a level of farm income and wages that permit the satisfaction of the needs of the rural population, at a similar level to the non-farm sector. Such a level of income should permit also a rate of saving and investment to promote a rate of growth of the farm sector similar, or superior, to that of the non-farm sector.

(vi) To increase the income of the lower agricultural groups faster than the high income groups, so as to reduce the range in income distribution.

(vii) To give full employment to the rural population, permitting their

* Ecuador

transference to the urban sector, whenever the level of technique allows it, and in accordance with the labour demand of the urban sector.

(viii) To conserve and improve natural resources so that they can be more useful in the future.

Even though there could be contradictions between these objectives within a general agricultural policy, it is also possible that with due regard to priorities – according to the social, economic and political conditions of each country – such objectives could be complementary to each other. The priorities and emphasis given by each country to those objectives will depend on the power structure in its government at the time concerned.

3. PRINCIPAL CONSTRAINTS TO ACHIEVING THE OBJECTIVES

3.1. *Land tenure*

To achieve the first four objectives related to production of food and raw materials for domestic consumption and for export, and the increase in production and productivity, requires optimum combination of land, labour, capital and management at a higher technical level to maximise revenue. Such an optimum combination will depend essentially on the size of farm units. In other words, the relation of man to land is a key factor in the rate of development of the agricultural sector. That relationship, unfortunately, is inadequate in Latin American countries, with few exceptions. Furthermore, the size of farms is tied to inflexible private or public ownership and so the problem cannot be solved in the short-run, and perhaps not in the long-run, except through a revolutionary change.

To illustrate the situation, in Latin America some 47% of the total number of farmers hold farm units of less than 5 hectares each and occupy only 0·7% of the total farm land. On the other hand, only 1% of farmers (approximately 100 000) hold 60% of farm land in farm units of over 1000 hectares.*

Farm units of less than 5 hectares present the following obstacles for development:

(i) There is an excess supply of farm labour which is offered to other farmers or to urban areas or simply remains as under-employed.

(ii) Such a low income is generated that it does not permit an adequate satisfaction of needs, especially of food, and this lowers the working capacity of farmers. Savings for future improvements are not possible, either.

(iii) The use of machinery to free labour from farming is not practicable on such small farms.

(iv) Farmers cannot go for higher education to accelerate farm labour mobility to urban areas.

(v) New subdivisions of land by inheritance continues the vicious circle of poverty from one generation to another.

(vi) Small farms make any programme of technical and financial assistance

* *Estructura Agraria de las Naciones Americanas* – Instituto Interamericano de Estadistica – 1961.

costly and ineffective, because a large number of people cannot improve their working capacity on such small holdings.

On the other side of the problem, big land holdings of over 1000 hectares present the following hindrances to development:

(i) The maximum use of land is hampered because of lack of capital. It is difficult for one family to accumulate capital needed to farm intensively 1000 has.

(ii) It ties farm development to the family cycle, causing stagnant and declining periods during the old age of the owner.

(iii) It make for an uneven distribution of income among farm owners and farm labour.

(iv) It impairs the social, cultural and economic development of farm labour working for the big owners.

(v) It discourages any utilization of new techniques, especially capital intensive ones, because there is ample labour at low wages.

(vi) It gives no encouragement to new inventions because owners are usually absentees in urban areas and are not sensitive to production problems.

The minifundia and latifundia problem was inherited from Indian and Spanish cultures and, since the colonial period, Latin America – with few exceptions – has been unable to solve this problem. Although during the 1960's almost all countries have passed agrarian reform laws, the land tenure systems remain almost unchanged with the exception of Cuba, Peru and Chile, which have undergone very important changes. Since the structural changes are slow and difficult, as history has taught us, we have to orientate agricultural research, extension services and credit facilities in the light of our deficient land tenure system which is a hindrance to any real effort for rural and economic development.

3.2. *Power structure*

A second important factor in achieving sound agricultural policy objectives is a well balanced power structure in each government. Policy-making groups should be constituted from representatives of farmers, merchants, manufacturers, consumers, and workers, in such a way that their decision power should be the synthesis of all national interests. In addition, governments should follow a sound method for reaching policy decisions, which should include the following steps:

(i) discussion of policy objectives;

(ii) a search for alternative solutions to problems in achieving such objectives;

(iii) a search for resources needed for the resolution of those problems; and

(iv) to be able to control such resources for use according to the decision taken.

Using such a method, in a well balanced policy decision group, the resulting policy ought to be in the interest of the majority, setting up priorities to achieve a sound economic growth in the long-run.

In that way, the method could guide the setting up of a framework of:

(i) stability, in the sense that the laws and regulations which determine the rules of the game are well known and more or less permanent;

(ii) flexibility, to adapt such rules to changes in the social and economic conditions;

(iii) efficiency, so that it permits a thorough knowledge of each problem, taking into consideration all the variables involved in its discussion and resolution – using the procedure set out above;

(iv) effectiveness, to take opportune action within the time limit the population requires.

Unfortunately, the power structure in Latin American countries is conditioned by the land tenure and land distribution and that of other capital goods related to industry, mines, trade, etc. Therefore, the average Latin American country does not enjoy a well balanced democracy where the policy decison groups are representative of the interests of all sectors in harmonious proportions. Almost all Latin American countries are going through an institutional crisis; Colombia and Venezuela are the only democracies in South America. The rest are governed by military dictators with professional rather than national interests. They do not take into consideration the real motivations, needs and desires of the majority who are participating in the daily work of the farm sector. Consequently, we should take into consideration the deficiencies in the power and government structure as a given factor which cannot be modified in the near future.

3.3. *Other incentives for farmers*

It is convenient now to single out some basic requirements if farmers are to be stimulated to accepting programmes of rural development, based on new techniques, extension services and credit facilities.

Farmers require security, especially for long-run farm family plans; security in respect of outcome of investment, production and prices in holding land and in the ownership of other capital goods. Risk and uncertainty are deterrents to farm investment. Such security is far from being present in most Latin American countries. Agrarian reforms that never end are always threatening large, medium and even small farmers. Most of the countries do not have a price support programme. There is a lack of storage facilities. Transport facilities are not adequate for perishable products. There are no adequate price information systems to guide farmers' decisions on when and what to produce. Farmers are in general isolated, depending on their own resources, and are often misled by absentee big landlords who do not know the real problems of the medium and small farmer.

4. NEW TECHNIQUES

Agricultural research in Latin America, as in most developing countries, has followed the pattern of research in North America and Europe. The research methods are experimental, using induction and statistical methods to generalise results.

The Latin American agricultural researchers, with some exception, have not focused their research towards their own real conditions. The problems of what crops require research, how and why, have been almost forgotten in many experimental stations, because they have followed the European or North American patterns. Experimental stations have usually been under the Ministry of Agriculture, and the Ministry most often is associated with big landlords and most of the researchers come from medium and high income groups with professional interest rather than the idea of service to the people at large. Thus the selections of groups to be improved and the techniques of research have been mostly exercised in the interests of big landlords, merchants, exporters and not for the well-being of the domestic consumer and producer.

Using the agricultural policy objectives already set out, the following criteria could be used to select an order of priority for the crops to be researched.

(i) Crops produced domestically that supply most food for domestic consumption.
(ii) Crops which most farmers are engaged in producing.
(iii) Crops with the largest area under cultivation.
(iv) Export crops which earn the largest amount of foreign exchange.

The history of experimental stations in Latin America leads to the conclusion that the last two criteria have been the basis of priorities. In the past, attention has been given in the main to coffee, cocoa, sugar cane, and bananas – all export products – rather than to beans, green peas, broad beans, maize or rice, which are basic food crops for our people. Only in the last two decades has more attention been given to basic food crops, such as corn and rice. Research on wheat and barley has been mainly in developed countries. Thus one of the most important features in research is the careful selection of crops whose improved techniques could benefit the pockets of the majority of the country's domestic producers and consumers.

The solution, as we have indicated, will depend on the power structure in government; and as we have said, the prevailing structure is not the best. By coincidence one crop may be the main food supplier and at the same time most farmers may have been engaged in its production, and it has also occupied more land – but it is not a usual case. More often export crops like coffee or cocoa are not for general consumption.

The second problem in agricultural research is the selection of techniques which are going to be researched for teaching to farmers. There are techniques aimed at substituting labour for machinery; others to increase land productivity; another to get more production out of labour. Extensive farming with low use of capital and labour is characteristic of big holdings, and in many experimental stations more attention has been given to techniques applicable to kinds of pasture, banana, sugar cane, beef and dairy cattle, rather than to techniques on horticulture, fruitculture, seeding distances, intercropping, rotations, etc., that are labour intensive techniques. Little attention has been paid to indigenous crops of the Andean region exceptionally rich in nutrients such as quinoa,* chocho,† and melloco which are almost unknown in experimental stations.

Another problem regarding research is that of coordination with extension services, credit facilities and, especially, with the main interest of farmers. Scientists often spend their time in research for research's sake, or for personal prestige, rather than for the wellbeing of those who are paying their salaries through taxes collected by government.

In summary, I consider that selection techniques should follow the main objectives of the agricultural policy in giving priority to:

(i) the practices which could be most easily understood and applied by most farmers.

(ii) those techniques that are less costly and that produce higher returns for most farmers, and

(iii) More sophisticated techniques for the use of big enterprises.

5. AGRICULTURAL EXTENSION SERVICES

In a generally accepted concept Extension is considered as "an informal, out-of-school, educational service for training and influencing farmers and their families to adopt improved practices in crop and livestock production, management, conservation and marketing".‡

According to this concept it is not only concerned with teaching and securing the adoption of a particular improved practice, but with changing the outlook of the farmer to the point where he will be receptive and, on his own initiative, will continuously seek means of improving his farm business at home. In rather marked contrast to this, is the concept of "Extension as an organization responsible for all Ministry of agriculture activities at the field level, including regulation, the provision of services, the collection of statistics, and with very little attention paid to the educational activity". In addition, there is the view that Agricultural Extension must be conducted as part of the broad extension efforts, going beyond the scope of the Ministry of Agriculture and involving health, education and other factors in the life of farm populations. To us, the main purpose of extension must be the dissemination of information derived from agricultural science and technology and it must be seen as part of an educational process directed to each farmer to improve his production and raise his standard of living.

The effectiveness of extension is measured by the ability of extension activities to change the static situation which prevails in rural areas into a dynamic one. One of the problems extension must overcome is the tendency of rural communities to resist changes. This tendency accounts for indifference, if not the hostility, farmers may show to any pressure of progress.

* Species of goosefoot

† Lupine.

‡ Science and technology for development, United Nations (1963).

Successful application usually requires changes in people's knowledge, attitudes, skills and habits and in the way people and their work are organized. It is true that relatively little has been done in the way of sociological, anthropological and psychological research on methods that may be effective in securing the adoption of new methods by farmers in various cultural and economic-political environments.

Latin American countries are facing the problem of the target for extension services: should they be directed to more than four million smallholders of less than 5 has. each, to 100 000 big landlords, or to the rest of medium owner operators. The teaching techniques required are different for each sector of the public. The big owners are urban people with medium and high educational levels; the smallholders are rural residents – illiterate or, at most, with primary schooling. Perhaps the best alternative would be to concentrate attention on medium size owner operators on farms larger than 5 has. and smaller than 200 has., and leave agrarian reform to solve the problem of latifundia and minifundia in the future. The alternative of giving more attention to medium size farms is convenient as the number of extension agents is quite insufficient – less than one for 4 000 farmers* in most of the Latin American countries.

In the face of those facts, and considering the agricultural policy objectives set out above, extension services should take into consideration the following suggestions:

(i) to teach groups and not individual farmers, in order to lower operational costs;

(ii) to use mass communication media, especially radio, to reach as many farmers as possible at the same time;

(iii) to select simple and more profitable techniques which can be easily understood and followed by farmers;

(iv) to prefer medium size owner operators;

(v) to use local leaders as extension agents to get acceptance of new ideas and facilitate communication between farmers and technical people; and

(vi) extension agents should teach small farmers how to organize in cooperatives or any other societies to facilitate introduction of new techniques.

6. FARM CREDIT

6.1. *Sources of finance*

Farmers' own savings finance most of the inputs, machinery, cattle, land improvements and many other investments in farming. Banking credit, or loans from dealers, finance only a small part of farm investments. In the case of Perú and Ecuador, for instance, banking credit does not finance more than 20% of total annual farm investment. I doubt that in any developing country farm credit finances more than 40% of total investment.

In spite of the small proportion financed by banking credit, it plays a very important role in rural development, for the following reasons:

* *Estado de la Agricultura y la alimentación*, FAO (1961).

(i) It is a complement to savings and investments and so credit accelerates the rate of capital formation and technical improvement.

(ii) Banking credit is usually selective as to farmers as well as to crops and techniques, therefore, the marginal efficiency of this capital is higher than that of farmers' own investments.

(iii) Loan conditions in terms of amount, maturity and repayment, are usually according to farmers's repayment capacity from farm revenue, generating additional income.

(iv) Interest rates are often lower than the free money market and loans are a kind of subsidy to farming. Inflation makes such subsidy higher and, in many cases, the interest rate becomes negative, with the exception of countries like Chile, Argentina, Brasil, where there is an automatic adjustment of debts.

7. THE MAIN CONSTRAINTS AND AGRICULTURAL CREDIT PROBLEMS

Providing farm credit is a costly and risky operation, and private banks, in many developing countries, are not interested in serving such a sector. For this reason most Latin American countries have state development banks specializing in agricultural credit. These banks face the same problem that researchers and extension agents face in selecting their clientele from among large numbers of small farmers and big landlords. To solve such problems some banks have established two types of credit: supervised credit for the small and some medium family farm owners, and banking credit as such, for large and medium size farm owners.

Supervised credit for the small farmer is a combination of technical and financial assistance. It finances integral farm plans to introduce permanent improvements, taking into consideration the farm family. Conditions for such loans regarding amount, security, interest rates and repayment periods are usually softer than for banking credit. Supervised credit programmes were set up in some South American countries during the 1960's. In Ecuador, for instance, a programme began in 1966, and the results measured by three evaluations show a success in terms of the amount invested, income received, and repayment, which were higher than what was planned.

Another problem which the development banks are facing is credit orientation, not only to small and big land owners but to specific areas and crops. Such orientation depends on the political structure of each country. Often those banks are governed by pressure groups of big farmers who direct the loans towards large commercial farms and for export crops.

To obtain resources for agricultural farm credit is perhaps the main problem for the state banks financed usually through state budget. When governments take the decision to finance farming by subsidising interest rates, it is difficult to get savings in the free money market for such purpose. Resources go to activities with a high rate of return such as housing, trade, industry; farm credit being costly and risky cannot compete in that market. Resources from international financial institutions have been a very interesting source of farm credit, especially for solving the foreign money scarcity, but unfortunately the amount for disposal is very limited.

To attack this problem countries like Columbia, for instance, have established a system for capturing public savings through saving and current accounts. Some 60% of total bank savings are channelled by Caja Agraria Industrial y Minera in some 650 banking offices throughout the country. Ecuador set up a similar system through the Banco Nacional de Fomento and public savings increased from US 10 million dollars in 1968 to US 100 million dollars in 1975. In the same period private banks increased their deposits only three times. Those examples show the possibility of using more savings and current accounts in state banks for financing agriculture.

Another of the serious problems which the state development banks are facing is how to obtain coordination between different agencies in charge of the execution of agricultural policy. The decisions on what crops to research and what techniques to teach are usually taken by Agricultural Ministries' personnel; but the decisions on what products to finance and what farmers get loans lie with bank personnel. Besides, farmers request loans not for the purposes set by Planning Boards or Agricultural Ministries, but for the crops which farmers consider to be most profitable. Therefore, after the completion of each agricultural national plan every evaluation of its achievement has shown how far the actual production figures are from what were supposed to be reached under the plan.*

Extension agents have been in charge of promoting some crops and techniques in some countries, but usually the result has been insignificant unless the preaching of extension agents has been accompanied by increasing prices or good weather conditions. This statement does not mean that I am in favour of the laws of demand and supply ruling agricultural production. Planning is a necessity for any approach to balanced agricultural development; but planning should be more on the side of the policy-decision-making process rather than in the pure planning figures.

To avoid the problems which arise from the participation of many organisations in policy decisions and their execution, the latest idea has been to organize so called "package programmes". This is discussed elsewhere in the conference but from what I saw in Ethiopia, package programmes are not the solution for planning and agricultural development at national level. Those packages are bad substitutes for adequate government organization in farm planning and policy decision systems. Consequently, agricultural economists should pay more attention to power structure and the policy decision process, rather than to planning techniques alone. In this respect it is convenient to set out some ideas for discussion here.

(i) In agricultural planning more attention should be paid to general policy objectives than to specific production figures.

(ii) Farmers, marketing firms, consumers and government officials, representing the main executive agencies, and research and extension services and

* If we are going to be serious we should write as a final footnote in every economic plan "Any similarity to real life is pure coincidence".

farm credit agencies, should participate in discussing general agricultural policy objectives.

(iii) Basic data for planning purposes should be collected and prepared by a central specialized planning agency.

(iv) In programming agricultural development each objective should be accompanied by a set of policies to stimulate or otherwise steer farmers to follow the guide lines set up in the plan. The policies and the plan should be well-known to farmers, marketing firms, agroindustrialists, and all institutions operating in the farm sector.

(v) The functions of researchers, extension agents and development banks should be clearly defined, avoiding all interference among them. Overlapping of functions, especially between technical assistance and farm credit, is a source of friction, misunderstanding and distrust between farmers and government officials.

(vi) In dealing with credit and technical assistance, this last concept should be divided in two aspects: technical assistance in farm management for investment purposes – which should be taught by banking personnel – and the farm practices – which should be taught by extension people at group level. In this way all the responsibility of credit will remain in the banks' hands and it will not be shared by extension agents who have a different philosophy with emphasis on agricultural development. They are usually most interested in production increase, whereas bank staff are more profit-minded, in line with the farmers' interest.

Finally, the more important problem in getting adequate policy decisions is the method and process employed. Method was discussed earlier, now it is convenient to set up criteria for the process in policy decisions.

(i) One of the most important tasks for an agricultural economist is to single out the issue and to present the problem as clearly as possible so that it is fully appreciated by the pressure groups concerned.

(ii) All the variables involved in the problem should be analysed pointing out the behaviour of each of them in order to make the problem understandable to everybody.

(iii) The participation of all the pressure groups should be secured in the discussion of the issue, making sure that they know the essence of the problem.

(iv) Alternatives for the solution of the problem should be presented and thoroughly discussed.

(v) The discussion should be steered towards the most adequate resolution of the problem, by evaluating the results of the possible alternatives and showing the way in which the interest of the pressure groups would be affected by it.

Unfortunately, this scheme can only be applied in a democratic organization. In military dictatorships something could be done in the above sense – although within very narrow limits. Therefore, more attention should be paid to power structure and politics, to get done what agricultural economics theory teaches us.

REFERENCES

Bićanić, R. (1967), 'Agriculture and the Political Scientist', *International Journal of Agrarian Affairs*, Vol. V, No. 2, May, pp. 157–182.

Hardin, L. S. and Collins, N. R. (1974), 'International Agricultural Research: Organising Themes and Issues', *Agricultural Administration*, Vol. 1, No. 1, Jan., pp. 13–22.

Howse, C. J. (1974), 'Agricultural Development without Credit', *Agricultural Administration*, Vol. 1, No. 4, Oct., pp. 259–262.

Hunter, G. (1974), 'The Implementation of Agricultural Development: Toward Criteria for the Choice of Tools', *Agricultural Administration*, Vol. 1, No. 1, Jan., pp. 51–72.

Mosher, A. T. (1971), *To Create a Modern Agriculture Organization and Planning*, Agricultural Development Council, New York.

Simantov, A. (1973), 'Economic Social and Political Priorities in Agricultural Policy Formulation in Industrialized Countries', *Proceedings of the 15th Conference of the International Association of Agricultural Economists, São Paulo, Brazil*, pp. 76–100.

Spiro, H. (1970), *Politics as the Master Science from Plato to Mao*, Harper & Row, New York.

Tiffen, M. (1974), 'Timing as a Factor in the Success of Extension Programmes: A Nigerian Case Study', *Agricultural Administration*, Vol. 1, No. 2, April, pp. 125–140.

United Nations (1963), *Science and Technology for Development – Vol. III – Agriculture*, Report on the United Nations Conference on the application of Science and Technology for the benefit of less developed areas.

DISCUSSION OPENING – M. Ngeze, *Tanzania*

In this paper the writer shows us a deplorable current state of affairs as regards land tenure system in Latin-American countries, agricultural system, administration of agricultural extension and credit facilities, agricultural research, power structure, etc., as instruments of economic development, and then presents suggestions, which, if accepted by governments and all concerned, he believes would speed up economic, and especially rural, development.

Dr. Benalcazar divides the Latin-American countries into two main groups; those which he refers to as being under dictatorial rule, examplified by Cuba, Peru and Chile (which I prefer to call revolutionary countries) and those with democratic rule, namely Columbia and Venezuala.

My impression from reading the paper is that the writer does not believe in, or sympathise keenly with, socialist aspiration.

The main challenge to those in government in developing countries is to choose between (i) continuing to assist those few who are already developed and enjoying a good level of living and (ii) paying more attention to the majority of the population who are still poor. This is the same thing as asking what is the purpose of rural development? Should governments in developing countries continue to develop a few individuals, who are already developed, living in towns but being called farmers? The grounds presumably would be that they own farms of over 1,000 hectares and employ hundreds of labourers who are being paid the minimum government wages of, say 150/- per month. Or should the purpose be to "to improve the living standards of the mass of the low income population residing in rural areas and making the process of their development self-sustaining"? (this is to borrow Uma Lele's definition of rural development). The choice will depend on the political ideologies in each country, the socialist countries will choose the latter and the capitalist countries the former.

With the limited resources available to developing countries, we cannot expect that the level of living of the masses will be improved through employing new techniques of agricultural production, agricultural extension, and credit facilities, so long as (a) these things continue to be accorded to those who do not need them very much, (b) virtually the land is in the hands of few individuals (c) there are urban dwellers who are farmers only in the sense that they own large farms worked by hired labour.

What is the future of those thousands of lowly paid farm labourers? These are the very ones who know very little about family planning. Are they not part of the rural population which rural development is all about? Is it not natural for these people to pray for a revolution? Could they also not be farmers and masters of themselves if assisted by government?

These are some questions not found in the paper. There is, for instance, no suggestion that there might be some form of co-operative farming system (such as block-farming) as one way to solve the problems of small farms. Although the author lists several disadvantages of large farms under absentee landlords he does not suggest what action is needed in relation to them. The paper does not suggest how to go about solving the problems caused by a few people owning the whole land and the majority being landless. To focus on essentials, –

(i) The writer tells the disadvantages of having small farms of less than 5 hectares and even also the disadvantages of big farms of over 1,000 hectares. He should tell us what he thinks is the optimum size for the different categories of farmers in Latin American countries.

(ii) The writer does not tell us the size of a farm which a normal family can manage.

(iii) He does not tell us the amount or resources which are available to the average family farmer, e.g. tractors, ploughs, and if every one can afford to hire labour.

In concluding my points I would like, once again, to praise the paper and mention that in developing countries, besides the suggestions he has made, there must be a revolution in land tenure system. The land should belong to the state and each citizen who can make economic use of it should have a right to it. Secondly, the purpose of development in these countries should be MAN – not a few men, but the people.

DISCUSSION OPENING – P. F. Philipp, *U.S.A.*

As Dr. Benalcazar mentioned, he changed his topic from "The agricultural extension services, new techniques and credit facilities as instruments of rural development", to "New Techniques, agricultural extension services and credit facilities as instruments of economic development, with special reference to Latin America." I want to open by referring to some consequences of the substitution of the words "economic development" with "rural development" in the paper.

Economic development deals with all income groups in a country while rural development, according to recent definitions – particularly by World Bank representatives – has a more limited target group. For example, Uma Lele, in *The design of rural development:lessons from Africa,* (John Hopkins University Press, 1975) uses the following description for integrated rural development: "Improving living standards of the mass of the low-income population residing in rural areas and making the process of their development self-sustaining." According to M. Yudelman's paper this morning rural development also implies "giving the target group the opportunity to enjoy the benefits of development; improved education, health, and nutrition, among other things".

Dr. Benalcazar's change of his topic from rural to economic development greatly affects policy conclusions. Take for example his question "to whom to direct the extension services in Latin-American countries"? Should the extension service serve the more than four million minifundia farmers operating less than five hectares each, who make up 47 per cent of the total number of farmers and hold only 0.7 per cent of the total farm land, or should it serve the 100,000 latifundia farmers with 1000 hectares or more each, who comprise only 1 per cent of all farmers but hold 60 per cent of all the farm land? Dr. Benalcazar comes to the conclusion in his written paper "To concentrate attention to medium-size owner operators on farms larger than 5 hectares and lower than 200 hectares", and let the future solve the problem of latifundio and minifundio. Had Dr. Benalcazar dealt with rural rather than with economic development, he would have had to conclude that the extension service needs to direct its attention primarily to the smallholders who clearly represent the mass of the low-income population. Similarly he would have had to reach different results with regard to new techniques.

Let me now contribute to Dr. Benalcazar's statement in his concluding sentence "to get done what agricultural economics teaches us in theory". I would like to discuss some of the practical issues and problems of agricultural extension, new techniques, and credit as instruments of rural development. Not withstanding the broad definition of rural development to which I previously referred, when we talk about the effect of extension, new techniques and credit on rural development, we obviously think primarily of the impact of these inputs on improving production.

The author seems to have some reservations about some aspects of the so-called "package program" in rural development, but he would probably agree that a complete package is needed to bring about improved output. This package must include such varied components as knowledge of the physical, biological, human, cultural and social conditions on the farm and in the village; farming know-how, markets, transportation, etc.

To expect a substantial increase in production from extension, new techniques and credit, these inputs need to be acceptable to the farmer. Let us look at extension first. What is agricultural extension? Dr. Benalcazar gives a good review of various definitions of extension. I prefer the following slightly different and simple definition which is common in the United States "Agricultural extension means helping farmers to help themselves".

To be of any use, an extension agent must be acceptable and not potentially harmful to the farmer. For example, in some parts of developing Asia, the extension agent is, as Dr. Benzlcazar puts it "responsible for all Ministry of Agriculture activities at the field level *including* regulation.....". To many an Asian farmer such a bureaucrat is a potentially harmful person – not unlike the tax collector. Add to this that the extension agent is often young, which is against him in many Asian cultures, or that he may extract payment from the farmer for his services. Finally, the farmer may ask himself whether such a city-trained extension agent, who frequently does not even want to get his hands dirty, really knows the practical problems of what he tries to teach. In short, is such an extension worker really acceptable to the farmer?

How about the new techniques which the extension agent tries to introduce? Are they really profitable at the farm level? True, the yields were much higher on the far-away experiment station, but production methods, conditions, know-how and risks are very different on the farmer's own farm.

And how about credit? Scarcity of subsidized credit, mentioned by Dr. Benalcazar, may not always be such a limiting factor. If a new technique has proved profitable in the village, the farmer would probably be willing to pay an unsubsidized rate of interest. However, will the promised government credit actually be available to the farmer in the village, on time and at the government rate, or will bureaucratic inefficiency and corruption make a shambles of the government system?

In summary, agricultural extension, new techniques and credit facilities are, of course, potentially vital instruments of rural development, but let us use realistic coefficients and not omit relevant variables in our production function models.

RAPPORTEUR'S REPORT – H. J. Plunkett, *Australia*

The general discussion, in effect, posed a range of questions in the process of clarifying issues.

One group of issues concerned political will, in one form or another. Thus the possibilities of development might be limited by insufficient political will to give concessional credit to appropriate target groups. Similarly a solution depended on the will to undertake land reform and to carry it through.

Possibly political will was concerned in the discussion of why there should be conflict over the needs of disadvantaged groups under military dictatorships if the latter only arose when there were domestic problems not solved by the existing political system. In these respects the author stressed that his concern was primarily with the role of other agricultural economists in identifying problems and proposing alternatives for action within the operative political system.

The proposed scope and priorities of research and extension were seen by several speakers as unduly narrow or unbalanced. For instance, the paper tended to give research priority to crops and agricultural technology without adequate attention to the social needs of farmers. Similarly it was questioned why extension was confined to agriculture and omitted health, education, etc. Indeed, doubts about the concentration of extension on middle sized farmers might fall under the same heading. It was noted that the arguments for some priority to this group was that they had the most readily realised potential.

The usefulness of seeking greater application of mechanised techniques in order to release labour was widely doubted in a country with high and expanding population. Prestige tended to overweight the attraction of mechanisation.

Finally, in relation to credit in addition to restraints from political hesitancy and generalised administrative inefficiency, the effect was noted of those administering it themselves probably wanting credit and being reluctant to help others to get it on concessional terms.

JOSEPH J. RICHTER*

*International Trade Models and Product Studies as a Base for Worldwide Decision-Making in Agriculture**

I

Decision-making in agriculture is of a nature so complex that for lack of knowledge of the impact of alternative policies, one is inclined to adhere to the rule that only measures which have consequences one can control with certainty should be applied. This attitude leads toward rather narrow concepts and limited horizons. Most agricultural policies consequently remain in the framework of a national state. This attitude is also reflected in agricultural economics and particularly in analysing and forecasting international trade.

The objectives of decision-making in agriculture on a worldwide scale are intricate and must be comprehensive because they affect very diverse facets of social and economic policies designed to improve the state of nutrition in all nations. As one country or region differs from others in endowments, resources for agricultural development, and production technology (which is expressed in a concept of comparative advantage), commodity trade should assume a predominant role in decision-making in worldwide agriculture. However, in the past, trade and commercial policies have suffered from the absence of economic tools to guide decision-makers. What intelligence was available rested on rather short run observation of markets and prices. International trade models and product studies were designed to remedy precarious situations and to provide a better foundation for policies, commercial as well as agricultural, for all countries.

The praxis of nineteenth century international commodity markets took it for granted that the instruments developed to facilitate trade functioned perfectly in the short run as well as in the foreseeable future. Whenever and wherever a commodity exchange was allowed to function and interference in the automatic system of adjustment was absent, trading provided for conditions best described as a dynamic equilibrium.

The three main components of the trade instruments on commodity exchanges were:

(a) the gold standard as the basis for international adjustment of the value of national currencies,

* University of Alberta, Edmonton, Alberta, Canada.

(b) the interaction of spot and futures trading activities, and

(c) the perfection of a worldwide communication system and wire service and newspaper reporting of prices and quantities traded.

Closer scrutiny of commodity exchanges in the past and in the present may lead us to question whether commodity exhanges have ever functioned without "product studies" and without concepts implied in "international trade models", albeit economists and traders in the past did not use the nomenclature nor the methodologies of research and analysis dealt with in this paper. However, there are other notions which merit special mention here. In the past, trading was regarded as an affair of the private sector of the economy. Governments did not or should not have interfered in the autonomous adjustments and mechanisms of international trade. It is, nevertheless, surprising that just about the time that the belief in the perfect market concept prevailed (in the second part of the last century and after the repeal of the Corn Laws in England), the need for economic research in support of commodity trading was felt much more than during the period of restrictive trade practices in the mercantilist era.

In the 1880's, the United States established an office in London to report on commodities, production, prices and trade. Today's United States Department of Agriculture commodity and country reports have their forerunners in these reports. Together with crop reports for the country itself (acreage and expected yields), USDA reports constitute a most important source of intelligence which affects price quotations on commodity markets in Chicago, New York, and other places in the United States. Most of the other countries participating in international trade feel compelled to observe, record, and forecast production and prices for commodities of special interest to them. Communication between commodity exchanges (Chicago, Minneapolis, Kansas City, London, Amsterdam, and Singapore are only a few examples) provides a network of international contacts which remains indispensable for the present and foreseeable future.

One cannot claim that the marketing system described lacks an economic and scientific foundation. Prices of commodities are the main feature in the information package. They allocate production as well as income and thus affect decision-making in agriculture all over the world. This applies to all countries regardless of their internal organisation of production processes (free market economies or centrally planned countries).

What is to be discussed in this paper is why the apparently perfect system of commodity exchanges did not and does not work satisfactorily and why it does not fully reflect the reality of the economic life and behaviour of participants. Even greater attention will be given later to the discussion of models and studies which could support worldwide decision-making in agriculture in the long run. This is particularly necessary because although the system of commodity exchanges (spot and futures trading) provides satisfactory information for short run conditions, it does not provide a data base for long run investment criteria nor for future structural adjustments in either production of agricultural commodities or the infrastructure for marketing food and fibres.

A second, often more important reason for rethinking the problem is the fact that an increasing percentage of international movements of agricultural products and food can be observed outside the traditional marketing system, bypassing commodity markets and exchanges as we know them and thus following criteria we might have neglected in the system of intelligence in support of trade and investment. Criteria which determine P.L. 480-based operations, considerations governing U.N. Food Aid, and a host of bilateral agreements are valid examples of "non-market", not price determined, movements of commodities in international trade.

A third feature in commodity trade not accounted for in the past are events which have fundamentally changed the scene, such as political decisions which have influenced regionalisation (EEC and COMECON; the OECD; the block of petroleum producing countries; or developing nations). In addition, economists have not considered such fundamental changes as the commodity crises of 1973/74 and the impact of considerations related to the protection of the environment. All these are events after which one hears more and more views that the agricultural economist's tool of economic analysis "throws but a partial light on the issues which are the most critical in the planning of world food supplies". At the same time there is convincing evidence that tools or techniques based on neoclassical theory are valid if only one can examine and project in an economic framework which includes factors which have been neglected in the past.[1]

II

International commodity trade studies fundamentally tend toward spatial equilibrium models either of the nature of a comprehensive analysis or partial analyses by commodity or by specific region. The underlying idea of this approach is that the interaction of demand and supply in the long run will be subject to forces which will neither produce surpluses nor leave people starving. This allows for interpretations which are somewhat Malthusian but at the same time express expectations that the food needs of the world population will be satisfactorily met.

Demand (as a rule, a simplified version of population and income growth projection) is mostly an extrapolation of time series. Therefore, two fundamental difficulties arise. The economist has to rely on statistical data gathered in a period of extraordinarily high growth rates. As already indicated, the first comprehensive projections of population growth prepared by the U.N. after World War II should be revised. The probability is great that projections should be scaled down considerably, especially for the more developed countries, the EEC and the COMECON. A second rather doubtful assumption that has crept into the demand analyses is that higher per caput income will be associated with somewhat uniform consumption patterns all over the world.

Even stronger are the reservations expressed in regard to the general pattern of supply and its impact on international trade. Countries and regions are so diverse so different when we assess the relationship between population and endowment in terms of land, water and other resources that generalized extrapolations of past trends in production probably have proven more misleading than helpful in explaining future production and trade patterns for agricultural commodities. For example, the USSR apparently decided to improve its nutritional standards and to participate in international trade as a buyer of wheat when the vagaries of its climate (something forecasters of agricultural production also failed to observe) caused production to fall 25 percent below trend. Other cases of such unforeseen but vital elements are the global shortage of fertilizer and the price increase of oil and gas. It is certainly possible to increase the number of observations and to account for greater amplitudes of year-to-year changes in factors such as climate, as well as socio-economic and political factors, which affect demand and supply at the international level. Modern computers can easily accommodate the additional data input.

One may claim that such an expansion of the data input may render the results of the analysis less reliable because of the widening of the range of probable solutions. Minima and maxima could easily be outside reasonable probabilities. But one is reminded that the accuracy of a forecast should not be the ultimate analytical problem. The closer to the present, the more accurate forecasts will be. The farther into the future, the more forecasting assumes a special nature. It is becoming a more dynamic indicator of possibilities and choices the decision-maker will welcome in support of his own socio-economic or political preferences. Simulation becomes in this case a more reasonable approach to economic analysis.

In the field of agriculture, it is of greatest importance for a country to account for cycles in production and to incorporate its cyclical production patterns to meet international market requirements. This is so because movements of agricultural commodities across the border of a country deal primarily with food products which represent residual quantities (often called surplus, or excess over domestic needs, or deficinecies, or residual quantities of food a country was not able to produce at all or was not able to produce at reasonable prices).

Macro-economic considerations (for example, a cyclical decline in economic activities and the subsequent relative decline of per caput real income) affect imports and exports more than the other sectors of the national income and expenditure accounts of an economy. Not even custom unions, free trade areas or zero-tariff conditions will prevent a decline in trade due to cyclical movements because of the protection built into the institutionalized arrangements of national economic infrastructure, prevailing consumption habits and political pressures exercized by predominant social groups.

III

The objectives of exporters and importers are to improve trading practices and to stabilize trade among nations and thus assure viability of investment in agriculture and the infrastructure related to the industry (particularly storage, transportation and food manufacturing facilities). Longer term forecasting serves these purposes and is being carried out by individual firms, by government and international organisations. The scope of economic analyses and forecasting varies so much that it is difficult to clearly designate the work as studies of trade. Classification is as difficult. In general terms, one can distinguish three types which denote the emphasis:[2]

(1) Country reports;
(2) Commodity studies; and
(3) Trade models.

(1) Country reports deal with trade problems and forecasts only in the case where exports (seldom imports) of agricultural commodities are of vital importance to the development of the industry. Some of the most elaborate country reports can be found in publications dealing with geographic or economic regions or an economic association.[3] Most country reports apply methodologies which yield either predictive or normative models for all sectors of the country's economy, not only the agricultural. Thus the farming sector is usually integrated into the network of a country's economic system. In more developed countries, where most of the economic analyses are done, the agricultural sector comprises but a small part of the economy and agricultural trade is less important. All newer studies are based on time series recursive models and/or cross-sectional analyses and apply computer programs.

From a rather simple notion that production in excess of domestic demand will find a market somewhere, country reports tend to underestimate the importance of foreign trade of agricultural products. This does not apply to country reports within the framework of a specific region where trade balances play a great role and where price trends are known or manipulated (as in the case of EEC agricultural trade). In such cases, elasticity coefficients have become a very reliable tool in the methodology for analysing trade.

(2) Commodity studies distinguish themselves because of the greater degree of accuracy, the well defined objectives and the rather longer time series upon which findings rest. What contributes most to the success of commodity studies applied to forecasting is the homogeneity of the statistics used. In many ways these studies serve also as outstanding sources for economic development planning as they contain an already standardized set of basic data and relationships (coefficients and elasticities) connected to demand and production functions.

Techniques of calculations are now often set in what is called "scenario" fashion or in variants. State assumptions and variables and their interactions are set out in considerable detail (in the case of a comprehensive feed grain study by the USDA, eight "cases" are specified). In reviewing commodity studies, one notices that many lack the essential element of economic analysis – the impact of prices on demand and supply. Why has this aspect been omitted in most of the studies? It appears that for agricultural commodities, market forces as measured by the impact of prices on commodity movements could not be quantified with a high degree of reliability for long periods after World War II because a major part of the food moving from country to country was not being traded through commercial transaction, thus putting it outside the effective price-making process. Other contributors to this difficulty were market distortions due to quantitative restrictions, quota systems and other government induced trade constraints and policies. In spite of these handicaps, computer models for trade of specific commodities were worked out to simultaneously determine levels of trade flows. There is sufficient indication that models with prices included and elasticities calculated could reflect real life conditions better than models without price variables. The techniques allow also for regionalisation according to specific criteria set by the researcher. Thus, one stage may include market economies; a second, centrally planned countries; a third, the less developed world; or the criteria can be geographic.

In commodity studies in general, the overriding advantages are the relatively high degree of homogeneity of the product and the relatively low substitutability which contribute to the transparency of relationships and high accuracy in forecasting.

(3) Comprehensive trade models appear to be the ultimate form of scientific work in support of worldwide decision-making in agriculture in spite of the rather hesitant acceptance of the first large-scale attempts at providing countries participating in commodity trade with such models.

IV

(1) Of all trade models, the ones which appear to be best known are the FAO Indicative World Plan trade component, the OECD longer term forecasting method and models prepared by the U.S. Department of Agriculture.[4]

The objective of FAO's Indicative World Plan for Agricultural Development was "to define the scale and nature and to assess long term prospects for world trade in major agricultural commodities".[5] Major variables affecting demand and supply were defined and parameters of prospective changes were sufficiently well quantified to provide for projections upon which individual member countries of the organisation could plan their own agricultural development strategy. In the "Outlook for Trade", nevertheless, the work proved hardly more than a general indication of major trends to 1975 and 1985. This was the result of the limited scope of the methodology employed.

The commodity coverage included all basic farm products. Major food items appeared in aggregated form (e.g., meat) as well as disaggregated, and main components were listed separately (e.g., beef and veal, mutton and lamb, pork, poultry and "other"). Cereals were subdivided into wheat, rice and "coarse grain". Sometimes the functional relationship between commodities were recognized and accounted for (as was the case with butter which was aggregated with "fats and oils" instead listing it as an animal product).

The country coverage was comprehensive (99 countries which accounted for 98 percent of world population), though difficulties arose from the procedure of grouping them in 24 sub-regions and only 3 zones (developed countries, centrally planned countries, and developing countries). The last grouping proved the least satisfactory because it did little for the better understanding of trade relations and often blurred the picture one could get from the analyses of sub-regions or individual countries.

Basic assumptions underlying the demand projections (population growth and change in GDP) follow the traditional line of U.N. studies. In addition, the FAO prepared "standardized food balance sheets" from which data were used to prepare (or correct) the rather simple income elasticities which measured the effect of per caput change of GDP on consumption of individual food items. In retrospect, one can observe that this procedure proved to be too theoretical, although it carried the seed for further work and improvement.

Production projections were based on past trends. This was probably the least satisfactory part of the study because it neglected the complexity of interrelationships of such factors as climate and to a great extent it also omitted the human factor, structural adjustment and government functions which affect the rate of adjustment of production technologies in agriculture.

Demand functions applied[6] provided for coefficients of income elasticities which were adequate where autonomous market and price-oriented economic conditions prevailed. Wherever this was not the case, the results were less satisfactory (in the case of government induced consumption, prices fixed by governments or non-price distribution of food). No adequate data were available for the calculation of price elasticities. For this reason, the inclusion of price elasticities in the model was postponed to a later date. The IWP projections were made uniformly at 1961–1963 prices.

If projected demand and production are brought together – calculated for every commodity in each country and projected to 1975 and 1985 – the only important information one gets is the extent to which a country (or a region) is or will be self-sufficient or the extent to which production will be in excess of domestic requirement. Thus the model provides "commodity balances" for respective years in the future. The balancing quantities (negative denotes import requirements, positive indicates export availabilities) "are not a projection of net imports or exports; they are only intended to provide an element for assessing trade prospects in 1975 and for analyzing possible changes in prices and in commodity policies".[7]

FAO's model clearly reveals that the magnitudes of residuals vary very much from year to year and from country to country. This accounts for the instability of agricultural commodity trade. planning and programming trade flows thus pose great difficulties. One recognizes also that there are limitations in the accumulation of "surpluses" as well as for "shortages".

In retrospect, and in conclusion of our observations in regard to the important work of the FAO, one should mention that the IWP projections were proven invalid in the early seventies, even before the revolutionary price changes which coincided with the cartelization of oil supply. The methodology of the study must have missed some most relevant correlations and disregarded some of the truly important factors of global production functions (for example, the consideration of cyclical changes of factors related to the climate of certain regions in the USSR which affect production functions for cereals).

Simplification of economic models through the use of parameters too complex in nature proved to be of little value. In the time series analysis, the change in per caput consumption calculated on the basis of a sole index proved less than satisfactory. Is the change due to per caput increase in disposable income, or is the greater contributor the change in price, the availability or change in the price of substitute goods, or a result of other factors?

Another problem in trade projections of this kind is the notion that consumption patterns depend on the level of per caput disposable income, and that this pattern is universally applicable. From what we know one cannot challenge the general rule, but the standard deviation might be very pronounced. Differences in consumption patterns between countries with market-oriented economies were very great. They were even more pronounced in centrally planned economies. Therefore, viable models which can usefully serve government and the public as a tool in decision-making include parameters not accounted for in the early studies (such as the FAO's IWP). More realistic regionalisation is also a mandatory requirement.

A most damaging attitude was shown in the early work on trade models. Economists were thinking in terms of the classical school where all conditions were perfect and the criteria of comparative advantage applied universally. The world is certainly not homogeneous as far as agricultural trade is concerned. Instead, there is a multiplicity of structures, of conduct and behaviour. Among countries there are associations of varying degrees and forms. Membership in associations change and depend on commodities, on regional arrangements based on geographic criteria and on internal as well as external policies. Such considerations are often more important than the grouping applied by FAO, particularly the one which segregated developed from developing countries.

A methodology which can be universally applied is an ideal the economist cherishes very much. Nevertheless, a valid question is: To what extent could modifications account for factors less easily quantified, as, for example, attitudes toward nutrition or toward trade deficits? In other words: Could the economist introduce parameters of a socio-economic nature or features attributed to the realm of "political economy"? How does one account for trade restrictions due to a variety of factors such as the level of unemployment, the rate of adjustment from predominantly rural employment to manufacturing and service industries, the balance of payment problem, the minimum rate of self-sufficiency due to political or defense considerations, etc? The inclusion of such factors in a model would contribute to a more meaningful IWP and to better forecasts of trade. In such a model the concept that trade deals only with residual quantities could be dropped and foreign trade could become an integral part of real balances on a worldwide scale. Also, a new "Perspective Study of World Agricultural Development" could depart from the central idea of determining degrees of self-sufficiency and would benefit from a realistic appraisal of what international trade could do for the world's population.

(2) Work on the IWP was a first attempt at a comprehensive projection of

agricultural trade possibilities. Other international organisations interested in trade (GATT/UNCTAD, OECD, International Wheat Council and other similar commodity-oriented institutions) also contributed to the development of the methodology of long term forecasting.

The OECD Agricultural Projections "are a complement to FAO's Indicative World Plan, which is primarily concerned with less developed countries".[8] Because member countries of the OECD have more detailed and longer run statistical data, this study was much more precise. Nevertheless, it did not pretend to forecast production, consumption and trade of agricultural products and chose to call itself a "projection exercise".

What distinguished this study was that specific assumptions underlying the projections were stated in considerable detail and the data system was used for reasonable extrapolations. In spite of this caution built into the methodology, a clear statement was made that "in several aspects the results are clearly unrealistic".[9]

It was recognized that where supply and demand were projected independently, residual items were large. Based on studies of individual countries, residuals were either net export or net import requirements. They were not projected trade!

Of particular value to researchers is a comparison with the FAO IWP findings. A review shows the weaknesses of the models:

(a) Demographic studies were out of date.

(b) The concept of "normal production" is a construct too far removed from reality in agricultural production because cyclical movements are dominant.

(c) The "composite elasticity" and the elasticities and correlation coefficients derived[10] proved to be applicable over narrow ranges only and do not accommodate conditions observed on commodity markets.

(d) The "net trade" concept tends to mask the acutal commodity movement (as each country is likely an exporter as well as an importer of the same commodity).

(e) The estimates of "unit value" where imports were recorded c.i.f. and exports f.o.b. poses questions of reconciliation for which there is apparently no ready answer available.

Compared with the IWP, the OECD study shows considerable improvement but a comparison is not fair because the OECD had a much more complete data system on its side. In the field of studies of methodologies, the OECD may likely proceed toward more dynamic recursive trade models.[11]

(3) In support of its agricultural and commodity trade policies, in the early sixties the EEC conducted rather extensive economic analyses and projections of demand and production of agricultural commodities, first for 1966 and then for the target year 1977. The methodology applied did not differ in principle from the one explained previously. Growth of population and per caput consumption were the basis for projecting domestic demand. Production increases were expected because of improved agricultural technologies

and were calculated by extrapolating past trends. Greater domestic supply affected the general increase in the degree of self-sufficiency which narrowed the gap between requirement and availability from domestic sources, a gap to be filled by imports from "third" countries.

It should be mentioned that the EEC presented for discussion two fundamental methodological studies before projections of demand and production were made.[12] Instead of looking for new avenues in projection, these studies emphasized the applicability of models to the conditions in the Community and its member countries. In these studies one notices emphasis on equilibrium models, on spatial equilibrium in particular. Guided by the basic philosophy of a market-oriented economic system, the role of prices and price function was emphasized. The work reflects also the dual foundation – market orientation as well as policies toward restructuring of the agricultural sectors of the member countries – of the CAP. International trade did not receive special consideration as greater attention was directed to the discussion of intra-Community trade.

In the EEC preference has been given to the application of spatial models to international trade with the objective of finding equilibrium solutions. Following Takayama and Judge,[13] a model based on linear demand and supply functions and including prices and other factors could logically lead to solutions as long as price competition in international trade is permitted. If conditions as set out are not met, this method could at least serve as a starting point and provide a model against which the impact of factors not easily quantified can be assessed. A next step would be to include such criteria as quantitative restrictions applied in trade, minimum level of self-sufficiency, "voluntary quota" agreements and other trade restricting practices.[14]

An inventory of research conducted in the member countries of the EEC in 1970 shows how little was done by agricultural economists to project international commodity trade. Reservations expressed by leading scholars as to the applicability of known methodologies probably led to a reluctance on the part of the administrators of research activities to proceed with further work. As a consequence, apparently no research input beyond the short term analyses and year-to-year projections indicated in the annual reports and budgets of the CAP and the Guarantee Fund (FEOGA) went into decision-making regarding agricultural trade.

The rather pessimistic attitude toward what the methodology could accomplish is reflected in a set of studies commissioned by the EEC's Directorate General for Agriculture. The studies were published over a number of years under the title "Projektionen ueber Erzeugung und Verbrauch landwirtschaftlicher Erzeugnisse – 1977" (individually for each member country and a summary volume for the six original EEC members).[15] Notes on the methodology used indicate that agricultural trade was treated solely as an actual or potential contributor to the balance of foreign trade. Thus trade was treated only as a residual and no indication of the direction of trade with "third" countries was included.

In general, the studies for 1977 indicate a much higher degree of self-sufficiency in all major commodities (for all cereals excluding rice, 99·3 percent for the six original EEC members and 91·4 percent for The EEC-nine;

corn, 80·7 percent for EEC-six and 63·9 percent for the EEC-nine; all meat, 93·9 percent for the EEC-six and 97·7 percent for the EEC-nine). This means that traditional suppliers to this market may have to find new outlets for their "surplus" production. Some of the expected increases in total availability appear disturbing to trade. For example, in the case of wheat in the target year 1977, the EEC-six might produce 14·1 percent more than their domestic requirement. However, at the same time, this "success" might ease the situation in the world supply and enable wheat deficient countries among the less developed nations to obtain this basic food with greater facility.

These studies were more comprehensive in coverage and interpretation when the degree of self-sufficiency was already high or when projections indicated that the EEC would approach a high level of self-sufficiency. Whenever supply depended to a great degree on imports (as in the case of citrus fruit or vegetable oil), the coverage was limited.

The EEC and individual member countries (particularly France, the Netherlands and the United Kingdom) were more successful in researching and projecting international trade in a series of so-called "commodity studies".[16]

An attempt by Dr. Rainer Schmidt to prognosticate international trade of milk products, a study commissioned by the EEC, is an example of a product study and will be reviewed here briefly.[17] The study listed twenty-eight exogenous variables for the model applied to determine import requirements for milk products to the United Kingdom. Among major items one recognizes some familiar features of expected change – a slight increase in population, an increase in per caput disposable income, a relative decline in the price of milk products against a relative increase in the price of beef, and an increase in the price of cheese and other milk products. Also included were features of adjustment to accommodate the trend in observed technological improvements (for example, higher yields of hay per acre). Unfortunately, no attempt was made to go beyond the calculation of import requirements. The study proved useful in assessing the impact of the United Kingdom's accession to the Common Market.

Income and price elasticities for milk and milk products, as well as the very important cross-elasticities, show extreme variations. Aggregations and regionalisation of demand are therefore nearly impossible. To eliminate errors from calculations of supply due to the fact that the milk cow is in most countries simultaneously the major source of beef is another facet of the complex problem. The animal is only a converter of domestically produced or imported feed. This accounts for further difficulty in providing information regarding the extent to which the milk industry contributes to foreign trade. Simple supply functions proved an insufficient base for forecasting trade in Schmidt's study.

On the positive side and in favour of his study, one should mention that it was demonstrated that different data systems from different countries required adjustments of the methodologies as was shown in the case of the report on the milk industry in the German Democratic Republic, the Philippines and Venezuela.

(4) In support of the work of the Club of Rome, discussion of the "Multi-level Computer Model of World Development System" is of interest to the study of agricultural trade.[18] This is a highly aggregative model which used the data base from the FAO studies, had regionalisation on similar principles and included FAO's Food Balance Sheet concept. A substantial addition to the FAO models was the inclusion of prices (per unit cost of land, livestock and other productive factors related to crops, as well as monetary values related to various categories of agricultural products). With such additions a comprehensive concept, the "agro-ecosystem", was established. Based on "sub-models" (land use and food production) and population, regional "surpluses" and "deficiencies" in food were predicted. Unfortunately, no true balances of trade were calculated.

Among "Scenarios" elaborated by the Club of Rome, as exemplified by a special note on South East Asia, completely unrealistic outcomes result. No reasonable trade could accommodate the discrepancies and food aid was considered at a level the magnitude of which is incomprehensible. In spite of such inconsistencies, the methodology is fascinating and promises to open doors for more detailed and more realistic studies. The basic concept – "agro-ecosystem" (to which economics and ecology contribute) – provides for parameters and insight into functional relationships often neglected in earlier studies.

V

FAO, OECD, and GATT/UNCTAD are international organisations and emphasize global economic and social aspects and problems. For various reasons these organisations have not appeared to place sufficient emphasis on balancing supply and demand of agricultural commodities on a worldwide scale. This is probably so because in the FAO priority was accorded to improving nutrition in less developed countries, increasing the degree of self-sufficiency and, consequently, improving agricultural techniques.

UNCTAD emphasized development to which agriculture can contribute, but this was only one of many avenues toward its goals. In GATT, the problem of agriculture did not play a major role as the agricultural sector was specifically designated as being outside its competence. In the work of the OECD one detects probably the most advanced data system and analyses in support of decision-making in agriculture, although its mandate is "regional" and thus not wide enough to affect decision-making on a worldwide scale.

The EEC was preoccupied with internal problems of coordination of agricultural development toward which intra-Community trade stabilization provided one, if not the most important, path. Intra-Community trade stabilization became one of the cornerstones of the Community. Imports from third countries were deliberately reduced to supplying residual quantities of agricultural products required solely in case domestic production was not sufficient to satisfy the demand. EEC exports to "third" countries often required subsidies and government supported marketing activities. Variable levies (in the case of imports) and export subsidization resulted in a situation where the participation of the EEC in commodity trade could not have been accommodated in methodologies of economic research which had as their major component prices of products and were generally thought of as complying with the classical principles of comparative advantage.[19]

The COMECON, the major economic association of centrally planned economies, with the USSR as its primary contributor to the trade of agricultural commodities, is the most difficult "region" to accommodate and to account for in a model of international trade or product study. The reason for this is not only because all prices in these countries are administered. Statistics upon which to calculate demand as well as production functions are not available. In spite of the difficulties arising from the inclusion of data from these economies, trade models and product studies which do not account for the components related to the COMECON and China will be highly deficient.

VI

The discussion related to economic regions as exemplified by the EEC and COMECON requires further elaboration because the observed trend toward more regionalized agricultural trade was in no way interrupted by the events of 1973/74. Established regional trade patterns proved viable trade arrangements (as was the EEC and its associates or the COMECON). Therefore trade patterns determined by geographic proximity and political affiliation must be recognized in trade models and product studies because a better knowledge of such patterns could significantly improve the forecasting of trade.

Some doubt must be expressed as to the applicability of the standard nomenclature which reflects the division of countries into "market economies" and "centrally planned economies". Also, identification according to institutional arrangements is not applicable. Not all market economies are members of the EEC or OECD, not even market economies in close geographical proximity. On the other hand, many countries distant from the EEC (geographic, political) benefit from association with EEC as it renders trade more of the kind denoted as "intraregional" or "intrazonal". Associates of the EEC enjoy privileged or free access to markets, long term trade arrangements and an increased flow of investment funds for the expansion of agricultural production. Links between participating countries are thus strengthened and trade flow is more stabilized in associations of this kind. Similarly, the COMECON is such a region or zone. Not all centrally planned economies are members. Its expansion (as in the case of Cuba) results first in a reorientation but afterwards in a stabilization of trade patterns. The United States has formed the core of such a region as well.

These observations should direct the attention of agricultural economists towards distinguishing between forecasting in the frame of a region or zone where the probability is high that predictions will be valid and forecasting agricultural trade on a worldwide basis. What is required is differentiation in methodologies and adjustments in accommodating the differences in data systems and in economic and market behaviour. This departure from looking at agricultural trade solely from the "global" aspect could provide for more realistic forecasting. Multilevel decision-making in agriculture is prevailing today. Free trade does not apply. A methodology of economic analysis, trade models and product studies will have to cope with the reality of the complex twentieth century.

There is justifiable hope for a change to better and easier forecasting because one can observe a trend toward greater homogeneity of economic conditions in individual regions, particularly where the regional set-up at the international level is institutionalized (EEC or COMECON). The gradual abolition of trade restrictive practices allows comparative advantage to affect allocation of production according to economic criteria. Simultaneously, some of the more damaging features on international commodity trade, particularly the high degree of uncertainty, tend to diminish. This is so because in the sixties and the beginning of the seventies, intraregional agricultural trade grew at an annual rate which was twice that of interregional or world trade. The sudden increase in interregional trade in 1973/74 may not fundamentally alter the relationship between intraregional and interregional or world trade. The purchase of huge quantities of cereals by the USSR and the general upheaval in trade in the recent past has probably strengthened the trend toward greater reliance on orderly and foreseeable relationships in commodity markets. This should also give direction to research and forecasting and to the methodologies to be applied.

Established research institutions and academic economists have already achieved a high degree of sophistication in forecasting trade, but the applicability and usefulness of models and commodity studies to government and industry require substantial resources and a degree of continuity only few countries can afford. For this reason international organisations will have to play an even more important role than in the past.

NOTES

[1]Allen, G. R. (1976) 'Some Aspects of Planning World Food Supplies', *Journal of Agricultural Economics*, Vol. XXVII, No. 1 (January, 1976), pp. 97–119.

[2]United States Department of Agriculture, ERS, *United Kingdom: Projected Levels of Demand, Supply, and Imports of Farm Products in 1965 and 1975*, ERS Foreign Bulletin 19 (Washington, D.C.: January, 1962);
United States Department of Agriculture, ERS, *Growth in World Demand for Feed Grains Related to Meat and Livestock Products and Human Consumption of Grain*, Foreign Agricultural Economic Report No. 63 (Washington, D.C.: July 1970);
United States Department of Agriculture, ERS, *The World Food Budget 1970*, Foreign Agricultural Economic Report No. 19 (Washington, D.C.: October 1964).

[3]Instituto Nacional de Technologia Agropecuaria, *Argentina: Proyecciones de la Oferta y la Demanda de Productos Agricolas seleccionados hasta 1982* (Buenos Aires: undated);
Institute for International Agriculture, Michigan State University, *The Impact on U.S. Agricultural Trade of the Accession of the United Kingdom, Ireland, Denmark and Norway to the European Economic Community* (East Lansing, Michigan: 1971).

[4]The work of the staff of UNCTAD will not be considered in this context because agricultural commodities constitute only a small part of a much more comprehensive work dealing with primary products.

[5]Food and Agriculture Organisation, *Agricultural Commodities – Projections for 1975 and 1985*, Vol. I (Rome, 1967), p. iii.

[6]Functions were: (a) logarathmic, $\log_e y = a + b \log_e x$; (b) semi-logarithmic,

$y = a + b \log_e x$; (c) log-inverse, $\log_e y = a - (b/x)$; (d) log–log-inverse, $\log_e y = a - (b/x) - c \log_e x$.

[7]Food and Agriculture Organisation, *Agricultural Commodities – Projections for 1975 and 1985*, Vol. II (Rome, 1967), p. xxxi.

[8]OECD, *Agricultural Projections for 1975 and 1985: Europe, North America, Japan, Oceania. Production and Consumption of Major Foodstuffs* (Paris, 1968).

[9]OECD, *op. cit.*, p. 9.

[10]OECD, *op. cit.*, p. 73. Rules: (a) positive elasticity coefficient from double log functions; (b) coefficient higher than 1·0 from log-inverse functions; (c) negative coefficient less than − 1·0 from semi-log functions.

[11]OECD, *Agricultural Projection Techniques. Papers Presented at OECD Meeting of Experts, 11th–14th October 1971 and Other Contributions* (Paris, 1972).

[12]EWG, *Landwirtschaftliche Vorausschaetzungen, Methoden, Techniken, und Modelle*, Hausmitteilungen ueber Landwirtschaft Nr. 48 (Brussel: September 1969); and
EWG, *Landwirtschaftliche Vorausschaetzungen Moeglichkeiten der Anwendung bestimmter Modelle, Methoden und Techniken in der Gemeinschaft*, Hausmitteilungen ueber Landwirtschaft Nr. 63 (Brussel, Oktober 1970).

[13]Takayama, T. and Judge, G. G. (1964) 'Equilibrium among Spatially Separated Markets: A Reformulation', *Econometrica*, Vol. 32.

[14]Marin Curtaud, B. (1965) 'Les Modeles prêvisionels des réseaux d'échanges internationaux et leur structure', *Bulletin du CEPREL* 5–6 (October, 1965).

[15]EWG, *Projektionen ueber Erzeugung und Verbrauch lanwirtschaftlicher Erzeugnisse – '1977' VI. Ergehnisse fuer die Europaeische Gemeinschaft*, Hausmitteilungen ueber Landwirtschaft Nr. 129 (Brussel, April 1974).

[16]CEPREL and CEDES in France; The Institute for Research in Agricultural Economics at Oxford University in the United Kingdom; The Research Institute in The Hague, Netherlands.

[17]Schmidt, Rainer (1971) *Analyse und Prognose der Importe von Milcherzeugnissen ausgewachtler Laender mit Hilfe oekonomischer Modelle*, Kieler Studien Nr. 117, Tuebingen.

[18]Contributors were: Clapham, W. B., Shook, T., Warshaw, M. W., Mesarovic, M. D., and Richardson, J. M., Jr.

[19]Haberler, Gottfried (1963) *Trends in International Trade*, GATT, Geneva.

WILHELM H. G. SCHEPER *

The Importance of World Models for Agricultural Policy

1. INTRODUCTION

This paper deals with three classes of world models

(a) international agricultural commodity models of the FAO-Gulbrandsen-type
(b) global world models of the Meadows-, Pestel/Mesarovič-, Linneman-type
(c) transmission models (LINK project)

The history of these classes of models is very different and also their contribution to decisions in agricultural policy. The global world models and the LINK models have been developed in recent years. The construction of models of these types can be expected to be one of the fastest growing disciplines in economics. The influence of the published models on decisions in agricultural policy is still rather low but this does not mean that the political influence of forthcoming models also will be low. Therefore, this paper also emphasises the future pattern in these fields. International agricultural commodity analysis has a longer history and, therefore, has had more influence on agricultural decisions.

The importance of economic models for decisions in agricultural policy is difficult to assess, especially for international agricultural models and decisions on international agricultural policy. Many world models would be much more useful for a centralised world government than they are for the decentralised national decision structure in the real world. Most of the decisions which determine the international agricultural scene are still done by national governments. National governments try to maximize the advantages for their countries – or for themselves. Their demand for information concerning the international agricultural scene differs from country to country. Countries which have rather small foreign trade compared to their GNP and total foreign

* University of Kiel. Federal Republic of Germany.
The author thanks his collaborator Gert SCHMIDT for assistance and critical comments.

trade find it rather easy to use various measures of national protection to avoid undesired influences from the agricultural world market. Countries which are main agricultural importers or exporters have much more interest in international agricultural developments. Their interests in information vary significantly with their import and export structures. Small countries are more or less price takers; large countries act like oligopolists or monopolists. The demand for information in western market economies differs also from the demand in socialist countries. To sum up, the need for information differs widely from country to country and model-builders should put more emphasis on this point than they have done in the past.

Most world models analyse long-run aspects but most political decisions deal with short-run or medium-run aspects. Therefore, this paper looks for possibilities to overcome this disparity. Transmission models seem to be a rather hopeful development. Various transmission techniques allow the linkage of complex national models to a more or less consistent international system. In future, it should be possible to build up transmission models which explain the interdependence between national economies on a rather high-aggregated level and also the interdependence between the national agricultural sectors on a rather low level of aggregation within one system, i.e. a system which includes also disaggregated models of national economies.

Transmission models so far have mainly dealt with short- and medium-run problems. The global world models of the Pestel/Mesarovič-, Bariloche-, and Linneman-type mainly create scenarios for the very long-run.

2. INTERNATIONAL AGRICULTURAL COMMODITY MODELS

2.1. *Surplus/deficit models*

The traditional and most used approach is the surplus/deficit approach. For each country and for each commodity the domestic produciton and the domestic demand are predicted separately, usually by simple trend methods. By countrywise aggregation of demand and supply projections future worldwide or regional deficits or surpluses can be estimated. In other words, the models point out expected shortages and surpluses for various countries and regions for the medium- and long-run future. In some models domestic demand is a simple function of the population and the per-capita income. Improvements of the supply predictions have been tried using simple supply functions which explain supply as a function of product and factor prices. Population, per-capita income and prices are exogenous variables. Models of this type have been built especially by FAO, OECD, and USDA[1] and have influenced the opinions of politicians and the public in general. The models do deliver a worldwide surplus/deficit scenario which enables us to locate critical points in food provision and they also give a useful starting point for the consideration of long-term market prospects. Secondly, the predicted trends help in discussing deviation problems together with stock statistics and also make a contribution to the analysis of short- and medium-run problems.

Critics of the trend models very often do not realize that accuracy is not the only criterion in valuing predictions. Valuation depends closely on the purpose for which the prediction will be used. In the short-run they allow separation of trends and deviations to a certain extent; in the long-run they provide at least a starting point for the discussion of possible patterns of development and critical points.

The criticism of surplus/deficit models should be concentrated in two fields. Firstly, domestic supply and domestic demand are not sufficiently explained by economic variables. Secondly, the models do not explain how the difference between domestic supply and domestic demand can be closed by international trade or by variation of buffer stocks.

2.2. *Equilibrium and international trade models*

There are some extensions of surplus/deficit models which try to explain the difference between domestic production and domestic demand by world market models. The well-known Gulbrandsen model[2] is a simultaneous model with the following characteristics:—

– countrywise for each product there is a supply function and a demand function.

– the supply, as well as the demand, function show microeconomic characteristics. Supply of each commodity is a log-linear function of agricultural product prices, and demand for each commodity is also a log-linear function of all agricultural product prices.

– the price of the commodity j in the region i is by definition equal to the world market price of this good plus the price support of this good which is given by the nation i, plus transportation costs. Transportation costs per unit commodity are the same for each nation.

– Transportation costs and national price supports are exogenous variables.

By iteration the model can be solved in a way that results in a simultaneous price and quantity system which fulfils the functions and definition equations described above. In addition, the solution fulfils the market clearance condition for each commodity – that is, world import equals world export. The solution does not include a world trade matrix, so there is no information on what quantity of a commodity is delivered from one country to another. The model cannot give this information because there is no matrix of transportation costs and no other market share approach. The value of the model depends among others on the quality of the national demand and supply functions. Gulbrandsen includes these functions in a very schematic way. The microeconomic approach, and especially the supply and demand elasticities with respect to the agricultural product prices, are rather problematic. We will come back to this point.

The oversimplified transportation costs assumption reduces the value of the model, especially for those commodities which have high transportation costs or can only be stored to a limited extent. The crucial transportation costs assumption can be replaced by the so-called "transportation costs minimizing approach". Here an exogenous given transportation costs matrix, in addition to the described equation system, allows the building of an optimisation model[3]. The solution fulfils the condition that world imports equal world exports for each commodity and, in addition, minimizes the transportation costs. Furthermore, this solution includes a world trade matrix. The formulation and solution of the model is rather easy for linear functions. In other cases the problem can only be solved by iteration.

The Gulbrandsen model, and also the spatial equilibrium models, are competitive models. Their national trade constraints are considered as exogenous variables, i.e., the price differences at the border are model input and not model output. In reality, however, many international trade constraints, especially the EEC levies, are functions of the difference between the desired domestic price and the price on the world market. The link between domestic prices and the world market prices is not as simple as it is assumed in the models described above. Furthermore, many exports and imports are the results of special government decisions which are out of accord with the principles of a competitive market. Irregularities in national buffer stock policy and concessional trade are also non-competitive elements. Especially in the agricultural sector it is unrealistic to make no distinction between price determination within a country or on the world market. Most countries want to achieve stability and continuity of prices and quantities within the national borders. This policy very often causes instability and non-competitive trade on international markets. The price equilibrium and trade models considered here given some hidden information which should be recognized. The models cover the main parts of international agricultural trade and provide a consistent price and quantity system. On the basis of this information it is quite easy to quantify deficits or surpluses in the national balance of payments caused by agricultural exports and imports.

2.3. *Extensions and modifications*

There are many models which are similar to the models described on the previous pages. Some concentrate on a smaller group of commodities or on just one single commodity[4]. Other models intensively work out the position of one specific country[5]. In the following we will concentrate on the problems of national supply and demand.

Many agricultural economists – as has been pointed out already – consider that high aggregated supply functions (estimated independently for each commodity by traditional methods of multiple regression analysis) only give very limited information. They believe that a satisfying explanation of the agricultural supply situation only can be worked out on the basis of disaggregated models which include farm structure, investment behaviour, expectations and decision functions of the farmers. The possibilities for explaining the supply in a better way vary from commodity to commodity and from country to country. To make use of these possibilities it is necessary to work with various approaches. Of course, this will make the models more complex and less comparable. But these more complex models can include also the relationship between the production of various commodities (e.g., joint production and joint investment, intermediate goods in the agricultural sector). Especially the relationship between agricultural supply and domestic demand in developed countries becomes more complicated from year to year, i.e., the processing and distribution of agricultural goods becomes more and more important. The relationship between agricultural producer prices and domestic food prices changes in the short-run as well as in the long-run. There is also the problem of input/output coefficients. To overcome all these difficulties, it is necessary to formulate the so-called integrated agricultural development models which describe the total national agricultural sector in a consistent way. Models of this type have been presented by various authors[6]. Their core is usually a linear optimisation model which simulates the short-run decisions of the farmers. The medium- and long-term development of the agricultural production potential is included by various approaches; the nature of these approaches is either positive or normative. Many recursive and behaviouristic elements help to increase the stability of the model and to introduce common-sense arguments. There are various possibilities in setting up these models:

(i) the whole national agri-business can be included or only a part of it,
(ii) the degree of aggregation can be very different,
(iii) the models can be constructed with special regard to short term, medium term or long term problems.

Thus there is enough flexibility within the model for its structure to be adjusted to problems of special interest. From country to country the integrated agricultural development models are faced with different questions and facts (different economic systems, different production structures, different structure and intensity of foreign trade, different stages of economic development, etc.).

Integrated agricultural development models must be linked to the other national sectors and to international markets. The agricultural output sector has to be linked with various groups of demanders. Major problems arise in explaining exports and imports and the demand for storage. So it is necessary to incorporate the institutional frame and also some government variables.

Investment and labour input in the agricultural sector depend to a large extent on the situation in the other sectors of the economy. Supply of capital goods and the prices of capital goods are not only a function of the general domestic economic development but especially in developing countries, they depend rather strongly on foreign relationships.

Fertilizer input and fertilizer prices can only be explained when the international fertilizer situation is included in the model.

The sector of other inputs deals with four different input groups:

(i) commodity flows which are similar to intermediate commodity flows in the agricultural sector (for example, feed concentrates with ingredients produced by the domestic agricultural sector),
(ii) current inputs delivered from the agricultural sectors of other nations,
(iii) non-agricultural current inputs with low dependence on international economic relationships (local goods),
(iiii) non-agricultural current inputs with high dependence on international economic relationships (for example, fuel).

The inputs of group (i) and of group (iii) can be incorporated in the model rather easily. The inputs of group (ii) and group (iiii) can only be endogenised when national and international models are included in the analysis.

Government policy in agriculture is hard to include in integrated agricultural models in a way other than as a set of exogenous variables. However, it is necessary to introduce government variables in the model in such a way that alternative behaviour of the government can be simulated.

The discussion of the borders between integrated agricultural development models and other national and international economic sectors shows that the information arising from these models can be increased substantially when the linkage to the surrounding sectors can be done successfully. Therefore, it is necessary to investigate the possibilities of linking the national agricultural development model to a macroeconomic model which includes the whole economy and to international economic models. It is difficult to define the most important limitations. In short-run analyses questions of investment and labour input are not so important, in long-run analyses the determination of the production potential plays the most important role. Very often more attention is given to the output markets than to the input markets. For many countries there are no reasons to look less carefully at the input side than at the output side. Especially in developing countries which import a major share of the agricultural input, the input side of the agricultural sector is at least as important as the output side.

The importance of national agricultural development models for the explanation of international agricultural product and factor markets has been analysed to a very small extent. This is understandable when we look at the history of these models. It begins with the analysis of the national farm sector on the basis of short-run linear programming models. Successively the models were extended to more complex models. It just seems to be a question of time before models are really linked with macroeconomic national models and international models.

In simple terms we have, at present, the following situation. On the one hand there are agricultural world models which explain in an oversimplified way the domestic supply and demand of agricultural commodities for most countries. On the other hand there are national agricultural development models which do not include relationship with other national sectors and to the world markets in a satisfying way. In the future it is necessary to connect both these developments in integrated schemes. In the following paragraphs we will discuss the possibilities and limitations of some approaches which may be useful starting points in this direction.

3. THE NEW SCOPE: GLOBAL WORLD MODELS

The Meadows model[7] initiated by the Club of Rome opened a new era in model-building. It produced rapid increasing interest in global world models concerning the future development of mankind. There are many reasons for this increasing interest and many of them have little to do with the professional economics. Nevertheless, the models of Meadows, Mesarovič-Pestel[8], the Bariloche group[9], and Linneman[10] are very important models, just because important individuals and important social groups regard them as important.

3.1. *The Meadows model*

Meadows presents a high aggregated model with 6 stock variables and 4 flow variables.

Stock variables: reproducible capital, effective capital units per capita, population, land, exhaustible natural resources, level of pollution;

Flow variables: non-food production per capita, non-food consumption per capita, gross investment per capita, food production per capita.

These variables are elements of a very complicated non-linear dynamic system. Besides the definitive equations there are many behaviour and technology functions of different types. Meadows solved the non-linear system on the computer year by year up to the year 2100. He calculated a lot of alternative parameter sets. From the mathematical point of view the Meadows system is mainly a system of difference equations of a rather high order. Besides some very special cases, these systems show a large degree of instability. That is why a mathematician is not surprised at Meadows' results. Most of his results tell us that there will be a sudden break-down of the world economy, and a tremendous decrease of population in the next 100 years. On the other hand Meadows, and in more detail the Sussex group, have shown that there are special parameter constellations which lead to a continuous growth of the economy or asymptotically to a stationary state. These parameters do not look very unrealistic but on the other hand the model is rather sensitive even to small variations of the parameters.

The agricultural sector has a major place in Meadows' model. It is always dangerous to interpret a system in a very simple recursive way when the system has also simultaneous and feed-back elements. However, the conclusion is allowed that Meadows' agricultural sector is a highly simplified approach. The agricultural production function is quite difficult to understand and there is no explicit investment function. The assumed relationship between agricultural inputs and negative effects on land fertility is just naïve. Furthermore, pollution elements and the phenomenon of technical progress are rather weak points in the analysis. The main instability is caused by the following – growing population and growth of other factors lead to an expansion of arable land and to higher yields per acre. However, negative effects are caused with a certain delay – higher intensity and increasing pollution lead to a decrease in land fertility and to land erosion. From this a very sharp decrease in agricultural production results.

Many of the weaknesses in Meadows' model can be easily avoided. It is possible to improve the model substantially on a similar aggregated level[11]. Using the tools of modern growth and investment theory it is possible to set up a highly sophisticated model which could fit perfectly in the parade of the neoclassical growth models of the Samuelson–Solow-type. It is possible to analyse the relationships between the variables of interest in a neoclassical model which is much more transparent than the Meadows model. Introduction of a production possibility frontier with the outputs pollution, food, non-agricultural goods and services and know-how, and the inputs labour, capital, land, and consumption of natural resources would be a large improvement. This introduction could help to discuss the partial influence of the determinants of long-run growth. These determinants are various types of technological progress, decreasing returns, various elasticities of substitution etc. To get a complete growth model there have to be added some dynamic functions (investment function, population function, labour supply function, accumulation function for technical know-how, and pollution reduction function). This model enables theoretical well-trained economists to analyse the conditions of balanced growth and to discuss various aspects of investment and technical progress policy. In general, these models are useful elements to train economic scholars in better understanding of the fundamental relationships in long run economics. Better understanding of these relations helps to avoid mistakes in applied economics analysis. However, the way from theoretical growth models to special predictions of long-term development is quite long. Looking at Meadows' model there are three crucial points: data problems, various questions of spatial distribution, the introduction of political and social elements.

Even a substantially improved Meadows' model will be rather sensitive to parameter variations. This results from the antagonism between technical progress and decreasing returns and from the various influences of assumptions on elasticities of substitution. That is why the result of model simulations depend heavily on the data input. Collection and definition of the data are rather difficult for the following reasons:

(i) in addition to other difficulties, there are the same difficulties which arise in the measurement of macroeconomic data on the national level,

(ii) there is little experience in collection and definition of high-aggregated world data,

(iii) for the set-up of the Meadows model, there are some unusual variables which must be defined and quantified (e.g., pollution).

(iv) it is difficult to estimate for the long run the exogenous variables of the model.

Nordhaus characterizes Meadows' analysis very concisely as "measurement without data". For various reasons it seems impossible to solve the data problem for high-aggregated models like the Meadows model in a satisfying way.

Models with an aggregation level similar to the Meadows model do not help to analyse the worldwide distribution problems; this is a serious disadvantage. The future of mankind is to a large extent a question of distribution:

(i) spatial distribution of world population,

(ii) spatial distribution of physical capital and know-how,

(iii) regional distribution of factor incomes,

(iv) world wide horizontal income and finance transfer, including capital transfer and foreign aid.

Political elements and social structures are not incorporated in the Meadows model. Models of the Meadows type offer only small possibilities of including government policy parameters. Perhaps it is possible to extend models of the Meadows type to normative models with the tools provided by the theory of optimal growth. Such an extension may be a useful addition to economic theory but there is no hope of their yielding further information which can be used in a direct way for political decisions.

3.2. *The Pestel/Mesarovič model*

Extensive discussion of the Meadows model encouraged the Club of Rome to sponsor another global world model. Pestel and Mesarovič tried to set up a model with the following improvements

(i) disaggregation of the world economy into ten regions,
(ii) disaggregation of the growth process by explicit introduction of more sub-sectors of production,
(iii) extension of the model by introduction of decision-functions and learning and information processes.

The Pestel and Mesarovič group put much emphasis on the collection of worldwide data. As far as possible, they made the data comparable and integrated them in a large system of definitive equations. The importance of this statistical work should not be underestimated; I have the feeling that it would be worthwhile to continue it.

Pestel and Mesarovič tried to present a global and disaggregated world model. This target has not been achieved. They have not presented a model which simulates the worldwide growth process and the growth process in the various world regions simultaneously. To achieve this a very complex model must be available and such a model cannot be built in two or three years as Pestel and Mesarovič tried to do.

These authors present partial analysis; they analyse the energy problem, the problem of foreign aid, the problem of some non-renewable resources, and the food problem. The food problem analysis will be described in the following.

The methods in this analysis are rather similar to the methods Pestel and Mesarovič used to analyse the other problems mentioned above. They take the line that, in the long run, the food problem will be most serious in South Asia. Therefore, they concentrate their analysis on this region. Food needs per capita and population growth are estimated for the long run by common methods. The food supply from South Asian agriculture is explained by an implicit production function; technical progress, current inputs, land and capital stock per acre are the main variables. The estimation of the production function is based on production curves and developments which have been observed in the developed world in the past. The linkage between the agricultural model and the non-agricultural sectors is less intensive than in the Meadows model. The Pestel/Mesarovič model does not have an extremely complicated feedback structure and therefore the growth paths of this model are much more stable. Commonsense arguments have major importance. Pestel and Mesarovič simulate some alternative developments.

The first scenario is, to a large extent, a simple trend analysis. It shows that in the period dated from year 2000 up to year 2015 about 500 millions of people in South Asia will die because of lack of food. The second scenario shows that the problem of hunger in South Asia cannot be solved even when the developing world gives maximal food aid to South Asia, if no other change of policies is assumed. The third scenario shows what happens when the countries in South Asia re-allocate their resources from the non-agricultural to the agricultural sector. If there are no other changes in economic policy this re-allocation only improves the food situation in the medium run. In the long run the reduced growth in the non-agricultural sector causes reduction of input in agriculture and this reduces food supply. Other scenarios investigate the influence of alternative reductions of the fertility rate and combinations of various methods. Finally Pestel and Mesarovič present – in their own opinion – the "best" scenario. In this scenario the developed countries give their foreign aid to South Asia mainly in the form of capital goods. These capital goods improve the agricultural and non-agricultural productivity in South Asia. Food production and export/import potential is increased in such a way that the problem of hunger can be solved.

There are some important characteristics of the Pestel/Mesarovič model. The selection and handling of the problems happen in a very pragmatic way. They concentrate their work on a few relationships which seem to be important. They investigate only selected possible development paths without using complicated methods. They select the "best one". This procedure seems to be very subjective and there is the danger of neglecting other important factors. Economists have some trouble when they try to make statements about the usefulness of the Pestel/Mesarovič scenarios. On one hand there is the feeling that most of the results they present are not new, at least not new in terms of quality. On the other hand, the quantitative results of Pestel and Mesarovič are quite impressive. The problem is the extent to which we can trust these quantifications. The publications of the Pestel and Mesarovič group do not give a sufficient basis for checking the quality of the published results. There are good reasons for doubts.

Firstly, the group must have had the same problems as other economists in estimating the food supply in South Asia. To achieve substantial improvements in supply analysis it is necessary to set up rather complex national agricultural models. This has been mentioned earlier in this paper. Secondly, the Pestel/Mesarovič group had to solve very difficult trade-off problems. The estimation of the trade-off between the agricultural and the non-agricultural sector in the form of a production possibility function has not been solved in a satisfying way up to now. Furthermore, Pestel and Mesarovič have been faced with the relationship between import/export structures and sector productivities. Let us assume that they have solved all these problems better than other economists before them. The question then arises to what extent the best Pestel/Mesarovič scenario influences economic and agricultural policy. Most of the politicians have no general objections to this scenario and favour

(i) more foreign aid for South Asia,
(ii) foreign aid for South Asia mainly as transfer of capital goods and know-how,
(iii) food aid for South Asia only in cases of emergency,
(iv) measures which help the developing countries to achieve larger market shares for industrial goods on the world market.

The real problem is to realize these general aims, that is, to find operational procedures. Pestel and Mesarovič do not suggest those procedures. Their scenarios cannot give the necessary answer, because of the global and extremely long-run orientated character of their model. These scenarios are more helpful as a basis for general discussions than for detailed political decisions.

3.3 *The Bariloche model*

A research group in Latin America is working on a global world model called the "Bariloche model". The work is still going on and there is not much published information about the model and the results. The following remarks are based on a discussion from Nordhaus[12] and some other information.

The Bariloche model puts large emphasis on the situation in the developing world. The world economy is divided into one developed region and three developing regions (South Asia, Latin America, and Africa). Little attention is paid to the international economic relationships between these regions. For each region there exists one model. The characteristics are the following. The model contains much more normative content than the previous global world models. The, so-called, "basic needs" play the leading part. There are four categories of basic needs

(i) calorie intake per capita must exceed 3000 per day,
(ii) at least 98% of the population between 6 to 12 years must be enrolled in schools,
(iii) each family must be provided with a house of minimum quality,
(iv) other substantial consumption goods (e.g., health service)

The per capita utility is described by a very unorthodox utility function. The variables of the utility function are life expectancy and surplus consumption, defined as the consumption which exceeds the basic needs. The production possibilities of the region depend on labour inputs, capital stock, and natural resources. The production technology is described by sectoral Cobb–Douglas functions. Technical progress as defined in modern production theory is excluded. There are five production sectors. Four of them produce consumption goods and one produces investment goods. The utility function is maximized under the constraints given by the technology and the available inputs year by year. It seems that there are further constraints in this maximization which predetermines the level of investment. Anyway, the model shows an investment rate of 20–25%. Population growth is an endogenous variable. It is assumed that the population growth rate decreases when per capita income increases. At least, this assumption is correct when the basic needs of the population are satisfied.

The complete model allows two alternative classes of growth paths depending on the parameter constellation. If there is sufficient growth of the surplus consumption in the initial phase, the population growth rate decreases. Then the total regional demand for basic needs grows less fast than the GNP and this process continues in the direction of a developed country (balanced growth or an asymptotic approach to a final stationary state). If, in the initial phase, surplus consumption decreases or does not increase enough, the population growth rate increases. Then the surplus consumption decreases continuously to zero. The downward movement does not stop here because the per capita income decreases still further. Finally, the process leads to a Malthusian situation. The available information can lead to the conclusion that the Bariloche group has formulated the model for central planning economies. In most of the developed countries the provision of the basic needs can only be achieved by central planning means which enable a very nearly equal income distribution to be reached.

The model runs show that Asia and Africa will have a decrease in per capita income in the way described above. Only if the developed countries give substantial foreign aid in the way of capital goods to these regions will there be a take-off. The importance of the Bariloche model for agricultural policy cannot be analysed due to the lack of information. However, it seems to be a heroic approach to explaining the world food situation without taking into consideration the worldwide trade in agricultural goods.

3.4. *The Linnemann model*

Sponsored by the Club of Rome a research group in the Netherlands (since 1972) has worked on a project called "Food for a Doubling World Population". The final results of the project are to be published in 1976 ("MOIRA – A Model of International Relations in Agriculture", North Holland Company), The project leader is Dr. Linnemann and, therefore, the project is known as the Linneman model. It deals only with the world food problem. Nevertheless, it is useful to put this model into the class of global world models. Results from this project are not yet published; the basis for the following comments are some unpublished papers presented by the Linneman group. Their main aim is to analyse the world food problem up to the year 2010. It is assumed that up to the year 2000 the world population will double. Linneman analyses domestic food supply and food demand country by country. The quantity of food supply and demand is measured in terms of protein. Linnemann believes that this index is a better measurement than the usual grain equivalent approach. Per capita consumption of food is a linear function of per capita income. Linnemann takes into account various income classes; the national demand function for food is derived from the demand functions of various social groups. National income distribution and the growth of population are exogenous data. The main emphasis is put on the supply side. Linnemann tries to estimate the upper limit of food production for each country; roughly speaking, this is the level which cannot be exceeded without revolutionary new basic inventions. This upper limit provides 40 billion people with food under the condition that food is equally distributed and the basic needs are satisfied. The national upper limit of food production is an important element in the national production function for food. The production function furthermore includes the variables labour, land, capital, and fertilizers. Linnemann has estimated the parameters for the production function on the basis of cross-sectional data. The information I had did not tell me to what extent Linneman incorporated product and factor prices in his analysis and what are the determinants of these prices. In his summary paper Linnemann gives the impression that he has incorporated an endogenous world-wide price system and important action parameters of governments also.

The Linnemann group presents in the summary paper some results of the model, e.g., some tables which describe food production and food demand for the regions North America, EEC, Latin America, Tropical Africa, Middle East, Southern Asia, and the world in total , for the period from 1975 up to the year 2010. Each table describes a development which will take place when certain assumptions are fulfilled. The main determinant for the development is the economic growth in the non-agricultural world. High non-agricultural growth leads in all cases to a higher level of food consumption per capita in almost all regions. For the assumption of high non-agricultural growth as well as for the assumption of low non-agricultural growth Linneman analyses the influence of various strategies of agricultural policy. He considers unchanged policy, moderation of food consumption in rich countries, food aid on a large scale by the rich countries, high and low world market price levels, and some combinations. Surprisingly, the strategy "unchanged policy" compared to other strategies leads to a rather good world-wide food situation. No strategy really improves the per capita consumption in South Asia. The results which are described in the summary paper presented by the Linneman group seem not to be based solely on the tables referred to. Anyhow, the existing information gives the impression that they have analysed the advantages and disadvantages of various strategies very carefully. To what extent the interpretations are really outputs of the Linnemann model and to what extent other considerations are important, is not absolutely clear. In the following we repeat some of the Linnemann theses:

(i) A decrease of economic growth in the developed world and also a decrease of food consumption in these areas has negative effects on the food situation in the developing world.

(ii) If the developed countries spend 0.6% of their GNP every year for food aid for the developing countries hunger in the world can be reduced to nearly zero up to the year 2010, at least when distribution is managed in an optimal way.

(iii) A liberalization of international agricultural trade has negative effects on the food situation in the developing world. The reason for this is a world-wide reduction of food production caused by the liberalization.

(iv) A stabilization of world market prices at a high level achieved by measures operated by the developed countries is an important improvement compared with the current policy in the opinion of Linnemann.

We have to wait until the Linneman group proves that their considerations for long-run policy measures are based on excellent investigations.

3.5. *Some remarks on the meaning of long-run global world models*
The history of long-run global world models is too short for us to be able to come to a final conclusion about the meaning of these models. However, already today some limitations are obvious. The effort to present a world model which incorporates all main economic activities in the world on a very high aggregated level has failed. At least, these models are not relevant for actual economic policy. That is why Pestel/Mesarovič and, especially, Linnemann analyse partial worldwide problems on a lower level of aggregation. They do not put as much emphasis on inter-sectoral linkages or bottlenecks caused by non-renewable resources and pollution as Meadows did. But their models also do not include real political and institutional elements; especially national problems are not part of the models.

A large part of the model analysis is concentrated on the deduction of long-run growth limits in the technologic field. These analyses seem to be rather useful. They can give incentives to governments and other institutions to formulate or reformulate long-run programs in research and development. For various reasons allocation of resources is much more imperfect in the social sense in the so-called "technical progress production" than in the ordinary production sectors (higher degree of monopoly, more externalities, higher degree of uncertainty, larger time differences between input and output). The models do not give many hints concerning the international division of labour. There is a large demand for these hints from politicians throughout the world. It should be clear that the factor capital and the factor labour, too, are not distributed in an optimal way, at least not in a way which leads to a maximization of world production. International trade, too, is not optimal. It is really necessary to get more insights into the limitations and possibilities of international capital and technology transfer. Probably most of the limits arise from institutional constraints. It would be useful to show the social opportunity costs of their nationalistic economic policy to some politicians. A large international capital and know-how transfer and the transfer of consumption goods, too, are related closely to questions of transfer of international finance. In this field it is necessary to develop new world-wide scenarios which include also institutional variables.

4. TRANSMISSION MODELS

4.1. *The LINK project*
Up to now we have discussed world models dealing with long-run problems. It seems to be impossible to extend these models to short-run and medium-run models. It is difficult to set up a world-wide disaggregated short- and medium-run model. There are suggestions that we should start with a world-wide input/output matrix[13] and develop a model on this basis. However, in practice the set-up of a world-wide Walrasian model seems to be an illusion. On the other hand there is a growing interest in the so-called "linkage models" in

the past year. World-wide linkage models consist of national models linked together. These models have several advantages:

(i) many national models are already available,

(ii) the national models can differ to a certain extent from country to country. This gives a chance to incorporate country-specific situations.

(iii) the world models can be set up in a decentralized way. Every nation presents its own model and the world model-builders link them together.

The interdependence between national business cycles has especially been forceful in convincing many economists that it is necessary to link national models together. Many countries have macroeconomic models for the short- and medium-run. These models differ in many aspects (level of aggregation, time horizon, quarterly or annual data, incorporation of the monetary sector). Exports and imports are explained by export and import functions. In most cases the export quantities are functions of the world trade and the relation between domestic prices and world market prices. The amount of imports is a function of national income or similar variables and of the relation between domestic prices and the world market prices. World market prices and world trade are exogenous variables in the national models. Exports and imports, and also the prices calculated and predicted by the national models, are not consistent in the world-wide sense. The consistency in the world-wide sense is given when the sum of all world imports equals world trade, the sum of all exports equals world trade and prices expressed in dollars and calculated on f.o.b. basis are the same from nation to nation. The LINK group[14] is working to achieve this world-wide consistency. They have developed some iteration techniques which lead to this consistency. In the iteration processes world trade is varied until consistency is achieved. Some iteration techniques secure only the consistency of imports, others the consistency of exports, imports, and prices. Some approaches result in countrywise consistent exports and imports, but no consistent results are achieved for the trade flows between different countries. Other iteration techniques calculate a world trade matrix on the basis of a constant market share approach. "Perfect" linkage techniques are not yet available, but the considerable progress in developing new techniques in the last few years gives us hope for the availability of much better techniques in the near future.

The LINK group tested its model for the first time in 1971 to get preliminary results. An extensive analysis was presented in 1975. The linkage of national economic models as it is done by the LINK group in Philadelphia, has attracted much attention from experts who are involved in employment and monetary policy. Up to now, the LINK model has delivered simulations for four years. But, as Klein pointed out, it is possible to do simulations for a period of 6 to 7 years. Such a time horizon is long enough for many decisions in economic policy. The LINK project is a long-term one; in the coming years some bottlenecks have to be overcome. National models must be improved; especially the monetary sector must be incorporated. This incorporation of the monetary sector allows the construction of a linkage model which explains capital flow and balance of payments and exchange rate problems. The LINK models have a rather high aggregation level. There are only four production sectors; one is the agricultural sector. It is well-known that it is rather difficult to build up a satisfying multi-sectoral business cycle model, but on the other hand we can expect that in the next years models with 4 or 5 production sectors will be available for many countries. The incorporation of normative elements and policy parameters has to be pushed forward, too.

The success of the LINK group mainly depends on three components

(i) development of linkage techniques,
(ii) development of national models,
(iii) administrative and financial constraints.

The development of new linkage techniques seems not to be the main constraint. The development of national models takes a lot of time. These models must be available at a central institute to be calculated there. This central institute needs all exogenous data as early as possible. This is necessary to ensure that the calculations can be done in time. To a certain degree the work of the national model builders has to be coordinated. However, the basic principle of the LINK approach is that the nations themselves know their problems better than a central institution and for this reason they should build their own models.

4.2. *LINK approach for the agricultural sector*

For the near future there is little chance to disaggregate the national short- and medium-run models in such a way that many production sectors can be considered explicitly. In the future, we will perhaps have national economic models with 4 or 5 production subsectors for many countries – and one of these subsectors will be the agricultural sector. That is why the linkage of national economic models can deliver only rather global insights in the world-wide agricultural relationships. For more detailed insights there is a need for a disaggregated world-wide agricultural model. For a world-wide linkage of disaggregated agricultural models with national models it is necessary to work with a multiple LINK approach. Figure 1 shows the basic idea. Every nation presents a national economic model with 4 or 5 sectors, the agricultural sector in this model is the result of an aggregation of a more detailed agricultural model called "disaggregated national agricultural model". This disaggregated national agricultural model can be of the Day/Heidhues/de Haen type. The national economic models can be different from country to country. They can be linked together by using one of the available linkage techniques. In this way, we get solutions of global national models which are consistent on a world-wide scale. At this stage it is possible to introduce some corrections and to solve the system again (i.e., modifications of policy parameters). The final solution of the national models also delivers aggregated national models which are of world-wide consistency. These models are the starting point for a disaggregated analysis of the international agricultural situation. The regional disaggregated national agricultural model is solved within the framework which is given by the solution of the aggregated national agricultural model. This has to be done for all national agricultural models. The disaggregated

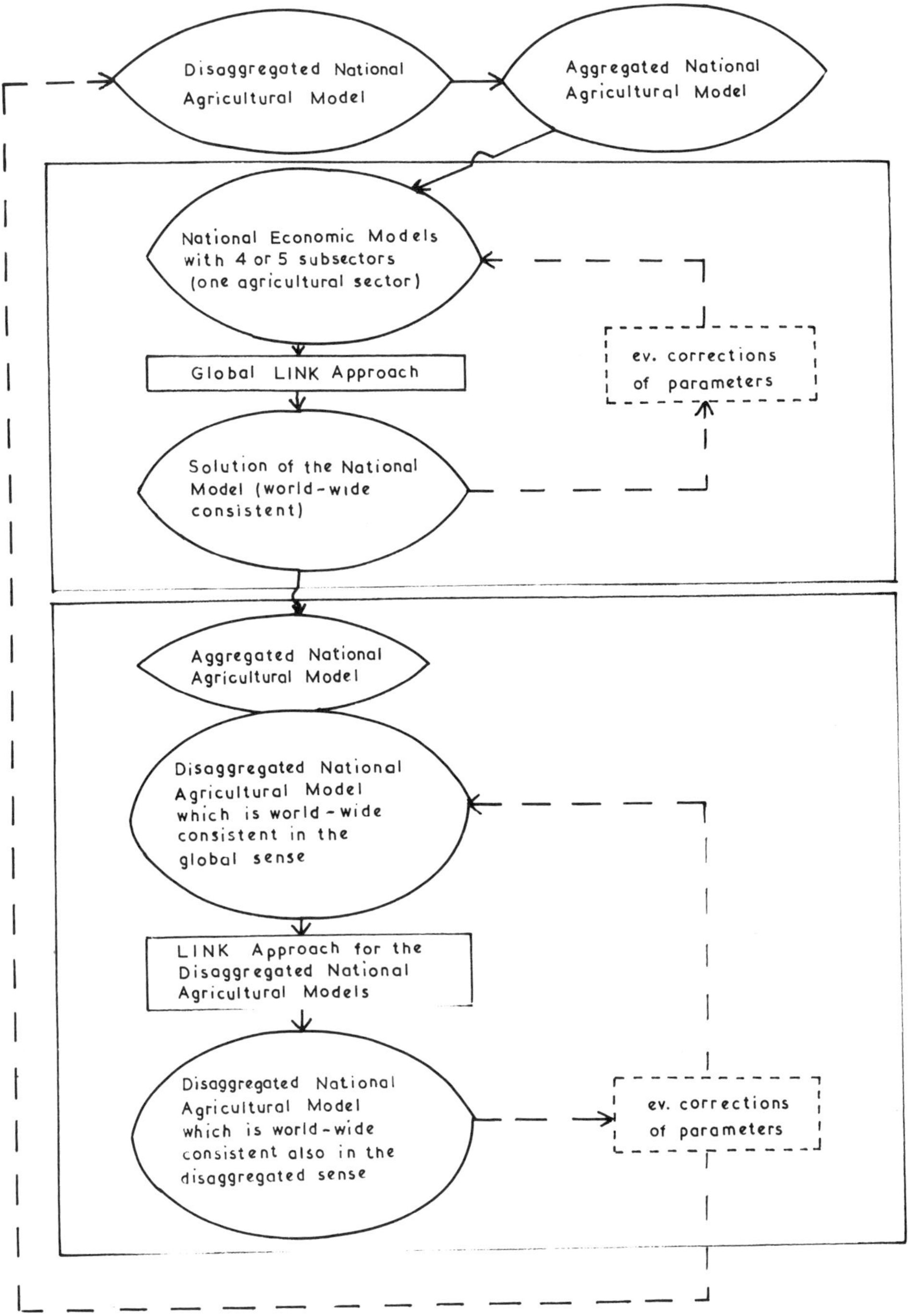

FIG. 1. Multiple LINK approach.

national agricultural models calculated in this way are consistent world-wide with respect to aggregated exports and imports but not to disaggregated exports and imports. To get consistency on the disaggregated level, too, the LINK approach is used. The results are disaggregated national agricultural models which are also consistent in detail. The solution of these models can be considered as the final solution of the multiple LINK approach. However, further calculations may be useful. It is possible, that various parameter variations, including government policy variables, may look quite unrealistic. Two corrections can be made. Firstly, it is possible to solve the disaggregated models again for the new parameters without attention to the relationships with the non-agricultural sector. This approach may be sufficient in many cases. Only in cases in which we have to expect that parameter corrections in the agricultural sector will have an important influence on the whole economy it is necessary to repeat all stages of the multiple LINK approach. The LINK approach for the disaggregated agricultural models in principle is confronted with the same difficulties which arise by the global LINK approach. However, there are some additional problems. More microeconomic elements have to be incorporated (price elasticities, special characteristics of the agricultural markets, etc.). Also government regulations must be incorporated in a more detailed way. The outlined scheme is rather flexible in the way that models can be modified in different ways according to the problems which are of interest.

The multiple LINK approach looks rather complex. However, current developments in model building are going in this direction. Most economists agree that it is necessary, and possible, to continue the development of national economic models which include 4 or 5 production sectors. Most economists, too, favour the set-up of integrated agricultural development models for all countries. The collaboration between national model-builders and agricultural economists has not been very close in the past but the situation is beginning to improve. Recent discussions have shown that a closer cooperation in model-building is necessary and is desired.

Of course, there are differences from country to country. In highly industrialized countries it is possible to explain major features of the economic development without a detailed consideration of the agricultural sector. In developing countries the opposite case will be relevant. The global LINK approach will be used to a much larger extent and the interest in world-wide consistent agricultural models is also growing. Therefore, the prospects for the multiple LINK approach are rather good.

The contributions of multiple LINK models to agricultural policy, as described above, can take place in various ways. Short- and medium-run national aggregated agricultural models which are linked properly to the other national sectors and also to the world-wide scene will gain more and more interest for institutions which give governmental advices. In recent years decision-makers in agricultural policy have realized that it is impossible to set up an optimal agricultural policy without considering the relationships to other parts of the national economy. Furthermore, international agricultural problems influence national agricultural policy at present more than in the past. Especially developing countries with a high intensity of agricultural foreign trade have to take into consideration world market situations when they make up their national development plans. The patterns of exchange rates, balance of payments, and business cycles influence strongly the prospects of their agricultural sector and their national food supply, especially in the developing world. Most failures in agricultural policy have arisen because decision-makers in agriculture do not have enough informations about the surrounding sectors. It should be possible to present the results of the multiple LINK approach in such a way that the decision-makers in agriculture get more and more interested in intersectoral and international relationships. This is a necessary condition for an impact of multiple LINK models on political decisions. Learning processes will take place and, finally, decisions will be influenced by these models. The advantage of the multiple LINK approach is that some of its elements, especially the national parts, have already an important influence on political decisions. The new parts are just extensions and therefore, learning step by step is possible.

REFERENCES

[1] FAO (Ed.) (1971), *Agricultural Commodity Projections 1970–1980*, Vol. I and II, Rome; OECD (Ed.) (1968), *Agricultural Projections for 1975 and 1985, Europe, North America, Japan, Oceania. Production and Consumption of Major Foodstuffs*, Paris; USDA (Ed.) (1973) *Growth Potential of Corn Production in Western Europe through 1975 and 1985*, Foreign Agricultural Report No. 88, Washington D.C.

[2] FAO (1971), *A World Price Equilibrium Model*, Projections Research Working Paper, No. 3, CCP 72/WP. 3, November.

[3] Samuelson, P. A. (1952), *Spatial Price Equilibrium and Linear Programming, American Economic Review*, Vol. 42; Fox, K. A. (1958), *Econometric Analysis for Public Policy*, Iowa State College Press.

[4] Labys, W. C. (1975), *Quantitative Models of Commodity Markets*, Cambridge Mass.

[5] USDA, *World Demand Prospects for Grain in 1980, with Emphasis on Trade by the Less Developed Countries*. Foreign Agricultural Economic Report No. 75.

[6] de Haen, H. *et al.* (1975), Integrierte Entwicklungsplanung. Eine Bestandsaufnahme, *Volkswirtschaftliche Schriften*, Heft 228, Berlin; Day, R. H. and Singh, J. (1971), *A Microeconometric Study of Agricultural Development*, Workshop Series 7120, Social Systems Research Institute, Madison (Wisc.).

[7] Meadows, D. H. *et al.* (1972), *The Limits to Growth*, New York; Meadows, D. L. *et al.* (1974), *Dynamics of Growth in a Finite World*, Cambridge, Mass.

[8] Mesarovič, M. and Pestel, E. (1974), *Mankind at the Turning Point*, New York.

[9] Bariloche Foundation, *Latin American World Model*, Preliminary report, presented at the International Institute for Applied Systems Analysis, Laxenburg (Austria) October 1974.

[10] De Hoogh, J., Keyzer, N. A., Linnemann, H. and van Heemst, H. D. J. (1976), *Food for a Growing World Population*, Economic and Social Institute, Free University, Amsterdam.

[11] Nordhaus, W. D. (1973), World Dynamics: Measurement without Data, *The Economic Journal*, Vol. 83, December; Scheper, W. (1974), Discussion of models of the Meadows, Pestel–Mesarovič type. In: H. Giersch (Ed.), *Das Umweltproblem in ökonomischer Sicht*, Symposium 1973, Tübingen.

[12] Nordhaus, W. D. (1975), *World Modelling from the Bottom up*, International Institute for Applied Systems Analysis, Laxenburg.

[13] Bottomley, A. *et al.*, *A World Agricultural Model: An Input–Output Research Proposal*, unpublished paper.

[14] Ball, R. J. (Ed.) (1973), *The International Linkage of National Economic Models*, Amsterdam, London, New York; Klein, L. R. (1976), *Philosophy and Methodology of Linkage*, paper presented at the Third International Conference of Applied Econometrics, European Community Commission, Brussels, Feb. 4–6.

DISCUSSION OPENING – C. Thomsen, *Denmark*

Professor Richter's paper is instructive and down-to-earth in an area which is so wide and complicated that it is easy to be carried away into highly sophisticated considerations. The complexity of the subject may be illustrated by the remark that world-wide decision-making in agriculture is a rather nebulous concept and international trade models are hard to find, to use an analogy applied to the New Economic Order.

Among the different levels of decision-making dealt with at the Conference, the one at the international level may be considered to be of the highest order. But I agree, that there is a tendency to limit decisions to areas and measures which can be controlled in the framework of a national policy, or a common policy as in the EEC. Unfortunately such decisions often have repercussions for other countries and for international trade, i.e. the so-called beggar-thy-neighbour policies. There is thus a growing need for concern about decision-making at the international level, and the subject seems worthy of more attention than it has received at the Conference.

The complexity of decision-making in the EEC has been well illustrated in Dr. Tracy's paper, and anybody who attended the UN World Food Conference in 1974 should agree that it would be extremely valuable to have an analysis of the decision-making that took place on that occasion. Although behavioural science is on the border-line of the discipline of economics, there is an obvious challenge in its application to international governmental organisations. Some work has already been done, but with the advent of the New Economic Order the need for a better understanding of the machinery involved will be even more pressing.

As regards examples of international trade models, Professor Richter has referred to the OECD projections and the FAO perspective studies. But in my view these studies could not be called trade models proper. As the paper points out, they only deal with international trade as residuals or balances resulting from projections of aggregate production and consumption in individual countries and regions. In a trade model worthy of the name the trade issue, with attention to prices should have a central position in the analysis. The paper makes reference to the deficiency of the studies in this respect and to the difficulties in obtaining reliable price elasticities both for supply and demand. But these difficulties could not diminish the importance of prices and price signals in any model dealing with trade. Without going into the other difficulties in connection with the setting up of international trade models referred to by Professor Richter, it seems worth making the point that some of the major criticism of the operation of the world market in agricultural products has come from quarters which have, themselves, contributed to the worsening of the situation by limiting its scope of operation still further.

With regard to the interrelationship between product studies, country studies and trade models, Professor Richter has emphasized that the problem of the quality of data becomes the more serious the more aggregated is the model. But if we want to work out realistic models, we cannot escape the need to integrate commodity studies into agricultural sector models, both nationally and internationally. The same applies to the integration of agricultural sector models into models for the total economy, and finally to the eventual integration of countries and regions into a world model. In my view, this can best be done through a step-by-step procedure, the same way as macro-models gain from being based on micro-models, and I find the exploration of possibilities of linking various models together very promising. Also in this context, prices will have a central role to play in the process of linking models together and providing for the necessary feed-back between them. This subject is described in the following paper by Professor Scheper.

My final point is concerned with the possibility of, in fact, using trade models as a base for world-wide decision-making. Although there is a growing need for this the experience of the past in international organisations leads to a great amount of scepticism. The predictive value of existing models has proved to be very limited and although the suggestions of Professor Richter for the inclusion of additional non-quantifiable parameters and factors, i.e. of institutional and political character, may improve their ability to explain the past I cannot see that this could improve their predictive value. The stochastic character of some of these variables, referred to earlier in the Conference, should support this view.

However, although I am sceptical about the direct value of these models for decision-making, I find more reason for optimism about their educational value. Apart from their value in university education, I also believe that they can play an important role in the education of the politicians. Here lies a challenge to agricultural economists and research institutes, both at the national and international level. International organisations no doubt also have an important part to play in the education of politicians but, so far as model-building as such is concerned, this task should preferably be entrusted to the independent research institutes.

DISCUSSION OPENING – A. S. Watson, *Australia*

Dr. Richter's indecision about the usefulness of formal models in predicting world trade in agricultural products makes it difficult to comment on this paper. In the early part he is sceptical about such models and there appear to be good reasons for this, particularly with respect to models like those of FAO which he discusses. More recent development in trade modelling may overcome some of the problems, but there are still good reasons for agreeing with the scepticism of Dr. Richter's introduction.

When only a small proportion of world production of a commodity is traded, it is inevitable that fluctuations in supply will change the status of countries, or groups of countries, that aim at self-sufficiency, from being net importers and net exporters. Quantities entering world trade, and world prices, therefore, are inherently unstable.

In the conclusion to his paper Dr. Richter suggests that international organisations put more resources into formal forecasting. This must reflect a judgment that efforts so far have been worthwhile, either for prediction or for qualitative analysis of trends revealed by formal models. In my opinion, the latter use of models is more acceptable.

My comments concentrate on specific points, but I would like, first, to refer to an important deficiency of the paper. There is no mention of the need to revise the approach to trade modelling in a world of flexible exchange rates. It is no longer sensible to separate agriculture from the rest of world trade and the world economy in model-building. To me, at least, this is a serious weakness of the models he has described; the current generation of trade models is considerably in advance of the models to which he has referred.

International trade in agricultural products will not, in itself, solve the world food problem. Expanded trade would assist in the process of income generation necessary for capital accumulation in developing countries. With respect to the food problem, trade in grains can help meet shortfalls in domestic production. The remainder of agricultural trade is either in commodities like sugar, coffee and tea – which are not generally described as food – or in agricultural products like the natural fibres which are industrial raw materials. In addition there is the trade in livestock products which is essentially restricted to developed countries because of the substantial investment required in the marketing infrastructure.

At the start of his paper Dr. Richter asserts that it was taken for granted in the 19th century that the instruments developed to facilitate trade functioned perfectly. This is a considerable over-simplification of economic history – to put it mildly. Later Dr. Richter refers to the fact that nowadays a greater percentage of world trade takes place outside traditional marketing systems. Most bilateral arrangements cannot be called "non-market" as they require information provided by commodity markets.

Dr. Richter further states: "In the field of agriculture, of greatest importance is to account for cycles in production and to incorporate one's own cyclical production patterns to meet international market requirements". I think this misleading, especially with respect to trade in grains which is essentially generated by shifts in supply in importing and exporting countries. So far as I know, there is no respectable evidence of cycles in weather, which makes it doubtful whether formal modelling would ever be useful for predictive purposes, although this is not the only reason for modelling.

In sections III and IV of his paper Dr. Richter discusses the sorts of models that can be applied to agricultural trade and describes the methods adopted by FAO, OECD and the EEC. I do not wish to attempt to add much to this review of previous work, except to say that the traditional method adopted by FAO of predicting average production forward using technological information, and consumption forward from estimated income elasticities and population assumptions, and then bringing the two elements together in a projection of price trends, leaves a great deal to be desired. It is arguable whether these are economic models at all.

"Gap analysis", as this approach has been called, ignores the self-correcting of imbalance by economic mechanisms. The trends revealed by such studies are likely to be reversed by adjustment to production, consumption and exchange rates that negate the forecast. Predictions, however carefully guarded their terms and even if they are presented as mere projections under stated assumption, frequently affect outcomes. I am not asserting that the only way predictions can be assessed is by comparing actual outcomes with predicted incomes: predictions or projections can be useful if they give market participants credible information in adjusting production and consumption.

However, official forecasts have occasionally brought about a convergence of expectations with uncritical acceptance of the authority of economists working in international agencies. Time is lacking to quote examples from recent history but official forecasting sometimes has made the situation worse rather than better.

One safeguard that would improve this situation would be for international agencies to observe standard academic conventions in reporting their empirical work. This has not been done on all occasions. It means their work cannot be evaluated properly by outsiders.

Dr. Richter concludes on an optimistic note. He refers to the "gradual abolition of trade restrictive practices allowing comparative advantage to affect allocation of production to economic criteria." I only wish I could share this hope.

DISCUSSION OPENING – G. L. Johnson, *U.S.A.*

I find Professor Scheper's written paper and verbal comments very helpful. In my view, his paper correctly stresses the significance of the advances now taking place in modelling approaches and techniques.

In surveying international commodity, global, and what he calls transmission models, Professor Scheper concentrates mainly on what I call subject-matter and disciplinary, as contrasted to problem-solving, models. Some of the subject-matter and disciplinary models reviewed are of questionable relevancy; at least the *set of problems* and the *decision-making* units facing them are not specified by the researchers involved. For instance, Professor Scheper points out vis-a-vis global models that they "would be much more useful for a centralized world government." Important as it is to be able to relate models to problems and decision-making units, it is also important to note that problem-solving, global models are virtually impossible because their makers cannot obtain information by interacting iteratively (Dr. Petit would say adaptively) with global decision-makers. Thus, the informal components involving administrators (referred to as essential by both Dr. Petit and myself earlier in the Conference) cannot be established in developing global problem-solving models.

Inability to construct global problem-solving models for a non-existent world government does not preclude, however, the development of relevant global subject-matter and disciplinary models. Such models can be germane to sets of problems faced by well-identified, existing national and supranational decision-makers, both public and private, as discussed by Michael Tracy and Guiseppe Barbero.

The Club of Rome models discussed by Professor Scheper are of questionable relevance and value either as problem-solving or subject-matter models, at least, in my view. The Meadow's model, among its other shortcomings, treats the world as homogeneous – like a cask of fermenting wine in which the yeast population "explodes" to either consume all available energy or to kill itself by polluting its own environment with alcohol. The Pestel–Mesarovic model does disaggregate the world into 10 separate wine casks. They attempt both to improve the world data base and to integrate the 10 regional models, but, according to Professor Scheper, have not yet succeeded. I believe the data and linkage problems they face are surmountable for their models, as presently formulated. However, I believe that, to be usable, such models must be made stochastic with respect to the impacts of weather, at least, so as to permit Monte Carlo simulations through time to determine variances in the consequences resulting from alternative national and supranational policies. We need to know more about the variances in regard to who is hurt and benefited, how and in what way, where, and when? We also need to be able to study the relationship between equality, knowledge and control (through government), on the one hand, and global stability, on the other.

I do not know the Bariloche model but Professor Scheper's description of it indicates that it is more specialized on economics than the other two efforts. It seems to lack the technological dimensions of the Meadows and Pestel–Mesarovic efforts and to share their lack of attention to the political and military realities of the world. A recent issue of Harper's magazine* castigated the Club of Rome and the Pestel–Mesarovic model, in a conservative way, for political naivete and for lack of objectivity in dealing with normative and prescriptive knowledge.

The Linnemann model is also incomplete. The tentative conclusion presented to us by Dr. Scheper suggests thet the Linneman group may have incorporated constraints into their models to reflect some reality with response to the ownership of market, political and military power in the world scene.

The LINK project discussed by Professor Scheper is not familiar to me though it is in charge of L. Klein with whom I have collaborated in doing past econometric research. What is envisioned by Klein is the linking of national econometric models into global models. It is the same linkage problem which bothers the Pestel–Mesarovic team but then their more complicated models deal with technology as well as economics. Klein's linkage problems are likely

* Florman, Samuel C., "Another Utopia Gone", *Harper's*, Harper's Magazine Co., New York, New York, August 1976.

to be fairly easily handled compared to those of Pestel–Mesarovic. Methods can be used which are roughly similar to those used when we combine enterprise components into firm models and subsector components into sector models. Klein's linkage problems would multiply geometrically for stochastic models not specialized on economics and dealing with technological, institutional and human change while giving attention to the reality of existing and changing world-wide distributions of market, political and military power. Until we learn to build some of these elements into our global models, even subject-matter global models will be of questionable relevance.

Professor Scheper's discussion of the LINK approach for the agricultural sector is interesting and sensible. Linkage is essential but so are good model components to be linked. I feel that good relevant model components for linkage would accumulate more rapidly if we modelled:

(a) the multidisciplinary domains of practical problems before real world decision-makers. We should do this in a philosophically flexible way without specialization on techniques or disciplines. As many decisions of national and supranational agencies are executed in the world arena, this would help us develop relevant components for global subject-matter models;
(b) subject-matter areas such as food and nutrition, employment generation, energy in agriculture, trade, technology adoption, national agricultural accounts models, environmental impacts, etc. This should be done in a multidisciplinary, philosophically flexible way, unspecialized on techniques. Much is to be gained in doing this work from defining the sets of practical problems and specifying the corresponding decision-makers for which a subject-matter model is relevant;
(c) disciplinary (economic) aspects of the domains of practical problems. We do much good and relevant work of this nature. The results are valuable components of both subject- and problem-solving models. One thing for us to be careful about in doing such work is not to oversell it. After all it is specialized on economics in a world of dynamic technology, institutions and people. We seldom solve problems as economists alone. We do not sell out to government or become hostages to it when we recognize that we have not handled the political, military and other dimensions of a practical problem; instead, we win credibility for recognizing our limitations while doing that which we do best – economics.

Before closing, I would like to return to Professor Scheper's excellent paper and ask him if he would be in position to comment for us briefly on such national models as Heady's Thailand model, Leroy Quance's NIRAP model in the United States Department of Agriculture, the IBRD Mexican model as an example of a GSSSA model such as Halter's Venezuelan model or the Nigerian simulation model. The Nigerian model accounted for key projections of 62 out of 536 pages of the Nigerian government's document on "Agricultural Development, 1973–1985". I know Dr. Scheper was pressed for space and time in his paper and presentation and feel therefore that he might appreciate an opportunity to extend his review beyond the surplus/deficit, equilibrium and trade models he touched on in the first part of his paper. All but one of the models I have just mentioned are complete or in advanced stages.

In closing, just a word about my optimism concerning models – and the GSSSA approach. My optimum is so realistic with respect to difficulties that it is hard to distinguish it from the pessimism of an equally realistic pessimist. Perhaps, my following discussant will be a realistic pessimist.

DISCUSSION OPENING – M. Daad, *U.K.*

Let me say at the outset that I was glad Professor Scheper noted that "the influence of published models on decision in agricultural policy is still rather low". I agree that this is a correct appraisal of the situation up to now, although one should also note that some of the earlier models, particularly Limits to Growth, have provided a useful contribution to public awareness and to public debate about finite resources and the balance of food production and world population. I would say we are still in a preliminary stage of theoretical development of world models.

I outline briefly four reasons why I believe these models have had only a limited direct impact on agricultural policy so far.

(a) The aggregation of the models working so far has been at much too high a level. For example, in spite of a tendency to disaggregate countries in more recent models into a number of geo-political economic groups, it still remains very difficult for individual countries to assess the optional course of action open to them. Indeed, even if the EEC were to be considered as one such group, individual EEC member states would still wish to be able to analyse the implications of the estimated results themselves. There is also a problem of the level of aggregation of agricultural commodities. I would suggest that most of the policy decision on which government economists advise relate to priorities within the agricultural sector; for example, the relative emphasis which should be placed on arable and livestock, between feed grains and food grains, between feed based and forage based livestock systems. Certainly for these sorts of decision one needs information about the quantities involved and, most importantly, about the relative prices of agricultural commodities in the future. This has certainly been our experience in the assessments for the U.K. policy document "Food from our own resources". The problems of allocation of resources between, say, agriculture and the rest of the economy are arguably not addressed so often by policy-makers and are, in any case, very much more difficult to assess quantitatively.

(b) I would like to raise a related problem about the modelling of the main institutional structures into world models. I am particularly thinking of trade arrangements which can, of course, feed back into production patterns. Reference to milk products and sugar readily illustrates the relatively high proportion of world trade which can be covered by trading agreements. Clearly, for the model to give useful results for policy purposes these institutional arrangements need to be considered.

(c) The next reason why I would suggest agricultural policy-makers and their economic advisers have not yet been generally convinced of the usefulness of results of these models, is that the models have not been sufficiently validated. Professor Scheper has perhaps not given enough attention in his necessarily brief paper to the problems of validation of these large models. Certainly, the fact that most of these models are iterative, projecting the outcome of $t + 1$ on the basis of the outcome of $t = 0$ gives serious reason to reflect upon the possible confidence limits at $t + 10$ or $t + 20$. Perhaps I can illustrate the problem of validation with reference to the Pestel–Mesarovic model. The central structure of the inter-relationships is so complex that few, if any, apart from the modelling team themselves have successfully unravelled it. No-one has, to my knowledge, repeated the Mesarovic modelling runs to check the results. This is clearly a highly undesirable state of affairs, because it seems that for the time being at least, we have to take these results on trust. Policy-makers are, for the most part, quite reasonably not prepared to do this.

(d) There is the problem of the relationship of the policy-decision-maker to the model. I believe, one of the main criticisms of the Forrester World II model, and some of the other early models, is that it assumes a total lack of anticipation by the decision-makers who are implicitly contained within the model structure. All decisions within the World II model are based on the current values of variables. There is no possibility of decision-makers modifying their decisions to take into account the likely consequences of current course of action. Policy-makers, I would suggest, do not act like lemmings marching to the edge of the cliff and falling off, even though, in our more cynical moments, we might feel that they have to look over the edge first before turning back. Certainly the Systems Analysis Research Unit Model currently being developed by a U.K. government team, with which I should add I am not associated, like the Linnemann model, programmes the decision-maker as a rational being guided by self-interest. The alternative open to policy-makers also raises important and difficult questions about the appropriate treatment of R and D in models, particularly whether it should be treated exogenously or endogenously.

Finally, I should like to raise the question of the future place of these models in the national and international decision-making process. We have already heard a number of speakers at this conference express doubts as to the usefulness of mathematical models in agricultural decision-making. For my part, I believe that it is still too early to tell how useful these models might be, and that in the meantime we should continue to give support to this area, provided that the models remain simple and problem orientated. Professor Scheper says in his paper "The construction of global world models and link models will be one of the fastest growing disciplines in economics". If this is the case, and I am not in a position to question his judgement, I wonder whether he would indicate whether he believes this is an optimal allocation of our own scarce professional resources.

Report of the general discussion

Part of the discussion consisted of contributions on more or less factual matters. In respect of Dr. Richter's remarks about FAO activities in the modelling field it was stated that FAO's main purpose in the Indicative World Plan was to build up a *normative* framework for the development of world agriculture for a period of about 25 years, not a trade equilibrium model. Within this, the emphasis was on the feasibility of expanding agriculture in the LDC's consistently *inter-alia*, with their macro-economic goals and their nutritional needs. It was also stated that, contrary to the statement in the paper, the supply figures were not simply based on trends but stemmed from a determination for each developing country of a feasible development path and strategy, taking account of the physical and economic potential, and technological improvements. Discrepancies between the supply figures in the IWP and experience are of the nature of divergences between IWP guidance and government action. Turning to recent FAO activities the speaker said that the new Perspective Study of World Agricultural Development which FAO was just starting would try to improve on the IWP, for example, by developing supply figures by agro-ecological zones and by more integral handling of the nutrition and trade/self-sufficiency objectives.

As regards results from FAO's work in these fields, it was argued that, in addition to the activities of the various inter-Governmental Commodity groups, decisions at the World Food Conference indicated the usefulness of the IWP analyses – for example, those in connection with International Undertaking on World Food Security; the work on the International Fund for Agricultural Development, the International Adjustment Strategy, minimum aid targets and developments generally on World Food Strategy.

One or two speakers regretted various omissions or underemphasis on matters which they saw as essential elements in decision-making at international level. These included treatment of the role of trade in development, the implied perpetuation of the pre-existing international division of labour, a framework for analysing appropriate commercial policies for backward countries, consideration of DC's trade policies towards LDC's, the farm sector policies of advanced countries, international price stabilisation schemes and UNCTAD's successes and failures.

Professor Richter was pressed on his definition of comparative advantage in view of the results of research in several countries, for example, by Hayami and Ruttan, Evenson, and Valentini, suggesting that land might not be the fundamental element in it but rather investment in adaptive research and rural education.

In so far as the need for data can be divorced from the models and regarded as factual, Professor Scheper's implication that model simulations were limited by the difficulty of giving long-run estimates of exogenous variables was challenged on the basis that the majority of the variables in the models being discussed were endogenous. Though there was emphasis on the need for more data on the endogenous variables to anticipate seasonal fluctuations, the force of the comments on the lack of data seemed to relate to the uncertainties in forecasts arising from lack of data on exogenous variables. In this connection doubt was expressed whether institutional and political variables could be omitted.

On the theoretical side the need for a fully valid foundation theory if extrapolations were not to lead to models at variance with facts, underlay various contributions, sometimes associated with grave doubts whether such foundation was available.

The positive types of model were seen by one speaker as inspired by the approach of natural scientists but to be consistent they must fulfil three conditions (i) the system must be closed, (ii) there must be complete knowledge of the state of the system and the relations within it and (iii) the law of conservation – but these are not met. (For example, see Northrop *The Logic of the Sciences and Humanities*.) So, in the speaker's view, there is no basis for reliable predictions.

Professional interest tended to press in the direction of greater research on international trade generally and more sophisticated models and interlinked complexes of models of individual commodities or subsectors. There were, however, voices arguing for more pragmatic analysis – it was reported that OECD had been working on an analysis of world food prospects which was not a projection exercise but gave a very useful guide to probable developments. It is, after all, important not only to know something of the trend but also of factors making for departure from the trend.

Though not expressing any doubts about the usefulness of "orthodox" model-building, some feeling was voiced – for example, in connection with the Meadows–Forrester approach – that a mystique was implied whereas in fact we were dealing simply with a useful tool which anyone could learn to handle in a few weeks.

Professor Scheper's support for the Day–Heidhues–de Haen type models for international linkage were criticised on the grounds that they often needed the introduction of rather *ad-hoc* flexibility constraints. The Duloy–Norton type of agricultural sector model was urged instead, since it explicitly includes price responsive primary factor supply and product demand relations. The model then simulates a competitive equilibrium, determining all factor and product prices endogenously, subject to policy and institutional constraints. It was stated that models were already being developed by the World Bank and others for a number of countries – for example, Mexico and the Philippines. They are open economy models and should lend themselves to the kind of linkages Professor Scheper suggests.

In response to Professor Scheper's pressure for more linkage of agricultural with non-agricultural sector models the PROLOG prototype modelling framework developed by Roger Norton *et al* at the World Bank Development Research was presented as an attempt at general equilibrium modelling in which interactions of agricultural and other sectors are explicitly included. There was the possibility that there would soon be enough price endogenous national agricultural models worked out to make it possible to begin on the linkage work which Professor Scheper had suggested.

Participants in the discussion included: A. Abdullah, *Bangladesh*; J. P. Bhattacharjee, *FAO*; J. Dubois, *France*; S. Midenberg, *France*; H. van Miltenburg, *Netherlands*; P. C. van den Noort, *Netherlands*; K. Prasad, *India*; R. Thomson, *U.S.A.*; M. Tracy, *EEC Belgium*.

J. J. Richter (in reply)

The complexity of international trade is such that it forces the researcher to single out items which he can safely investigate, document and present to a meeting like this. It was very helpful to have, from one of our friends, a catalogue of questions concerned with international trade problems.

Regarding the usefulness of models, I do not feel that I have to justify the plain fact that economic theorising has its own merit and justification, but the quantification methods we use in agricultural economics have to be applied so that we receive, at least in some degree, explanatory as well as predictive or indicative answers to the questions we pose. I was very careful in the paper to say that, so applied, these models are an aid to the decision-making process, contributing to the intelligence with which we handle international trade problems. Models are not substitutes for policy-decisions but they can be, if wisely prepared, an aid to the decision-making processes.

We are so very reluctant to accept that outside the economic fields there are decision-making processes which affect agricultural economics in spite of the fact that they are so incredibly difficult to quantify. In regard to international relationships, I find very often that it would be helpful if we borrowed from our colleagues in the other social sciences and political sciences. We have basically utilized only "switch on, switch off" dummy variable systems without taking advantage of the very advanced knowledge of social scientists in other disciplines.

I agree that the question of exchange rates in agricultural trade is of the greatest importance. I singled them out in my original draft, but came to the conclusion that they were too complex a problem to be handled in the paper. In relation to our experience, I would only point out that the European Economic Commission has been working in this field for many years. The Commission is seeking better solutions, but to date we must use the tools available. A very similar situation applies to the COMECON. Originally, international or regional trade in the COMECON was based on fixed prices, but they learned by experience that adjustments over time were needed. I certainly want to give credit to our friends in the East for developing serviceable models in this field. I personally do not believe that the predictive truths we have developed are worse than the previous ones and therefore I think that some contribution to this can be derived from models.

W. Scheper (in reply)

As I mentioned in my paper, it is very hard to solve the data problem for models of the Meadows type. A discussant argued against it and said that the important variables of the Meadows Models are indigenous and calculated from year to year. That is correct. However, there are many parameters which have to be estimated exogenously and I tried to say that it is very difficult to forecast the values of these parameters. I agree that it is a weakness of my paper to neglect national agricultural models of the World Bank Mexico type. These models are indeed a good starting point for international linkage approaches. I am in line with Professor Johnson's argument that model-builders should put more emphasis on possible variations of parameters in future and the consequences which arise from these variations. Dr. Daad pointed out that the interpretation of models is very important. To have an influence on agricultural policy, economists must present results and assumptions of their models in a way which politicians can understand and, furthermore, in a way which provides a foundation on which the politician can base trust in the economist.

I felt that Dr. Daad's argument is one of the most important arguments in this session. Useful world models will be very complex and, therefore, necessarily there will arise the Black Box problem. This means that the model-builder cannot explain everything in very simple terms and, therefore, the politician has to trust the economist to a certain extent. The construction of complex models certainly takes a long time. They are never ready – in the sense that no further improvement and adjustment is possible. Further improvements and adjustments of existing models very often seems more useful than starting with a very different new model.

Finally, in my opinion the discussion gave too little attention to link problems. I feel that in the coming years we should invest a lot of work to link agricultural models with highly integrated macroeconomic national and international models.

HARRY E. WALTERS*

Towards a World Food Policy Progress and Problems in Decision-making

The title of this paper implies that there is a world food policy and that we can measure progress towards reaching the goals of that policy. But while the nature of the world food problems is now fairly clear, as are many of the solutions, and a world food strategy was outlined at the World Food Conference in November 1974, the policies through which the goals of that strategy can be achieved are not well established. When the national implications of the policies required to achieve these goals become clear to decision-makers, their willingness to undertake them dwindles rapidly in the absence of well developed programmes to show how they can be achieved. For world food problems are the result of larger economic, social and political realities and not, as is too often suggested, simply the result of inadequate technology or resources.

The basic factors explaining the food problem have been analysed in great detail over the past few years – both in the early Sixties and again during 1972–74. There is a large measure of agreement that:

(a) unless food production is greatly increased in the developing countries those countries will face impossible food import requirements in the coming decade and even more impossible import requirements thereafter;

(b) the food "surpluses" experienced by the developed countries during the two decades leading up to 1972 were not only the result of productive capabilities but also of policies which in various ways supported food production or prices at levels above what would otherwise have been sustained;

(c) consumers with low incomes consume inadequate diets, largely dependent on cereals, starchy roots or other inferior food sources, while consumers with high incomes consume more food than would be required for an adequate diet;

(d) the nature of the elasticity of the demand for food and the unstable supply of food can lead, in the absence of stock policies, to dramatic price changes, both up and down, and that these changes if upward, will adversely affect the poorest consumers, and if downward will reduce the incentive to produce food; and

* Assistant Executive Director, World Food Council.

(e) food aid has had both beneficial and detrimental effects on food production in developing countries.

If it is possible to agree on these general elements of the food problem, why is it that the policies that would be needed to bring about solutions to these problems have not come into effect.

I think that there are at least four reasons.

First, although serious studies may agree that the above are the essential problems, there are a number of issues which capture the attention of a great many people, and may deflect attention from problems which are less dramatic and may be much more intractable. Among these I would include the possibility of major weather changes, resource exhaustion, the fear that food demand may permanently outstrip supply and the concern that it may be necessary to determine who shall have food and who shall not. I do not suggest that these are not worthy concerns in themselves but they are not the most essential problems affecting food now or in the relatively predictable future.

Second, while there is widespread agreement about the essential problems and their solutions, these solutions have not been developed so that decision-makers can see how to bring them about within the context of their own economic, political and social realities. The Second Development Decade, the Indicative World Plan, the Strategy for International Agricultural Adjustment, the International Undertaking on World Food Security, and the studies and reports surrounding the World Food Conference, all point essentially to the same conclusions and they all embody the same proposals for action. But this has been about as far as things have gone. The fact that these strategies, plans and undertakings have been agreed to in principle, but have resulted in limited action suggests that while there is broad general agreement about what should be done, there is little practical agreement about how to do it, and especially who should make what decisions.

A third, and closely related explanation, is that in times of crisis there is widespread support for food-focussed policies. Food is seen to be "special" and in some way separate from other economic, social and political problems. As soon as the "crisis" has passed, however, these other considerations resume their dominant position and decisions affecting food are easily overshadowed.

A fourth factor is the existence of different socio-economic systems. No single world food policy will apply or be acceptable to all systems.

I would like to turn now to what I think is meant by a world food policy – the combination of policies that emerge out of the 22 Resolution of the World Food Conference – and consider the progress and problems that decision-makers have faced in implementing this strategy. Inevitably this will reflect much of the experience of the World Food Council, the institution set up by that Conference to oversee the implementation of its resolutions.

In essence the strategy which emerged from the World Food Conference had the following major elements.

First, it called for a major increase in food production in the developing countries. To achieve this, eight resolutions were adopted which were directly relevant, dealing with agricultural inputs, research, seeds, soils, water and others. Among these, the most significant was Resolution XIII calling for the establishment of an International Fund for Agricultural Development (IFAD). All of these resolutions flowed from the recognition that, over the past two decades, food production in the developing countries was falling progressively behind demand and that, as a result, food imports by developing countries were rising sharply. Of all the decisions of the World Food Conference this one received the broadest agreement.

Second, it called for the development of a system of world food security which centred on the establishment of a cereals reserve to replace the grain surpluses that had offset previous shortfalls in production or shifts in trade but had been eliminated during 1972–73. The need for such a system was a recognition of the detrimental effects on food prices of shortfalls or surpluses in production when stocks were not available.

Third, it called for an improved system of food aid. This was in response to the abrupt reduction in food aid which took place in 1973 as a result of the absence of surpluses and high commercial demand for grain.

Fourth, it called for policies and programmes to improve nutrition in recognition of the widespread prevalence of mal- or undernutrition affecting an estimated 460 million people in developing countries.

Fifth, it called for changes involving international trade, stabilization and agricultural adjustment; and

Sixth, it set up a series of institutions to ensure that these resolutions were implemented. Overlaying this strategy was the Declaration on the Eradication of Hunger and Malnutrition which was adopted by the Conference. That declaration includes the statement that "society today already possesses sufficient resources, organizational ability and technology and hence the competence to achieve this "objective". This echoes the statement of the United States Secretary of State at the Conference that the world possesses the capability to ensure that within a decade "no child should go to bed hungry, no family should want for its daily bread".

1. WHAT PROGRESS HAS BEEN MADE SINCE THE WORLD FOOD CONFERENCE?

A critical analysis would have to conclude that progress toward the implementation of the world food policy outlined at the World Food Conference has been very slow. One positive achievement has been the recognition of the need to increase food production. This is reflected in the all-out efforts of the major food producing developed countries to expand production and the substantial increase in food production in many developing countries in 1975. How much of this can be attributed to the World Food Conference and how much to weather and the response to food shortages and high prices in 1973–4 is a matter of judgement.

Another positive achievement has been the creation of IFAD, the contributions to which are near the one billion dollar target and the agreement to which was signed by participating countries in June.

Progress on the other resolutions, especially the crucial ones dealing with food security, food aid and nutrition, has been extremely limited, however. At the present time grain stocks are emerging or are likely to emerge only where unexpected surpluses arise. Food aid has gradually increased to over 9 million tons, but major efforts to restructure such aid have not been undertaken. Favourable improvements in nutrition have come about, if indeed they have, only because of lower food prices and more plentiful supplies. Developments in food trade are, at best, the reflection of slow progress in trade negotiations in general.

If the World Food Conference created the essentials of a world strategy which was broadly agreed to, why have the policy decisions which should have emerged been so slow in coming. An easy answer would be that decision-makers lack the will to make the policy decisions that would be needed or that they have become preoccupied with other matters. To a certain extent this is true, but a more important explanation is that most countries can identify with and support these objectives in principle, but in practice they imply profound decisions involving the internal policies of developed and developing countries alike. In the absence of a new crisis which can be used as a wedge to force changes in policies, substantial progress toward a world food policy cannot be expected until the feasibility of such a policy is much better demonstrated and until the political compromises needed can be reached.

It might be instructive to consider how action has progressed on the major elements of the World Food Conference strategy and in that way appraise the problems and progress of decision-makers.

2. INCREASING FOOD PRODUCTION IN THE DEVELOPING COUNTRIES

To achieve the major increase in food production in developing countries the Conference called on all countries to give a high priority to food production and to increase the resources devoted to food production within developing countries and in the form of external assistance. This objective was reinforced by the creation of the World Food Council which has given its highest priority to bringing such increases about and by the creation of the Consultative Group on Food Production and Investment in developing countries (CGFPI) whose job it is to stimulate an increase in the flow of external resources for food production and their better direction.

At the World Food Conference it was relatively easy to agree that there was a need to overcome a projected deficit of 85 to 100 million tons of cereals in the developing countries by 1985. But these agencies concerned have found it a far more complex matter to ensure that production increases take place where they are most needed. This presumes the determination of priorities which immediately raise issues of serious concern to recipient and donor countries alike. It also brings one face to face with the problem of determining the kinds of policy changes that a country, or others, may need to make, in order to stimulate production in a given country. Choosing among competing wants in the allocation of resources may be a familiar problem to the economist, but it is not a comfortable subject for the politician or international diplomat. And the internal policies of every country are strongly defended as matters of exclusive concern to that country.

Another problem that emerges is that there is a surprising absence of firm, agreed information about how individual developing countries can increase their food production. Determining by extrapolations of past trends or other aggregative methods that there will be a food deficit of a given magnitude in a given country is of little help in demonstrating how and at what costs that deficit could be overcome in a decade.

The World Food Council proposed at its second session in June 1976 that a special effort be launched to determine just how and at what cost food production could be increased in those countries which appear to have the most severe present, or expected, food problems and also appear to have the least nearly adequate resource endowment to achieve the necessary increases. Combined with this effort, which would be initiated by the countries themselves in cooperation with one or more of the international agencies, would also be a determination of the policy issues and changes that might be involved. At best this will not be an easy task.

These problems reflect the fact that there is a big step between the generalization that production increases are needed in the developing countries to the stage of determining – in sufficient detail to gain the support of the country itself, international agencies and donor countries – just what needs to be done and who is to do it. But this step is crucial if we are to move from anxiety and recrimination to positive and effective action.

3. WORLD FOOD SECURITY AND GRAIN RESERVES

While there may be general agreement that a system of world food security including grain reserves is necessary, the stage for discussion has shifted from the anxiety of the World Food Conference to negotiations at the International Wheat Council, the Committee on World Food Security, GATT and UNCTAD. In these negotiations, the anxiety over the precarious condition of a world without grain reserves is overshadowed by the trade and other policy concerns of grain exporting and importing countries. These are issues about which decision-makers are aware on a daily basis. While their constituants may share a general concern for the existence of food reserves, they also have other and maybe conflicting concerns with other aspects of these negotiations.

The World Food Council at its second session advanced for consideration by these negotiating bodies a compromise proposal which combined elements of a grain reserve, food aid and emergency food assistance, but it is apparent that achieving a workable compromise among the various national interests will be a difficult matter.

4. FOOD AID

The World Food Conference called for the return to a minimum annual level of 10 million tons of food aid on a three year forward planning basis as well as other substantial changes in food aid policies and programmes. While there has been a gradual growth of food aid to over 9 million tons, there are very significant disagreements among governments over the usefulness of food aid. Governments who provided such aid in the past are reluctant to do so at previous levels when the demand for grain is high and surpluses no longer exist. Similarly there is a recognition of the adverse effects food aid may have had on food production efforts in the past and many governments are reluctant again to contribute to such results. Moreover, changes in food aid policies and programmes immediately involve issues of internal policies in recipient and donor countries which are not easily resolved.

5. NUTRITION AND THE ERADICATION OF HUNGER AND MALNUTRITION

Progress in nutrition within the context of the objectives of the World Food Conference has yet to be tackled in a serious way and there has been a tendency to treat the objective of eradicating hunger and malnutrition as at best a necessary humanitarian goal but of little practical significance as a programme to be seriously undertaken.

6. CONCLUSIONS

The World Food Conference is too close and the institutions it created too new to draw any hard and fast conclusions about their effectiveness in overcoming the world's food problems. But it does seem that enough time has elapsed to draw certain important conclusions.

If the objective is to develop a world food policy that will in fact eliminate or greatly ameliorate the major food problems which are now apparent and will, in a decade, achieve the aspirations of the Declaration on the Eradication of Hunger and Malnutrition, then additional steps are needed. Among them:

(a) Food will have to be elevated to a position in international affairs where governments are willing to take policy decisions affecting food which they might not be willing to take for other economic problems.

(b) If hunger and malnutrition are to be eradicated or greatly reduced within a decade this will involve some measure of income and food redistribution in addition to the best efforts to increase food production.

(c) If effective policies are to be developed to increase food production, to develop a system of world food security, and to improve the usefulness of food aid, much more effort will have to be devoted to developing policies that can be accommodated within the existing economic, social and political realities of the multiplicity of nations which will have to agree to these policies. This will require a willingness to compromise in favour of food and malnourished people and these compromises are likely only when there is the conviction that the desired result will be achieved.

The long series of events which culminated in the World Food Conference, not just the events of 1972–74, but the many conferences that preceded it, and the general agreement with the aspirations enunciated in the resolutions adopted by that Conference, suggest that there is a serious international desire to take these additional steps. They must be taken because the resolutions of the World Food Conference only tell us where we think we should be, we must still work out how we are going to get there.

Report of the general discussion

Part of the discussion, in effect, reflected doubts about the effectiveness of the kind of international action proposed by the World Food Conference. It took the form, on the one hand, of uneasiness that creation of further international agencies invited overlapping and doubts whether an International Fund for Agricultural Development would put more resources to use than would the present agencies. On the other, the usefulness of alternative approaches was emphasized and contributions ranged over the scope for farmers' organisations to influence policy; the assertion that, without international action, food production had been promoted by farmers' efforts, by the investment of the agribusiness sector and research agencies and the technical assistance activities of charitable agencies and national governments.

Attention was also drawn to the slight attention given in the paper to features of the New International Economic Order considering the importance of some of its features in helping low income countries to solve their food problems.

Another group of comments urged closer scrutiny of issues than the paper afforded. More precise indications were sought about the quantities of food needed to achieve specified stabilization objectives; the decision methods and operating procedures for an international stabilization agency; the decision methods for national buffer stock holdings; the costs of such schemes and the development which might be promoted with the money which they would undoubtedly absorb and the success of the world crop surveillance scheme which the World Food Conference entrusted to the World Meteorological Organization (Geneva) and FAO (Rome).

Finally, the need to treat food and population problems as a single whole was stressed.

Participants in the discussion included: J. H. Dalton, *Ethiopia*; G. von der Goltz, *Kenya*; D. Paarlberg, *U.S.A.*; K. Prasad, *India*; G. Schmidt, *Fed. Rep. of Germany*; M. J. Schultheis, *Tanzania*; D. Tomic, *Yugoslavia*; H. C. Trelogan, *U.S.A.*; N. C. Westermark, *Finland*.

H. E. Walters (in reply)

Regarding the justification for creating new administrative machinery, I should emphasize that in mid-1976 the World Food Council staff consisted of about four people struggling to prepare for the meeting of the ministers.

The new institutions are not large institutions and are not designed to be. Duplication does not therefore arise. Some of the ministers will be ministers of agriculture, others of planning, of finance, of foreign affairs. Food is larger than FAO, than the World Bank, than UNICEF. Moreover, the world food problem is a national government problem; and it is also an international agency problem with a great many elements. It was felt that a World Council was needed to provide a place where ministers with responsibilities bearing on food – but not necessarily involved in food – could fight through towards decisions. Experience suggests that it will be quite a long time before highly desirable solutions are achieved. Whether the World Food Council will, or should, succeed is up to governments, but, for the reasons given, I think it should. Meaningful large scale international decisions will not be made until they are approached like this.

With respect to buffer stocks, following in the pattern of what the World Food Council was supposed to do, many of these questions are to be left to other people. The World Food Council is charged with trying to sort out the political issues. Responsibility for many matters concerned with buffer stocks lies elsewhere.

At its meeting in June the World Food Council did consider a set of proposals for its Secretariat which, while not an ideal solution to the world's food security problem, were considered to be a compromise that governments might be willing to agree to now, in lieu of working out a sounder and more satisfactory solution through other agencies. The proposals for the size of the buffer stock range, as far as I know, from the levels of ten to 12 million tons up to international stocks of 60 million tons or so. We proposed a buffer stock in the neighbourhood of about 20 to 25 million tons operated on an international level, in a way somewhat similar to the one discussed yesterday here at the meeting on models. It would dampen the tendency for prices to go to extreme levels in either direction. This proposal was combined with other aspects of a world food security system so that there would be at least a floor in terms of emergency relief, food aid and buffer stocks. It was very clear at the Council meeting that a major gulf separates those who think of a reserve as a physical pile of grain against a crisis and those who see a reserve in larger terms. For the former it is hard to understand why it cannot simply be decided to have a large pile of grain somewhere. Those who have experienced the effect of a large grain stock on the world food situation in the last twenty years, are concerned that the existence of surpluses of grain, as such is not a desirable thing in itself and can have adverse effects on production and distribution of food.

FAO, within its present arrangements, is doing a number of specific studies in specific countries with respect to desirable sizes of reserves and their costs. Certainly the World Food Council and other institutions might be in a position to acquire additional funds for those purposes if a convincing argument can be made for them.

In principle, the International Fund for Agricultural Development was to be set up to accumulate additional funds for increasing food production.

I think it is clear that the initial contributions to that fund are additional in most cases but it is the next time around where the concerns I was mentioning in the paper come into play. Governments will have to have food still very high on their priorities list and see to it that they are not shuffling funds from one place to another. The International Fund for Agricultural Development is being set up as essentially a place where funds are accummulated and where the major decisions on their use will be made. But the Fund itself will rely on the existing institutions for the basic implementation and working out of the kinds of projects it intends to support.

Briefly, about the current surveillance of crop prospects and developments, FAO has made a number of improvements in the early-warning system and its general surveillance. Most people seem to feel that there is a better balanced set of perceptions on the issue than there was a few years ago.

I do not think we can create a long-term Scientific World Food Policy – in fact, I am not even sure that is the issue. The need for increased food production, for food security and improvement in nutrition – all these are, in a sense, marginal issues. As I think Dr. Paarlberg put it once a long time ago "one of the problems of food is that a little too much is a lot too much and little too little is a great deal too little". The efforts that will be needed to raise the levels of production and improve consumption patterns involve a complex mass of changing political and social decisions in a number of countries. It is my feeling that the need is for a mechanism for constantly raising these issues when these adjustments get out of line, and constantly reminding people of what they need to do and, hopefully, being able to convince them of how they can do it. In the long run it is not a Scientific World Food Policy that will last for ever that we are searching for but a mechanism that will bring us face to face with the problems we need to solve, two, three or five years before they become problems that we are not in a position to solve right then.

I can only agree that the population problem is an important aspect. However, I have been concerned with the solution of problems that face us in the next five to ten years and we cannot do much to change the population effect on the demand for food in that period. I am in wholehearted agreement with those who argue that we need to find not a proper balance between population and food, but we need to find a proper balance between population and the way of life that people would like to have. Food is a part of that. So are medicine, education, housing and so on.

DENIS K. BRITTON*

Research Priorities Synoptic View

I feel rather like a headmaster of a school who calls his pupils together at the end of term to tell them how well they have done, to warn them that there is a need for still greater efforts next time but to reassure them that the holiday has now come and the school rules can be relaxed and a certain amount of frivolity is permitted once again. Perhaps we have become too serious. Anyway if I use some of these next few minutes in a rather light-hearted and even fanciful way, I hope I will be forgiven and that you will all enter into the spirit of this end-of-term occasion.

For one thing, we have the ladies with us, and we owe it to them not to be serious or too professional in their company. We all know that the ladies get on and do something while the men just sit and talk; we therefore have some obligation to justify all our sitting and talking. I am reminded of the story of the lady who was asked by an admiring visitor the secret of her happy marriage. "Oh, it's quite simple" she said, "my husband and I share all the planning and decision-making equally between us; he does all the planning and I do all the decision-making".

We have had a vast number of words delivered during the past ten days and what I have tried to do is to dip my fishing net into this pool of words. Much material was so refined that it slipped right through my net, but I am reporting on those creatures which I have retained. My filter is no doubt somewhat different from yours and I am not going to give you a ready-made report to take home which your bosses will accept as a satisfactory account of how you have used your time.

I may not be able to mention by name all those from whom I have fished some of my ideas and I apologise if I do not give your full academic titles to those whom I shall mention.

Let me say, first of all, what is my own attitude to agricultural economics. I believe that when we strip away all the wrappings with which we like to encase this subject and get right to the core of it, we do not find a mathematical equation. We do not find a set of systems. We find at the centre a human family. This is my starting point and I recall what Schultz said on our first

* Wye College, University of London, U.K.

day: "It's what farm people do or what farm people will be able to do which holds the key to our story". I shall be saying a little more about econometric models but first I just want to say this. Econometric models, and indeed all the other models we use, must in the end be buildings in which men and women could feel at home and recognize their identity. They must not be too clinical, or so inhuman that people are thought of just as labour units. The more model builders polish the marble floors of their buildings, the more slippery and uninviting they become; and men and women may not want to make their habitation there.

Now I suppose I should ask, have we in this Conference been, to any extent, off course? Have we wandered at all from our proper function? I suggest you might refer to our Constitution, and find Article II (Objectives) which reads as follows: "To foster the application of the science of agricultural economics in the improvement of the economic and social conditions of rural people and their associated communities". That is the first point. I think we can all recognize the hand of Leonard Elmhirst in drafting that section. The Article on Objectives continues: "To advance knowledge of agricultural processes and the economic organisation of agriculture". That last phrase, by the way, is the title of one of Schultz's books. "And to facilitate communication and exchange of information among those concerned with rural welfare throughout the world". If we are not personally pursuing any of those aims I think we should ask, ought we to be here at all? Or do the objectives of the Association need changing, because after fifty years our world and our ways of looking at it have changed? In passing it is interesting to note that our French colleagues describe themselves not as agricultural economists but as rural economists. Perhaps this is an indication that they have remained closer to the original objectives of the Association. Even so, at the present state of Michel Petit's work on the rôle of models in the decision process I do not find it to be noticeably rural in character.

It has been suggested to me by some people to whom I have talked during the Conference, that our papers have been too general. Do we want, at a conference of this kind, general or specific papers? The contributed papers which we received yesterday did seem to me closer to research and to the scientific method than some of our other papers have been. I think, for example, of Kenneth Shapiro's paper on "Efficiency Differentials in Peasant Agriculture and their Implications for Development Policies". That paper re-examines a proposition previously formulated by another research worker, in this case none other than Professor Schultz, concerning the high efficiency of traditional agriculture, given the available inputs and technology. Shapiro's paper produces data purporting to show that the previously held proposition is not supported, and postulates that many farmers could operate much more efficiently than they do with the resources they now have. Finally, the paper suggests that further research is needed to determine *why* this is so. Now that kind of scientific criticism, debate, and reassessment, seems to me to be of the very essence of research and I feel sure that Karl Popper, the leading thinker in the philosophy of science who was quoted earlier in this Conference, would thoroughly approve of that kind of paper. He might be much more dubious about the papers of a general stock-taking kind. Of course we cannot please everybody. If our Conference consisted mainly of the discussion of reports on highly specific and localised case-studies or small-scale surveys, a number of disappointed participants would say, "I found there was very little of relevance to my situation". That number would probably increase if we invited mainly local research papers. Each attempt to generalise or to make a comparative analysis enables more people to identify with the question under discussion, but at too high a level of generality the depth of such identification can become very superficial and even totally impersonal.

Next I want to turn to Glenn Johnson's outline of the triple rôle of agricultural economists. After that I want to underline some points from which I think we could derive satisfaction, then some points giving cause for concern, then a word about models and policy-makers and finally a programme for action.

Now the first part of the triple rôle of agricultural economists, says Glenn Johnson, is to participate in decision-making, the eventual object being to help to prescribe a solution. An inter-action is needed between the decision-makers and agricultural economists. The agricultural economists will be looked to particularly for maximising decisions or indications, and Johnson reminded us that great philosophic flexibility, as he called it, is required on the part of agricultural economists in that context. Capstick suggested that maybe we should switch some more of our resources in agricultural economics into this first category of Glenn Johnson's. On the other hand Ohkawa clearly prefers that we confine attention mainly to the second and third and not get too involved in direct participation.

The second category was to carry out subject-matter research, putting together information of a specific subject where the immediate objective is not to prescribe policy but to prepare the components of models. Here I would recall Uma Lele's request for more accurate data on such matters as yields, acreages, prices and market behaviour; and Tracy's suggestion that we should do more outlook-projection work before the next crisis appears, rather than analysing it after it has happened.

The third category was to carry out disciplinary research, where we attempt to improve theories and quantitative techniques. As Glenn Johnson said, often the practical problems encountered by agricultural economists in their first rôle go unsolved because the necessary disciplinary research has not been done. Here Johnson underlined that we need better techniques for working with non-monetary values and Richter also said, when he was talking about world models, that these must take account of some factors not easily quantified – socio-economic parameters, as he called them – which belong more to political economy.

Now I will move on to points from which I think perhaps we can derive some satisfaction. I think we could congratulate ourselves on our intellectual stamina and alertness throughout these long days; I wonder though if it has always been so. I can recall some earlier conferences at which I believe that a participant could sometimes have got away – perhaps at four o'clock in the afternoon at the fifth plenary session – with some most outrageous statements. They would solemnly have got into the record and no one would have taken any notice. I think we are much more alert today.

One of our colleagues said to me in the corridor, "I feel very happy here, because we have experience a *planetary dimension*". I know what he feels and I share his view. I think discussion here has been more systematic, more relevant, less episodic, than at previous conferences. Too often in the past we have had to listen to members who have used valuable discussion time to read long prepared statements which took no account of the course of the debate and seemed to be introduced at random points. There has been less of that this time and I am very happy about it. Perhaps our preoccupation with the development of rigorous conceptual systems has had that effect.

Now if I may turn, Mr. President, to points raised in this Conference which should give us some cause for concern. The first one I want to mention is – are we, perhaps, not doing enough empirical research? Are we losing our appetite for, or even our belief in, research? So many of our papers consist of commentaries upon current trends in the economy and in society – for example the trend towards decentralisation mentioned by Yudelman – or express a concern with the analysis of concepts and systems without any specific content. I wondered sometimes whether we had done more at this Conference than to design and assemble a set of empty boxes. Is it the case that our large institutions such as the World Bank, are still learning more by trial and error than by systematic research? These are questions I am asking. I remember one comment made by Mr. Martens at one of the discussion sessions, pointing out that we should not suppose that the content of our Conference papers, or of our journal articles, reflects fairly the bulk of empirical work which is going on but which is not widely reported. I think that is a very good point. But perhaps it is a point which editors of journals should note, not to over-weight the methodological side of the subject and not to restrict themselves to philosophical articles as though there was nothing of interest to report in actual research outcomes.

I wonder whether the apparent lack of research content in some of our papers is due to the age composition of those attending our conference. Have we got too many administrators amongst us who have little time for research and indeed are not expected to do research? Anyway I think I am going to suggest to my new Executive Committee that a certain number of papers, and a certain proportion of discussion time, should be allocated to members under 35 years of age, and only if such young contributors are not forthcoming would their share of the time be surrendered to the elder statesmen. Does the world, perhaps, pay more attention and pay more money to those who advise (often on the basis of knowledge much of which was gained a quarter of a century ago) than to those who are engaged in the process of discovery? Where are we to look today for the fresh springs of vigorous intellectual activity on our subject which are replenishing our discipline and its body of tenable propositions?

Schultz noticed an increasing opposition to economics and the abasement of economics by governments. If economists have gone wrong, I wonder where they have gone wrong? I think the Conference has shown, Mr. President, that we may have gone wrong in three main ways. First, we overlook social and political factors in our own sector. We still think we can identify purely economic problems when really they are socio-politico-economic problems. This point was stressed, Mr. President, in your own address at the opening of our Conference, and it was also very effectively expressed in Mbithi's paper. I was not able to be present at the session when he delivered that paper, but it seemed to me to be a very challenging paper about decision-making in the farmhouse household. The farmer's social decision-making environment is prescribed by his culture of which he is a product and its complex values, cognitions, beliefs, and experiences are important components of his decision-making frame of reference. Technical and economic factors in *some* decisions are not really critical. Mbithi went on to list some of the false assumptions made by professional extension workers. First false assumption: that the rural farm can be assumed to be a responsive, expectant production firm which can be organised. Second false assumption: that the farm is operating in a knowledge vacuum and that historical, social and environmental factors, as well as the personality of the farmer, can be ignored. Third false assumption: that the change agent or extension worker is himself operating with all the necessary information. Fourth false assumption: that the farmer understands and appreciates the technical nature of the expert prescription and has the resources to implement it. Mbithi utterly rejects the "trickle-down-effect" theory which he says leads to rigid barriers in access to national farm resources and perpetuates unacceptable income disparities. I was very impressed by that line of argument and I think we all have to review our analysis of economic problems against the background of those criticisms.

The second possible failure of agricultural economists is that we do not sufficiently integrate our studies in the agricultural sector with events in the rest of the economy. This was underlined by Ohkawa in his paper. Vos also emphasised that there is a clear and strong feed-back between the rates of growth in the two sectors of the economy, and that this is of vital significance when it comes to the choice of the strategy for the economic development of the country. Tracy pointed out in connection with this matter of relating our own sector with other sectors, that in the same way we should integrate our studies of regional agricultural policy with the world situation more than we have done; and I did notice that when Tracy was giving his paper about European agriculture it so happened that there were no Asians or Africans present. I know that there were other regional group discussions going on at the same time, but it does illustrate the tendency for our regional studies to become isolated from each other.

The third factor which I think has emerged as important in the criticism of our attitude is that we do not take enough account of the changing exogenous factors which can so often upset the agricultural picture. We seem to be sometimes unwilling to accept that the world is dynamic. Nussbaumer underlined this. The pace of change is probably accelerating. We cannot stop the system so as to take a long hard look at it; we have got to adapt our studies to the changing system. Van den Noort said that we have to accept that the decision-making process is affected by the fact that the conditions, the economic machinery, the objectives, even the general welfare function are not constants but are changing over time. All this leads to the point, I think, that we need more multi-disciplinary studies. There has been insufficient awareness of the findings of our colleagues in other social sciences and so our models lack realism. Giving another angle on this, Paarlberg underlined that we must concern ourselves with the question "who makes the decisions, what is decided and what is the outcome of the decision?". I find a reflection of this same pointing to involvement with decision-makers when Scheper pointed out, when he was talking about model building, that our real problem is to find operational procedures. That is the main problem, not to analyse situations but to find operational procedures, and I think this is what Paarlberg was getting at. Again, Folkesson said in another session, "If we are interested in agricultural policy we should spend more time on the analysis of the policy formulation process itself". Bublot drew our attention to the operations of pressure groups, bodies, social classes, and we cannot ignore that these are real factors in the decision-making process, Ohkawa said that preferences of political parties are decisive and preferences of administration cannot be ignored. Tracy recommends us to ask, when we are making our analysis, where does power lie? Similar points were made by Anthonio and Sebestyen. Another comment here is that we are telling ourselves that we have not been sufficiently dis-aggregated in our approach. We need to aim programmes at specific groups of needy people, not just ensure that the global supply keeps pace with population increase.

Now after discussing decision-making for ten days, I suppose it could be said, have we made any decisions? I do not think it would be a fair question, because the decision-makers we are talking about are not agricultural economists, they are mostly politicans or farmers. We are seeking to help them, but not to become the decision-makers ourselves. It is, however, perhaps worth asking, did we come here to talk to each other or to the world outside? In the speech by Vice President Arap Moi he clearly challenged us to speak to the world outside, to give prescriptions, especially for his own country. I still think that it is perhaps not fair to suggest that we came here for that purpose. I think we have come here mainly for a kind of professional stock-taking, to remind ourselves or to inform ourselves of what different people are doing in our subject, to discuss what is the current conventional wisdom, to discuss what has been discarded since we last met or over recent years and to discuss what is emerging. All this will not necessarily have immediate relevance to actual problems in particular countries; we are mostly further back down the line than that.

Now I want to move on to a few words about models and policy-making. I start with Dr. Dadd's comment that policy-makers will not take our models on trust; they have got to be validated. We have to show that if the models had been used in the past they would have come up with reasonably good indications of what has actually happened. Dr. Kim said that we have got to sell these models step by step to those to whom we think they will be useful; not oversell them, or they will just be rejected, but sell them step by step, get people used to the idea that they are an instrument for policy-making, though only one instrument. Nielsen said in one of the discussion groups that models should be developed in response to a problem which someone has asked us to solve, not for their own sake. And Paarlberg said if we want to advance in relevance we may have to do this at some expense of rigour. I thought that summed it up rather well. Petit was honest enough to admit some of the limitations of model-building, when he said towards the end of his paper that the usefulness of models obviously depends upon the quality of the new knowledge which they help to produce. The social relevance of this new knowledge must ultimately be judged from the point of view of the decision-making processes to which it contributes. In other words it is a pragmatic approach in the end, not just theoretical. Petit said that we must not overlook that in society conflicts are normal and recognised as such. We have to recognise that the adaptive behaviour model does not tell us how conflicts are solved. This is another point at which we need to touch people in neighbouring disciplines; adaptive behaviour models do not even tell us which conflicts are important and as such they fall short of being a global theory of social change. I think that could be a vital weakness, if all the problems today are socio-politico-economic and not just economic.

For these and other reasons, Mr. President, I have my own personal doubts about taking the model-building approach too far. I believe that decision-making will always be an art, at least as much as a science. Can we by analysis reveal the real essence of decision-making? Can we make optimum decision-making reproducible and available to all? I doubt it. I think there is an element of intuition and an element of improvisation in decision-making which will always elude us, but I am willing to accept that model-building is not a retreat from reality but the construction of a spring-board from which we can approach reality more effectively. But whatever attitude members may take on this question I would say, do not forget to read the forthcoming issue of the *European Review of Agricultural Economics* in which Petit, Renborg, Glenn Johnson and others will have a series of papers all about the validity of the adaptive behaviour model approach. Meanwhile some agricultural economists may prefer always to work with informal models to which they are accustomed; and Ohkawa pointed out that too sophisticated methods may not help much as tools for reasonable decision-making, a proposition with which Dr. Kim agreed.

Now I will turn to our programme for action. The very first point I would like to mention under this heading is that we need to give more attention to the clarity of our presentation. Mr. Arap Moi rubbed this in quite strongly. He challenged us to translate our professional sophistication into clear, precise and communicable language, to give a clear analysis of alternatives and the consequences of each alternative of action. I think we can do more along these lines and Capstick suggested that it is up to us economists to project our ideas to the policy-makers. If we wait until we are asked for our views we may wait a very long time, he said. I wonder whether our papers both here and in our journals are not sometimes written in much too impersonal a style. Why do we never use the first person singular in our papers? I will always welcome some papers of the kind in which someone will come to us and say

"I have made an investigation of the following kind and I have produced the following results. I will explain why I think these results may have some significance for conditions wider than those for which they were observed, and I invite your comments on my methods and my results". Why do we always say, for instance, "A sample survey of farm businesses was undertaken using interviewing techniques", when we mean "I went and talked to quite a lot of farmers"? Papers from the socialist countries, if I may say so, are particularly characterised by a highly impersonal and generalised style. I look forward to the day when a research worker from the Soviet Union will come to our conference and tell us what he and his immediate colleagues are doing, rather than making general statements about organisations and methods as they apply throughout the Union. But that is only one example; I am not suggesting that it is only the socialist countries which use the impersonal style. We seem to think it is more professional to use the impersonal style.

I think we have also got to be ready to meet ignorance of economics in high places with patience and with restraint. My experience is that not only do the plants and animals not read what we have written, as Schultz reminded us, but nine out of ten administrators are entirely unconvinced that any agricultural economists are worth reading or are readable anyway! Econometric models are regarded as too esoteric to be understandable and too artificial or historical to be relevant. They are seen in relation to the issues of the day rather as crossword puzzles might be seen in relation to the main news pages of the newspaper. Administrators want their economics encapsulated in immediately digestible tablets. They do not ask the questions which economists are accustomed to answering. They ask, for example, "Which is the best size of farm?" rather than asking "what is the relationship between size and the other economic characteristics of farming?" We have to be ready for that kind of attitude.

What will be the most important research needs for the future? The first I would mention, I think, is the general subject of how can we identify and ensure a better distribution of the national agricultural product? That is, we should study more closely the economic aspects of social justice for rural people. How should the benefits of growth be distributed? How much should success be rewarded? We have got to consider here what Ohkawa calls the trade-off between growth and equity. Bhattacharjee spoke of a dearth of analysis of policy options in this area. Nazarenko pointed out that we cannot use *only* marginal equilibrium theory in relation to the agricultural sector, because this sector is socially so important and so different. Do we not agree? Hence the USSR is doing a great deal to improve the rural infrastructure and so on. Can we sustain success in rural development without increasing inequality? We have also to fill some of the elementary gaps in technical knowledge indicated, for example, by Petit, when he said that some production models had to be abandoned because of lack of the necessary biological data about such things as the relationship between rainfall and the growth of grass. Again this is an indication of multi-disciplinary study being needed. Bhattacharjee asked the question: "Given a labour-intensive strategy of growth in many countries, which crop and livestock mixes should be included and at what stages of the evolution of production should different crop and livestock mixes be encouraged?" This is an area for research, I feel sure. In general the recognition of people's multiple goals, not just their economic goals, means that we need more of the trade-off type of analysis. Yudelman drew our attention to urgent research areas through which to investigate the reasons for differences in the rate of take-up of new technology. These need to be better understood, including the elements of risk associated with new technology in different situations. This bring in agronomy and sociology as well as economics. He also pointed the finger to welfare measurement: we need to find practicable methods to measure what happens to rural welfare when we alter the relative weights of the components of development projects. We need to know more about the principles and patterns of behaviour of small-scale producers, and we need to monitor development projects much more effectively so as to help us to design new systems. This, Yudelman said, was a challenge to the academic community at large. So there is plenty to be done in all those areas.

How hopeful can we be about completion? I liked Campbell's quotation from Samuel Johnson: "where there is no hope there can be no endeavour". It reminded me of a remark I heard Leonard Bernstein, the musician, make when he was giving a televised lecture about a symphony and about music teaching generally. He said "all teaching is an expression of faith in the future"; and I think we might say in answer to those who question the validity of our research efforts, that all our research is an expression of faith that in due time we may bring order out of chaos. I believe that there will be enough time, and that we should not be too impatient about the apparent lack of progress. I have calculated that if each one of us could add only one small but significant increment to knowledge in a professional lifetime of forty years, then we could look forward collectively to receiving one such contribution every month.

I come now to the end of my talk, Mr. President. As I settle down into my seat in the aeroplane on the way back to London I think perhaps I shall remember some words of Shakespeare which he put into the mouth of Prospero in "The Tempest". (Please, dear interpreter, do not try to translate Shakespeare; I cannot ask you to do that.)

> Our revels now are ended. These are actors,
> (As I foretold you) were all spirits, and
> Are melted into air, into thin air . . .
> The cloud-capp'd towers, the gorgeous palaces,
> The solemn temples, the great globe itself,
> Yea, all which it inherit, shall dissolve,
> And, like this insubstantial pageant faded,
> Leave not a rack behind: we are such stuff
> As dreams are made on; and our little life
> Is rounded with a sleep.

But then, Mr. President, as I twist and turn in my seat and dawn comes up over the Mediterranean, I shall remember with pleasure one thing that

Shakespeare could not know, namely that *we* have the 16th Volume of Proceedings to look forward to, edited by Theodore Dams and Kenneth Hunt. So long as they promise not to publish this speech, I salute them. And, my friends, I salute you all.†

† For the welfare of readers we did not promise!. Edit.

INDEX

This index is intended mainly to complement the Contents List. For locating the main subjects of the conference and the ways in which they were treated the contents list is likely to be the better aid. The index is likely to be the more useful in locating references to items which appeared in a number of papers or discussion sessions and, to a lesser extent, to items which arose more or less incidentally to the particular line of discussion being followed. Where a subject is frequently referred to in a paper or report each individual reference is not given. Names of countries are included only if they appear in connection with some major activity; entries in statistical tables are generally not indexed. Names of paper readers and openers of discussions are indexed. Other names are included only if this is convenient for reference because they have become associated with some concept, research area, etc. (for example, Keynes, Meadows (model), etc.).